Stochastic Komatu-Loewner Evolutions

Stochastic Komatu–Loewner Evolutions

Zhen-Qing Chen
University of Washington, USA

Masatoshi Fukushima
Osaka University, Japan

Takuya Murayama
Kyushu University, Japan

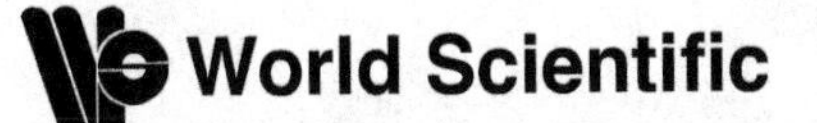

NEW JERSEY • LONDON • SINGAPORE • BEIJING • SHANGHAI • HONG KONG • TAIPEI • CHENNAI • TOKYO

Published by

World Scientific Publishing Co. Pte. Ltd.
5 Toh Tuck Link, Singapore 596224
USA office: 27 Warren Street, Suite 401-402, Hackensack, NJ 07601
UK office: 57 Shelton Street, Covent Garden, London WC2H 9HE

Library of Congress Cataloging-in-Publication Data
Names: Chen, Zhen-Qing, author. | Fukushima, Masatoshi, 1935– author. | Murayama, Takuya, author.
Title: Stochastic Komatu-Loewner evolutions / Zhen-Qing Chen (University of Washington, USA), Masatoshi Fukushima (Osaka University, Japan), Takuya Murayama (Kyushu University, Japan).
Description: New Jersey : World Scientific, [2023] | Includes bibliographical references and index.
Identifiers: LCCN 2022038167 | ISBN 9789811262784 (hardcover) | ISBN 9789811262791 (ebook) | ISBN 9789811262807 (ebook other)
Subjects: LCSH: Brownian motion processes--Mathematical models. | Markov processes--Mathematical models. | Stochastic analysis--Mathematical models.
Classification: LCC QA274.75 .C454 2023 | DDC 519.2/33--dc23/eng20221107
LC record available at https://lccn.loc.gov/2022038167

British Library Cataloguing-in-Publication Data
A catalogue record for this book is available from the British Library.

For any available supplementary material, please visit
https://www.worldscientific.com/worldscibooks/10.1142/13038#t=suppl

Desk Editors: Nimal Koliyat/Lai Fun Kwong

Typeset by Stallion Press
Email: enquiries@stallionpress.com

Preface

In this monograph, we aim at extending the celebrated theory of stochastic Loewner evolution (SLE) from a simply connected planar domain to a multiply connected planar domain in a unified manner by using Brownian motion with darning (BMD) that was coined in [Chen and Fukushima (2012), Chapter 7] in relation to the boundary theory of symmetric Markov processes.

Let $\mathbb{H}$ denote the upper half-plane, viewed as an open subset of the complex plane $\mathbb{C}$, and $\gamma = \{\gamma(t) : 0 \leq t < t_\gamma\}$ be a Jordan arc with $\gamma(0) \in \partial\mathbb{H}$ and $\gamma(0, t_\gamma) \subset \mathbb{H}$. For each $t \in (0, t_\gamma)$, there exists then a unique Riemann map g_t from $\mathbb{H} \setminus \gamma(0, t]$ onto $\mathbb{H}$ satisfying the hydrodynamic normalization at infinity:

$$g_t(z) = z + \frac{a_t}{z} + o(1/|z|) \quad \text{as } z \to \infty.$$

Under the reparametrization of γ to make $a_t = 2t$ (called the half-plane capacity parametrization), $t \mapsto g_t(z)$ satisfies the following ordinary differential equation

$$\frac{dg_t(z)}{dt} = -2\pi\Psi_{\mathbb{H}}(g_t(z), \xi(t)) \quad \text{with } g_0(z) = z \quad \text{for } z \in \mathbb{H} \setminus \gamma(0, t],$$

up to some positive time $\widetilde{t}_\gamma$. The above equation is called a (chordal) *Loewner equation*. Here

$$\Psi_{\mathbb{H}}(z, \zeta) = -\frac{1}{\pi}\frac{1}{z - \zeta}, \quad z \in \mathbb{H}, \ \ \zeta \in \partial\mathbb{H},$$

and $\xi(t) = g_t(\gamma(t)) = \lim_{z \in \mathbb{H}\setminus\gamma(0,t],\, z \to \gamma(t)} g_t(z)$, which is a continuous function in t taking value in the boundary $\partial\mathbb{H}$ of $\mathbb{H}$. The Loewner equation was initially formulated for simply connected planar domains such as the unit disk $\mathbb{D}$ and is effectively utilized in solving the Bieberbach conjecture (cf. [Löwner (1923); Conway (1995)]).

Conversely, given any real-valued continuous function $\{\xi(t),\ t \geq 0\}$, let $\{g_t(z) : 0 \leq t < t_z\}$ be the unique solution of the chordal Loewner equation with the right maximal interval $[0, t_z)$ of existence (cf. [Hartman (1964)]) and define the set $F_t = \{z \in \mathbb{H} : t_z \leq t\}$ for each $t > 0$. Then $\overline{F}_t$ is a compact subset of $\mathbb{C}$, $F_t = \overline{F}_t \cap \mathbb{H}$ and $\mathbb{H} \setminus F_t$ is simply connected. Such a set is called an $\mathbb{H}$-*hull* or simply a *hull*. Furthermore, g_t becomes a conformal map from $\mathbb{H} \setminus F_t$ onto $\mathbb{H}$. The increasing family $\{F_t\}$ of $\mathbb{H}$-hulls will be called a *Loewner evolution driven by* $\{\xi(t)\}$.

In 2000, Oded Schramm [Schramm (2000)] made a remarkable breakthrough on possible scaling limits of various two-dimensional lattice models in statistical physics. He found that for these lattice models at criticality, which have a certain domain Markov property and are conjectured to have conformal invariance, their scaling limits can be described by Loewner evolutions driven by continuous random processes ξ having stationary independent increments and zero mean. Accordingly, each ξ has to be a constant multiple of the one-dimensional standard Brownian motion B. He called the Loewner evolution driven by $\xi(t) = \sqrt{\kappa}B_t$, where κ is a positive constant, the *stochastic Loewner evolution*, which is denoted by SLE_κ. This evolution is now often called the *Schramm-Loewner evolution* in honor of Oded Schramm. One of the significant contributions of Schramm is that SLE_κ provides explicit limiting objects for many two-dimensional lattice models at criticality that were conjectured to be conformal invariant, some of which have been established rigorously later including the loop-erased random walk, Ising model, percolation and uniform spanning trees.

Remarkable features of SLE_κ have then been revealed including the *locality* of SLE_6; see, for example, [Lawler, Schramm and Werner (2001, 2003); Rohde and Schramm (2005)]. We refer the reader to [Lawler (2005)] and [Werner (2004)] for systematic accounts of the SLE theory. In the SLE theory, an important role has been played by the planar Brownian motion, particularly, by the *absorbed Brownian motion* (ABM) $\mathbf{Z}^{\mathbb{H}}$ on $\mathbb{H}$. Notice that the kernel $\Psi_{\mathbb{H}}(z,\zeta)$ appearing on the right-hand side of the chordal Loewner equation can be called the *complex Poisson kernel* of $\mathbf{Z}^{\mathbb{H}}$ because its imaginary part $\Im\Psi_{\mathbb{H}}(z,\zeta) = \dfrac{1}{\pi}\dfrac{y}{(x-\xi)^2+y^2}$ for $z = x+iy$, $\zeta = \xi + i0$, is nothing but the classical Poisson kernel giving an integral representation of harmonic functions for $\mathbf{Z}^{\mathbb{H}}$.

We now give a brief outline of the contents of the present monograph. Consider a planar domain $D = \mathbb{H} \setminus \bigcup_{j=1}^N C_j$ for $N \in \mathbb{N}$, where $\{C_j : 1 \le j \le N\}$ are mutually disjoint horizontal line segments in $\mathbb{H}$. Such a domain is called a *standard slit domain* and is known to be one of the canonical multiply connected planar domains. Denote by $D^* = D \cup \{c_1^*, \dots, c_N^*\}$ the quotient topological space obtained from $\mathbb{H}$ by regarding each slit C_j as a singleton c_j^*. Let m be the Lebesgue measure on D being extended to D^* by setting $m(c_j^*) = 0$ for $1 \le j \le N$. The *Brownian motion with darning* (BMD) on D^* is by definition an m-symmetric diffusion process on D^* with no killing on $\{c_1^*, \dots, c_N^*\}$ whose part process on D is identical in law with the ABM on D.

In Chapter 1, we establish the existence and uniqueness of the BMD on D^* (denoted by $\mathbf{Z}^*$) by using its Dirichlet form characterization. We then present the distinguished property of $\mathbf{Z}^*$ that the restriction to D of any $\mathbf{Z}^*$-harmonic function admits a harmonic conjugate up to an additive constant. This property allows us to introduce the *complex Poisson kernel of the BMD* $\Psi_D(z,\zeta)$, $z \in D, \zeta \in \partial\mathbb{H}$, an analytic function in z whose imaginary part is the Poisson kernel of the BMD giving an integral representation of harmonic functions of $\mathbf{Z}^*$.

In Chapter 2, we first present a self-contained proof of the existence and uniqueness of the Riemann map g_F^0 from $\mathbb{H}\setminus F$ to $\mathbb{H}$ with $g_F^0(\infty)=\infty$ for any $\mathbb{H}$-hull F. We then construct a conformal map g_F but with $\mathbb{H}$ being replaced by any standard slit domain. We next consider a Jordan arc $\gamma=\{\gamma(t):0\le t<t_\gamma\}$ in a standard slit domain D with $\gamma(0)\in\partial\mathbb{H}$. For each $t\in(0,t_\gamma)$, there exists then a unique conformal map g_t from $D\setminus\gamma(0,t]$ onto a unique standard slit domain D_t satisfying the hydrodynamic normalization at infinity. We show in Section 2.5 that g_t satisfies, under the half-plane capacity reparametrization, the ODE

$$\frac{dg_t(z)}{dt}=-2\pi\Psi_{D_t}(g_t(z),\xi(t)),\quad z\in D\setminus\gamma(0,t],\quad g_0(z)=z.$$

The above equation is called a chordal *Komatu-Loewner equation* (KL equation in abbreviation). The kernel $\Psi_{D_t}(z,\zeta)$ on the right-hand side is the BMD-complex Poisson kernel for the image domain D_t.

It was Yusaku Komatu who first extended the Loewner equation to multiply connected planar domains, especially to an annulus [Komatu (1943)] and then to a circularly slit annulus [Komatu (1950)]. This is the reason that we call the above equation as a KL equation. In Section 2.1, we can make complex analytic considerations analogous to [Komatu (1950)] in obtaining the above ODE, but only in a left derivative sense which does not uniquely characterize the solution. Before strengthening it into a genuine ODE in Section 2.5, we need to investigate through Sections 2.2–2.5 the continuity of various related quantities in t, by using an expression of $\Im g_t(z)$ in terms of BMD $\mathbf{Z}^*$ and the Lipschitz continuity of the BMD-complex Poisson kernel $\Psi_D(z,\zeta)$ under the perturbation of the standard slit domain D.

The Jordan arc γ induces not only the continuous motion $\xi(t)$ on $\partial\mathbb{H}$ but also the motion of the horizontal slits $\mathbf{s}(t)$ of the image domain $D_t=g_t(D)$. In Sections 2.6 and 3.1, an equation (see (3.1.3)) describing the horizontal slits $\mathbf{s}(t)$ will be derived that involves $\xi(t)$, obtained by taking suitable trace of the KL equation on the slits.

Conversely, given any pair $(\xi(t),\mathbf{s}(t))$ such that

(i) $\xi(t)$ is a real-valued continuous function in t,
(ii) $(\xi(t),\mathbf{s}(t))$ satisfies the slit equation (3.1.3),

we can substitute it into the right-hand side of the KL equation for $g_t(z)$. The unique solution $g_t(z)$ of the resulting equation gives rise to a family of growing hulls $\{F_t\}$ contained in D, which will be called a *Komatu-Loewner evolution* (KLE) driven by $\xi(t)$ or by $(\xi(t),s(t))$. Basic properties of KLE are presented in Sections 3.2 and 3.3.

In Sections 4.1 and 4.2, we randomize the Jordan arc γ in D and assume that it enjoys in distribution a domain Markov property and the invariance in law under linear maps in an analogous manner to [Schramm (2000)]. The random process

$\mathbf{W}_t = (\xi(t), \mathbf{s}(t))$ induced by the random Jordan arc γ will then be shown to satisfy a system of stochastic differential equations (4.2.10) and (4.2.11). The equation (4.2.11) for the slit motion $\mathbf{s}(t)$ is the same as (3.1.3). The diffusion and drift coefficients α and b appearing in the SDE (4.2.10) for the driving process $\xi(t)$ are certain homogeneous functions. When $D = \mathbb{H}$, namely, when no slit C_j is present, $\mathbf{W}_t = \xi(t)$ is reduced to $\xi(t) = \xi(0) + \alpha B_t$ for the one dimensional standard Brownian motion B_t and a non-zero constant α, namely, to the case of SLE_κ with $\kappa = \alpha^2$. Given functions α, b as above with a Lipschitz continuity, we can solve the SDE (4.2.10) and (4.2.11) in $(\xi(t), \mathbf{s}(t))$ and consider the KLE $\{F_t\}$ driven by it. This family $\{F_t\}$ of random growing hulls is called the *stochastic Komatu-Loewner evolution* driven by $(\xi(t), \mathbf{s}(t))$ and denoted by $\mathrm{SKLE}_{\alpha,b}$. The adaptedness of the random conformal mapping $g_t(z)$, its inverse $g_t^{-1}(z)$ and the resulting random hulls $\{F_t\}$ with respect to the filtration $\{\mathcal{G}_t\}$ generated by the Brownian motion B_t appearing in (4.2.10) are carefully investigated in Section 4.3. These adaptedness results are presented here for the first time and are crucial in applying stochastic calculations to investigate transforms of $\mathrm{SKLE}_{\alpha,b}$ in Chapter 5.

Chapter 5 deals with transformations of KLE and $\mathrm{SKLE}_{\alpha,b}$ by a certain class of univalent maps. Using the concept of kernel convergence of unbounded domains formulated in Section 5.1, a geometric characterization of KLE is given in Theorem 5.2.7 that does not explicitly involve the KL equation. This characterization will enable us to establish in Section 5.3 the invariance of KLE under a class of univalent maps in the following sense:

Suppose that D and $\widetilde{D}$ are either standard slit domains or the upper half-plane $\mathbb{H}$, and the degree of the connectivity of $\widetilde{D}$ is equal to or smaller than that of D. We consider a specific class of univalent functions h that maps a subdomain of D into $\widetilde{D}$. Then, for any KLE $\{F_t\}$ on D, the family $\{h(F_t)\}$ of the image hulls by h is again a KLE on $\widetilde{D}$ up to an appropriate reparametrization of t and up to the random time that $\{F_t\}$ leaves the subdomain of D; see Proposition 5.3.7 and Lemma 5.3.8.

Section 5.4 studies the transformations of $\mathrm{SKLE}_{\alpha,b}$ by h. A semi-martingale decomposition of the driving function (process) of the image hulls $\{h(F_t)\}$ is derived in (5.4.4). A local use of Itô's formula is made to derive the identity (5.4.4). Formula (5.4.4) contains a function on $\partial\mathbb{H}$ defined by

$$b_{\mathrm{BMD}}(\zeta, D) = 2\pi \lim_{z\to\zeta}(\Psi_D(z,\zeta) - \Psi_{\mathbb{H}}(z,\zeta))$$

which indicates a discrepancy of the standard slit domain D from the upper half-plane $\mathbb{H}$ relative to BMD. We call $b_{\mathrm{BMD}}(0, D)$ the *BMD domain constant.*

As consequences, the conformal invariance of $\mathrm{SKLE}_{\sqrt{6},-b_{\mathrm{BMD}}}$ and the equivalence of $\mathrm{SKLE}_{\sqrt{\kappa},b}$ to SLE_κ are established in Section 5.5.

In this monograph, we only treat the chordal case, namely, the upper half-plane and its standard slit subdomains. As for investigations of extensions of Loewner equation and Loewner evolution to other canonical multiply connected domains like annulus, circularly slit annulus and circularly slit disk, we refer the readers to [Fukushima and Kaneko (2014); Chen, Fukushima and Suzuki (2017)] and references therein.

Our research on SKLE was strongly influenced by the works of R. O. Bauer and R. M. Friedrich [Bauer and Friedrich (2006, 2008)] who initiated a comprehensive approach to SKLE and, by the works of G. F. Lawler [Lawler (2006)] and S. Drenning [Drenning (2011)] who used an excursion reflected Brownian motion in their study of SKLE. We are most grateful to Steffen Rohde and Hiroyuki Suzuki for collaborations and discussions on SKLE. Thanks are also due to Ms. Lai Fun Kwong of World Scientific Publishing Co. for her constant and truly helpful cooperation.

Zhen-Qing Chen, Masatoshi Fukushima and Takuya Murayama

About the Authors

Zhen-Qing Chen is a Professor of Mathematics at the University of Washington, Seattle. He obtained his PhD from Washington University in St. Louis in 1992. His research interests include Dirichlet form theory, stochastic analysis, heat kernels, stochastic processes in deterministic as well as random environments, and potential theory. He is a Fellow of the American Mathematical Society and a Fellow of the Institute of Mathematical Statistics. He is a recipient of the Itô Prize in 2019, together with Masatoshi Fukushima.

Masatoshi Fukushima is a Professor Emeritus at Osaka University, Japan, who is an author of three books on Dirichlet forms and Markov processes. He received the 2003 Analysis Prize from the Mathematical Society of Japan, and also 2019 Itô Prize from the Bernoulli Society (together with Z.-Q. Chen) for the paper [Chen and Fukushima, 2018].

Takuya Murayama is currently an Assistant Professor at Kyushu University, Japan. He received a doctoral degree from Kyoto University in March 2021 and was a postdoctoral research fellow of Japan Society for the Promotion of Science in 2021.

Contents

Preface v

About the Authors xi

1. Multiply connected planar domain and Brownian motion 1

1.1 Absorbed Brownian motion (ABM), Green function and harmonic function . . . 1
1.2 Brownian motion with darning and analytic function . . . 16
1.3 Existence and uniqueness of BMD . . . 19
1.4 Localization properties . . . 35
1.5 Zero flux characterization of L^2-generator of BMD . . . 37
1.6 BMD harmonic function and zero period property . . . 38
1.6.1 BMD harmonic function: Continuity and locality . . . 39
1.6.2 Zero period property of BMD-harmonic function . . . 41
1.6.3 Harmonic conjugate . . . 43
1.7 Green function and Poisson kernel of BMD . . . 44
1.8 Conformal invariance of BMD and its Green function . . . 49
1.9 Complex Poisson kernel $\Psi(z, \zeta)$ of BMD on a standard slit domain . . . 54

2. Chordal Komatu-Loewner differential equation and BMD 59

2.1 Komatu-Loewner left differential equation generated by a Jordan arc . . . 59
2.1.1 $\mathbb{H}$-hulls and canonical maps for multiply connected domains 59
2.1.2 The map $g_{t,s}$ and an expression of $a_t - a_s$ for a Jordan arc 64
2.1.3 Chordal Komatu-Loewner left differential equation for a Jordan arc . . . 68
2.2 Probabilistic representation of canonical map and half-plane capacity . . . 69
2.2.1 Expression of $\Im g_F$ by BMD and ABM . . . 69
2.2.2 Representing a_F by BMD and uniform bound of $g_F(z) - z$ 73
2.2.3 Representing a_F^0 by ABM and uniform bound of g_F^0 . . . 76

2.3 Continuity of various objects generated by Jordan arc 77
2.3.1 Joint continuity of $\Im g_t(z)$ and equi-continuity of $\{g_t(z)\}$ . . 77
2.3.2 Continuity of $g_{t,s}(z)$, a_t, D_t and $\xi(t)$ 81
2.4 BMD complex Poisson kernel under perturbation of standard slit domain 85
2.4.1 Small perturbation of standard slit domain 85
2.4.2 Lipschitz continuity of $\Psi_D(z, \xi)$ 89
2.5 Joint continuity of $\Psi_t(z, \zeta)$ and chordal Komatu-Loewner differential equation 101
2.6 Chordal Komatu-Loewner differential equation for slit motions 103

3. Komatu-Loewner evolution (KLE) 111

3.1 Slit motion $\mathbf{s}(t)$ induced by a motion $\xi(t)$ on $\partial\mathbb{H}$ 111
3.2 KLE driven by $\xi(t)$ and its basic properties 114
3.3 Half-plane capacity for Komatu-Loewner evolution 124

4. Stochastic Komatu-Loewner evolution (SKLE) 129

4.1 Random Jordan arc and induced process $(\xi(t), \mathbf{s}(t))$ 129
4.1.1 Domain Markov property and conformal invariance 129
4.1.2 Markov property of $\mathbf{W}_t = (\xi(t), \mathbf{s}(t))$ 132
4.1.3 Scaling property and homogeneity in horizontal direction of $\mathbf{W}$ 137
4.2 Stochastic differential equation for $(\xi(t), \mathbf{s}(t))$ 140
4.3 Stochastic Komatu-Loewner evolution $\mathrm{SKLE}_{\alpha,b}$ 145
4.3.1 SDE with homogeneous coefficients 145
4.3.2 $\mathrm{SKLE}_{\alpha,b}$ and its basic properties 147

5. KLE and its transformation 157

5.1 Kernel theorem for subdomains of $\mathbb{H}$ 157
5.2 Continuously growing hulls and KLE 163
5.3 Invariance of KLE under univalent map 172
5.4 Transformation of $\mathrm{SKLE}_{\alpha,b}$ 180
5.4.1 BMD domain constant b_{BMD} 180
5.4.2 Transformation of driving process of $\mathrm{SKLE}_{\alpha,b}$ 181
5.5 Invariance of $\mathrm{SKLE}_{\sqrt{6},-b_{\mathrm{BMD}}}$ and equivalence of $\mathrm{SKLE}_{\sqrt{\kappa},b}$ to SLE_κ 188

Appendix 193

A.1 Dirichlet problem and α-order exit distribution 193
A.2 Conformal invariance of ABM 195
A.3 Construction of BMD starting at every point of D^* 196

A.4 Probabilistic representation of $\Im g_F$ 202
A.5 Proper maps and their degree 207
A.6 Green function of ABM under perturbation of standard slit domain 209
A.7 BMD Poisson kernel under small perturbation of $\mathbb{H}$ 218
A.8 Comparison of half-plane capacities 225

Notes 231

Bibliography 233

Index 237

Chapter 1

Multiply connected planar domain and Brownian motion

1.1 Absorbed Brownian motion (ABM), Green function and harmonic function

Let $\mathbb{C}$ be the complex plane and $\mathbf{Z} = \{Z_t, t \geq 0; \mathbb{P}_z, z \in \mathbb{C}\}$ be the *planar Brownian motion*, where $\mathbb{P}_z$ denotes the law of the Brownian motion starting from z, namely, $\mathbb{P}_z(Z_0 = z) = 1$ for $z \in \mathbb{C}$. In this book, we use $:=$ as a way of definition. For $z \in \mathbb{C}$ and $r > 0$, we use $B_r(z)$ to denote the open disk in $\mathbb{C}$ centered at z with radius r; that is, $B_r(z) := \{w \in \mathbb{C} : |w-z| < r\}$. Denote by $\mathcal{B}(\mathbb{C})$ the collection of all Borel subsets of $\mathbb{C}$. For any open or a closed set $A \subset \mathbb{C}$, we use $C(A)$ and $C_b(A)$ to denote the space of continuous functions on A and the space of bounded continuous functions on A, respectively. For any open set $D \subset \mathbb{C}$ and integer $k \geq 1$, $C^k(D)$ stands for the space of continuous functions on D that have continuous kth derivatives, while $C^\infty(D)$ is the space of continuous functions on D that are infinitely differentiable. We use $C_c(D)$ for the space of continuous functions on D having compact support. Similarly, $C_c^\infty(D)$ is the collection of infinitely differentiable functions on D having compact support.

First of all, we state several basic properties of planar Brownian motion. Planar Brownian motion $\mathbf{Z}$ is a Markov process on $\mathbb{C}$ with the following transition probability (see, e.g., [Port and Stone (1978), Chapter 1]): for any $z \in \mathbb{C}$, $t > 0$ and $B \in \mathcal{B}(\mathbb{C})$,

$$\mathbb{P}_z(Z_t \in B) = \int_B n(t, z - \zeta) d\zeta, \tag{1.1.1}$$

where $d\zeta := dudv$ for $\zeta = u + iv$ and $n(t,z) := \dfrac{1}{2\pi t} e^{-|z|^2/(2t)}$. Planar Brownian motion $\mathbf{Z}$ is an irreducible recurrent strong Markov process with continuous sample paths. The expectation of a random variable relative to the law $\mathbb{P}_z$ will be designated by $\mathbb{E}_z$. For $B \in \mathcal{B}(\mathbb{C})$, the *hitting time* σ_B of B by Brownian motion $\mathbf{Z}$ is a random time defined by

$$\sigma_B = \inf\{t > 0 : Z_t \in B\}$$

with $\inf \emptyset := \infty$, while the *exit time* τ_B from B by $\mathbf{Z}$ is defined by

$$\tau_B = \inf\{t \geq 0 : Z_t \notin B\}.$$

A set $B \in \mathcal{B}(\mathbb{C})$ is called *polar* if $\mathbb{P}_z(\sigma_B < \infty) = 0$ for any $z \in \mathbb{C}$. If $B \in \mathcal{B}(\mathbb{C})$ is non-polar, then $\mathbb{P}_z(\sigma_B < \infty) = 1$ for any $z \in \mathbb{C}$ owing to the irreducible recurrence of $\mathbf{Z}$ ([Fukushima, Oshima and Takeda (2011), Theorem 4.7.1, Exercise 4.7.1]). A point $z \in \mathbb{C}$ is called a *regular point* for a set $B \in \mathcal{B}(\mathbb{C})$ if $\mathbb{P}_z(\sigma_B = 0) = 1$. We use B^r to denote the collection of all regular points of $B \in \mathcal{B}(\mathbb{C})$.

As a probabilistic counterpart of a Lebesgue's theorem, it holds that (cf. [Port and Stone (1978), Theorem 2.7.2], see also [Itô and McKean (1965), §7.11])

(Z.1) if $B \in \mathcal{B}(\mathbb{C})$ is non-empty and no connected component of B reduces to a singleton, then B is non-polar and $z \in B^r$ for every $z \in B$.

A Borel measurable function u defined on an open set $U \subset \mathbb{C}$ is called *harmonic* if it is bounded on each compact subset of U and satisfies the *mean value property*: for any $z \in U$ and $r > 0$ with $\overline{B_r(z)} \subset U$,

$$u(z) = \mathbb{E}_z \left[u(Z_{\tau_{B_r(z)}}) \right].$$

Note that the exit distribution of Brownian motion starting from z from the disk $B_r(z)$ is the uniform distribution on the circle $\partial B_r(z)$. Hence the above display is equivalent to

$$u(z) = \frac{1}{2\pi} \int_0^{2\pi} u(z + re^{i\theta}) d\theta.$$

It is well known (cf. [Port and Stone (1978), Proposition 4.1.2]) that a function u is harmonic in an open set U in the above sense if and only if $u \in C^\infty(U)$ and its Laplacian $\Delta u = 0$ on U.

Let U be a non-empty open subset of $\mathbb{C}$ whose complement is non-polar. For a bounded measurable function f on ∂U, define

$$\mathcal{H}_U f(z) := \mathbb{E}_z \left[f(Z_{\tau_U}) \right], \quad z \in U.$$

It then holds that

(Z.2) $\mathcal{H}_U f$ is bounded and harmonic on U, and further $\lim_{z \to w, z \in U} \mathcal{H}_U f(z) = f(w)$ whenever $w \in \partial U \cap (\mathbb{C} \setminus U)^r$ and f is continuous in a neighborhood of w.

This is a special case of $\alpha = 0$ of Theorem A.1.1 formulated in Section A.1 of the Appendix. See also [Port and Stone (1978), Proposition 4.2.1]. Note that the harmonicity of $\mathcal{H}_U f$ is an immediate consequence of the strong Markov property of $\mathbf{Z}$.

In this section, we investigate the relationship of the 0-order resolvent density of the absorbed Brownian motion on a planar domain D with the classical Green function, and study some behaviors of Green functions and harmonic functions on D especially when D is unbounded.

Let D be a fixed domain such that $\mathbb{C} \setminus D$ is non-polar. The *absorbed Brownian motion* (abbreviated by *ABM*) $\mathbf{Z}^0 = (Z_t^0, \zeta^0, \mathbb{P}_z^0)_{z\in D}$ on D is the part of the planar Brownian motion $\mathbf{Z} = (Z_t, \mathbb{P}_z)_{z\in\mathbb{C}}$ on D defined by

$$Z_t^0 = \begin{cases} Z_t, & 0 \le t < \tau_D \\ \partial, & t \ge \tau_D, \end{cases} \qquad \mathbb{P}_z^0 = \mathbb{P}_z,\ z \in D, \quad \zeta^0 = \tau_D,$$

where ∂ is cemetery point added to D as its one-point compactification. The random time ζ^0 is called the *lifetime* of $\mathbf{Z}^0$. Denote by $p_t^0(z,\zeta)$, $z, \zeta \in D$, the transition probability density of the ABM $\mathbf{Z}^0$ on D with respect to the Lebesgue measure. It follows from (1.1.1) and the *first passage time relation* for $\mathbf{Z}$ ([Itô (1960), §2.6]) that

$$p_t^0(z,\zeta) = n(t, z-\zeta) - \mathbb{E}_z[n(t-\tau_D, Z_{\tau_D} - \zeta); \tau_D < t]. \tag{1.1.2}$$

For $\alpha \ge 0$, the *α-order resolvent density* $G_\alpha^0(z,\zeta)$ of $\mathbf{Z}^0$ is defined to be

$$G_\alpha^0(z,\zeta) = \int_0^\infty e^{-\alpha t} p_t^0(z,\zeta)dt, \quad \text{for } z,\ \zeta \in D. \tag{1.1.3}$$

When $\alpha = 0$, the 0-*order resolvent density* will simply be denoted as $G^0(z,\zeta)$. To emphasize their dependence on the underlying domain D, we shall occasionally adopt the notations $\mathbf{Z}^D = (Z_t^D, \zeta^D, \mathbb{P}_z^D)$, $p_t^D(z,\zeta)$, $G_\alpha^D(z,\zeta)$ and $G^D(z,\zeta)$ in place of $\mathbf{Z}^0 = (Z_t^0, \zeta^0, \mathbb{P}_z^0)$, $p_t^0(z,\zeta)$, $G_\alpha^0(z,\zeta)$ and $G^0(z,\zeta)$, respectively.

The integral $\int_0^\infty n(t,z)dt$ is infinite but the re-centered integral

$$\begin{aligned}\int_0^\infty (n(t,z) - n(t,\mathbf{e}_1))\,dt &= \int_0^\infty \frac{1}{2\pi t}\left(e^{-|z|^2/2t} - e^{-1/2t}\right)dt \\ &= \frac{1}{2\pi}\int_0^\infty \left(\int_{|z|^2 t}^t e^{-s}ds\right)\frac{dt}{t} = -\frac{1}{\pi}\log|z| \quad \text{for} \quad z \in \mathbb{C},\end{aligned} \tag{1.1.4}$$

is finite where $\mathbf{e}_1 = (1,0) \in \mathbb{C}$. Since $n(t,z) \ge n(t,\mathbf{e}_1)$ when $|z| \le 1$ and $n(t,z) < n(t,\mathbf{e}_1)$ when $|z| > 1$, in fact we have from (1.1.4) that

$$\int_0^\infty |n(t,z) - n(t,\mathbf{e}_1)|dt = \frac{1}{\pi}\,|\log|z|| \qquad \text{for } z \in \mathbb{C}. \tag{1.1.5}$$

We now make use of the *fundamental identity for the logarithmic potential* due to S.C. Port and C.J. Stone stating that

$$G^0(z,\zeta) < \infty \quad \text{and} \quad \mathbb{E}_z\left[|\log|Z_{\tau_D} - \zeta||\right] < \infty \quad \text{for every distinct } z,\zeta \in D, \tag{1.1.6}$$

and

$$G^0(z,\zeta) = -\frac{1}{\pi}\log|z-\zeta| + \frac{1}{\pi}\mathbb{E}_z \log|Z_{\tau_D} - \zeta| + W_D(z), \quad z,\ \zeta \in D, \quad z \ne \zeta, \tag{1.1.7}$$

where W_D is a non-negative locally bounded function on $\mathbb{C}$ vanishing on $(\mathbb{C} \setminus D)^r$. This was shown in [Port and Stone (1978), Theorem 3.4.2] by using the identity (1.1.4) among others.

From (1.1.2), one has

$$\begin{aligned}p_t^0(z,\zeta) = n(t, z-\zeta) - n(t,\mathbf{e}_1) - \mathbb{E}_z[n(t-\tau_D, Z_{\tau_D} - \zeta) - n(t-\tau_D, \mathbf{e}_1); \tau_D < t] \\ + n(t,\mathbf{e}_1) - \mathbb{E}_z[n(t-\tau_D, \mathbf{e}_1); \tau_D < t].\end{aligned}$$

Hence (1.1.5) and (1.1.6) imply that, for distinct $z, \zeta \in D$,

$$\int_0^\infty |n(t, \mathbf{e}_1) - \mathbb{E}_z[n(t-\tau_D, \mathbf{e}_1); \tau_D < t]|dt$$
$$\leq G^0(z,\zeta) + \frac{1}{\pi}|\log|z-\zeta|| + \frac{1}{\pi}\mathbb{E}_z\left[|\log|Z_{\tau_D} - \zeta||\right] < \infty.$$

Moreover, (1.1.4) and (1.1.7) imply that

$$W_D(z) = \int_0^\infty \mathbb{E}_z\left[n(t,\mathbf{e}_1) - n(t-\tau_D, \mathbf{e}_1)\mathbf{1}_{\{t>\tau_D\}}\right] dt, \quad z \in D. \tag{1.1.8}$$

It follows from (1.1.8) that

$$\begin{aligned} W_D(z) &= \lim_{N\to\infty} \int_0^N \mathbb{E}_z\left[n(t,\mathbf{e}_1) - n(t-\tau_D, \mathbf{e}_1)\mathbf{1}_{\{t>\tau_D\}}\right] dt \\ &= \lim_{N\to\infty} \mathbb{E}_z\left[\int_0^N n(t,\mathbf{e}_1)dt - \int_0^{(N-\tau_D)^+} n(s,\mathbf{e}_1)ds\right] \\ &= \lim_{N\to\infty} \mathbb{E}_z\left[\int_{(N-\tau_D)^+}^N n(t,\mathbf{e}_1)dt\right] \geq 0. \end{aligned} \tag{1.1.9}$$

Here for $a \in R$, $a^+ := \max\{a, 0\}$. From the last display, we get the following monotonicity property of W_D:

$$\text{if } D_1 \supset D_2, \text{ then } W_{D_1}(z) \geq W_{D_2}(z) \text{ for every } z \in D_2. \tag{1.1.10}$$

Denote by $\mathbb{H}$ the upper half plane $\{z = x + iy \in \mathbb{C} : y > 0\}$ and by $G^{\mathbb{H}}(z,\zeta)$ the 0-order resolvent density of the ABM $\mathbf{Z}^{\mathbb{H}}$ on $\mathbb{H}$. One can easily deduce from the reflection principle of Brownian motion that

$$p_t^{\mathbb{H}}(z,\zeta) = n(t, z-\zeta) - n(t, z-\bar{\zeta}) \quad \text{for } z, \zeta \in \mathbb{H} \tag{1.1.11}$$

(cf. [Itô (1960), §2,6]), which combined with (1.1.4) yields that

$$G^{\mathbb{H}}(z,\zeta) = \frac{1}{\pi}\left(\log|z-\bar{\zeta}| - \log|z-\zeta|\right), \quad z, \zeta \in \mathbb{H}. \tag{1.1.12}$$

Lemma 1.1.1. *$W_{\mathbb{H}}(z) = 0$ for $z \in \mathbb{C}$.*

Proof. For $z = x+iy \in \mathbb{H}$, by the reflection principle for one-dimensional Brownian motion, it is well known (see, e.g., [Karatzas and Shreve (1998), (8.4) on p. 96]) that

$$\mathbb{P}_z(\tau_{\mathbb{H}} < t) = \sqrt{\frac{2}{\pi}} \int_{y/\sqrt{t}}^\infty e^{-r^2/2} dr \quad \text{for } t \geq 0. \tag{1.1.13}$$

Thus, we have by (1.1.9) that for any $\varepsilon \in (0,1)$,

$$\begin{aligned}
W_{\mathbb{H}}(z) &= \lim_{N\to\infty} \mathbb{E}_z\left[\int_{(N-\tau_{\mathbb{H}})^+}^N n(t,\mathbf{e}_1)dt\right] \\
&\le \limsup_{N\to\infty}\left(\mathbb{E}_z\left[\int_{(1-\varepsilon)N}^N \frac{1}{2\pi t}e^{-1/(2t)}dt;\tau_{\mathbb{H}}\le \varepsilon N\right]\right. \\
&\qquad \left. +\mathbb{E}_z\left[\int_0^N \frac{1}{2\pi t}e^{-1/(2t)}dt;\tau_{\mathbb{H}}> \varepsilon N\right]\right) \\
&\le \frac{1}{2\pi}\log\frac{1}{1-\varepsilon} + \limsup_{N\to\infty}\frac{1}{2\pi}\left(\int_0^1 t^{-1}e^{-1/(2t)}dt+\log N\right)\mathbb{P}_z(\tau_{\mathbb{H}}>\varepsilon N) \\
&\le \frac{1}{2\pi}\log\frac{1}{1-\varepsilon} \\
&\quad + \limsup_{N\to\infty}\frac{1}{\sqrt{2}\pi^{3/2}}\left(\int_0^1 t^{-1}e^{-1/(2t)}dt+\log N\right)\int_0^{y/\sqrt{\varepsilon N}} e^{-r^2/2}dr \\
&\le \frac{1}{2\pi}\log\frac{1}{1-\varepsilon} + \limsup_{N\to\infty}\frac{1}{\sqrt{2}\pi^{3/2}}\left(\int_0^1 t^{-1}e^{-1/(2t)}dt+\log N\right)\frac{y}{\sqrt{\varepsilon N}} \\
&= \frac{1}{2\pi}\log\frac{1}{1-\varepsilon}.
\end{aligned}$$

As the above holds for every $\varepsilon \in (0,1)$, taking $\varepsilon \downarrow 0$ yields that $W_{\mathbb{H}}(z)=0$ for every $z\in\mathbb{H}$ and hence for every $z\in\mathbb{C}$. □

It follows from Lemma 1.1.1 and the monotonicity (1.1.10) that $W_D=0$ on $\mathbb{C}$ for any domain $D\subset\mathbb{H}$ and

$$G^0(z,\zeta) = -\frac{1}{\pi}\log|z-\zeta| + \frac{1}{\pi}\mathbb{E}_z\log|Z_{\tau_D}-\zeta|, \quad z,\zeta\in D, \quad z\neq\zeta. \tag{1.1.14}$$

In other words, the 0-order resolvent density of $\mathbf{Z}^0$ coincides with the classical Green function up to a constant multiple $1/\pi$ whenever the domain D is contained in a half plane. For this reason, we call the 0-order resolvent density of the ABM $\mathbf{Z}^0$ on D also its *Green function*. When D is bounded, D can be a subset of $\mathbb{H}$ by a suitable translation and the law of the planar Brownian motion is translation invariant, we also have the identity (1.1.14). But we note that W_D can be non-trivial for some unbounded D. For example, when $D=\{z\in\mathbb{C}:|z|>r\}$ for $r>0$, $W_D(z)=\frac{1}{\pi}\log(|z|/r)$ on D (cf. [Port and Stone (1978), Proposition 3.4.9]).

We also notice that, as $p_t^0(z,\zeta)$ is symmetric in $z,\zeta\in D$ ([Fukushima, Oshima and Takeda (2011), Lemma 4.1.3]), so is the Green function of $\mathbf{Z}^0$:

$$G^0(z,\zeta)=G^0(\zeta,z), \quad z,\ \zeta\in D. \tag{1.1.15}$$

For $\rho>0$, define $B_\rho=\{z\in\mathbb{C}:|z|<\rho\}$. The following proposition for $\rho=1$ has appeared in [Lawler (2005), (2.12)].

Proposition 1.1.2. *Let $\mathbf{Z}^{\mathbb{H}}=(Z_t^{\mathbb{H}},\zeta^{\mathbb{H}},\mathbb{P}_z^{\mathbb{H}})$ be the ABM on the upper half plane $\mathbb{H}$.*

For $0 \le \theta_1 < \theta_2 \le \pi$, *let* $\Gamma_{\theta_1,\theta_2} = \{\rho e^{i\theta} : \theta \in (\theta_1, \theta_2)\} \subset \mathbb{H} \cap \partial B_\rho$. *It then holds that*

$$\mathbb{P}_z^{\mathbb{H}}\left(Z^{\mathbb{H}}_{\sigma_{\mathbb{H}\cap\partial B_\rho}} \in \Gamma_{\theta_1,\theta_2}\right) = \frac{2}{\pi}\eta \sin\vartheta \int_{\theta_1}^{\theta_2} (1 + c(\theta,\vartheta,\eta)\eta) \sin\theta d\theta \qquad (1.1.16)$$

for $z \in \mathbb{H} \setminus \overline{B}_\rho$. *Here*

$$z = re^{i\vartheta} \quad (r > \rho), \qquad \eta = \frac{\rho}{r} \in (0,1),$$

and $c(\theta,\vartheta,\eta)$ *is a real valued continuous function of* $(\theta,\vartheta,\eta) \in (0,\pi)^2 \times (0,1)$ *satisfying, for any* $\kappa \in (0,1)$,

$$c_\kappa := \sup_{\theta\in(0,\pi),\ \vartheta\in(0,\pi),\ \eta\in(0,\kappa]} |c(\theta,\vartheta,\eta)| < \infty. \qquad (1.1.17)$$

Proof. $\psi(z) = z + \rho^2 z^{-1}$ defines a conformal map from $\mathbb{H}_\rho := \mathbb{H} \setminus \overline{B}_\rho$ onto $\mathbb{H}$ that sends the part $\partial B_\rho \cap \mathbb{H}$ of the boundary of $\mathbb{H}_\rho$ onto the interval $(-2\rho, 2\rho) \subset \partial\mathbb{H}$ by $\psi(\rho e^{i\theta}) = 2\rho\cos\theta$, $\theta \in (0,\pi)$. We make use of the conformal invariance of the ABM $\mathbf{Z}^{\mathbb{H}_\rho} := (Z_t^{\mathbb{H}_\rho}, \zeta^{\mathbb{H}_\rho}, \mathbb{P}_z^{\mathbb{H}_\rho})$ on $\mathbb{H}_\rho$ according to Theorem A.2.1 as follows: Let $\check{\mathbf{Z}}^{\mathbb{H}_\rho} := (\check{Z}_t^{\mathbb{H}_\rho}, \check{\zeta}^{\mathbb{H}_\rho}, \mathbb{P}_z^{\mathbb{H}_\rho})_{z\in\mathbb{H}_\rho}$ be the time change of $\mathbf{Z}^{\mathbb{H}_\rho}$ by means of a strictly increasing continuous additive functional $A_t = \int_0^t |\psi'(Z_s^{\mathbb{H}_\rho})|^2 ds$, namely $\check{Z}_t^{\mathbb{H}_\rho} = Z^{\mathbb{H}_\rho}_{A_t^{-1}}$, $\check{\zeta}^{\mathbb{H}_\rho} = A_{\zeta^{\mathbb{H}_\rho}}$. Then, for any $z \in \mathbb{H}_\rho$,

$$\left(\psi(\check{Z}_t^{\mathbb{H}_\rho}), \check{\zeta}^{\mathbb{H}_\rho}, \mathbb{P}_z^{\mathbb{H}_\rho}\right) \text{ is identical in law with } (Z_t^{\mathbb{H}}, \zeta^{\mathbb{H}}, \mathbb{P}^{\mathbb{H}}_{\psi(z)}). \qquad (1.1.18)$$

Accordingly, if we denote by $p_\Gamma(z)$ the left-hand side of (1.1.16), then $p_\Gamma(z) = \mathbb{P}_z^{\mathbb{H}_\rho}\left(Z^{\mathbb{H}_\rho}_{\zeta^{\mathbb{H}_\rho}-} \in \Gamma\right)$ for $\Gamma = \Gamma_{\theta_1,\theta_2}$, and

$$p_\Gamma(z) = \mathbb{P}_z^{\mathbb{H}_\rho}\left(\psi(\check{Z}^{\mathbb{H}_\rho}_{\check{\zeta}^{\mathbb{H}_\rho}-}) \in \psi(\Gamma)\right) = \mathbb{P}^{\mathbb{H}}_{\psi(z)}\left(Z^{\mathbb{H}}_{\zeta^{\mathbb{H}}-} \in (2\rho\cos\theta_2, 2\rho\cos\theta_1)\right).$$

The last expression admits the Poisson integral representation so that

$$p_\Gamma(z) = \frac{1}{\pi}\int_{\xi_1}^{\xi_2} \frac{v}{(u-\xi)^2 + v^2} d\xi.$$

with $u = \Re\psi(z) = x(1 + \frac{\rho^2}{|z|^2})$, $v = \Im\psi(z) = y(1 - \frac{\rho^2}{|z|^2})$ for $z = x + iy = re^{i\vartheta} \in \mathbb{H}_\rho$ and $\xi_1 = 2\rho\cos\theta_2$, $\xi_2 = 2\rho\cos\theta_1$.

We substitute these and $\xi = 2\rho\cos\theta$ into the above by setting $\eta = \frac{\rho}{r}$. We then have $(u-\xi)^2 + v^2 = r^2(1 + d(\theta,\vartheta,\eta)\eta)$ with

$$d(\theta,\vartheta,\eta) = \eta^3 + 2\eta(\cos^2\vartheta - \sin^2\vartheta) - 4(1+\eta^2)\cos\vartheta\cos\theta + 4\eta\cos^2\theta,$$

so that

$$\frac{1}{\pi}\frac{v}{(x-\xi)^2 + v^2}d\xi = -\frac{2}{\pi}\eta\sin\vartheta(1 + c(\theta,\vartheta,\eta)\eta)\sin\theta d\theta,$$

with

$$c(\theta,\vartheta,\eta) = -\frac{d(\theta,\vartheta,\eta) + \eta}{1 + d(\theta,\vartheta,\eta)\eta}.$$

As $\eta \in (0,1)$, we have $|d(\theta,\vartheta,\eta)| < 15$ and accordingly, for $\eta \le 1/16$,

$$|c(\theta,\vartheta,\eta)| \le \frac{|d(\theta,\vartheta,\eta)|+1}{1-|d(\theta,\vartheta,\eta)|\eta} \le \frac{16}{1-15/16} = 16^2.$$

Since $(u-\xi)^2+v^2$ is positive unless $\eta = 1$, we see for $(\theta,\vartheta,\eta) \in (0,\pi)^2 \times (0,1)$ that $1 + d(\theta,\vartheta,\eta)\eta$ does not vanish and consequently, $c(\theta,\vartheta,\eta)$ is continuous. In particular, for any $\kappa \in (1/16,1)$, $\sup_{\theta\in(0,\pi),\ \vartheta\in[0,\pi),\eta\in[\frac{1}{16},\kappa]} |c(\theta,\vartheta,\eta)|$ is finite, yielding (1.1.17). □

Here, we state the well known maximum principle for harmonic functions in a way convenient to be used for unbounded domains. For a locally compact Hausdorff space E, denote by $E \cup \{\partial\}$ the one-point compactification of E and by $C_\infty(E)$ the space of all continuous functions on $E \cup \{\partial\}$ vanishing at ∂.

Lemma 1.1.3. *Let u be a harmonic function on a domain $D \subset \mathbb{C}$.*

(i) *If u is non-negative on D, then u is either identically 0 on D or positive everywhere on D.*
(ii) *If $u \in C_\infty(D)$, then u is identically 0 on D.*

Proof. (i) We have by the mean value property (1.1.1) of u that, if $u(z) = 0$ for some $z \in D$, then u vanishes on the circle $\partial B_r(z)$ for any $r > 0$ with $\overline{B_r(z)} \subset D$, and hence on the disk $B_r(z)$ contained in D. Consequently, the set $U = \{z \in D : u(z) = 0\}$ is open and closed so that U is either D or empty.

(ii) If $a = \max_{z\in D\cup\{\partial\}} u(z)$ is positive, then a is attained at some point of D and, by the same reasoning as above, $u = a$ identically on D, contradicting the assumption. In the same way, we see that u cannot be negative. □

A closed connected subset of $\mathbb{C}$ containing at least two points will be called a *continuum*. We fix a planar domain E such that either $E = \mathbb{C}$ or $\mathbb{C} \setminus E$ is a continuum. Consider, for $N \in \mathbb{N}$, the domain $D \subset E$ expressed as

$$D = E \setminus K, \qquad K = \bigcup_{j=1}^{N} A_j, \tag{1.1.19}$$

where $\{A_j\}_{1\le j\le N}$ are mutually disjoint compact continua contained in E. In the case that $E \ne \mathbb{C}$, such a domain D is called an $(N+1)$-*connected domain.*

A typical example of an $(N+1)$-connected domain is a *standard slit domain* which is by definition a domain D expressed as above with $E = \mathbb{H}$ and with $\{A_i,\ 1 \le i \le N\}$ being mutually disjoint line segments parallel to the x-axis contained in $\mathbb{H}$. From Section 1.9 of this chapter throughout the rest of the book except for Sections 2.1.1 and 2.2.1, we will work on standard slit domains. See [Conway (1995)], [Fukushima (2020)] and [Chen, Fukushima and Suzuki (2017)] for other types of canonical multiply connected planar domains.

Given an $(N+1)$-connected domain (1.1.19), Let $\mathbf{Z}^E = (Z_t^E, \zeta^E, \mathbb{P}_z^E)_{z\in E}$ be the ABM on E. and let

$$\varphi^{(i)}(z) = \mathbb{P}_z^E\left(Z_{\sigma_K}^E \in A_i,\ \sigma_K < \infty\right), \quad z \in E,\ 1 \le i \le N. \tag{1.1.20}$$

We call the family $\{\varphi^{(j)}|_D : 1 \le j \le N\}$ the *harmonic basis* for the multiply connected domain D.

Let γ be a C^1-smooth simple closed curve surrounding A_j, namely, $\gamma \subset D$, $\mathrm{ins}\gamma \supset A_j$, $\overline{\mathrm{ins}\gamma} \cap A_k = \emptyset$ for every $k \ne j$. Here, $\mathrm{ins}\gamma$ denotes the bounded component of $\mathbb{C} \setminus \gamma$ and is called the interior of γ. For a harmonic function u defined on $U \cap D$ for some neighborhood U of A_j, the value

$$\int_\gamma \frac{\partial u(\zeta)}{\partial \mathbf{n}_\zeta} s(d\zeta), \tag{1.1.21}$$

is independent of the choice of such curve γ with $\mathbf{n}$ denoting the unit normal vector pointing toward A_j and s the arc length of γ. This value is called the *period* of u around A_j.

Lemma 1.1.4. *Let D be a domain defined by* (1.1.19).

(i) *For the function $\varphi^{(i)}$ defined by* (1.1.20), *$\varphi^{(i)}\big|_D$ is harmonic and strictly positive on D. It is continuously extendable to $\overline{E}$ with value* 1 *on A_i and* 0 *on $\partial E \cup (\bigcup_{1\le j\le N, j\ne i} A_j)$. It further satisfies*

$$\lim_{z\to\infty, z\in D} \varphi^{(i)}(z) = 0, \qquad 1 \le i \le N. \tag{1.1.22}$$

(ii) *Assume that $E \ne \mathbb{C}$. For any bounded continuous function f on ∂E, the function defined by*

$$\psi_f(z) := \mathbb{E}_z^0\left[f\left(Z_{\zeta^0-}^0\right); Z_{\zeta^0-}^0 \in \partial E\right], \quad z \in D, \tag{1.1.23}$$

in terms of the ABM $\mathbf{Z}^0 = (Z_t^0, \mathbb{P}_z^0)$ on D is harmonic on D. It is continuously extendable to $\overline{E}$ with value f on ∂E and 0 *on $\bigcup_{j=1}^N A_j$. If f is further of compact support, then*

$$\lim_{|z|\to\infty, z\in D} \psi_f(z) = 0. \tag{1.1.24}$$

Proof. (i) In terms of the planar Brownian motion $\mathbf{Z} = (Z_t, \mathbb{P}_z)$, $\varphi^{(i)}(z)$, $z \in D$, can be expressed as $\varphi^{(i)}(z) = \mathbb{P}_z\left(Z_{\sigma_K} \in A_i, \sigma_K < \sigma_{\partial\mathbb{H}}\right) = \mathcal{H}_D g(z)$, where g is a continuous function on $\partial D = \partial K \cup \partial E$ taking value 1 on ∂A_i and 0 on $\partial D \setminus \partial A_i$. The stated harmonicity and continuous extendability follow from this and the properties (**Z**.1), (**Z**.2) of **Z**. Its strict positivity is a consequence of Lemma 1.1.3 (i).

Choose $\rho > 0$ such that $B_\rho \supset K$. Then, for $z \in \mathbb{H} \setminus \overline{B}_\rho$, $\varphi^{(i)}(z)$ is dominated by the left hand side of (1.1.16) with $\theta_1 = 0, \theta_2 = \pi$ because the path $Z_t^{\mathbb{H}}$ must cross ∂B_ρ before hitting the set K. So (1.1.22) follows from (1.1.16). We note that $\varphi^{(i)}$ can be also expressed as $\varphi^{(i)}(z) = \mathbb{P}_z^0\left(Z_{\zeta^0-}^0 \in A_i\right)$, $z \in D$, in terms of the ABM $\mathbf{Z}^0 = (Z_t^0, \zeta^0, \mathbb{P}_z^0)$ on D.

(ii) The first statement can be shown similarly to (i). For the second statement, it suffices to choose $\rho > 0$ such that B_ρ contains the support $S \subset \partial E$ of f to get (1.1.24). □

In the proof of the following proposition, we need to consider a smooth $(N+1)$-connected bounded domain D (1.1.19) whose boundary ∂D consists of mutually disjoint closed analytic Jordan curves

$$\Gamma_0 = \partial E, \quad \Gamma_j = \partial A_j, \quad 1 \le j \le N.$$

Here, an *analytic Jordan curve* is a Jordan curve that is a finite union of open analytic arcs, and an *analytic arc* is by definition the image $\psi((-1,1))$ of the interval $(-1,1) \subset \mathbb{C}$ under a one-to-one analytic map ψ defined on a neighborhood of $(-1,1)$.

Let $G^0(z,\zeta)$ be the Green function of the ABM $\mathbf{Z}^0$ on such a bounded smooth domain D. According to [Garnett and Marshall (2005), Theorem II.2.5], $G^0(z,\zeta)$ can be extended to be harmonic in ζ on a neighborhood of ∂D with

$$\frac{\partial G^0(z,\zeta)}{\partial \mathbf{n}_\zeta} < 0, \quad \text{for any} \quad \zeta \in \partial D, \tag{1.1.25}$$

where $\mathbf{n}_\zeta$ denotes the unit outward normal vector at $\zeta \in \partial D$. Furthermore, any function $u \in C(\overline{D})$ that is harmonic on D admits the *Poisson integral formula*:

$$u(z) = -\frac{1}{2}\int_{\partial D} \frac{\partial G^0(z,\zeta)}{\partial \mathbf{n}_\zeta} u(\zeta)s(d\zeta), \quad z \in D, \tag{1.1.26}$$

by making use of Green's second formula.

Proposition 1.1.5. *Let $D = \mathbb{H}\setminus K,\ K = \bigcup_{i=1}^N A_i$, be an $(N+1)$-connected domain and $G^0(z,\zeta)$ be the Green function of the ABM $\mathbf{Z}^0$ on D. Define $\varphi^{(i)}$ by (1.1.20) with $E = \mathbb{H},\ 1 \le i \le N$.*

(i) $\lim_{z\to\infty, z\in D} G^0(z,\zeta) = 0$. *The function $z \mapsto G^0(z,\zeta)$ is positive and harmonic in $D \setminus \{\zeta\}$, and can be continuously extended to $\overline{\mathbb{H}} \setminus \{\zeta\}$ by setting its value on $(\bigcup_{j=1}^N A_j) \cup \partial\mathbb{H}$ to be zero.*

(ii) *Let $\mathbf{Z}^{\mathbb{H}} = \{Z_t^{\mathbb{H}}, t \ge 0; \mathbb{P}_z^{\mathbb{H}}, z \in \mathbb{H}\}$ be the ABM on $\mathbb{H}$ and $K_{\mathbb{H}}(z,\zeta) = \frac{1}{\pi}\frac{\Im z}{|z-\zeta|^2}$, $z \in \mathbb{H}$, $\zeta \in \partial\mathbb{H}$, be the Poisson kernel of $\mathbf{Z}^{\mathbb{H}}$. Then*

$$-\frac{1}{2}\frac{\partial}{\partial \mathbf{n}_\zeta} G^0(z,\zeta) = K_{\mathbb{H}}(z,\zeta) - \mathbb{E}_z^{\mathbb{H}}\left[K_{\mathbb{H}}(Z_{\sigma_K}^{\mathbb{H}},\zeta); \sigma_K < \infty\right], \quad z \in D,\ \zeta \in \partial\mathbb{H}. \tag{1.1.27}$$

The left-hand side of (1.1.27) is positive harmonic in $z \in D$.

(iii) *For each $z \in D$ and $1 \le i \le N$, $-2\varphi^{(i)}(z)$ equals the period of the harmonic function $G^0(z,\cdot)$ around A_i.*

(iv) *For any bounded continuous function f on $\partial\mathbb{H}$,*

$$\mathbb{E}_z^0\left[f(Z_{\zeta^0-}^0); Z_{\zeta^0-}^0 \in \partial\mathbb{H}\right] = -\frac{1}{2}\int_{\partial\mathbb{H}} \frac{\partial G^0(z,\zeta)}{\partial \mathbf{n}_\zeta} f(\zeta)s(d\zeta), \ z \in D, \quad (1.1.28)$$

where s is the Lebesgue measure on $\partial\mathbb{H}$.

Proof. (i) $G^0(z,\zeta)$ is strictly positive because of (1.1.3) and Lemma 1.1.3. The first assertion follows from $0 < G^0(z,\zeta) \le G^{\mathbb{H}}(z,\zeta)$ and (1.1.12). The property $\lim_{z\in D, z\to\xi} G^0(z,\zeta) = 0$ for $\xi \in \partial\mathbb{H}$ also follows from this and the symmetry (1.1.15). The same property for $\xi \in \partial K$ follows from the expression (1.1.14) of $G^0(z,\zeta)$ and the identity in terms of the planar Brownian motion

$$\mathbb{E}_z\left[\log|Z_{\tau_D} - \zeta|\right] = \mathbb{E}_z\left[\log|Z_{\sigma_K} - \zeta|; \sigma_K < \sigma_{\partial\mathbb{H}}\right] + \mathbb{E}_z\left[\log|Z_{\sigma_{\partial\mathbb{H}}} - \zeta|; \sigma_K > \sigma_{\partial\mathbb{H}}\right].$$

The first term of the right-hand side tends to $\log|\xi - \zeta|$ as $z \in D$, $z \to \xi \in \partial K$ by the same consideration as in the proof of Lemma 1.1.4 (i). The square of the second term is dominated by the product $\mathbb{E}_z\left[(\log|Z_{\sigma_{\partial\mathbb{H}}} - \zeta|)^2\right]\mathbb{P}_z(\sigma_{\partial\mathbb{H}} < \sigma_K)$. The first factor is locally bounded in $z \in \mathbb{H}$, while the second factor tends to zero as $z \in D$, $z \to \xi \in \partial K$, completing the proof.

(ii) Since $\mathbf{Z}^0$ is the part of $\mathbf{Z}^{\mathbb{H}}$ on D,

$$G^0(z,\zeta) = G^{\mathbb{H}}(z,\zeta) - \mathbb{E}_z^{\mathbb{H}}\left[G^{\mathbb{H}}(Z_{\sigma_K}^{\mathbb{H}},\zeta)\right], \quad z,\zeta \in D, \ z \ne \zeta. \quad (1.1.29)$$

By (1.1.12), we have

$$-\frac{1}{2}\frac{\partial}{\partial \mathbf{n}_\zeta} G^{\mathbb{H}}(z,\zeta) = \frac{1}{2}\frac{\partial}{\partial\eta} G^{\mathbb{H}}(z,\xi+i\eta)\big|_{\eta=0} = K_{\mathbb{H}}(z,\zeta), \quad z \in \mathbb{H}, \ \zeta = \xi + i0. \ (1.1.30)$$

and

$$0 < \frac{1}{2}\frac{\partial}{\partial\eta} G^{\mathbb{H}}(x+iy,\xi+i\eta) \le \frac{1}{\pi\delta} \quad \text{whenever } z \in K \text{ and } \eta > 0, \quad (1.1.31)$$

where $\delta = \mathrm{dist}(K,\partial\mathbb{H}) > 0$. (1.1.27) follows from (1.1.29), (1.1.30) and (1.1.31).

The left-hand side of (1.1.27) is non-negative harmonic in $z \in D$ by (i) and (1.1.15). As $\lim_{z\to\zeta} K_{\mathbb{H}}(z,\zeta) = \infty, \zeta \in \partial\mathbb{H}$, it is positive for $z \in D$ by (1.1.27) and Lemma 1.1.3.

(iii) Suppose $h(\zeta)$ is a non-constant harmonic function on a simply connected region $U \subset \mathbb{H}$. Then h is the imaginary part of a certain analytic function f on U. The set of points $\zeta \in U$ with $\nabla h(\zeta) = 0$ (called critical points of h) must be finite on each compact subset of U because otherwise $f' \equiv 0$ on U and so h is constant on U.

In what follows, we fix $z \in D$. Consider the set $D_a = \{\zeta \in D : G^0(z,\zeta) > a\}$ for $a > 0$. For a sufficiently small $a > 0$, D_a is a bounded $(N+1)$-connected subdomain of D. By taking into account the above observation on critical points of $G^0(z,\cdot)$, one can choose a sequence $\{a_n\}$ decreasing to 0 such that ∂D_{a_n} avoids critical points and consists of mutually disjoint closed curves $\Gamma_0^{(n)}, \Gamma_1^{(n)}, \ldots, \Gamma_N^{(n)} \subset \mathbb{H}$ with $\mathrm{ins}\Gamma_0^{(n)} \supset \bigcup_{j=1}^N \Gamma_j^{)n)}$, $\mathrm{ins}\Gamma_j^{(n)} \supset A_j$, $1 \le j \le N$, for each $n \ge 1$, in view of (i) and (1.1.15).

Each $\Gamma_j^{(n)}$ is an analytic Jordan curve, $0 \le j \le N$. Indeed, one can take a simply connected bounded open set U with $\Gamma_j^{(n)} \subset U \subset D \setminus \{z\} \setminus \bigcup_{k\neq j} \text{ins}\Gamma_k^{(n)}$ such that U contains no critical point of $G^0(z,\cdot)$. There exists then an analytic function f on U with $\Im f(\zeta) = G^0(z,\zeta)$, $\zeta \in U$. Since $f'(\zeta) \neq 0$, $\zeta \in U$, f maps U onto $V = f(U) \subset \mathbb{C}$ univalently. Let $\Lambda = \{w \in V : \Im w = a_n\}$ and $\Psi = f^{-1}$. Then $\Psi(\Lambda) = \{\zeta \in U : G^0(z,\zeta) = a_n\} = \Gamma_j^{(n)}$ as was to be proved.

We let $K^{(n)} = \bigcup_{j=1}^N \Gamma_j^{(n)}$ and

$$\varphi_n^{(i)}(z) = \mathbb{P}_z^{\mathbb{H}}\left(\sigma_{K^{(n)}} < \infty,\ Z_{\sigma_{K^{(n)}}} \in \Gamma_i^{(n)}\right), \quad z \in \mathbb{H} \setminus \bigcup_{j=1}^N \overline{\text{ins}\Gamma_j^{(n)}},\ 1 \le i \le N,$$

which is harmonic on $\mathbb{H} \setminus \bigcup_{j=1}^N \overline{\text{ins}\Gamma_j^{(n)}}$, continuous on its closure, and taking value 1 on $\Gamma_i^{(n)}$ and 0 on $\partial\mathbb{H} \cup \bigcup_{1\le j\le N, j\neq i} \Gamma_j^{(n)}$ by Lemma 1.1.4 with $\overline{\text{ins}\Gamma_j^{(n)}}$ in place of A_j, $1 \le j \le N$. Since the Green function for D_{a_n} equals $G^0(z,\zeta) - a_n$, we get from (1.1.26)

$$\varphi_n^{(i)}(z) = -\frac{1}{2}\int_{\Gamma_i^{(n)}} \frac{\partial G^0(z,\zeta)}{\partial \mathbf{n}_\zeta} ds(\zeta) - \frac{1}{2}\int_{\Gamma_0^{(n)}} \frac{\partial G^0(z,\zeta)}{\partial \mathbf{n}_\zeta} \varphi_n^{(i)}(\zeta) ds(\zeta). \tag{1.1.32}$$

The first term on the right hand side equals $-\frac{1}{2}p_i$, where p_i denotes the period (1.1.21) of $G^0(z,\cdot)$ around A_i which is independent of $n \ge 1$.

Notice that $\varphi_n^{(i)}(\zeta) \le \varphi_1^{(i)}(\zeta)$, $\zeta \in D \setminus \text{ins}\Gamma_i^{(1)}$. Since $\lim_{\zeta\to\infty} \varphi_1^{(i)}(\zeta) = 0$ by Lemma 1.1.4 and $\lim_{y\downarrow 0} \varphi_1^{(i)}(x+iy) = 0$ uniformly in $x \in \mathbb{R}$, there exist, for any $\varepsilon > 0$, a large $L > 0$ and a small $\delta > 0$ such that $\varphi_1^{(i)}(\zeta) < \varepsilon$ for any $\zeta \in Q_{L,\delta}$, where

$$Q_{L,\delta} = (\mathbb{H} \setminus \overline{B}_L) \cup R_{L,\delta}, \quad R_{L,\delta} = \{\zeta = x+iy : -L < x < L,\ 0 < y < \delta\}. \tag{1.1.33}$$

We take $\delta > 0$ such that $R_{L,\delta} \cap \bigcup_{j=1}^N \Gamma_j^{(1)} = \emptyset$. Further, we can choose $n_0 = n_0(L,\delta) \in \mathbb{N}$ with

$$\Gamma_0^{(n)} \subset Q_{L,\delta}, \quad \text{for any} \quad n \ge n_0. \tag{1.1.34}$$

Indeed, it suffices to take n_0 with $a_{n_0} < \alpha$ for

$$\alpha = \inf\{G^0(z,\zeta) : \zeta \in B_L \cap \mathbb{H} \setminus \bigcup_{j=1}^N \overline{\text{ins}\Gamma_j^{(1)}} \setminus R_{L,\delta}\} (> 0).$$

Since it follows from (1.1.32), (1.1.34), (1.1.25) and (1.1.26) for $u = 1$ that

$$0 \le \varphi_n^{(i)}(z) + \frac{1}{2}p_i \le \int_{\Gamma_0^{(n)}} \frac{\partial G^0(z,\zeta)}{\partial \mathbf{n}_\zeta} ds(\zeta) < \varepsilon, \quad n \ge n_0,$$

it only remains to show that

$$\lim_{n\to\infty} \varphi_n^{(i)}(z) = \varphi^{(i)}(z). \tag{1.1.35}$$

Define $V = \bigcup_{j=1}^N \text{ins}\Gamma_j^{(1)}$. It is a neighborhood of K with compact closure. Consider the harmonic measures

$$\mu_n^z(d\eta) = \mathbb{P}_z^{\mathbb{H}}\left(\sigma_{K^{(n)}} < \infty, Z_{\sigma_{K^{(n)}}} \in d\eta\right) \quad \text{and} \quad \mu^z(d\eta) = \mathbb{P}_z^{\mathbb{H}}\left(\sigma_K < \infty, Z_{\sigma_K} \in d\eta\right)$$

with $z \in \overline{V}$. For any $f \in C(\overline{V})$, let $\mathbf{H}_K f(\zeta) = \mathbb{E}_\zeta^{\mathbb{H}}\left[f(Z_{\sigma_K}); \sigma_K < \infty\right]$, $\zeta \in \overline{V}$. We then have $\mathbf{H}_K f \in C(\overline{V})$ and $\mathbf{H}_K f\big|_K = f$ in the same way as the proof of Lemma 1.1.4. On the other hand, we can see in a similar way to the proof of (1.1.34) that, for any neighborhood U of K, the support $K^{(n)}$ of μ_n^z is contained in U from some n on. Therefore

$$\lim_{n\to\infty} \int_{\overline{V}} (\mathbf{H}_K f(\eta) - f(\eta))\mu_n^z(d\eta) = 0.$$

Since

$$\int_{\overline{V}} \mathbf{H}_K f(\eta)\mu_n^z(d\eta) = \mathbb{E}_z^{\mathbb{H}}\left[\mathbf{H}_K f(Z_{\sigma_{K^{(n)}}})\right] = \mathbb{E}_z^{\mathbb{H}}\left[f(Z_{\sigma_K}); \sigma_K < \infty\right] = \int_K f(\eta)\mu^z(d\eta),$$

$\{\mu_n^z\}$ converges weakly to μ^z on $\overline{V}$. In particular, for $f \in C(\overline{V})$ with $f = 1$ on $\mathrm{ins}\Gamma_i^{(1)}$ and $f = 0$ on $\mathrm{ins}\Gamma_j^{(1)}$ for $1 \le j \le N, j \neq i$,

$$\lim_{n\to\infty} \varphi_n^{(i)}(z) = \lim_{n\to\infty} \int_{\overline{V}} f(\eta)\mu_n^z(d\eta) = \int_{\overline{V}} f(\eta)\mu^z(d\eta) = \varphi^{(i)}(z),$$

arriving at (1.1.35).

(iv) We make use of the sets D_{a_n} considered in the above for a fixed $z \in D$. Let f be a continuous function on $\partial\mathbb{H}$ with support contained in a finite open interval $I \subset \partial\mathbb{H}$. We first assume that f is a restriction to $\partial\mathbb{H}$ of a C^1-function F on $\mathbb{C}$.

Let $\psi_f(\zeta)$, $\zeta \in D$, be defined by (1.1.23). By Lemma 1.1.4, there exists, for any $\varepsilon > 0$, a large $L > 0$ with

$$|\psi_f(\zeta)| < \varepsilon, \qquad |\zeta| > L. \tag{1.1.36}$$

We choose L satisfying further $\overline{I} \subset (-L, L)$, and define the sets $R_{L,\delta}$ and $Q_{L,\delta}$ by (1.1.33) for a $\delta > 0$ chosen as before. By (1.1.34), the level curve $\Gamma_0^{(n)}$ is contained in $Q_{L,\delta}$ for any $n \ge n_0(L,\delta)$. We may also assume that the C^1-function F vanishes on $\overline{\mathbb{H}} \setminus B_L$.

We define

$$H^n F(w) = \mathbb{E}_w^{D_{a_n}}\left[F(Z_{\zeta-}); Z_{\zeta-} \in \Gamma_0^{(n)}\right], \quad w \in D_{a_n},$$

in terms of the ABM $(Z_t, \mathbb{P}_w^{D_{a_n}})$ on the set D_{a_n}. Similarly to ψ_f, $H^n F$ is harmonic on D_{a_n} and extends continuously to $\overline{D}_{a_n}$ taking value F on $\Gamma_0^{(n)}$ and 0 on $\bigcup_{j=1}^N \Gamma_j^{(n)}$. The Poisson integral formula (1.1.26) applied to the domain D_{a_n} and the function $u = H^n F$ then gives

$$H^n F(z) = -\frac{1}{2}\int_{\Gamma_0^{(n)} \cap R_{L,\delta}} \frac{\partial G^0(z,\zeta)}{\partial \mathbf{n}_\zeta} F(\zeta)ds(\zeta), \quad z \in D_{a_n},\ n \ge n_0(L,\delta) \tag{1.1.37}$$

In the same way, we have from (1.1.25), (1.1.26) and (1.1.36)

$$\left| H^n(\psi_f)(z) + \frac{1}{2}\int_{\Gamma_0^{(n)} \cap R_{L,\delta}} \frac{\partial G^0(z,\zeta)}{\partial \mathbf{n}_\zeta} \psi_f(\zeta)ds(\zeta) \right| < \varepsilon \quad \text{for } n \ge n_0(L,\delta). \tag{1.1.38}$$

Denote $\tau_{D_{a_n}}$ by τ_n for simplicity. We can use the strong Markov property of the planar Brownian motion $\mathbf{Z} = (Z_t, \mathbb{P}_z)$ to obtain

$$\begin{aligned} H^n(\psi_f)(z) &= \mathbb{E}_z\left[\mathbb{E}_{Z_{\tau_n}}\left[f(Z_{\tau_D}); Z_{\tau_D} \in \partial\mathbb{H}\right]; Z_{\tau_n} \in \Gamma_0^{(n)}\right] \\ &= \mathbb{E}_z\left[f(Z_{\tau_D}); Z_{\tau_D} \in \partial\mathbb{H}\right] = \psi_f(z). \end{aligned} \tag{1.1.39}$$

Since both F and ψ_f are continuous extensions of f from $[-L, L] \subset \partial\mathbb{H}$ to $R_{L,\delta}$, we can choose, for any $\varepsilon > 0$, a small $\delta > 0$ with $|F(\zeta) - \psi_f(\zeta)| < \varepsilon$ for any $\zeta \in R_{L,\delta}$. Therefore, it follows from (1.1.37), (1.1.38) holding for $n \geq n_0(L, \delta)$ and (1.1.39) that

$$\psi_f(z) = \lim_{n\to\infty} H^n F(z). \tag{1.1.40}$$

We have seen that $G^0(z, \zeta)$ is harmonic in $\zeta \in D \setminus \{z\}$ admitting the limit 0 at each $\zeta \in \partial\mathbb{H}$ so that it is extendable to be harmonic to a neighborhood of $\partial\mathbb{H}$. As F is assumed to be a C^1-function on $\overline{R}_{L,\delta}$, (1.1.37) and Green's first formula lead us to

$$H^n F(z) + \frac{1}{2}\int_{\partial\mathbb{H}} \frac{\partial}{\partial \mathbf{n}_\zeta} G^0(z,\zeta) f(\zeta) s(d\zeta) = \frac{1}{2}\int_{R_{L,n}} \nabla_\zeta G^0(z,\zeta)\cdot\nabla F(\zeta) du dv, \tag{1.1.41}$$

where $\zeta = u + iv$ and $R_{L,n}$ is a subregion of $R_{L,\delta}$ bounded by $\Gamma_0^{(n)}$, $\partial\mathbb{H}$ and $\{\zeta \in \mathbb{H} : \Re\zeta = \pm L\}$. By letting $\delta \downarrow 0$ and accordingly $n \to \infty$, the last integral tends to zero. Consequently, (1.1.40) and (1.1.41) yield (1.1.28) for any continuous function f on $\partial\mathbb{H}$ with compact support that is the restriction to $\partial\mathbb{H}$ of a C^1-function $\mathbb{C}$.

For a non-negative continuous function f on $\partial\mathbb{H}$ with compact support, we first extend it to a bounded continuous function F on $\mathbb{C}$ and consider convolutions $\varphi^\varepsilon * F$ with mollifiers φ^ε. Then (1.1.28) holds for $\varphi^\varepsilon * F\big|_{\partial\mathbb{H}}$. It holds for f by letting $\varepsilon \downarrow 0$. (1.1.28) for any non-negative bounded continuous function on $\partial\mathbb{H}$ also holds as an increasing limit of functions of the same type with compact support. □

Proposition 1.1.6. *Let D be an $(N+1)$-connected domain as in Proposition* 1.1.5. *Suppose that f is a continuous function on $\overline{D}$ that is harmonic on D taking a constant value f_i on each ∂A_i, vanishing on $\partial\mathbb{H}$ outside a compact set and satisfying*

$$\lim_{|z|\to\infty,\ z\in D} f(z) = 0. \tag{1.1.42}$$

Then for every $z \in D$,

$$f(z) = \sum_{k=1}^N f_k \varphi^{(k)}(z) - \frac{1}{2}\int_{\partial\mathbb{H}} \frac{\partial G^0(z,\zeta)}{\partial \mathbf{n}_\zeta} f(\zeta) ds(\zeta). \tag{1.1.43}$$

It also holds that for each $1 \leq i \leq N$,

$$\text{the period of } f \text{ around } A_i = \sum_{k=1}^N f_k a_{ki} + \int_{\partial\mathbb{H}} \frac{\partial \varphi^{(i)}(\zeta)}{\partial \mathbf{n}_\zeta} f(\zeta) ds(\zeta), \tag{1.1.44}$$

where a_{ki} denotes the period of $\varphi^{(k)}$ around A_i.

Proof. Let h be the difference of the functions on both sides of (1.1.43). Then h is a harmonic function on D taking value 0 continuously on $(\cup_{i=1}^N A_i) \cup \partial\mathbb{H}$ by virtue of (1.1.28) and Lemma 1.1.4. By condition (1.1.42) and Lemma 1.1.4, $\lim_{|z|\to\infty, z\in D} h(z) = 0$. Hence we can conclude that $h = 0$ identically on D by Lemma 1.1.3.

Let γ be a smooth simple curve surrounding A_i. We shall show that

$$-\frac{1}{2}\int_\gamma \frac{\partial}{\partial \mathbf{n}_z}\left[\int_{\partial\mathbb{H}} \frac{\partial}{\partial \mathbf{n}_\zeta} G^0(z.\zeta) f(\zeta) ds(\zeta)\right] ds(z) = \int_{\partial\mathbb{H}} \frac{\partial}{\partial \mathbf{n}_\zeta}\varphi_i(\zeta) f(\zeta) ds(\zeta). \quad (1.1.45)$$

Then (1.1.44) is obtained by evaluating the period of the functions around A_i on both sides of (1.1.43).

We can find disks $U_k = B_r(z_k)$, $r > 0$, $z_k \in \gamma$, $1 \le k \le n$, satisfying $\gamma \subset \bigcup_{k=1}^n U_k \subset D$, and disjoint portions γ_k, $1 \le k \le n$, of γ with $\bigcup_{l=1}^n \gamma_k = \gamma$ and $\overline{\gamma}_k \subset U_k$, $1 \le k \le n$. Since $G^0(z,\zeta)$ is harmonic in $z \in D \setminus \{\zeta\}$,

$$G^0(z,\zeta) = \int_{\partial U_k} p_k(z,z') G^0(z',\zeta) ds(z'), \quad z \in U_k, \ \zeta \neq z. \quad (1.1.46)$$

where $p_k(z,z')$ is the Poisson kernel of the disk U_k.

Notice that

$$M_k = \sup_{z\in\gamma_k, z'\in\partial U_k} \left|\frac{\partial}{\partial \mathbf{n}_z} p_k(z,z')\right| \quad \text{is finite.}$$

As (1.1.31) holds for any compact subset of $\mathbb{H}$ in place of K, we get from (1.1.29) that

$$0 < \frac{1}{\eta} G^0(z', \xi + i\eta) \le \frac{2}{\pi d_k} \quad \text{whenever } z' \in \partial U_k \text{ and } \eta > 0, \quad (1.1.47)$$

where $d_k = \mathrm{dist}(\partial U_k, \partial\mathbb{H})$. Hence the identity (1.1.46) yields for $z \in U_k$ and $\zeta \in \partial\mathbb{H}$,

$$-\frac{1}{2}\frac{\partial}{\partial \mathbf{n}_z}\frac{\partial}{\partial \mathbf{n}_\zeta} G^0(z,\zeta) = -\frac{1}{2}\int_{\partial U_k} \frac{\partial}{\partial \mathbf{n}_z} p_k(z,z') \frac{\partial}{\partial \mathbf{n}_\zeta} G^0(z',\zeta) ds(z'). \quad (1.1.48)$$

By integrating the both sides of (1.1.48) in z over γ_k and taking (1.1.47) into account, we get

$$\begin{aligned}
&-\frac{1}{2}\int_{\gamma_k} \frac{\partial}{\partial \mathbf{n}_z}\frac{\partial}{\partial \mathbf{n}_\zeta} G^0(z,\zeta) ds(z) \\
&= -\frac{1}{2}\frac{\partial}{\partial \mathbf{n}_\zeta}\int_{\gamma_k}\int_{\partial U_k} \frac{\partial}{\partial \mathbf{n}_z} p_k(z,z') G^0(z',\zeta) ds(z') ds(z) \\
&= -\frac{1}{2}\frac{\partial}{\partial \mathbf{n}_\zeta}\int_{\gamma_k} \frac{\partial}{\partial \mathbf{n}_z}\int_{\partial U_k} p_k(z,z') G^0(z',\zeta) ds(z') ds(z) \\
&= -\frac{1}{2}\frac{\partial}{\partial \mathbf{n}_\zeta}\int_{\gamma_k} \frac{\partial}{\partial \mathbf{n}_z} G^0(z',\zeta) ds(z), \quad \zeta \in \partial\mathbb{H}. \quad (1.1.49)
\end{aligned}$$

Furthermore, (1.1.48) and (1.1.28) yield

$$\int_{\gamma_k\times\partial\mathbb{H}} \left|\frac{\partial}{\partial \mathbf{n}_z}\frac{\partial}{\partial \mathbf{n}_\zeta} G^0(z,\zeta)\right| |f(\zeta)| ds(z) ds(\zeta) \le 4\pi r M_k \sup_{\zeta\in\partial\mathbb{H}} |f(\zeta)|.$$

Hence, the Fubini theorem applies and we obtain using (1.1.49)

$$
\begin{aligned}
&-\frac{1}{2}\int_{\gamma_k}\frac{\partial}{\partial \mathbf{n}_z}\left[\int_{\partial\mathbb{H}}\frac{\partial}{\partial \mathbf{n}_\zeta}G^0(z.\zeta)f(\zeta)ds(\zeta)\right]ds(z)\\
&=-\frac{1}{2}\int_{\gamma_k\times\partial\mathbb{H}}\frac{\partial}{\partial \mathbf{n}_z}\frac{\partial}{\partial \mathbf{n}_\zeta}G^0(z.\zeta)f(\zeta)ds(\zeta)ds(z)\\
&=-\frac{1}{2}\int_{\partial\mathbb{H}}\left[\int_{\gamma_k}\frac{\partial}{\partial \mathbf{n}_z}\frac{\partial}{\partial \mathbf{n}_\zeta}G^0(z.\zeta)ds(z)\right]f(\zeta)ds(\zeta)\\
&=-\frac{1}{2}\int_{\partial\mathbb{H}}\left[\frac{\partial}{\partial \mathbf{n}_\zeta}\int_{\gamma_k}\frac{\partial}{\partial \mathbf{n}_z}G^0(z.\zeta)ds(z)\right]f(\zeta)ds(\zeta).
\end{aligned}
$$

Summing up in k, we see that the above identity holds for γ in place of γ_k, yielding (1.1.45) by virtue of Proposition 1.1.5 (iii) and the symmetry of $G^0(z,\zeta)$. □

Lemma 1.1.7. *For $a>0$, consider the rectangle $R_a := \{z=x+iy: -a<x<a,\ 0<y<a\}$ and put $\Sigma_a=\partial R_a\setminus\partial\mathbb{H}$. Let h be a harmonic function on $\mathbb{H}\setminus R_{\ell_0}$ for some $\ell_0>0$ such that*

$$\lim_{z\in\mathbb{H}\setminus R_{\ell_0}\ z\to\infty}|h(z)|=0 \quad \textit{and} \quad h(x+i0+)=0 \quad \text{for any } x \text{ with } |x|>\ell_0. \tag{1.1.50}$$

Then by setting $C=\sup_{z\in\mathbb{H}\setminus R_{\ell_0}}|h(z)|$, which is finite, we have

$$\int_{\Sigma_\ell}\left|\frac{\partial h(z)}{\partial \mathbf{n}_z}\right|ds(z)\le 8C \quad \text{for any } \ell>2\ell_0. \tag{1.1.51}$$

Proof. For a fixed $\ell_1>\ell_0$, let $\mathbb{H}^+_{\ell_1}=\{z\in\mathbb{H}:\Im z>\ell_1\}$. Using the kernel $p(\xi,\eta)=\frac{1}{\pi}\eta/(\xi^2+\eta^2)$, $\xi,\eta\in\mathbb{R}$, $\eta>0$, we then have the Poisson integral representation of h

$$h(x+iy)=\int_{-\infty}^{\infty}p(x-\xi,y-\ell_1)h(\xi+i\ell_1)d\xi. \quad x+iy\in\mathbb{H}^+_{\ell_1}. \tag{1.1.52}$$

To see this, define the function $\widetilde{h}(z)$, $z=x+iy\in\mathbb{H}^+_{\ell_1}$ by the right hand side of (1.1.52). By [Garnett and Marshall (2005), Theorem I.1.2], $\widetilde{h}$ is a harmonic function on $\mathbb{H}^+_{\ell_1}$ belonging to $C_\infty(\overline{\mathbb{H}}^+_{\ell_1})$ as $h\big|_{\partial\mathbb{H}^+_{\ell_1}}\in C_\infty(\partial\mathbb{H}^+_{\ell_1})$ by (1.1.50). By (1.1.50) again, $h-\widetilde{h}\in C_\infty(\mathbb{H}^+_{\ell_1})$ and consequently, (1.1.52) follows from Lemma 1.1.3 (ii).

Because of

$$|p_y(x-\xi,y-\ell_1)|_{y=\ell}\le\frac{1}{\ell-\ell_1}p(x-\xi,\ell-\ell_1), \quad \ell>\ell_1,$$

we have $|h_y(x+iy)|_{y=\ell}\le\dfrac{C}{\ell-\ell_1}$, $\ell>\ell_1$, and so $\displaystyle\int_{-\ell}^{\ell}|h_y(x+iy)|_{y=\ell}dx\le\frac{2C\ell}{\ell-\ell_1}$, $\ell>\ell_1$.

Under the condition (1.1.50), $h(x+iy)$, $x > \ell_0$, $y > 0$, can be extended to a harmonic function on $x > \ell_0$, $y \in \mathbb{R}$, by the reflection $h(\bar{z}) = -h(z)$. Similarly to the above, we have for $x > \ell_1$, $y > 0$,

$$h(x+iy) = \frac{1}{\pi}\int_{-\infty}^{\infty} \frac{x-\ell_1}{(y-y')^2+(x-\ell_1)^2} h(\ell_1+iy')dy', \tag{1.1.53}$$

and $\int_0^\ell |h_x(x+iy)|_{x=\ell}\, dy \le \frac{C\ell}{\ell-\ell_1}$ for $\ell > \ell_1$. Thus $\int_{\Sigma_\ell} \left|\frac{\partial h(z)}{\partial \mathbf{n}_z}\right| ds(z) \le \frac{4C\ell}{\ell-\ell_1}$ for $\ell > \ell_1$, arriving at (1.1.51). □

1.2 Brownian motion with darning and analytic function

It is well known that any harmonic function in a simply connected domain D can be realized as the real (or imaginary) part of an analytic function in D. This property is no longer true if D is not simply connected. In this book, we will consider conformal mappings on multiply connected domains. Given a multiply connected domain D in $\mathbb{C}$, we like to find a diffusion process on it such that its harmonic functions can all be realized as the real (or imaginary) part of an analytic function in D.

Let $D \subset \mathbb{C}$ be a connected open set. A function $f\colon D \to \mathbb{C}$ is *analytic* if the limit

$$f'(z) = \lim_{\Delta z \to 0} \frac{f(z+\Delta z)-f(z)}{\Delta z}$$

exists for every $z \in D$. Here are some basic properties of analytic functions in D:

(i) An analytic function is C^∞-smooth in D.
(ii) Let u and v be the real and imaginary parts, respectively, of an analytic function f in D; that is, $f(z) = u(z)+iv(z)$. By the definition of the derivative f', we have

$$f'(z) = u_x + iv_x = \tfrac{1}{i}u_y + i\cdot\tfrac{1}{i}v_y = v_y - iu_y.$$

Thus, u and v satisfy the *Cauchy–Riemann equation*:

$$\begin{cases} u_x = v_y \\ u_y = -v_x \end{cases} \quad \text{in } D.$$

This implies $\Delta u = \Delta v = 0$ in D; that is both u and v are harmonic in D. Conversely if u and v are any real-valued C^2-smooth functions in D satisfying the above Cauchy-Riemann equation, then $f(z) := u(z)+iv(z)$ is an analytic function in D.

Fix a base point $z_0 \in D$. For any $z \in D$, we take a smooth curve γ connecting z_0 to z with $\gamma(t) = (x(t), y(t))$, $0 \le t \le T$. Here, we identify a complex number

$x + iy \in \mathbb{C}$ with the two-dimensional vector $(x, y) \in \mathbb{R}^2$. For an analytic function $f = u + iv$ in D, we have

$$
\begin{aligned}
u(z) - u(z_0) &= \int_0^T \frac{d}{dt} u(x(t), y(t))\, dt = \int_0^T (u_x x'(t) + u_y y'(t))\, dt \\
&= \int_0^T (u_x, u_y) \cdot (x'(t), y'(t))\, dt = \int_0^T (v_y, -v_x) \cdot (x'(t), y'(t))\, dt \\
&= \int_0^T (v_x, v_y) \cdot (-y'(t), x'(t))\, dt \\
&= \int_\gamma \frac{\partial v}{\partial \mathbf{n}}\, d\sigma. \qquad (1.2.1)
\end{aligned}
$$

Here, $d\sigma = \sqrt{(x'(t))^2 + (y'(t))^2}\, dt$ is the arc length measure on γ, and $\mathbf{n}(t)$ is the unit normal vector of γ obtained by turning $90°$ clockwise of the tangent direction $(x'(t), y'(t))$. In particular, if γ is a closed smooth curve in D, then

$$\int_\gamma \frac{\partial v}{\partial \mathbf{n}}\, d\sigma = 0. \qquad (1.2.2)$$

Conversely, if v is a harmonic function in D satisfying (1.2.2) for any closed smooth curve γ in D, we can use (1.2.1) to define a function $u(z)$ as follows. Let $u(z_0) = a \in \mathbb{R}$. For any $z \in D$, let $\gamma = \{\gamma(t); 0 \le t \le T\}$ be a smooth curve in D so that $\gamma(0) = z_0$ and $\gamma(T) = z$. Define

$$u(z) = u(z_0) + \int_\gamma \frac{\partial v}{\partial \mathbf{n}}\, d\sigma.$$

By (1.2.2), the value of $u(z)$ is independent of the smooth curve γ that connects z_0 to z. By taking γ approach to z horizontally (respectively, vertically) we obtain $u_x = v_y$ and $u_y = -v_x$. Hence $f = u + iv$ is analytic in D. Note that u and hence f is uniquely determined by v up to an additive real constant.

We now fix a planar domain E. It will be assumed that either $E = \mathbb{C}$ or $\mathbb{C} \setminus E$ is a continuum. Let

$$D = E \setminus K, \quad K = \bigcup_{i=1}^N A_i, \qquad (1.2.3)$$

where $\{A_i, 1 \le i \le N\}$ is a finite family of mutually disjoint compact continua contained in E. It is easy to verify that (1.2.2) holds for any closed smooth curve Γ in D if and only if it holds for any closed smooth curve Γ in D whose interior contains exactly one A_i. We call this property as v having zero period around A_i for every $1 \le i \le N$. Thus a function v on D is the imaginary part of an analytic function in D if and only if v is harmonic in D having zero period around A_i for every $1 \le i \le N$.

The diffusion process that will be introduced below, called Brownian motion with darning enjoys the property that any of its harmonic function has not only the harmonicity on D in the ordinary sense but also zero period around A_i for

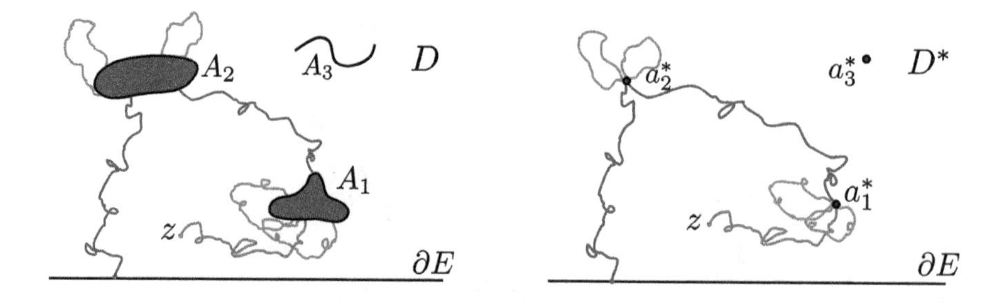

Fig. 1.1 The quotient space D^* of E and the BMD on it.

every $1 \leq i \leq N$ as will be demonstrated in Section 1.6. This is the reason why the Brownian motion with darning is of a great use in the analysis on multiply connected planar domains.

Let

$$D^* = D \cup K^*, \quad K^* = \{a_i^*, 1 \leq i \leq N\}, \tag{1.2.4}$$

and define a neighborhood U_i^* of each point a_i^* in D^* by $\{a_i^*\} \cup (U_i \setminus A_i)$ for some neighborhood U_i of A_i in E. That is, D^* is a space equipped with the quotient topology obtained from E by regarding each continuum A_i as one point a_i^*. Denote by m the Lebesgue measure in D that is extended to D^* by setting $m(K^*) = 0$. Intuitively speaking, Brownian motion with darning on D^* is an m-symmetric diffusion process on D^* obtained from the ABM on E by "darning" or "shorting" each set A_j into a single point a_j^*; see Figure 1.1 with $N = 3$.

Definition 1.2.1. *Brownian motion with darning* (BMD in abbreviation) $\mathbf{Z}^* = (Z_t^*, t \geq 0, \zeta^*; \mathbb{P}_z^*, z \in D^*)$ is an m-symmetric diffusion on D^* such that

(i) its subprocess killed upon leaving D has the same law as the ABM on D;
(ii) it admits no killing on K^* in the sense that

$$\mathbb{P}_x(Z_{\zeta^*-}^* \in K^*, \zeta^* < \infty) = 0 \qquad \text{for every } x \in D.$$

In general, a Markov process $\mathbf{X} = (X_t, \zeta, \mathbb{P}_x)$ on a topological state space M with lifetime ζ is called a *diffusion* if $\mathbf{X}$ is strong Markov and of continuous sample paths in the sense that

$$\mathbb{P}_x(X_t \text{ is continuous in } t \in [0, \zeta)) = 1 \quad \text{for every } x \in M.$$

A Markov process $\mathbf{X}$ is said to be *symmetric* with respect to a σ-finite positive measure ν on M if its transition function P_t defined by $P_t f(x) = \mathbb{E}_x[f(X_t)]$, $t > 0$, $x \in M$, satisfies

$$\int_M P_t f(x) g(x) \nu(dx) = \int_M f(x) P_t g(x) \nu(dx) \quad \text{for any } t > 0 \text{ and } f, g \in \mathcal{B}_+(M).$$

A ν-symmetric Markov process $\mathbf{X}$ with right continuous sample paths can be well studied by means of the Dirichlet form $(\mathcal{E},\mathcal{F})$ on $L^2(M;\nu)$ with which $\mathbf{X}$ is associated.

In Chapter 7 of [Chen and Fukushima (2012)], the BMD has been studied under a more general context of the *boundary problem for a symmetric Markov process* $\mathbf{X}^0$ on a state space M^0 whose boundary F consists of finitely or countably many points. The problem is to look for all possible symmetric Markovian extensions of $\mathbf{X}^0$ from M^0 to $M^0 \cup F$ and to establish and characterize all these possible extensions.

In this general context, the BMD $\mathbf{Z}^* = (Z_t^*, \zeta^*, \mathbb{P}_z^*)$ can be regarded as an m-symmetric diffusion extention of the ABM $\mathbf{Z}^0 = (Z_t^0, \zeta^0, \mathbb{P}_z^0)$ on D from D to $D^* = D \cup K^*$. In particular, Theorem 7.7.3 in [Chen and Fukushima (2012)] applied to this special case reads as follows:

The Dirichlet form $(\mathcal{E}^*, \mathcal{F}^*)$ on $L^2(D^*;m)$ of the BMD $\mathbf{Z}^* = (Z_t^*, \zeta^*, \mathbb{P}_z^*)$ can be uniquely described as

$$\mathcal{E}^*(u,v) = \frac{1}{2}\int_D \nabla u(z) \cdot \nabla v(z) m(dz)$$

and

$$\mathcal{F}^* = \text{linear span of } H_0^1(D) \text{ and } \{u_1^{(i)}\big|_D,\ 1 \le i \le N\},$$

where $u_1^{(i)}(z)$ is the 1-order hitting distribution of $\mathbf{Z}^*$ on $K^* = \{a_1^*, \ldots, a_N^*\}$ starting at $z \in D^*$:

$$u_1^{(i)}(z) = \mathbb{E}_z^*\left[e^{-\sigma_{K^*}}; Z_{\sigma_{K^*}} = a_i^*\right], \quad z \in D^*, \quad 1 \le i \le N.$$

Guided by this description, we shall establish in the next section the existence and uniqueness of BMD in a much more direct and self-contained fashion. See also Remark 1.3.11 and Remark 1.8.3.

1.3 Existence and uniqueness of BMD

In this section, we show under the setting in Section 1.2 that BMD always exists and is unique in law by making use of a relevant Dirichlet form. As mentioned in the above, BMD $\mathbf{Z}^*$ on D^* can be intuitively thought of as a diffusion process obtained from the ABM $\mathbf{Z}^E$ on E by "shorting" each A_j into a single point a_j^*. We start with a lemma concerning the α-order hitting distribution of the ABM $\mathbf{Z}^E$ on $K = \cup_{i=1}^N A_i$.

For $p > 0$, denote by $L^p(E)$ the L^p-space with respect to the Lebesgue measure on E. According to [Chen and Fukushima (2012), Example 3.5.9] or [Fukushima, Oshima and Takeda (2011), Example 4.4.1], the symmetric Dirichlet form $(\mathcal{E}^E, \mathcal{F}^E)$ on $L^2(E)$ of the ABM $\mathbf{Z}^E$ is given by

$$(\mathcal{E}^E, \mathcal{F}^E) = (\frac{1}{2}\mathbf{D}_E, H_0^1(E)), \quad \text{where} \quad \mathbf{D}_E(u,v) = \int_E \nabla u(x) \cdot \nabla v(x) dx. \quad (1.3.1)$$

$H_0^1(E)$ is the $\mathcal{E}_1^E$-closure of $C_c^\infty(E)$ in the Sobolev space $H^1(E) = \{u \in L^2(E) : |\nabla u| \in L^2(E)\}$. Here, for $\alpha > 0$, $\mathcal{E}_\alpha^E(u,v) := \mathcal{E}^E(u,v) + \alpha(u,v)_{L^2(E)}$.

$(\frac{1}{2}\mathbf{D}_E, H_0^1(E))$ is a symmetric regular Dirichlet form on $L^2(E;m)$ having $C_c^\infty(E)$ as its core. In the references cited above, the $\mathcal{E}_1^E$*-capacity* $\mathrm{Cap}^E(A)$ of an open set $A \subset E$ is defined by

$$\mathrm{Cap}^E(A) = \inf\{\mathcal{E}_1^E(u,u) : u \in \mathcal{F}^E,\ u = 1 \text{ a.e. on } A\} \tag{1.3.2}$$

and extended to any set $A \subset E$ by $\mathrm{Cap}^E(A) = \inf_{B:\text{open},\, B\supset A} \mathrm{Cap}^E(B)$. The terms $(\mathcal{E}^E$-)*quasi-everywhere* (q.e. in abbreviation) and $(\mathcal{E}^E$-)*quasi-continuous* are defined in terms of the $\mathcal{E}_1^E$-capacity Cap^E. Each function $u \in \mathcal{F}^E$ admits its quasi-continuous version denoted by $\widetilde{u}$. If a quasi-continuous function vanishes a.e. on E, then so does it q.e. on E. A set $A \subset E$ is called $\mathcal{E}^E$-polar if $\mathrm{Cap}^E(A) = 0$. The notion of the *energy measure* $\mu_{\langle u\rangle}$ for $u \in H_0^1(E)$ is also introduced and it holds that

$$\mu_{\langle u\rangle}(dx) = |\nabla u(x)|^2 dx.$$

The energy measure $\mu_{\langle u\rangle}$ is the same as its strongly local part $\mu^c_{\langle u\rangle}$ because the Dirichlet form $(\frac{1}{2}\mathbf{D}, H_0^1(E))$ is strongly local. Further, it holds that

$$\mathcal{E}^E(u,u) = \frac{1}{2}\mu^c_{\langle u\rangle}(E), \quad u \in H_0^1(E). \tag{1.3.3}$$

Using the ABM $\mathbf{Z}^E = (Z_t^E, \zeta^E, \mathbb{P}_z^E)$ on E, define, for $1 \le i \le N$ and $\alpha > 0$,

$$u_\alpha^{(i)}(z) = \mathbb{E}_z^E\left[e^{-\alpha\sigma_K}; Z^E_{\sigma_K} \in A_i\right], \quad z \in E. \tag{1.3.4}$$

Lemma 1.3.1. *We have the following for each* $1 \le i \le N$.

(i) $u_\alpha^{(i)} \in H_0^1(E)$, $u_\alpha^{(i)}(z) < 1$ *for any* $z \in D$, *and* $u_\alpha^{(i)}$ *is* α*-harmonic on* D *in the sense that* $u_\alpha^{(i)} \in C^\infty(D)$ *and* $(\alpha - \frac{1}{2}\Delta)u_\alpha^{(i)}(z) = 0,\ z \in D$.
(ii) $u_\alpha^{(i)}$ *is continuous on* E *with value* 1 *on* A_i *and* 0 *on* $\bigcup_{1\le j\le N,\, j\ne i} A_j$. *It is continuously extendable to* $\mathbb{C}$ *with value* 0 *on* $\mathbb{C}\setminus E$.
(iii) $u_\alpha^{(i)} \in L^1(E)$ *and* $\lim_{z\in E,\, z\to\infty} u_\alpha^{(i)}(z) = 0$.
(iv) $\|u_\alpha^{(i)}\|_{L^2(D)} > 0$ *and* $\mathrm{Cap}^E(A_i) > 0$.

Proof. (i) $u_\alpha(z) < 1$ for any $z \in D$ because $z \notin K^r$. Notice that

$$u_\alpha^{(i)}(z) = \mathbb{E}_z^E\left[e^{-\alpha\sigma_K} f(Z^E_{\sigma_K})\right] =: \mathbf{H}_K^\alpha f(z), \quad z \in E, \tag{1.3.5}$$

for a function $f \in H_0^1(E)$ with $f = 1$ on A_i and $f = 0$ on $\cup_{l\ne i}A_j$. The function $\mathbf{H}_K^\alpha f(z)$, $z \in E$, is the projection of $f \in H_0^1(E)$ on the $\mathcal{E}_\alpha^E$-orthogonal complement of $H_0^1(D)(\subset H_0^1(E))$. Consequently, $u_\alpha^{(i)} \in H_0^1(E)$ and $\mathcal{E}_\alpha^E(u_\alpha^{(i)}, v) = 0$ for any $v \in C_c^\infty(D)$, which implies that $u_\alpha^{(i)}$ is a solution of $(\alpha - \frac{1}{2}\Delta)u_\alpha^{(i)} = 0,\ z \in D$, in the distribution sense. By the well-known Weyl's lemma, there exists $\widetilde{u}_\alpha^{(i)} \in C^\infty(D)$ such that $(\alpha - \frac{1}{2}\Delta)\widetilde{u}_\alpha^{(i)} = 0,\ z \in D$, and $\widetilde{u}_\alpha^{(i)}(z) = u_\alpha^{(i)}(z)$ a.e. on D.

Let $D_i = D \cup A_i (= E \setminus \bigcup_{j \neq i} A_j)$ and consider the ABM $\mathbf{Z}^{D_i} = (Z_t^{D_i}, \zeta^{D_i}, \mathbb{P}_z^{D_i})$ on D_i and the function

$$w(z) = \mathbb{E}_z^{D_i}\left[e^{-\alpha\sigma_{A_i}}\right] = \mathbb{E}_z^{D_i}\left[e^{-\alpha\sigma_{A_i}}\,\mathbb{1}_{D_i}(Z_{\sigma_{A_i}}^{D_i})\right], \quad z \in D_i.$$

As $\mathbb{1}_{D_i}$ is α-excessive relative to $\mathbf{Z}^{D_i}$, so is the function w by [Chen and Fukushima (2012), Lemma A.2.4]. Observe that $u_\alpha^{(i)}\big|_D = w\big|_D$. Hence, $u_\alpha^{(i)}$ is α-excessive relative to the ABM $\mathbf{Z}^D$ on D in view of the Remark preceding Theorem 12.4 of [Dynkin (1965)].

The transition function p_t^D of $\mathbf{Z}^D$ is absolutely continuous with respect to the Lebesgue measure so that

$$u_\alpha^{(i)}(z) = \lim_{t\downarrow 0} e^{-\alpha t} p_t^D u_\alpha^{(i)}(z) = \lim_{t\downarrow 0} e^{-\alpha t}\mathbb{E}_z^D\left[\widetilde{u}_\alpha^{(i)}(Z_t^D)\right] = \widetilde{u}_\alpha^{(i)}(z),$$

for every $z \in D$.

(ii) For $z \in D = E \setminus K$, $u_\alpha^{(i)}(z)$ also admits the following expression in terms of the planar Brownian motion $\mathbf{Z} = (Z_t, \mathbb{P}_z)$,

$$u_\alpha^{(i)}(z) = \mathbb{E}_z\left[e^{-\alpha\tau_D} : Z_{\tau_D} \in \partial A_i\right] =: \mathcal{H}_D^\alpha g(z), \quad z \in D, \tag{1.3.6}$$

for the function g on ∂D with value 1 on ∂A_i and 0 on $\partial E \cup \left(\bigcup_{1\le j\le N,\, j\neq i} \partial A_j\right)$. The assertion (ii) follows from (i), (1.3.6), Theorem A.1.1 and property (**Z.1**) of the planar BM **Z**.

(iii) Take $\rho > 0$ with $B_\rho \supset K$ and define $v_\alpha(z) = \mathbb{E}_z\left[e^{-\alpha\sigma_{B_\rho}}\right]$, $|z| > \rho$, in terms of the planar BM $\mathbf{Z} = (Z_t, \mathbb{P}_z)$. Then $v_\alpha(z) \ge u_\alpha^{(i)}(z)$, $|z| > \rho$, $1 \le i \le N$. Since $v_\alpha(z) = v_\alpha(r)$, $r = |z|$, is decreasing in r and satisfies

$$\left(\alpha - \frac{1}{2}\Delta\right) v_\alpha(z) = \left(\alpha - \frac{d}{dm}\frac{d}{ds}\right) v_\alpha(r) = 0, \quad r > \rho,$$

for $ds = r^{-1}dr$, $dm = 2rdr$ (cf. [Chen and Fukushima (2012), p. 126]), we get

$$\alpha\int_\rho^R v_\alpha(r)dm(r) = \frac{dv_\alpha}{ds}(R) - \frac{dv_\alpha}{ds}(\rho) \le -\frac{dv_\alpha}{ds}(\rho), \quad R > \rho,$$

yielding the two properties of (iii).

(iv) If $\|u_\alpha^{(i)}\|_{L^2(D)} = 0$, $u_\alpha^{(i)}$ vanishes identically on D by (i), contradicting (ii). Let $e_\alpha^{(i)}(z) = \mathbb{E}_z^E\left[e^{-\alpha\sigma_{A_i}}\right]$, $z \in E$. Then $e_\alpha^{(i)} \in \mathcal{F}^E$ and $\mathrm{Cap}^E(A_i) = \mathcal{E}_1^E(e_1^{(i)}, e_1^{(i)})$ by [Chen and Fukushima (2012), Corollary 3.2.3]. As $e_1^{(i)}(z) \ge u_1^{(i)}(z)$, we get $\mathrm{Cap}^E(A_i) \ge \|u_1^{(i)}\|_{L^2(D)}^2 > 0$. □

Denote by m the Lebesgue measure on $D = E\setminus K$ being extended to $D^* = D\cup K^*$ by setting $m(K^*) = 0$. We shall construct a symmetric Dirichlet form $(\mathcal{E}^*, \mathcal{F}^*)$ on $L^2(D^*; m)$ relevant to the BMD on D^* in the following manner.

Define

$$\mathcal{F}^* = \text{ linear subspace of } H^1(D) \text{ spanned by } H_0^1(D) \text{ and } \{u_1^{(j)}|_D;\ 1 \le j \le N\}, \tag{1.3.7}$$

and for $u, v \in \mathcal{F}^*$,

$$\mathcal{E}^*(u, v) = \frac{1}{2}\int_D \nabla u(x)\cdot\nabla v(x)dx. \tag{1.3.8}$$

Theorem 1.3.2.

(i) *$(\mathcal{E}^*, \mathcal{F}^*)$ is a symmetric strongly local regular Dirichlet form on $L^2(D^*; m)$.*
(ii) *Each a_j^* has positive $\mathcal{E}_1^*$-capacity.*
(iii) *The part form of $(\mathcal{E}^*, \mathcal{F}^*)$ on D equals $(\frac{1}{2}\mathbf{D}_D, H_0^1(D))$.*

Proof. (i) For $u, v \in \mathcal{F}^*$, let

$$\mathcal{E}_1^*(u, v) := \mathcal{E}^*(u, v) + (u, v)_{L^2(D^*;m)} = \frac{1}{2}\mathbf{D}_D(u, v) + (u, v)_{L^2(D)}, \quad u, v \in \mathcal{F}^*.$$

By Lemma 1.3.1 (i),

$$H_0^1(D) \subset \mathcal{F}^* \subset H^1(D),$$

and, for any $1 \le j \le N$ and $f \in H_0^1(D)$, $\mathcal{E}_1^*(u_1^{(j)}, f) = \lim_{n\to\infty} \mathcal{E}_1^*(u_1^{(j)}, f_n) = 0$ where $\{f_n\}$ are functions in $C_c^\infty(D)$ that is $\mathcal{E}_1^*$-convergent to f. Therefore any $u \in \mathcal{F}^*$ admits the expression

$$\begin{cases} u = f + h, \quad \text{where } f \in H_0^1(D) \text{ and } h \in \mathcal{H} := \{\sum_{k=1}^N c_k u_1^{(k)},\ c_k \in \mathbb{R}\}, \\ \text{and } \ \mathcal{E}_1^*(u, u) = \mathcal{E}_1^*(f, f) + \mathcal{E}_1^*(h, h). \end{cases} \tag{1.3.9}$$

Clearly, $(H_0^1(D), \mathcal{E}_1^*)$ is closed, while $\mathcal{H}$ is a finite dimensional linear subspace of $H^1(D)$ so that it is a closed linear subspace of $H^1(D)$ with inner product $\frac{1}{2}\mathbf{D}^D(u, v) + (u, v)_{L^2(D)}$. Hence $(\mathcal{E}^*, \mathcal{F}^*)$ is a closed symmetric form on $L^2(D^*; m) = L^2(D)$.

Next observe that

$$\mathcal{F}^* = \widehat{\mathcal{F}}^E\big|_D \quad \text{for } \widehat{\mathcal{F}}^E = \{u \in H_0^1(E) : \widetilde{u} \text{ is constant } \mathcal{E}^E\text{-q.e. on each } A_j\}. \tag{1.3.10}$$

The constant value taken by $u \in \widehat{\mathcal{F}}^E$ on A_j is denoted by $\widetilde{u}(A_j)$. The inclusion $\subset$ is obvious. As $H_0^1(D) = \{u \in H_0^1(E) : \widetilde{u} = 0 \text{ q.e. on } K\}$ by [Chen and Fukushima (2012), §3.2], any $u \in \widehat{\mathcal{F}}^E$ admits the $\mathcal{E}_1^E$-orthogonal decomposition

$$\widetilde{u}(z) = f(z) + \mathbf{H}_K^1 \widetilde{u}(z) = f(z) + \sum_{j=1}^N \widetilde{u}(A_j) u_1^{(j)}(z) \tag{1.3.11}$$

on E with $f \in H_0^1(D)$. It follows that $\mathcal{F}^* \supset \widehat{\mathcal{F}}^E\big|_D$ and so (1.3.10) holds.

Property (1.3.10) implies that $v = 0 \vee u \wedge 1 \in \mathcal{F}^*$ for any $u \in \mathcal{F}^*$. Clearly, $\mathcal{E}^*(v, v) \le \mathcal{E}^*(u, u)$, namely, the unit contraction operates on $(\mathcal{E}^*, \mathcal{F}^*)$. This proves that $(\mathcal{E}^*, \mathcal{F}^*)$ is a symmetric Dirichlet form on $L^2(D^*; m)$.

Let $\mathcal{C}$ be the linear span of C_c^2 and $\{u_1^{(j)}; 1 \le j \le N\}$, that is,

$$\mathcal{C} = \left\{ u = f_0 + \sum_{k=1}^N c_k u_1^{(k)}(x) : \ f_0 \in C_c^2(D), c_k \in \mathbb{R} \text{ for } 1 \le k \le N \right\}. \tag{1.3.12}$$

In view of Lemma 1.3.1, the function $u_1^{(j)}$ can be viewed as a function in $C_\infty(D^*)$ with $u_1^{(j)}(a_i^*) := \delta_{ij}$. Thus by defining $u(a_j^*) = c_j$ for $u \in \mathcal{C}$ of (1.3.12), we can view $\mathcal{C}$ as a subspace of $C_\infty(D^*) \cap \mathcal{F}^*$. Since $\mathcal{C}$ is an algebra that separates points in D^*, by

Stone-Weierstrass theorem, $\mathcal{C}$ is uniformly dense in the space $C_\infty(D^*)$ of continuous functions in D^* vanishing at ∂ for the one-point compactification $D^* \cup \{\partial\}$ of D^*. By (1.3.9), $\mathcal{C}$ is $\mathcal{E}_1^*$-dense in $\mathcal{F}^*$. Hence $(\mathcal{E}^*, \mathcal{F}^*)$ is a symmetric regular Dirichlet form on $L^2(D^*; m)$ by [Chen and Fukushima (2012), Lemma 1.3.12]. It is clear that $(\mathcal{E}^*, \mathcal{F}^*)$ is strongly local.

(ii) Fix $1 \le j \le N$. Let O be a relatively compact open subset of $D^* \setminus (K^* \setminus \{a_j^*\})$ so that $O \cap K^* = \{a_j^*\}$. Let $(\mathcal{E}^{*,O}, \mathcal{F}^{*,O})$ be the part form of $(\mathcal{E}^*, \mathcal{F}^*)$ on O. It is well known (see, e.g., [Chen and Fukushima (2012), Theorems 3.3.8 and 3.3.9]) that $(\mathcal{E}^{*,O}, \mathcal{F}^{*,O})$ is a regular Dirichlet form on $L^2(O; m)$, and a set $A \subset O$ is $\mathcal{E}^*$-polar if and only if it is $\mathcal{E}^{*,O}$-polar. Let $D_1 \subset O$ be any open subset that contains a_j^*. For $f \in \mathcal{F}^{*,O}$ with $f = 1$ m-a.e. on D_1, by (1.3.9) and (1.3.10), there is $g = f_0 + \sum_{i=1}^N c_i u_1^{(i)} \in \widehat{\mathcal{F}}^E$ for some $f_0 \in H_0^1(D)$ and constants c_i, $1 \le i \le N$ so that $f = g|_D$. Denote by $\widetilde{g}$ the $\mathcal{E}^E$-quasi-continuous version of g on E. Since $f = 0$ m-a.e. on $D \cap (D^* \setminus O)$ and $f = 1$ m-a.e. on D_1, there exists an $\mathcal{E}^E$-polar set $\mathcal{N}_0 \subset E$ with $\widetilde{g}(x) = 0$ for $x \in D \cap (D^* \setminus O) \setminus \mathcal{N}_0$ and $\widetilde{g}(x) = 1$ for $x \in (D \cap D_1) \setminus \mathcal{N}_0$. Hence, due to the continuity of $t \mapsto \widetilde{g}(X_t^E)$ from [Chen and Fukushima (2012), Theorem 3.1.7] and the $\mathcal{E}_1^E$-orthogonal decomposition of $\widetilde{g}$, there exists a Borel $\mathcal{E}^E$-properly exceptional set $\mathcal{N} \subset E$ containing $\mathcal{N}_0$ (see right below for the meaning of a properly exceptional set) so that for every $x \in D \setminus \mathcal{N}$,

$$\sum_{i=1}^N c_i u_1^{(i)}(x) = \mathbf{H}_K^1 \widetilde{g}(z) = \mathbb{E}_x\left[e^{-\sigma_K} \widetilde{g}(Z_{\sigma_K}^E)\right] = \mathbb{E}_x\left[e^{-\sigma_K} \widetilde{g}(Z_{\sigma_K -}^E)\right] = u_1^{(j)}(x).$$

It follows that $f(x) = g(x) = f_0(x) + u_1^{(j)}(x)$ for m-a.e. $x \in D$ and so

$$\mathcal{E}_1^*(f, f) = \mathcal{E}_1^*(f_0, f_0) + \mathcal{E}_1^*(u_1^{(j)}, u_1^{(j)}) \ge \|u_1^{(j)}\|_{L^2(D)}^2.$$

Consequently,

$$\mathrm{Cap}^{*,O}(D_1) = \inf\left\{\mathcal{E}_1^*(u, u) : u \in \mathcal{F}^{*,O},\ u = 1 \ m\text{-a.e. on } D_1\right\} \ge \|u_1^{(j)}\|_{L^2(D)}^2.$$

As $\|u_1^{(j)}\|_{L^2(D)}^2 > 0$ by Lemma 1.3.1, we conclude that $\mathrm{Cap}^{*,O}(a_j^*) \ge \|u_1^{(j)}\|_{L^2(D)}^2 > 0$ and thus a_j^* has positive $\mathcal{E}_1^*$-capacity.

(iii) In view of [Chen and Fukushima (2012), (3.3.1)], (1.3.9) and (ii), the part form $(\mathcal{E}^{*,D}, \mathcal{F}^{*,D})$ of $(\mathcal{E}^*, \mathcal{F}^*)$ on D is given by

$$\mathcal{F}^{*,D} = \{u \in \mathcal{F}^* : \widetilde{u} = 0 \ \ \mathcal{E}^*\text{-q.e. on } K^*\} = H_0^1(D),$$

and

$$\mathcal{E}^{*,D}(u, v) = \mathcal{E}^*(u, v) = \frac{1}{2}\mathbf{D}_D(u, v) \quad \text{for } u, v \in H_0^1(D).$$

This completes the proof of the theorem. □

In order to construct a BMD on D^* from the symmetric strongly local regular Dirichlet form $(\mathcal{E}^*, \mathcal{F}^*)$ on $L^2(D^*, m)$ defined by (1.3.8) and (1.3.7), we first recall some general notions and facts from [Fukushima, Oshima and Takeda (2011)].

Let M be a locally compact separable metric space, ν a positive Radon measure on M of full support and $(\mathcal{E}, \mathcal{F})$ a regular Dirichlet form on $L^2(M;\nu)$. By [Fukushima, Oshima and Takeda (2011), Theorem 7.2.1], there exists a ν-symmetric Hunt process $\mathbf{X} = (X_t, \zeta, \mathbb{P}_x)$ on M with which $(\mathcal{E}, \mathcal{F})$ is associated in the following sense: the strongly continuous contraction semigroup $\{T_t; t > 0\}$ of symmetric operators on $L^2(M;\nu)$ determined by the transition function $\{P_t; t > 0\}$ of $\mathbf{X}$ does generate the form $(\mathcal{E}, \mathcal{F})$. The Hunt process $\mathbf{X}$ is then automatically *properly associated with* $(\mathcal{E}, \mathcal{F})$ in the sense that $P_t u$ is an $\mathcal{E}$-quasi continuous version of $T_t u$ for any $t > 0$ and $u \in \mathcal{B}_+(M) \cap L^2(M;\nu)$ (cf. [Fukushima, Oshima and Takeda (2011), Theorem 4.2.3]).

The Hunt process $\mathbf{X}$ is strong Markov and its sample path $X_t(\omega)$ has the properties that it takes the value in the one-point compactification $M_\partial := M \cup \{\partial\}$ of M with $X_t(\omega) = \partial$ for every $t \geq \zeta(\omega)$, satisfies the right continuity in $t \in [0, \infty)$ and admits the left limit $X_{t-}(\omega)$ for every $t \in (0, \infty)$.

A Borel set $\mathcal{N} \subset M$ is called *properly exceptional for* $\mathbf{X}$ if $\nu(\mathcal{N}) = 0$ and $M \setminus \mathcal{N}$ is $\mathbf{X}$-invariant in the sense that $\mathbb{P}_x(\sigma_{\mathcal{N}} < \infty) = 0$ for every $x \in M \setminus \mathcal{N}$. A set $\mathcal{N} \subset M$ is called *$\mathcal{E}$-polar* if $\mathrm{Cap}(\mathcal{N}) = 0$ where Cap denotes the capacity defined as (1.3.2) for $(\mathcal{E}, \mathcal{F})$. Any properly exceptional set for $\mathbf{X}$ is $\mathcal{E}$-polar and any $\mathcal{E}$-polar set is contained in some Borel properly exceptional set for $\mathbf{X}$ ([Fukushima, Oshima and Takeda (2011), Theorems 4.1.1 and 4.2.1]).

Suppose that the regular Dirichlet form $(\mathcal{E}, \mathcal{F})$ on $L^2(M;\nu)$ is strongly local. By [Fukushima, Oshima and Takeda (2011), Theorem 4.5.3] or [Chen and Fukushima (2012), Theorem 4.3.4], there exists then a properly exceptional Borel set $\mathcal{N} \subset M$ for $\mathbf{X}$ such that $\mathbf{X}\big|_{M\setminus\mathcal{N}}$ is a diffusion admitting no killing inside M in the following sense: for every $x \in M \setminus \mathcal{N}$,

$$\mathbb{P}_x(X_t \text{ is continuous in } t \in [0, \zeta)) = 1, \tag{1.3.13}$$

$$\mathbb{P}_x(X_{\zeta-} \in M,\ \zeta < \infty) = 0. \tag{1.3.14}$$

(1.3.14) means that, for $x \in M \setminus \mathcal{N}$, $\mathbb{P}_x(X_{\zeta-} = \partial,\ \zeta < \infty) = \mathbb{P}_z(\zeta < \infty)$, namely, X_t is continuous on M_∂ also at ζ whenever $\zeta < \infty$. Therefore $\mathbf{X}\big|_{M\setminus\mathcal{N}}$ enjoys the path continuity

$$\mathbb{P}_x(X_t \text{ is continuous on } M_\partial \text{ in } t \in [0, \infty)) = 1, \quad x \in M \setminus \mathcal{N}, \tag{1.3.15}$$

stronger than (1.3.13).

Let $D^*_\partial := D^* \cup \{\partial\}$ be the one-point compactification of D^*. Accordingly, there exist an m-symmetric Hunt process $\mathbf{Z}^* = (Z^*_t, \zeta^*, \mathbb{P}^*_z)$ on D^* associated with the strongly local regular Dirichlet form $(\mathcal{E}^*, \mathcal{F}^*)$ on $L^2(D^*;m)$ and a properly exceptional set $\mathcal{N}_0 \subset D^*$ for $\mathbf{Z}^*$ such that

$$\mathbb{P}^*_z(Z^*_t \text{ is continuous on } D^*_\partial \text{ in } t \in [0, \infty)) = 1 \quad \text{for every } x \in D^* \setminus \mathcal{N}_0. \tag{1.3.16}$$

As a^*_j is non $\mathcal{E}^*$-polar for $1 \leq j \leq N$, by Theorem 1.3.2,

$$\mathcal{N}_0 \subset D = D^* \setminus K^*.$$

Let $\mathbf{Z}^{*,0}$ be the part process of $\mathbf{Z}^*$ on D obtained from $\mathbf{Z}^*$ by killing upon its exit time τ_D from D. Denote by $p_t^{*,0}(z, B)$, $z \in D, B \in \mathcal{B}(D)$, the transition function of $\mathbf{Z}^{*,0}$. By [Fukushima, Oshima and Takeda (2011), Theorem 4.4.2], $\mathbf{Z}^{*,0}$ is m-symmetric and associated with the part form $(\mathcal{E}_D^*, \mathcal{F}_D^*)$ of $(\mathcal{E}^*, \mathcal{F}^*)$ on $L^2(D)$. Notice that a subset of D is $\mathcal{E}^*$-polar if and only if it is $\mathcal{E}_D^*$-polar in view of [Fukushima, Oshima and Takeda (2011), Theorem 4.4.3]. We also know from Theorem 1.3.2 that $(\mathcal{E}_D^*, \mathcal{F}_D^*) = (\frac{1}{2}\mathbf{D}_D, H_0^1(D))$.

Denote by $p_t^0(z, B)$, $z \in D$, $B \in \mathcal{B}(D)$, the transition function of the ABM $\mathbf{Z}^0$ on D which is associated with $(\frac{1}{2}\mathbf{D}_D, H_0^1(D))$.

Proposition 1.3.3.

(i) *There exists a Borel set $\mathcal{N} \subset D$ that is properly exceptional for both $\mathbf{Z}^*$ and $\mathbf{Z}^0$ so that*

$$p_t^{*,0}(z, B) = p_t^0(z, B), \quad t \geq 0, \ z \in D \setminus \mathcal{N}, \ B \in \mathcal{B}(D \setminus \mathcal{N}). \tag{1.3.17}$$

(ii) $\mathbf{Z}^*\big|_{D^* \setminus \mathcal{N}}$ *is a BMD starting at any point of $D^* \setminus \mathcal{N}$. It further satisfies the path-continuity* (1.3.16) *for every $z \in D^* \setminus \mathcal{N}$.*

Proof. (i) As $(\mathcal{E}_D^*, \mathcal{F}_D^*) = (\frac{1}{2}\mathbf{D}_D, H_0^1(D))$ is a regular Dirichlet form on $L^2(D)$, both $\mathbf{Z}^{*,0}$ and $\mathbf{Z}^0$ are properly associated with it so that, for a countable uniformly dense subset $\mathcal{C}$ of $C_c(D)$, there exists an $\mathcal{E}_D^*$-polar set $\mathcal{N}_1$ with $\mathcal{N}_0 \subset \mathcal{N}_1 \subset D$ such that

$$p_t^{*,0} f(z) = p_t^0 f(z), \quad z \in D \setminus \mathcal{N}_1,$$

for every $t \in \mathbb{Q}_+$ and $f \in \mathcal{C}$.

By noting that a_j^* is non-$\mathcal{E}_1^*$-polar for $1 \leq j \leq N$, we can find an increasing sequence $\{\mathcal{N}_n\}$ of Borel sets with $\mathcal{N}_1 \subset \mathcal{N}_n \subset D$ such that $\mathcal{N}_{2n}$ (resp. $\mathcal{N}_{2n+1}$) is properly exceptional for $\mathbf{Z}^0$ (resp. $\mathbf{Z}^*$). Then $\mathcal{N} = \bigcup_{n=1}^\infty \mathcal{N}_n \subset D$ has the desired property.

(ii) By virtue of (i), the part process $\mathbf{Z}^{*,0} = (Z_t^{*,0}, \zeta^{*,0}, \mathbb{P}_z^{*,0})$ of $\mathbf{Z}^*$ on D has the same law as the ABM $\mathbf{Z}^0 = (Z_t^0, \zeta^0, \mathbb{P}_z^0)$ on D for each $z \in D \setminus \mathcal{N}$. As $\mathcal{N}_0 \subset \mathcal{N}$, $\mathbf{Z}^*\big|_{D^* \setminus \mathcal{N}}$ enjoys the path continuity property (1.3.16) and consequently admits no killing inside $D^* \setminus \mathcal{N} (\supset K^*)$, so that it is a BMD starting at any point $z \in D^* \setminus \mathcal{N}$. □

From the regular Dirichlet form $(\mathcal{E}^*, \mathcal{F}^*)$ on $L^2(D^*; m)$, we have constructed a BMD $\mathbf{Z}^*\big|_{D^* \setminus \mathcal{N}} = (Z_t^*, \zeta^*, \mathbb{P}_z^*)_{z \in D^* \setminus \mathcal{N}}$ starting at any points of D^* outside some Borel properly exceptional set $\mathcal{N} \subset D$. Note that $K^* = \{a_1^*, \ldots, a_N^*\} \subset D^* \setminus \mathcal{N}$ by Theorem 1.3.2 (ii). The transition function of $\mathbf{Z}^*\big|_{D^* \setminus \mathcal{N}}$ will be denoted by $p_t^*(z, B)$:

$$p_t^*(z, B) = \mathbb{P}_z^*(Z_t^* \in B), \quad t \geq 0, \ z \in D^* \setminus \mathcal{N}, \ B \in \mathcal{B}(D^* \setminus \mathcal{N}). \tag{1.3.18}$$

Our next task is to construct a BMD starting at every point of D^* by joining together the ABM $\mathbf{Z}^0 = (Z_t^0, \zeta^0, \mathbb{P}_z^0)$ on D (starting at every point of D) with

$\mathbf{Z}^*\big|_{D^*\setminus\mathcal{N}}$ starting at a_j^*, $1 \le j \le N$, by means of a *piecing out procedure* described as follows: for every $z \in D$, we run Z_t^0 starting at z until its lifetime ζ^0; if $Z^0_{\zeta^0-} \in A_j$, we identify A_j with a_j^* and run a copy of $\mathbf{Z}^*\big|_{D^*\setminus\mathcal{N}}$ starting from a_j^*. Indeed this procedure can be carried out rigorously to get.

Theorem 1.3.4. *The procedure of piecing out from $\mathbf{Z}^0$ and $\mathbf{Z}^*\big|_{D^*\setminus\mathcal{N}}$ yields a BMD $\widetilde{\mathbf{Z}} = (\widetilde{Z}_t, \widetilde{\zeta}, \widetilde{\mathbb{P}}_z)$ on D^* which specifically enjoys the following properties: $\widetilde{\mathbf{Z}}$ is an m-symmetric strong Markov process. The part process of $\widetilde{\mathbf{Z}}$ on D is identical in law with the ABM on D. Its sample path $\widetilde{Z}_t$ takes value in the one-point compactification $D^*_\partial := D^* \cup \{\partial\}$ of D^* with $\widetilde{Z}_t = \partial$ for any $t \ge \widetilde{\zeta}$ and*

$$\widetilde{\mathbb{P}}_z^*\left(\widetilde{Z}_t^* \text{ is continuous on } D^*_\partial \text{ in } t \in [0,\infty)\right) = 1 \quad \text{for every } z \in D^*. \tag{1.3.19}$$

The proof of this theorem will be given in Section A.3 of the Appendix.

In order to establish a uniqueness theorem on BMD, we need to prepare the following extensions of some general theorems in Sections 3.1, 3.2 and 4.3 of [Chen and Fukushima (2012)] to a quasi-regular Dirichlet form and a properly associated symmetric standard process. These theorems were formulated for a regular Dirichlet form and an associated symmetric Hunt process.

See Chapter 1 of [Chen and Fukushima (2012)] for the definition of quasi-regular Dirichlet forms and all relevant basic notions.

Proposition 1.3.5. *Let M be a locally compact separable metric space, ν a positive Radon measure on it with full support, $(\mathcal{E}, \mathcal{F})$ a quasi-regular Dirichlet form on $L^2(M;\nu)$ and $\mathbf{X} = (X_t, \zeta, \mathbb{P}_x)$ an ν-symmetric standard process on M that is properly associated with $(\mathcal{E}, \mathcal{F})$. For $f \in \mathcal{F}$, denote by $\widetilde{f}$ its $\mathcal{E}$-quasi-continuous version.*

(i) *A subset N of M is $\mathcal{E}$-polar if and only if it is ν-polar in the sense that $\mathbb{P}_\nu(\sigma_{N_1} < \infty) = 0$ for some nearly Borel set $N_1 \supset N$. Any ν-polar set is contained in a Borel properly exceptional set.*
(ii) *For $B \in \mathcal{B}(M)$, let*

$$\mathcal{F}_{M\setminus B} = \{f \in \mathcal{F} : \widetilde{f} = 0 \ \mathcal{E}\text{-q.e. on } B\}.$$

For $\alpha > 0$, denotes by $\mathcal{H}_B^\alpha$ the orthogonal complement of $\mathcal{F}_{M\setminus B}$ in $(\mathcal{F}, \mathcal{E}_\alpha)$. Define $\mathbf{H}_B^\alpha$ by

$$\mathbf{H}_B^\alpha g(x) = \mathbb{E}_x\left[e^{-\alpha\sigma_B} g(X_{\sigma_B})\right], \quad x \in M, \ g \in \mathcal{B}_+(M).$$

For any $f \in \mathcal{F}$, $\mathbf{H}_B^\alpha|\widetilde{f}|(x) < \infty$ $\mathcal{E}$-q.e. $x \in M$ and $\mathbf{H}_B^\alpha \widetilde{f}$ is an $\mathcal{E}$-quasi-continuous version of the projection $P_{\mathcal{H}_B^\alpha} f$ of f on $\mathcal{H}_B^\alpha$.
(iii) *Let K be a compact subset of M such that the set $\mathcal{L}_{K,1} = \{f \in \mathcal{F}; \widetilde{f} \ge 1 \ \mathcal{E}\text{-q.e. on } K\}$ is non-empty. Then, for $\alpha > 0$, the α-order hitting probability $p_K^\alpha(x) = \mathbb{E}_x[e^{-\alpha\sigma_K}]$, $x \in M$, is an $\mathcal{E}$-quasi continuous function in $\mathcal{F}$.*

Proof. (i) is taken from the last statement of Theorem 3.1.13 of [Chen and Fukushima (2012)] which is formulated for a more general quasi-regular Dirichlet form and a properly associated symmetric right process. By taking (i) into account, we can see that the proof of all assertions in Section 3.2 of [Chen and Fukushima (2012)] works word for word for the present $(\mathcal{E}, \mathcal{F})$ and $\mathbf{X}$ and (ii) and (iii) correspond to Theorem 3.2.2 and Corollary 3.2.3 of [Chen and Fukushima (2012)], respectively. □

Proposition 1.3.6. *Under the same setting as in the preceding proposition, we have the following.*

(i) *Any semi-polar subset of M for $\mathbf{X}$ is $\mathcal{E}$-polar.*

(ii) *For each $u \in \mathcal{F}_b$, there exist a unique positive measure $\mu_{\langle u\rangle}$ on M charging no $\mathcal{E}$-polar set and satisfying*

$$\int_M f(x)\mu_{\langle u\rangle}(dx) = 2\mathcal{E}(u, uf) - \mathcal{E}(u^2, f) \quad \text{for every } f \in \mathcal{F}_b, \tag{1.3.20}$$

where the function f on the left-hand side is represented by its $\mathcal{E}$-quasi-continuous version.

(iii) *Assume that $\mathbf{X}$ is a diffusion on M admitting no killing inside M in the sense that*

$$\mathbb{P}_x(X_{\zeta-} \in M, \zeta < \infty) = 0 \quad \text{for } \mathcal{E}\text{-q.e. } x \in M.$$

Then the measure $\mu_{\langle u\rangle}$ in (ii) *further satisfies the following three properties:*

(P.1) $\mu_{\langle u\rangle}(M) = 2\mathcal{E}(u, u)$ *for $u \in \mathcal{F}_b$.*

(P.2) *For any $u \in \mathcal{F}_b$ and any Borel modification $\dot{u}$ of u, the push forward measure ν of $\mu_{\langle u\rangle}$ under the map $\dot{u}$ defined by $\nu(A) := \mu_{\langle u\rangle}(\dot{u}^{-1}(A))$, $A \in \mathcal{B}(\mathbb{R})$, is absolutely continuous with respect to the Lebesgue measure on $\mathbb{R}$. Here $\dot{u}^{-1}(A) := \{x \in M : \dot{u}(x) \in A\}$. This in particular implies that $\mu_{\langle u\rangle}$ does not charge on level sets of $\dot{u}$.*

(P.3) *Suppose $O \subset M$ is finely open and $u \in \mathcal{F}$ whose $\mathcal{E}$-quasi-continuous version is constant $\mathcal{E}$-q.e. on O. Then $\mu_{\langle u\rangle}(O) = 0$.*

Proof. We use the transfer method based on [Chen and Fukushima (2012), Theorem 1.4.3] which states that $(\mathcal{E}, \mathcal{F})$ is quasi-homeomorphic to a regular Dirichlet form $(\widehat{\mathcal{E}}, \widehat{\mathcal{F}})$ on $L^2(\widehat{M}, \widehat{\nu})$ for some locally compact separable metric space $\widehat{M}$ by a quasi homeomorphism j from $M_1 = \cup_{k=1}^{\infty} F_k (\subset M)$ onto $\widehat{M}_1 = \cup_{k=1}^{\infty} \widehat{F}_k (\subset \widehat{M})$. Here $\{F_k\}$ (resp. $\{\widehat{F}_k\}$) denotes an $\mathcal{E}$-nest (resp $\widehat{\mathcal{E}}$-nest), j is a homeomorphism from F_k onto $\widehat{F}_k$ for each k, and $\widehat{\nu}$ is the image measure of ν by j.

Let $\widehat{\mathbf{X}} = (\widehat{X}_t, \widehat{\zeta}, \widehat{\mathbb{P}}_x)$ be an $\widehat{\nu}$-symmetric Hunt process on $\widehat{M}$ associated with the regular Dirichlet form $(\widehat{\mathcal{E}}, \widehat{\mathcal{F}})$. In view of [Chen and Fukushima (2012), Theorem 3.1.13], there exist properly exceptional sets $\mathcal{N} \supset M \setminus M_1$ and $\widehat{\mathcal{N}} \supset \widehat{M} \setminus \widehat{M}_1$ of $\mathbf{X}$ and $\widehat{\mathbf{X}}$, respectively, such that the Hunt process $\widehat{\mathbf{X}}\big|_{\widehat{M}_1 \setminus \widehat{\mathcal{N}}}$ has the same distribution as the image process of the standard process $\mathbf{X}\big|_{M_1 \setminus \mathcal{N}}$ under the map j.

(i) If a set $B \subset M$ is semi-polar for $\mathbf{X}$, then so is the set $B_1 = B \cap (M_1 \setminus \mathcal{N})$ for $\mathbf{X}\big|_{M_1 \setminus \mathcal{N}}$ and hence, so is the set $jB_1 \subset \widehat{M}_1 \setminus \widehat{\mathcal{N}}$ for $\widehat{\mathbf{X}}\big|_{\widehat{M}_1 \setminus \widehat{\mathcal{N}}}$. Consequently jB_1

is $\widehat{\mathcal{E}}$-polar by virtue of [Chen and Fukushima (2012), Theorem 3.1.10]. In view of [Chen and Fukushima (2012), Exercise 1.4.2], B_1 and B are $\mathcal{E}$-polar.

(ii) For the regular Dirichlet form $(\widehat{\mathcal{E}}, \widehat{\mathcal{F}})$ on $L^2(\widehat{M}; \widehat{\nu})$, Let $\widehat{\mu}_{\langle \widehat{u} \rangle}$ be the energy measure on $\widehat{M}$ of $\widehat{u} \in \widehat{\mathcal{F}}$. According to [Chen and Fukushima (2012), Theorem 4.3.11] and [Fukushima, Oshima and Takeda (2011), §3.2], it is uniquely characterized by the equation

$$\int_{\widehat{M}} \widehat{f}(\widehat{x}) \widehat{\mu}_{\langle \widehat{u} \rangle}(d\widehat{x}) = 2\widehat{\mathcal{E}}(\widehat{u}, \widehat{u}\widehat{f}) - \widehat{\mathcal{E}}(\widehat{u}^2, \widehat{f}) \quad \text{for any } \widehat{u}, \widehat{f} \in \widehat{\mathcal{F}}_b,$$

where the function $\widehat{f}$ on the left-hand side is taken to be any $\widehat{\mathcal{E}}$-quasi-continuous version of it. $\widehat{\mu}_{\langle \widehat{u} \rangle}$ charges no $\widehat{\mathcal{E}}$-polar set.

For any $f \in \mathcal{F}$, define $\widehat{f} \in \widehat{\mathcal{F}}$ by $\widehat{f}(\widehat{x}) = f(j^{-1}\widehat{x})$, $\widehat{x} \in \widehat{M}_1$. Then the above identity reads

$$\int_{\widehat{M}_1} f(j^{-1}\widehat{x}) \widehat{\mu}_{\langle \widehat{u} \rangle}(d\widehat{x}) = 2\mathcal{E}(u, uf) - \mathcal{E}(u^2, f), \quad u, f \in \mathcal{F}_b,$$

where the function $f \in \mathcal{F}$ on the left-hand side is taken to be any $\mathcal{E}$-quasi-continuous version of it. This means that, if we let $\mu_{\langle u \rangle}\big|_{M_1}$ be the image measure of $\widehat{\mu}_{\langle \widehat{u} \rangle}\big|_{\widehat{M}_1}$ under the map j^{-1} and set $\mu_{\langle u \rangle}(M \setminus M_1) = 0$, then $\mu_{\langle u \rangle}$ charges no $\mathcal{E}$-polar set and satisfies (1.3.20).

To verify the uniqueness of $\mu_{\langle u \rangle}$, observe that, for any compact set K contained in F_k for some k, $\widehat{K} = jK \subset \widehat{F}_k$ is compact so that it admits a function $\widehat{f} \in \widehat{\mathcal{F}} \cap C_c(\widehat{M})$ with $\widehat{f}(\widehat{x}) \geq 1$, $\widehat{x} \in \widehat{K}$. Hence $f \in \mathcal{L}_{K,1}$ for f defined by $f(x) = \widehat{f}(jx)$, $x \in M_1$, and consequently, the α-order hitting probability p_K^α is an $\mathcal{E}$-quasi-continuous function in $\mathcal{F}$ by Proposition 1.3.5 (iii).

Now, for $u \in \mathcal{F}_b$, let μ be a positive measure on M charging no $\mathcal{E}$-polar set such that $\int_M f(x)\mu(dx)$ is equal to the right-hand side of (1.3.20) for any $\mathcal{E}$-quasi-continuous $f \in \mathcal{F}_b$. Then $\int_M p_K^\alpha(x)\mu(dx) = \int_M p_K^\alpha(x)\mu_{\langle u \rangle}(dx)$ for any compact set K contained in F_k for some k. As the semi-polar set $K \setminus K^r$ is $\mathcal{E}$-polar by (i), we get $\mu(K) = \mu_{\langle u \rangle}(K)$ by letting $\alpha \to \infty$, yielding the uniqueness of $\mu_{\langle u \rangle}$. Thus, we have

$$\mu_{\langle u \rangle}(A) = \widehat{\mu}_{\langle \widehat{u} \rangle}(j(A \cap M_1)) \quad \text{for any } A \subset \mathcal{B}(M). \tag{1.3.21}$$

(iii) **(P.1)** follows directly from [Chen and Fukushima (2012), (4.3.15)] for the regular Dirichlet form $(\widehat{\mathcal{E}}, \widehat{\mathcal{F}})$ and the identity (1.3.21).

Under the assumption of (iii), the process $\widehat{\mathbf{X}}\big|_{\widehat{M}_1 \setminus \widehat{\mathcal{N}}}$ considered in the beginning of the proof becomes of continuous sample path on $[0, \infty)$:

$$\widehat{\mathbb{P}}_{\widehat{x}}\left(t \in [0,\infty) \mapsto \widehat{X}_t \in \widehat{M} \cup \{\widehat{\partial}\} \text{ is continuous with } \widehat{X}_t = \widehat{\partial} \text{ for } t \geq \widehat{\zeta}\right) = 1,$$

for $\widehat{x} \in \widehat{M}_1 \setminus \widehat{\mathcal{N}}$, where $\widehat{M} \cup \{\widehat{\partial}\}$ is the one-point compactification of $\widehat{M}$. Hence, for any quasi-continuous function $\widehat{u} \in \widehat{\mathcal{F}}$, the additive functional $\widehat{A}_t^{[u]} = \widehat{u}(\widehat{X}_t) - \widehat{u}(\widehat{X}_0)$, $t \geq 0$, is continuous in $t \in [0, \infty)$, $\widehat{\mathbb{P}}_{\widetilde{x}}$-a.s. for any $\widetilde{x} \in \widehat{M}_1 \setminus \widehat{\mathcal{N}}$ on account of

[Chen and Fukushima (2012), Theorem 3.1.7] by choosing a larger properly exceptional set $\widehat{N} \subset \widehat{M}$ of $\widehat{\mathbf{X}}$ if necessary.

In Section 4.3 of [Chen and Fukushima (2012)], the energy measure $\widehat{\mu}_{\langle\widehat{u}\rangle}$ of $\widehat{u} \in \widehat{\mathcal{F}}$ and its strongly local part $\widehat{\mu}^c_{\langle\widehat{u}\rangle}$ are defined to be the Revuz measures of the predictable quadratic variations of the martingale part $\widehat{M}^{[\widehat{u}]}$ in the Fukushima decomposition of $\widehat{A}^{[\widehat{u}]}$ and its continuous part $\widehat{M}^{[\widehat{u}],c}$, respectively. As $\widehat{M}^{[\widehat{u}]} = \widehat{M}^{[\widehat{u}],c}$, we have $\widehat{\mu}_{\langle\widehat{u}\rangle} = \widehat{\mu}^c_{\langle\widehat{u}\rangle}$ for every $\widehat{u} \in \widehat{\mathcal{F}}$.

For $u \in \mathcal{F}$, take any Borel version $\dot{u}$ of u. For $A \in \mathcal{B}(\mathbb{R})$, let $B = \dot{u}^{-1}(A) \cap M_1 \in \mathcal{B}(M_1)$. Define $\widehat{u} \in \widehat{\mathcal{F}}$ by $\widehat{u}(\widehat{x}) = \dot{u}(j^{-1}\widehat{x})$, $\widehat{x} \in \widehat{M}_1$. Then $\widehat{u}^{-1}(A) \cap \widehat{M}_1 = jB \in \mathcal{B}(\widehat{M}_1)$ and by (1.3.21),

$$\nu(A) = \mu_{\langle u\rangle}(\dot{u}^{-1}(A)) = \mu_{\langle u\rangle}(B) = \widehat{\mu}_{\langle\widehat{u}\rangle}(jB) = \widehat{\mu}_{\langle\widehat{u}\rangle}(\widehat{u}^{-1}(A)) = \widehat{\mu}^c_{\langle\widehat{u}\rangle}(\widehat{u}^{-1}(A)).$$

The property **(P.2)** now follows from [Chen and Fukushima (2012), Theorem 4.3.8].

To show **(P.3)**, suppose $u \in \mathcal{F}$ is $\mathcal{E}$-quasi-continuous and equals a constant c $\mathcal{E}$-q.e. on O and define $\widehat{O} = j(O \cap M_1)(\subset \widehat{M}_1)$, $\widehat{u}(\widehat{x}) = u(j^{-1}\widehat{x})$ for $\widehat{x} \in \widehat{M}_1$. Then $\widehat{u} \in \widehat{\mathcal{F}}$ and $\widehat{u}(\widehat{x}) = c$ $\widehat{\mathcal{E}}$-q.e. $\widehat{x} \in \widehat{O}$. Hence by (1.3.21) and [Chen and Fukushima (2012), Proposition 4.3.1], $\mu_{\langle u\rangle}(O) = \widetilde{\mu}_{\langle\widetilde{u}\rangle}(\widetilde{O}) = 0$. □

We call the measure $\mu_{\langle u\rangle}$ appearing in Proposition 1.3.6 (ii) the *energy measure* of $u \in \mathcal{F}_b$ for the quasi-regular Dirichlet from $(\mathcal{E}, \mathcal{F})$.

Theorem 1.3.7. *BMD on D^* is unique in law. Its Dirichlet form on $L^2(D^*; m)$ is equal to $(\mathcal{E}^*, \mathcal{F}^*)$ defined by* (1.3.8) *and* (1.3.7).

Proof. Let $\mathbf{Z}^* = (Z_t^*, \zeta^*, \mathbb{P}_z^*)$ be a BMD on D^* and $(\mathcal{E}, \mathcal{F})$ its associated Dirichlet form on $L^2(D^*; m) = L^2(D)$. By [Chen and Fukushima (2012), Theorem 1.5.3], $(\mathcal{E}, \mathcal{F})$ is quasi-regular and $\mathbf{Z}^*$ is properly associated with it. We will prove by several steps that

$$(\mathcal{E}, \mathcal{F}) = (\mathcal{E}^*, \mathcal{F}^*). \tag{1.3.22}$$

Recall that for $f \in \mathcal{F}$, its $\mathcal{E}$-quasi-continuous version is denoted by $\widetilde{f}$.

Claim (i). Each a_i^* is non $\mathcal{E}$-polar, $1 \leq i \leq N$.

To see this, let $\mathbf{Z}^0 = (Z_t^0, \zeta^0, \mathbb{P}_z^0)$ be the ABM on $D = E \setminus K$. By Definition 1.2.1, $\mathbf{Z}^*$ is a diffusion on D^* with no killing on K^* whose part process on D is $\mathbf{Z}^0$. Therefore,

$$\mathbb{P}_z^*(\sigma_{a_i^*} < \infty) \geq \mathbb{P}_z^0(Z_{\zeta^0-} \in A_i,\ \zeta^0 < \infty) = \varphi^{(i)}(z), \quad z \in D,$$

which is positive by Lemma 1.1.4. Hence a_i^* is non m-polar for $\mathbf{Z}^*$ and consequently Claim (i) holds by Proposition 1.3.5.

The part form $(\mathcal{E}^D, \mathcal{F}_D)$ of $(\mathcal{E}, \mathcal{F})$ on the open set $D \subset D^* = D \cup K^*$ is defined by

$$\mathcal{F}_D = \{f \in \mathcal{F} : \widetilde{f} = 0\ \mathcal{E}\text{-q.e. on } K^*\}, \quad \mathcal{E}^D = \mathcal{E}\big|_{\mathcal{F}_D \times \mathcal{F}_D},$$

which is a Dirichlet form on $L^2(D;m) = L^2(D)$.

Claim (ii). $(\mathcal{E}^D, \mathcal{F}_D) = (\frac{1}{2}\mathbf{D}_D, H_0^1(D))$.

Denote by G_α^* and G_α^0 the resolvent of $\mathbf{Z}^*$ and $\mathbf{Z}^0$, respectively. Then

$$G_\alpha^* f(z) = G_\alpha^0 f(z) + \mathbf{H}_{K^*}^\alpha G_\alpha^* f(z), \quad z \in D^*, \ f \in \mathcal{B}(D^*) \cap L^2(D^*;m). \tag{1.3.23}$$

We can see as in the proof of [Chen and Fukushima (2012), Theorem 3.1.7 (ii)] that $G_\alpha^* f$ is $\mathcal{E}$-quasi continuous. Consequently, by Proposition 1.3.5 (ii) and Proposition 1.3.6 (i), $G_\alpha^0 f \in \mathcal{F}_D$ and

$$\mathcal{E}_\alpha(G_\alpha^0 f, v) = (f, v)_{L^2(D)}, \quad v \in \mathcal{F}_D,$$

which means that $(\mathcal{E}^D, \mathcal{F}_D)$ is a Dirichlet form on $L^2(D)$ generated by the strongly continuous contraction semigroup determined by the transition function of the ABM $\mathbf{Z}^0$. Since $(\frac{1}{2}\mathbf{D}_D, H_0^1(D))$ has the same property, we get Claim (ii).

Claim (iii). $\mathcal{F} = \mathcal{F}^*$.

As in the proof of Proposition 1.3.6, $(\mathcal{E}, \mathcal{F})$ is quasi-homeomorphic to a regular Dirichlet form $(\widehat{\mathcal{E}}, \widehat{\mathcal{F}})$ on $L^2(\widehat{D}; \widehat{m})$ for some locally compact separable metric space $\widehat{D}$ by a one-to-one map j from $D_1^* = \cup_{k=1}^\infty F_k (\subset D^*)$ onto $\widehat{D}_1 = \cup_{k=1}^\infty \widehat{F}_k (\subset \widehat{D})$. Here $\{F_k\}$ (resp. $\{\widehat{F}_k\}$) denotes an $\mathcal{E}$-nest (resp. $\widehat{\mathcal{E}}$-nest), j is a homeomorphism from F_k onto $\widehat{F}_k$ for each k and $\widehat{m}$ is the image measure of m by j,

Since each a_i^* is non $\mathcal{E}$-polar by Claim (i), $K^* = \{a_1^*, \ldots, a_N^*\} \subset D_1^*$. If we let $a_k = ja_k^*$, $1 \le k \le N$, $\{a_1, \ldots, a_N\}$ are distinct points in $\widehat{D}_1$. As $(\widehat{\mathcal{E}}, \widehat{\mathcal{F}})$ is regular, there exists, for each k, a function $\widehat{f}_k \in \widehat{\mathcal{F}} \cap C_c(\widehat{D})$ with $\widehat{f}_k(a_k) = 1$, $\widehat{f}_k(a_\ell) = 0$, $\ell \ne k$ (cf. [Chen and Fukushima (2012), Exercise 1.3.13]). Then the function f_k on D_1^* defined by $f_k(x) = \widehat{f}_k(jx)$ is an $\mathcal{E}$-quasi-continuous function in $\mathcal{F}$ taking value 1 at a_k^* and 0 at a_ℓ^* for $\ell \ne k$.

By Proposition 1.3.5 (ii), $\mathbf{H}_{K^*}^1 f_k\big|_D \in \mathcal{F}$, while

$$\mathbf{H}_{K^*}^1 f_k(z) = \mathbb{E}_z^*\left[e^{-\sigma_{K^*}}; Z_{\sigma_{K^*}}^* = a_k^*\right] = \mathbb{E}_z^E\left[e^{-\sigma_K}; Z_{\sigma_K}^E \in A_k\right] = u_1^{(k)}(z), \ z \in D.$$

Hence $u_1^{(k)}\big|_D \in \mathcal{F}$, $1 \le k \le N$, which together with Claim (ii) implies $\mathcal{F}^* \subset \mathcal{F}$. For any $f \in \mathcal{F}$, $f_0 := \widetilde{f} - \sum_{k=1}^N \widetilde{f}(a_k^*)\mathbf{H}_{K^*}^1 f_k$ is an $\mathcal{E}$-quasi continuous function in $\mathcal{F}$ vanishing on K^* so that $f_0\big|_D \in H_0^1(D)$ by Claim (ii). This shows that $f \in \mathcal{F}^*$, yielding $\mathcal{F} \subset \mathcal{F}^*$. Thus we have $\mathcal{F} = \mathcal{F}^*$.

Claim (iv). $\mathcal{E} = \mathcal{E}^*$.

Let $\mu_{\langle u \rangle}$ be the energy measure of $u \in \mathcal{F}_b$ for the quasi-regular Dirichlet form $(\mathcal{E}, \mathcal{F})$ of $\mathbf{Z}^*$. Since the ABM on D admits no killing inside D, $\mathbf{Z}^*$ is a diffusion on $D^* = D \cup K^*$ with no killing inside D^*. Consequently, we get from **(P.2)** of Proposition 1.3.6 that $\mu_{\langle u \rangle}(K^*) = 0$ and so from **(P.1)** that

$$\mathcal{E}(u,u) = \frac{1}{2}\mu_{\langle u \rangle}(D), \quad u \in \mathcal{F}.$$

For a bounded $u \in H_0^1(D)$, we have $\mathcal{E}(u,u) = \frac{1}{2}\mathbf{D}_D(u,u)$ by Claim (ii) so that the equation (1.3.20) for bounded $u, f \in H_0^1(D)$ implies

$$\mu_{\langle u \rangle}(dx) = |\nabla u(x)|^2 dx \quad \text{on } D. \tag{1.3.24}$$

For $u = u_1^{(k)} (\subset \mathcal{F}_b \cap C^\infty(D))$, $1 \leq k \leq N$, take any open set $O \subset D$ with $\overline{O} \subset D$, choose $v \in C_c^\infty(D) \subset H_0^1(D)$ with $v = u$ on O. Then $\mu_{\langle u-v \rangle}(O) = 0$ by (P.3) of Proposition 1.3.6 so that, by using the inequality stated in the above of Lemma 3.2.1 of [Fukushima, Oshima and Takeda (2011)], we have $\mu_{\langle u \rangle}(B) = \mu_{\langle v \rangle}(B) = \int_B |\nabla v(x)|^2 dx$ for any Borel set $B \subset O$, yielding (1.3.24) for such u. In the same way, (1.3.24) holds for any bounded $u \in \mathcal{F}^*$, yielding Claim (iv).

We have shown (1.3.22). The resolvent G_α^* of $\mathbf{Z}^*$ is then unique because $G_\alpha^* f(a_i^*)$, $1 \leq i \leq N$, are unique owing to Claim (i), and (1.3.23) reads $G_\alpha^* f(z) = G_\alpha^0 f(z) + \sum_{j=1}^N u_\alpha^{(j)}(z) G_\alpha^* f(a_j^*)$ for $z \in D$. □

In the rest of this section, we describe the extended Dirichlet space of $(\mathcal{E}^*, \mathcal{F}^*)$ and thereby deduce a specific global behavior of the BMD especially when $E = \mathbb{C}$.

For a general Dirichlet form $(\mathcal{E}, \mathcal{F})$, its extended Dirichlet space is denoted by $(\mathcal{F}_e, \mathcal{E})$. When $(\mathcal{E}, \mathcal{F})$ is transient, $\mathcal{F}_e$ is a real Hilbert space with inner product $\mathcal{E}$. The extended Dirichlet space $\mathcal{F}_e$ of $(\frac{1}{2}\mathbf{D}_U, H_0^1(U))$ on $L^2(U)$ for a domain $U \subset \mathbb{C}$ is denoted by $H_{0,e}^1(U)$. The extension of $\mathcal{E}$ from $H_0^1(U)$ to $H_{0,e}^1(U)$ is still given by $\frac{1}{2}\mathbf{D}_U$ due to a property of the space $\mathrm{BL}(U)$.

The 0-order resolvent (operator) G^0 of the ABM $\mathbf{Z}^0 = (Z_t^0, \zeta^0, \mathbb{P}_z^0)$ on D is defined by

$$G^0 f(z) = \mathbb{E}_z^0 \left[\int_0^{\zeta^0} f(Z_t^0) dt \right], \quad z \in D, \quad f \in \mathcal{B}_+(D).$$

Proposition 1.3.8. *Assume that $E \neq \mathbb{C}$.*

(i) *The Dirichlet form $(\mathcal{E}^*, \mathcal{F}^*)$ of $\mathbf{Z}^*$ on $L^2(D^*, m)$ and $(\frac{1}{2}\mathbf{D}_D, H_0^1(D))$ on $L^2(D)$ are both transient.*

(ii) *Let $\mathcal{H} = \left\{ \sum_{i=1}^N c_i \varphi^{(i)}\big|_D : c_i \in \mathbb{R} \right\}$. It then holds that*

$$\mathcal{F}_e^* = \{u = g + h : g \in H_{0,e}^1(D),\ h \in \mathcal{H}\}. \tag{1.3.25}$$

$$\mathcal{E}^*(u, v) = \frac{1}{2}\mathbf{D}_D(u, v), \quad u, v \in \mathcal{F}_e^*. \tag{1.3.26}$$

(iii) *It further holds that*

$$\mathcal{F}_e^* = \widehat{\mathcal{F}}_e^E\big|_D \text{ for } \widehat{\mathcal{F}}_e^E := \{u \in H_{0,e}^1(E) : \widetilde{u} \text{ is constant } \mathcal{E}^E\text{-q.e. on each } A_j\}. \tag{1.3.27}$$

Proof. (i) For the BMD $\mathbf{Z}^* = (Z_t^*, \zeta^*, \mathbb{P}_z^*)$ on $D^* = D \cup K^*$ and the ABM $\mathbf{Z}^0 = (Z_t^0, \zeta^0, \mathbb{P}_z^0)$ on D, we have

$$\mathbb{P}_z^*(\zeta^* < \infty,\ Z_{\zeta^*-}^* \in \partial E) \geq \mathbb{P}_z^0(\zeta^0 < \infty,\ Z_{\zeta^0-}^0 \in \partial E), \quad z \in D,$$

which can be seen to be positive by making a similar consideration to the proof of Lemma 1.1.4 (i) and using the assumption that $\mathbb{C} \setminus E$ is a continuum. Hence, $\bigcup_{\ell=1}^\infty \{z \in D^* : \mathbb{P}_z^*(Z_\ell^* \in D^*) < 1\} = D^*$ m-a.e., so that the Dirichlet form $(\mathcal{E}^*, \mathcal{F}^*)$

on $L^2(D^*;m)$ of $\mathbf{Z}^*$ is transient in view of [Fukushima, Oshima and Takeda (2011), Lemma 1.6.5].

The transience of $(\frac{1}{2}\mathbf{D}_D, H_0^1(D))$ can be proved in a similar way because $\mathbb{P}_z^0(\zeta^0 < \infty) > 0$ for any $z \in D$.

(ii) Clearly $H_{0,e}^1(D) \subset \mathcal{F}_e^*$. In view of [Chen and Fukushima (2012), Exercise 5.5.1], it holds that

$$G^0 u_1^{(i)}(z) = \varphi^{(i)}(z) - u_1^{(i)}(z) \le 1, \quad z \in D. \tag{1.3.28}$$

We get from (1.3.28) and Lemma 1.3.1 (iii) that

$$\int_D u_1^{(i)}(z) G^0 u_1^{(i)}(z) dz \le \int_D u_1^{(i)}(z) dz < \infty.$$

Therefore, by [Chen and Fukushima (2012), Theorem 2.1.12], $G^0 u_1^{(i)} \in H_{0,e}^1(D)$, $1 \le i \le N$, and accordingly $\varphi^{(i)} \in \mathcal{F}_e^*$, $1 \le i \le N$. We have shown the inclusion $\supset$ in (1.3.25).

To prove the opposite inclusion, express the family $\mathcal{F}^*$ as

$$\mathcal{F}^* = \left\{ u = f - \sum_{i=1}^N c_i\, G^0 u_1^{(i)} + \sum_{i=1}^N c_i \varphi^{(i)} : \ f \in H_0^1(D),\ c_i \in \mathbb{R},\ 1 \le i \le N \right\}.$$

As the function $g = f - \sum_{i=1}^N c_i G^0 u_1^{(i)}$ for $f \in H_0^1(D)$ belongs to $H_{0,e}^1(D)$ and the function $h = \sum_{i=1}^N c_i \varphi^{(i)} \in \mathcal{F}_e^*$ is harmonic on D, we have $\mathbf{D}_D(g,h) = 0$ and $\mathbf{D}_D(u,u) = \mathbf{D}_D(g,g) + \mathbf{D}_D(h,h)$ for $u = g + h \in \mathcal{F}^*$.

Take any $u \in \mathcal{F}_e^*$ and let $\{u_n = g_n + h_n\} \subset \mathcal{F}^*$ be an approximating sequence of u. u_n converges to u m-a.e. on D^*. Since $\{u_n\}$ is $\mathbf{D}_D$-Cauchy, so are the sequences $\{g_n\} \subset H_{0,e}^1(D)$ and $\{h_n = \sum_{i=1}^N c_{i,n}\varphi^{(i)}\} \subset \mathcal{F}_e^*$.

On account of (i), g_n (resp. h_n) is $\mathbf{D}_D$-convergent to some $g \in H_{0,e}^1(D)$ (resp. $h \in \mathcal{F}_e^*$). Notice that, as $(\mathcal{E}^*, \mathcal{F}^*)$ is regular and transient, $\varphi^{(i)}$, $1 \le i \le N$, are $\mathcal{E}^*$-quasi continuous functions in $\mathcal{F}_e^*$ by virtue of [Chen and Fukushima (2012), Theorem 3.4.2]. Hence, by the 0-order version of [Chen and Fukushima (2012), Theorem 1.3.3], there is a subsequence $\{n_k\}$ such that h_{n_k} converges to $\widetilde{h}$ $\mathcal{E}^*$-q.e. on D^*. Since each a_i^* is non $\mathcal{E}^*$-polar by Theorem 1.3.2, $h_{n_k}(a_i^*) = c_{i,n_k}$ converges to some $c_i \in \mathbb{R}$ as $k \to \infty$ and $\widetilde{h} = \sum_{i=1}^N c_i \varphi^{(i)}$. We have shown $u = g + \widetilde{h}$, namely, the inclusion $\subset$ in (1.3.25).

(iii) Notice that the Dirichlet form $(\mathcal{E}^E, \mathcal{F}^E) = (\frac{1}{2}\mathbf{D}_E, H_0^1(E))$ is also transient. Obviously $H_{0,e}^1(D) \subset \widehat{\mathcal{F}}_e^E$. Since $\varphi^{(i)}(z) = \mathbb{E}_z^E\left[f_i(Z_{\sigma_K}^E)\right]$, $z \in E$, for a function $f_i \in C_c^\infty(E) \subset H_{0,e}^1(E)$ taking value 1 on A_i and 0 on A_k for $k \ne i$, $\varphi^{(i)} \in \widehat{\mathcal{F}}_e^E$, $1 \le i \le N$, and any $u \in \widehat{\mathcal{F}}_e^E$ admits the $\mathcal{E}^E$-orthogonal decomposition $\widetilde{u} = f + \sum_{j=1}^N \widetilde{u}(A_j)\varphi^{(j)}$ for some $f \in H_{0,e}^1(D)$ by virtue of [Chen and Fukushima (2012), Theorem 3.4.2]. Hence (iii) follows from (ii). □

Proposition 1.3.9. *Assume that* $E = \mathbb{C}$. *The Dirichlet form* $(\mathcal{E}^*, \mathcal{F}^*)$ *on* $L^2(D^*;m)$ *(for* $D^* = D \cup K^*$ *with* $D = \mathbb{C}\setminus K$*) is recurrent and its extended Dirichlet* *space* $(\mathcal{F}_e^*, \mathcal{E}^*)$ *is*

$$\mathcal{F}_e^* = \widehat{\mathcal{F}}_e^{\mathbb{C}}\big|_D, \tag{1.3.28}$$

where $\widehat{\mathcal{F}}_e^{\mathbb{C}} := \{u \in H_e^1(\mathbb{C}) : \widetilde{u}$ is constant $\mathcal{E}^{\mathbb{C}}$-q.e. on each $A_j\}$, *and*

$$\mathcal{E}^*(u,v) = \frac{1}{2}\mathbf{D}_{\mathbb{C}}(u,v) \quad \textit{for } u, v \in \mathcal{F}_e^*. \tag{1.3.29}$$

Proof. Notice that, according to [Chen and Fukushima (2012), Theorem 2.2.13], the extended Dirichlet space $H_e^1(\mathbb{C})$ is identical with the Beppo Levi space $\mathrm{BL}(\mathbb{C}) = \{f \in L^2_{\mathrm{loc}}(\mathbb{C}) : |\nabla f| \in L^2(\mathbb{C})\}$ and any $\mathbf{D}_{\mathbb{C}}$-Cauchy sequence $\{v_n\} \subset \mathrm{BL}(\mathbb{C})$ admits $v \in \mathrm{BL}(\mathbb{C})$ and constants c_n such that v_n is $\mathbf{D}_{\mathbb{C}}$-convergent to v and $v_n + c_n$ is $L^2_{\mathrm{loc}}(\mathbb{C})$-convergent to v.

Now, for any $u \in \mathcal{F}_e^*$, take its approximating sequence $u_n \in \mathcal{F}^*$ so that $\mathbf{D}_D(u_n - u_m, u_n - u_m) \to 0$ as $n, m \to \infty$, and $u_n(z) \to u(z)$ as $n \to \infty$ for a.e. $z \in D = \mathbb{C}\setminus K$. By (1.3.10), $u_n = v_n|_D$ for some $v_n \in \widehat{\mathcal{F}}^{\mathbb{C}}$. By (1.3.3) and [Chen and Fukushima (2012), Theorem 4.3.8], $\mathbf{D}_{\mathbb{C}}(v_n - v_m, v_n - v_m) = \mathbf{D}_D(u_n - u_m, u_n - u_m)$, $m, n \geq 1$.

Hence there exist $v \in \mathrm{BL}(\mathbb{C})$ and constants c_n such that v_n is $\mathbf{D}_{\mathbb{C}}$-convergent to v and $v_n + c_n$ is $L^2_{\mathrm{loc}}(\mathbb{C})$-convergent to v. By taking a suitable subsequence if necessary, $\lim_{n\to\infty} c_n = c$ exists and $u = v - c$ a.e. on D. This means that $\{v_n\} \subset \widehat{\mathcal{F}}^{\mathbb{C}}(\subset H^1(\mathbb{C}))$ is an approximating sequence for $v - c \in H_e^1(\mathbb{C})$. It follows from [Chen and Fukushima (2012), Theorem 2.3.4] that, for a suitable subsequence $\{n_k\}$, $\widetilde{v}_{n_k}(z)$ converges to $\widetilde{v}(z) - c$ q.e. on $\mathbb{C}$ as $k \to \infty$. Consequently, $v - c \in \widehat{\mathcal{F}}_e^{\mathbb{C}}$, $u = (v - c)|_D$ and

$$\mathcal{E}^*(u,u) = \lim_{n\to\infty} \frac{1}{2}\mathbf{D}_D(u_n, u_n) = \lim_{n\to\infty} \frac{1}{2}\mathbf{D}_D(v_n, v_n) = \frac{1}{2}\mathbf{D}_D(v,v),$$

as v_n is $\mathbf{D}_{\mathbb{C}}$-convergent to v. We have shown that $\mathcal{F}_e^* \subset \widehat{\mathcal{F}}_e^{\mathbb{C}}|_D$.

On the other hand, by virtue of [Fukushima, Oshima and Takeda (2011), Exercise 4.6.4] and Lemma 1.3.1 (iv), any $u \in \widehat{\mathcal{F}}_e^{\mathbb{C}}$ admits a unique $\mathbf{D}_{\mathbb{C}}$-orthogonal decomposition $u = f + \sum_{j=1}^N \widetilde{u}(A_j)\varphi^{(j)}$, $f \in H^1_{0,e}(D)$. Since $H^1_{0,e}(D) \subset \mathcal{F}_e^*$ and $\varphi^{(j)} \in \mathcal{F}_e^*$, $1 \leq j \leq N$, as was shown in the first paragraph of the proof of Proposition 1.3.8 (ii), we get the converse inclusion $\mathcal{F}_e^* \supset \widehat{\mathcal{F}}_e^{\mathbb{C}}|_D$.

It follows from (1.3.28) and (1.3.29) that $1 \in \mathcal{F}_e^*$ and $\mathcal{E}^*(1,1) = 0$, yielding the recurrence of $(\mathcal{E}^*, \mathcal{F}^*)$ by [Chen and Fukushima (2012), Theorem 2.1.8]. □

Theorem 1.3.10. *Assume that* $E = \mathbb{C}$. *Let* $\mathbf{Z}^* = (Z_t^*, \zeta^*, \mathbb{P}_z^*)$ *be the BMD on* $D^* = D \cup K^*$ *with* $D = \mathbb{C} \setminus \bigcup_{j=1}^N A_j$.

(i) *The process* $\mathbf{Z}^*$ *is conservative, that is,*

$$\mathbb{P}_z^*(\zeta^* = \infty) = 1 \quad \textit{for every } z \in D^*. \tag{1.3.30}$$

(ii) *Let* B *be a Borel subset of* D *that is non-polar with respect to the planar Brownian motion. Then*

$$\mathbb{P}_z^*(\sigma_B < \infty,\ Z^*_{\sigma_B -} \in \overline{B}) = \mathbb{P}_z^*(\sigma_B < \infty) = 1 \quad \text{for every } z \in D^*. \tag{1.3.31}$$

Proof. Since $(\mathcal{E}^{\mathbb{C}}, \mathcal{F}^{\mathbb{C}})$ and $(\mathcal{E}^*, \mathcal{F}^*)$ share a common part form $(\mathcal{E}^D, \mathcal{F}^D)$, we know from [Fukushima, Oshima and Takeda (2011), Theorem 4.4.3 (i)] that a subset of D is $\mathcal{E}^{\mathbb{C}}$-polar if and only if it is $\mathcal{E}^*$-polar. Suppose that B is a Borel subset of D that is non-polar with respect to the planar Brownian motion. B is then non-$\mathcal{E}^{\mathbb{C}}$-polar and consequently non-$\mathcal{E}^*$-polar.

The Dirichlet form $(\mathcal{E}^*, \mathcal{F}^*)$ on $L^2(D^*; m)$ is recurrent by Proposition 1.3.9. As has been proved under a much more general setting by [Chen and Fukushima (2012), Lemma 7.7.2], it is irreducible as well so that we get from [Fukushima, Oshima and Takeda (2011), Theorem 4.7.1]

$$\mathbb{P}_z^*(\sigma_B < \infty) = 1, \tag{1.3.32}$$

for $\mathcal{E}^*$-quasi every $z \in D^*$. In particular, (1.3.32) holds for $z \in K^*$ as each a_j^* is non-$\mathcal{E}^*$-polar by Theorem 1.3.2.

Denote by p_t^0 the transition function of the ABM $\mathbf{Z}^0 = (Z_t^0, \zeta^0, \mathbb{P}_z^0)$ on D. Since the function $v(z) = \mathbb{P}_z^*(\sigma_B < \infty)$, $z \in D^*$, is excessive relative to $\mathbf{Z}^*$, $v(Z_t^*)$ is right continuous in $t \in [0, \infty)$ $\mathbb{P}_z^*$-a.s. for every $z \in D^*$ in view of [Chen and Fukushima (2012), Theorem A.2.2]. Hence, for any $z \in D$,

$$p_t^0 v(z) = \mathbb{E}_z^0[v(Z_t^0)] = \mathbb{E}_z^*[v(Z_t^*); t < \tau_D] \ \to \ v(z) \quad \text{as } t \downarrow 0.$$

As $p_t^0(z, \cdot)$ is absolutely continuous with respect to the Lebesgue measure on D, $p_t^0 v(z) = p_t^0 \mathbb{1}_D(z) \to 1$ as $t \downarrow 0$ for every $z \in D$. Consequently, (1.3.32) holds for every $z \in D$.

On the other hand, in view of [Fukushima, Oshima and Takeda (2011), Lemma 1.6.5, Exercise 4.5.1], the recurrence of $(\mathcal{E}^*, \mathcal{F}^*)$ implies that (1.3.30) is valid for $\mathcal{E}^*$-q.e. $z \in D^*$. By the same reasoning as above, (1.3.30) can be seen to be valid for every $z \in D^*$. In particular, Z_t^* is continuous on $[0, \infty)$ $\mathbb{P}_z^*$-a.s. for every $z \in D^*$, which yields the first identity of (1.3.31). □

Remark 1.3.11 (BMD and excursions around holes). In Theorem 1.3.4, the BMD on D^* is constructed by piecing together the ABM $\mathbf{Z}^0$ on D starting at any $z \in D$ and the process $\mathbf{Z}^*$ on D^* starting at a_j^*, $1 \le j \le N$. The latter is readily constructed due to the regularity of the Dirichlet form (1.3.7), (1.3.8) and the positivity of the associated capacity of a_j^*. But it is hard to see how its sample path looks like.

When $N = 1$, $\mathbf{Z}^*$ starting at a_1^* was actually constructed in [Fukushima and Tanaka (2005)] by piecing together the excursions around the hole A_1 according to an excursion-valued Poisson point process in Itô's sense ([Itô (1970)]) whose characteristic measure is uniquely determined by $\mathbf{Z}^0$. When $N \ge 2$, this procedure can be performed inductively as in [Chen and Fukushima (2012), Theorem 7.7.4] to produce $\mathbf{Z}^*$ starting at a_j^* $1 \le j \le N$. See Figure 1.1.

Remark 1.3.12. The construction and the Dirichlet form characterization of BMD shown by Theorem 1.3.4 and Theorem 1.3.7 in the above can be extended to a more general symmetric diffusion than the ABM $\mathbf{Z}^E$ on E we started with. We shall see this in Proposition 1.4.3 of the next section.

1.4 Localization properties

Just as in Section 1.2, we assume, for a planar domain E, that either $E = \mathbb{C}$ or $\mathbb{C} \setminus E$ is a continuum. Let $\{A_1, \dots, A_N\}$ be mutually disjoint continua contained in E, $D = E \setminus \bigcup_{j=1}^N A_j$, and $D^* = D \cup \{a_1^*, \dots, a_N^*\}$ be the space obtained from E by rendering each A_j into a singleton a_j^*. $\mathbf{Z}^*$ denotes the BMD on D^*.

Consider a subdomain E_1 of E such that $\mathbb{C} \setminus E_1$ is a continuum, $\bigcup_{j=1}^\ell A_j \subset E_1$ for some $\ell \le N$ and $E_1 \cap A_j = \emptyset$ for $\ell < j \le N$. Analogously to the above, one can consider the sets $D_1 = E_1 \setminus \bigcup_{j=1}^\ell A_j$ and $D_1^* = D_1 \cup \{a_1^*, \dots, a_\ell^*\}$.

Theorem 1.4.1. *The part process $\mathbf{Z}^{*,D_1^*}$ of $\mathbf{Z}^*$ killed upon leaving D_1^* is the BMD on D_1^*.*

Proof. Let $\mathbf{Z}^0$ be the part process on D of the BMD $\mathbf{Z}^*$. $\mathbf{Z}^0$ is the ABM on D. As the part process on D_1^* of the m-symmetric diffusion $\mathbf{Z}^*$, $\mathbf{Z}^{*,D_1^*}$ is a diffusion on D_1^* that is symmetric with respect to $m|_{D_1^*}$ admitting no killing on $\{a_1^*, \dots, a_\ell^*\}$. The part of $\mathbf{Z}^{*,D_1^*}$ on D_1 coincides with the part on D_1 of the ABM $\mathbf{Z}^0$ on D so that it is the ABM on D_1. Hence $\mathbf{Z}^{*,D_1^*}$ must be the BMD on D_1^* by Theorem 1.3.7.

We present an alternative proof of this theorem by using Dirichlet form characterization of BMD on D_1^* shown by Theorem 1.3.7 and its expression (1.3.10). Let $(\mathcal{E}^*, \mathcal{F}^*)$ and $(\mathcal{E}, \mathcal{F})$ be the Dirichlet forms of BMD in D^* and D_1^*, respectively. Then

$$\mathcal{F}^* = \{u|_D : \ u \in H_0^1(E), \ \widetilde{u} \text{ is constant } \mathcal{E}^E\text{-q.e. on each } A_j\},$$

and $\mathcal{E}^*(u, v) = \frac{1}{2}\mathbf{D}_D(u, v)$ for $u, v \in \mathcal{F}^*$. It is known that $\mathbf{Z}^{*,D_1^*}$ has Dirichlet form $(\mathcal{E}^*, \mathcal{F}^*_{D_1^*})$ on $L^2(D_1^*; m)$, where

$$\mathcal{F}^*_{D_1^*} := \{u \in \mathcal{F}^* : \widetilde{u} = 0 \ \mathcal{E}^*\text{-q.e. on } D^* \setminus D_1^*\}.$$

Since each a_j^* has positive capacity, we conclude that

$$\mathcal{F}^*_{D_1^*} = \Big\{u|_D : u \in H_0^1(E), \ \widetilde{u} \text{ is constant } \mathbf{D}\text{-q.e. on } A_j \text{ for } j = 1, \dots, l$$
$$\text{and } \widetilde{u} = 0 \ \mathcal{E}^E\text{-q.e. on } E \setminus E_1\Big\} = \mathcal{F}.$$

So $(\mathcal{E}^*, \mathcal{F}^*_{D_1^*}) = (\mathcal{E}, \mathcal{F})$, which proves that $\mathbf{Z}^{*,D_1^*}$ is the BMD on D_1^*. □

As an immediate consequence of Theorems 1.3.10 and 1.4.1, we obtain the following;

Theorem 1.4.2. *Suppose that E is a planar domain so that $\mathbb{C} \setminus E$ is a continuum. Let $\{A_1, \dots, A_N\}$ be mutually disjoint continua contained in E and $\mathbf{Z}^* = (Z_t^*, \zeta^*, \mathbb{P}_z^*)$ be the BMD on the quotient space $D^* = (E \setminus \bigcup_{j=1}^N A_j) \cup \{a_1^*, \dots, a_N^*\}$. Then*

$$\mathbb{P}_z^*(\zeta^* < \infty, \ Z^*_{\zeta^*-} \in \partial E) = \mathbb{P}_z^*(\zeta^* < \infty) = 1 \quad \text{for every } \ z \in D^*. \tag{1.4.1}$$

Proof. Consider the BMD $\mathbf{Z}^{\mathbb{C},*}$ on the quotient space $(\mathbb{C}\setminus\bigcup_{j=1}^N A_j)\cup\{a_1^*,\dots,a_N^*\}$. By Theorem 1.4.1, $\mathbf{Z}^*$ is the part process of $\mathbf{Z}^{\mathbb{C},*}$ on D^* being killed at the exit time τ_{D^*} which equals the hitting time $\sigma_{\mathbb{C}\setminus E}$. Therefore (1.4.1) follows from the property (1.3.31) of $\mathbf{Z}^{\mathbb{C},*}$. □

The next Proposition says that one can darn (or short) holes one by one.

Proposition 1.4.3. *Let* $\mathbf{Y}$ *be the BMD on* $D_1^* := (E\setminus\cup_{j=1}^{N-1}A_j)\cup\{a_1^*,\dots,a_{N-1}^*\}$ *obtained from the ABM on* E *by darning the first* $N-1$ *holes. Let* $\mathbf{Z}$ *be an* m*-symmetric diffusion on* D^* *obtained from* $\mathbf{Y}$ *by darning* A_N *into a single point* a_N^**, namely, an* m *symmetric diffusion extension of* $\mathbf{Y}$ *from* D_1^* *to* D^**. Such a diffusion* $\mathbf{Z}$ *exists and coincides with the BMD on* D^**.*

Proof. Let $D_1 = E\setminus\cup_{j=1}^{N-1}A_j$ and denote by $(\mathcal{E},\mathcal{F})$ the Dirichlet form of $\mathbf{Y}$ on $L^2(D_1^*;m)$. As was observed in the above,

$$\mathcal{F} = \left\{u|_{D_1} :\ u\in H_0^1(E),\ \widetilde{u}\text{ is constant }\mathcal{E}^E\text{-q.e. on }A_j\text{ for }j=1,\dots,N-1\right\}, \tag{1.4.2}$$

and $\mathcal{E}(u,v)=\frac{1}{2}\mathbf{D}_{D_1}(u,v)$.

The process $\mathbf{Z}$ is an m-symmetric diffusion extension of the BMD $\mathbf{Y}$ on D_1^* from D_1^* to $D^* = (D_1^*\setminus A_N)\cup\{a_N^*\}$, while the BMD on D^* is an m-symmetric diffusion extension of the ABM on D from D to $D^*=D\cup\{a_1^*,\dots,a_N^*\}$. Actually, by replacing the ABM on D with the BMD $\mathbf{Y}$ on D_1^* in the proof of Theorem 1.3.7 and the expression (1.3.10), one can readily obtain the following expression of the Dirichlet form $(\widehat{\mathcal{E}},\widehat{\mathcal{F}})$ on $L^2(D^*;m)$ of $\mathbf{Z}$:

$$\widehat{\mathcal{F}} = \{u|_D :\ u\in\mathcal{F},\ \widetilde{u}\text{ is constant }\mathcal{E}\text{-q.e. on }A_N\},\quad \widehat{\mathcal{E}}(u,v)=\mathcal{E}(u,v). \tag{1.4.3}$$

From (1.4.2) and (1.4.3), we get

$$\widehat{\mathcal{F}} = \left\{u|_D :\ u\in H_0^1(E),\ \widetilde{u}\text{ is constant }\mathcal{E}^E\text{-q.e. on }A_j\text{ for }j=1,\dots,N\right\} = \mathcal{F}^*,$$

and $\widehat{\mathcal{E}}(u,v)=\frac{1}{2}\mathbf{D}_D(u,v)$ for $u,v\in\mathcal{F}^*$.

The darned process $\mathbf{Z}$ starting at every point of D^* can be constructed as in the proof of Theorem 1.3.4 by noting that Lemma A.3.1 remains valid if the ABM on D is replaced by the BMD on D_1^*. □

Proposition 1.4.4. *Let* $K=A\cup B$ *be the union of two disjoint non-polar compact subsets of* E*. Let* $\mathbf{Y}$ *be BMD on* $(E\setminus A)^*$ *by darning* A*, and* $\mathbf{Z}$ *the diffusion with darning on* $(E\setminus K)^*$ *obtained from* $\mathbf{Y}$ *by darning* $A^*\cup B$*. Then* $\mathbf{Z}$ *is BMD on* $(E\setminus K)^*$ *by darning* K *into one single point.*

Proof. Let $(\mathcal{E},\mathcal{F})$ and $(\widehat{\mathcal{E}},\widehat{\mathcal{F}})$ be the Dirichlet forms for the processes $\mathbf{Y}$ and $\mathbf{Z}$ on $L^2((E\setminus A)^*;m)$ and $L^2((E\setminus K)^*;m)$, respectively. Note that

$$\mathcal{F} = \left\{u|_{E\setminus A} :\ u\in H_0^1(E),\ \widetilde{u}\text{ is constant }\mathcal{E}^E\text{-q.e. on }A\right\},$$

$$\mathcal{E}(u,v) = \frac{1}{2}\int_{E\setminus A}\nabla u(x)\cdot\nabla v(x)dx \qquad \text{for } u,v\in\mathcal{F},$$

while

$$\begin{aligned}\widehat{\mathcal{F}} &= \{u|_{E\setminus K} :\ \widetilde{u}\in\mathcal{F},\ u \text{ is constant } \mathcal{E}\text{-q.e. on } A^*\cup B\}\\ &= \{u|_{E\setminus K} :\ u\in H_0^1(E),\ \widetilde{u} \text{ is constant } \mathcal{E}^E\text{-q.e. on } K=A\cup B\}\\ &= \mathcal{F}^*,\end{aligned}$$

$$\widehat{\mathcal{E}}(u,v)=\mathcal{E}(u,v)=\frac{1}{2}\int_{E\setminus K}\nabla u(x)\cdot\nabla v(x)dx=\mathcal{E}^*(u,v) \qquad \text{for } u,v\in\widetilde{\mathcal{F}}.$$

Here $(\mathcal{E}^*,\mathcal{F}^*)$ is the Dirichlet form for BMD $\mathbf{Z}^*$ on $(E\setminus K)^*$. This proves that $\mathbf{Z}$ has the same distribution as BMD $\mathbf{Z}^*$ on $(E\setminus K)^*$. □

1.5 Zero flux characterization of L^2-generator of BMD

Under the setting of Sections 1.2 and 1.3, let $(\mathcal{A}^*,\mathcal{D}(\mathcal{A}^*))$ denote the L^2-*infinitesimal generator* of BMD $\mathbf{Z}^*$, or equivalently, of the Dirichlet form $(\mathcal{E}^*,\mathcal{F}^*)$ on $L^2(D^*,m)$. That is, $u\in\mathcal{D}(\mathcal{A}^*)$ if and only if $u\in\mathcal{F}^*$ and there is some $f\in L^2(D)=L^2(D^*;m)$ so that

$$\mathcal{E}^*(u,v)=-\int_D f(x)v(x)dx \qquad \text{for every } v\in\mathcal{F}^*. \tag{1.5.1}$$

We denote the above f as $\mathcal{A}^*u$. In view of (1.3.7) and (1.3.8), condition (1.5.1) is equivalent to

$$\frac{1}{2}\int_D\nabla u(x)\cdot\nabla v(x)dx=-\int_D f(x)v(x)dx \qquad \text{for every } v\in C_c^\infty(D) \tag{1.5.2}$$

and

$$\frac{1}{2}\int_D\nabla u(x)\cdot\nabla u_1^{(j)}(x)dx=-\int_D f(x)u_1^{(j)}(x)dx \qquad \text{for every } j=1,\ldots,N. \tag{1.5.3}$$

(1.5.2) says that Δu on D in the distribution sense is in $L^2(D)$ and $f=\frac{1}{2}\Delta u\in L^2(D)$. Let us define the *flux* $\mathcal{N}(u)(a_j^*)$ of u at a_j^* by

$$\mathcal{N}(u)(a_j^*)=\frac{1}{2}\int_D\nabla u(x)\cdot\nabla u_1^{(j)}(x)dx+\frac{1}{2}\int_D\Delta u(x)u_1^{(j)}(x)dx. \tag{1.5.4}$$

Then (1.5.3) is equivalent to

$$\mathcal{N}(u)(a_j^*)=0 \qquad \text{for every } j=1,\ldots,N. \tag{1.5.5}$$

Hence, we have established the following.

Theorem 1.5.1. *A function $u\in\mathcal{F}^*$ is in $\mathcal{D}(\mathcal{A}^*)$ if and only if the Laplacian Δu of u in the distribution sense is in $L^2(D)$ and u has zero flux at every a_j^*. Moreover, for $u\in\mathcal{D}(\mathcal{A}^*)$, $\mathcal{A}^*u=\frac{1}{2}\Delta u$ on D.*

Note that when $\partial D=(\cup_{j=1}^N\partial A_j)\cup\partial E$ is smooth and u is smooth in a neighborhood of $\overline{D}$, then by the Green-Gauss formula we have

$$\mathcal{N}(u)(a_j^*)=\frac{1}{2}\int_{\partial A_j}\frac{\partial u(x)}{\partial\mathbf{n}}\sigma(dx),$$

where $\mathbf{n}$ is the unit outward normal vector field of D on ∂D and σ is the surface measure on ∂D, because $u_1^{(j)}\big|_D$ extends continuously with value 1 on A_j and 0 on $(\cup_{k\neq j} A_k) \cup \partial E$.

We define the 0-order resolvent (operator) G^* of the BMD $\mathbf{Z}^* = (Z_t^*, \zeta^*, \mathbb{P}_z^*)$ by

$$G^* f(z) = \mathbb{E}_z^* \left[\int_0^{\zeta^*} f(Z_t^*) dt \right], \quad z \in D^*, \quad f \in \mathcal{B}_+(D^*). \tag{1.5.6}$$

Theorem 1.5.2. *Assume that E is bounded. For every $f \in L^2(D)(= L^2(D^*; m))$, $G^* f \in \mathcal{D}(\mathcal{A}^*)$ with*

$$\mathcal{A}^* G^* f = -f \quad \text{and} \quad \mathcal{N}(G^* f)(a_i^*) = 0, \quad 1 \leq i \leq N. \tag{1.5.7}$$

Furthermore, any $u \in \mathcal{D}(\mathcal{A}^*)$ admits the expression $u = G^* f$ for $f = -\mathcal{A}^* u (\in L^2(D))$.

Proof. As E is bounded, the well known Poincaré inequality holds (cf. [Fukushima, Oshima and Takeda (2011), (1.5.11)]), that is, there is $\lambda_1 > 0$ so that

$$\mathbf{D}(f, f) \geq \lambda_1 \int_E f(x)^2 dx \qquad \text{for } f \in H_0^1(E).$$

In view of Lemma 1.3.1 (i), (1.3.7) and (1.3.8), this in particular implies that

$$\mathcal{E}^*(u, u) \geq \frac{\lambda_1}{2} \int_D u(x)^2 dx \qquad \text{for } u \in \mathcal{F}^*. \tag{1.5.8}$$

It follows that $(\mathcal{E}^*, \mathcal{F}^*)$ is transient and, as $\mathcal{F}^*$ is already complete with the metric $\mathcal{E}^*$, $\mathcal{F}_e^* = \mathcal{F}^*$. Further, in view of [Chen and Fukushima (2012), (2.1.10)], we get from (1.5.8) that, for any non-negative $f \in L^2(D)$,

$$\int_D f(z) G^* f(z) m(dz) = \sup_{u \in \mathcal{F}^*} \frac{(|u|, f)_{L^2(D)}^2}{\mathcal{E}^*(u, u)} \leq \frac{2}{\lambda_1} \|f\|_{L^2(D)}^2.$$

By [Chen and Fukushima (2012), Theorem 2.1.12], this implies that $G^* f \in \mathcal{F}_e^*(= \mathcal{F}^*)$ and

$$\mathcal{E}^*(G^* f, v) = -(f, v)_{L^2(D)}, \quad \text{for} \quad f \in L^2(D), \quad v \in \mathcal{F}^*, \tag{1.5.9}$$

which means $G^* f \in \mathcal{D}(\mathcal{A}^*)$ and (1.5.7) by Theorem 1.5.1.

The second assertion follows from (1.5.1) and (1.5.9). □

1.6 BMD harmonic function and zero period property

Under the setting of Sections 1.2 and 1.3, denote by $\mathbf{Z}^E = (Z_t^E, \zeta^E, \mathbb{P}_z^E)$ and $\mathbf{Z}^D = (Z_t^D, \zeta^D, \mathbb{P}_z^D)$ the ABM on E and $D = E \setminus K$ for $K = \bigcup_{i=1}^N A_i$, respectively, and by $\mathbf{Z}^* = (Z_t^*, \zeta^*, \mathbb{P}_z^*)$ the BMD on the set

$$D^* = D \cup K^*, \quad K^* = \{a_1^*, \ldots, a_N^*\}$$

obtained from E by rendering each hole A_i as one point a_i^*.

1.6.1 *BMD harmonic function: Continuity and locality*

Definition 1.6.1. A Borel measurable real-valued function u on a connected open subset O of D^* is said to be $\mathbf{Z}^*$-*harmonic* or *BMD-harmonic* on O if for every relatively compact open subset O_1 of O,

$$\mathbb{E}_z^*\big[|u(Z^*_{\tau_{O_1}})|\big] < \infty \quad \text{and} \quad u(z) = \mathbb{E}_z^*\big[u(Z^*_{\tau_{O_1}})\big] \qquad \text{for every } z \in O_1. \qquad (1.6.1)$$

Here $\tau_{O_1} := \inf\{t \geq 0 : Z_t^* \notin O_1\}$.

Notice that $\mathbb{P}_z^*(\tau_{O_1} < \infty) = 1$ for every $z \in D^*$ on account of Theorem 1.3.10 when $E = \mathbb{C}$ and Theorem 1.4.2 when $\mathbb{C} \setminus E$ is a continuum.

The restriction to $O \cap D$ of any $\mathbf{Z}^*$-harmonic function on O is harmonic there in the classical sense (with respect to Brownian motion) and hence is continuous there.

Theorem 1.6.2. *Suppose that u is $\mathbf{Z}^*$-harmonic on a connected open set O of D^*. Then u is continuous on O.*

Proof. It suffices to show that u is continuous at each point of K^* contained in O. Choose a relatively compact open set $O_1 \subset O$ such that $O_1 \cap K^* = \{a_j^*\}$ for some $1 \leq j \leq N$. Denote the set $O_1 \setminus \{a_j^*\}$ by $\widehat{O}_1$. Since u is $\mathbf{Z}^*$-harmonic on O and the part process of $\mathbf{Z}^*$ on D has the same law as the ABM on D, we have for every $z \in \widehat{O}_1$,

$$\begin{aligned} u(z) &= \mathbb{E}_z^*\left[u(Z^*_{\tau_{O_1}})\right] = \mathbb{E}_z^*\left[u(Z^*_{\sigma_{\partial O_1}}); \sigma_{\partial O_1} < \sigma_{a_j^*}\right] + u(a_j^*)\mathbb{P}_z^*(\sigma_{\partial O_1} > \sigma_{a_j^*}) \\ &= \mathbb{E}_z^E\left[u(Z^E_{\sigma_{\partial O_1}}); \sigma_{\partial O_1} < \sigma_{A_j}\right] + u(a_j^*)\mathbb{P}_z^E(\sigma_{\partial O_1} > \sigma_{A_j}). \end{aligned}$$

Note that u is bounded on ∂O_1 and A_j is a continuum. By making use of the properties **(Z.1)** and **(Z.2)** of the planar Brownian motion $\mathbf{Z}$ in a similar way to the proof of Lemma 1.1.4 (i), we have $\lim_{z \in O_1 \setminus A_j, z \to w} u(z) = u(a_j^*)$ for every $w \in \partial A_j$ which means that $\lim_{z \in \widehat{O}_1, z \to a_j^*} u(z) = u(a_j^*)$ in the quotient topology. □

Theorem 1.6.3. *Suppose that D_1 and D_2 are two connected open subsets of D^* and that $D_1 \cap D_2 \neq \emptyset$. If u is $\mathbf{Z}^*$-harmonic in D_i for $i = 1, 2$, then u is $\mathbf{Z}^*$-harmonic in $D_1 \cup D_2$.*

Proof. Let O be a relatively compact open subset of $D_1 \cup D_2$ so that $O \cap D_1 \cap D_2 \neq \emptyset$. Let $\{U_k^{(i)}; k \geq 1\}$ be an increasing sequence of relatively compact open subsets whose union is D_i and $\partial U_k^{(i)}$ is a smooth subset in D for $i = 1, 2$. Since $\{U_k^{(1)} \cup U_k^{(2)}; k \geq 1\}$ forms an open cover for $\overline{O}$, there is some $k_0 \geq 1$ so that $\overline{O} \subset U_{k_0}^{(1)} \cup U_{k_0}^{(2)}$. For notational simplicity, denote $U_{k_0}^{(i)}$ by U_i for $i = 1, 2$. Note that $O_i := O \cap U_i$ is a relatively compact open subset of D_i for $i = 1, 2$ and $O_1 \cap O_2 \neq \emptyset$. We claim that $u(z) = \mathbb{E}_z^*\left[u(Z^*_{\tau_O})\right]$ for every $z \in O$.

Let $\{\theta_t; t \geq 0\}$ be the shift operator for BMD $\mathbf{Z}^*$ on D^*. We use $\{\mathcal{F}_t; t \geq 0\}$ to denote the minimal augmented natural filtration generated by $\mathbf{Z}^*$. Define a sequence of stopping times as follows. Let $T_0 = 0$,

$$T_{2k+1} := T_{2k} + \tau_{O_1} \circ \theta_{T_{2k}} \quad \text{and} \quad T_{2k+2} := T_{2k+1} + \tau_{O_2} \circ \theta_{T_{2k+1}} \quad \text{for } k \geq 0.$$

As was noted in the above, $\tau_O < \infty$ a.s. Clearly $T_k \leq \tau_O$ for every $k \geq 1$. Since u is $\mathbf{Z}^*$-harmonic in both D_1 and D_2, we have for every $z \in O$,

$$\mathbb{E}_z^* \left[u(Z^*_{T_{k+1}}) | \mathcal{F}_{T_k} \right] = u(Z^*_{T_k}) \quad \mathbb{P}_z^*\text{-a.s.} \qquad \text{for every } k \geq 0. \tag{1.6.2}$$

In fact, when k is even, the left-hand side of (1.6.2) equals $Y_k := \mathbb{E}_{Z^*_{T_k}} \left[u(Z^*_{\tau_{O_1}}) \right]$, $\mathbb{P}_z^*$-a.s. by the strong Markov property of $\mathbf{Z}^*$. If $Z^*_{T_k} \in O_1$, then $Y_k = u(Z^*_{T_k})$ by (1.6.1). If $Z^*T_k \in \partial O_2 \setminus O_1$, then $Y_k = \mathbb{E}^*_{Z^*_{T_k}} [u(Z^*_0)]$ which equals $u(Z^*_{T_k})$ again. When k is odd, we get (1.6.2) in a similar way by using (1.6.1) for O_2 in place of O_1. In other words, $\{u(Z^*_{T_k}); k \geq 0\}$ is a martingale with respect to the filtration $\{\mathcal{F}_{T_k}\}_{k \geq 0}$ under $\mathbb{P}_z$ for every $z \in O$.

Let $T := \lim_{k \to \infty} T_k$. We show that $T = \tau_O$. Clearly $T \leq \tau_O$ a.s. On $\{T < \tau_O\}$, $Z^*_T(\omega) \in O = O_1 \cup O_2$, say, $Z^*_T(\omega) \in O_2$. There is some large $k_0 = k_0(\omega)$ so that $Z^*_{T_k}(\omega) \in O_2$ for all $k \geq k_0$. This is impossible as for even $k \geq k_0(\omega)$, $Z^*_{T_k}(\omega) \notin O_2$. So we must have $T = \tau_O$ a.s.

By virtue of Theorem 1.6.2, u is continuous on $D_1 \cup D_2$ so that it is bounded on $\overline{O}$. Further $\tau_O < \infty$ a.s. Hence the continuity of the sample path of $\mathbf{Z}^*$ yields

$$u(z) = \lim_{k \to \infty} \mathbb{E}_z \left[u(Z^*_{T_k}) \right] = \mathbb{E}_z \left[u(Z^*_{\tau_O}) \right], \quad z \in O,$$

as was to be proved. □

Consider an open set $\widetilde{E}$ such that $\widetilde{E} \subset E$, $\mathbb{C} \setminus \widetilde{E}$ is a continuum and each compact continuum A_i is either contained in $\widetilde{E}$ or disjoint from $\widetilde{E}$. Let $\widetilde{K} = \widetilde{E} \cap K = \bigcup_{j=1}^{\ell} A_{i_j}$ for $1 \leq i_1 < \cdots < i_\ell$, $\ell \leq N$, and $\widetilde{D} = \widetilde{E} \setminus \widetilde{K}$. Denote by $\widetilde{\mathbf{Z}}^*$ the BMD on the set

$$\widetilde{D}^* = \widetilde{D} \cup \widetilde{K}^*, \quad \widetilde{K}^* = \{a^*_{i_1}, \ldots, a^*_{i_\ell}\},$$

obtained from the ABM on $\widetilde{E}$ by rendering each hole A_{i_j} as a one point $a^*_{i_j}$, $1 \leq j \leq \ell$. Then, by virtue of Theorem 1.4.1, $\widetilde{\mathbf{Z}}^*$ is identical in law with the part process $\mathbf{Z}^{*,\widetilde{D}^*}$ of $\mathbf{Z}^*$ on $\widetilde{D}^*$, namely, the subprocess of $\mathbf{Z}^*$ obtained by killing upon its exit time from $\widetilde{D}^*$. Let O be an open connected subset of $\widetilde{D}^* (\subset D^*)$. A function u on O is $\mathbf{Z}^*$-harmonic on O if and only if u is harmonic with respect to the part process $\mathbf{Z}^{*,\widetilde{D}^*}$ of $\mathbf{Z}^*$ on $\widetilde{D}^*$. Therefore, we have the following equivalence:

$$u \text{ is } \mathbf{Z}^*\text{-harmonic on } O \iff u \text{ is } \widetilde{\mathbf{Z}}^*\text{-harmonic on } O. \tag{1.6.3}$$

1.6.2 *Zero period property of BMD-harmonic function*

Let γ be a C^1-smooth simple closed curve surrounding A_j, namely, $\gamma \subset D$, ins$\gamma \supset A_j$, $\overline{\text{ins}\gamma} \cap A_k = \emptyset$, $k \neq j$. Here insγ denotes the bounded component of $\mathbb{C} \setminus \gamma$ and is called the interior of γ. For a harmonic function u defined in a neighborhood of A_j, by the Green-Gauss formula, the value

$$\int_\gamma \frac{\partial u(\zeta)}{\partial \mathbf{n}_\zeta} ds(\zeta),$$

is independent of the choice of such curve γ with $\mathbf{n}$ denoting the unit normal vector pointing toward A_j and s the arc length of γ. This value is called the *period* of u around A_j.

Theorem 1.6.4. *Let O be a connected open subset of D^*. A Borel measurable real valued function v in O is $\mathbf{Z}^*$-harmonic on O if and only if v is continuous on O, harmonic in $D \cap O$ and the period of v around a_i^* is 0 for every i such that $a_i^* \in O$.*

Proof. The assertion trivially holds if O does not contain any a_i^*. In view of Theorem 1.6.3 and the equivalence (1.6.3), without loss of generality, we may and do assume that E is bounded with $\mathbb{C} \setminus E$ being a continuum, $O = D^*$ and that D^* contains exactly one of the a_i^*'s, say, a_1^* (that is, K consists of exactly one compact continuum A_1 and $E = D \cup A_1$).

(i) Suppose that v is $\mathbf{Z}^*$-harmonic on D^*. It is continuous on $D^* = D \cup \{a_1^*\}$ by Theorem 1.6.2. Clearly v is harmonic in D. Let U_1 and U_2 be relatively compact open subsets of E so that $A_1 \subset U_1 \subset \overline{U}_1 \subset U_2 \subset \overline{U}_2 \subset E$. Take $\psi \in C_c^\infty(\mathbb{R}^2)$ such that $\psi = 1$ on U_1 and $\psi = 0$ on U_2^c. Define $f(x) = -\frac{1}{2}\mathbb{1}_D \cdot \Delta(\psi v)$. We first show that

$$(\psi v)(z) = G^* f(z) \quad \text{for every } z \in D^*, \tag{1.6.4}$$

where G^* is the 0-order resolvent of $\mathbf{Z}^*$ defined by (1.5.6).

Since f is bounded on D and $f = 0$ on $(D \cap U_1) \cup (D \setminus U_2)$, $G^* f \in \mathcal{F}^*$ by Theorem 1.5.2. Furthermore, $G^* f$ can be verified to be $\mathbf{Z}^*$-harmonic on $(D \cap U_1) \cup \{a_1^*\}$ by its expression (1.5.6) and the strong Markov property of $\mathbf{Z}^*$. Hence $w = \psi v - G^* f$ is also $\mathbf{Z}^*$-harmonic on this set. On the other hand, by the strong Markov property of $\mathbf{Z}^*$, we have for $z \in D$,

$$G^* f(z) = G^D f(z) + \mathbb{E}_z^* \left[G^* f(Z^*_{\sigma_{a_1^*}}) \right] = G^D f(z) + G^* f(a_1^*) \varphi(z),$$

where $\varphi(z) = \mathbb{P}_z^*(\sigma_{a_1^*} < \infty)$, $z \in D$. Notice that $G^*|f|(a_1^*)$ is finite because $G^*|f|(z) < \infty$ m-a.e., $\{a_1^*\}$ is of positive $\mathcal{E}_1^*$-capacity by Theorem 1.3.2 and so [Chen and Fukushima (2012), Theorem A.2.13] applies. $G^D f$ is bounded on D and $\int_D f(z) G^D f(z) m(dz) < \infty$ on account of the explicit expression (1.1.14) of the 0-order resolvent density $G^D(z, \zeta)$ for the bounded domain D. Hence $G^D f \in H^1_{0,e}(D) \cap L^2(D) = H^1_0(D)$ and

$$\frac{1}{2}\mathbf{D}_D(G^D f, g) = -\frac{1}{2} \int_D \Delta(\psi v)(z) g(z) m(dz) \quad \text{for any } g \in H^1_0(D),$$

by [Chen and Fukushima (2012), Theorem 2.1.12]. This means that $w = \psi v - G^* f = \psi v - G^D f - G^* f(a_1^*)\varphi$ is harmonic on D and consequently $\mathbf{Z}^*$-harmonic on D. Therefore, w is $\mathbf{Z}^*$-harmonic on D^* by virtue of Theorem 1.6.3.

Using the expression (1.1.14) along with the properties **(Z,1)**, **(Z.2)** of the planar Brownian motion $\mathbf{Z}$, we also see that $G^D f$ is continuously extendable to $\overline{D}$ with value 0 on $\partial D = \partial E \cup \partial A_1$. In view of Lemma 1.1.4 (i), w is therefore continuously extendable to $\overline{D}^*$ with value 0 on ∂E. Hence, for any $\varepsilon > 0$, we can find a relatively compact connected open subset O_1 of D^* with $a_1^* \in O_1$ and $|w(z)| < \varepsilon$, $z \in D^* \setminus O_1$. We then get $|w(z)| < \varepsilon$ for $z \in D^*$ from (1.6.1), yielding (1.6.4).

Let $u_1(z) := \mathbb{E}_z^E[e^{-\sigma_{A_1}}]$, which is smooth, strictly smaller than 1 on D, and continuous on E with value 1 on A_1 by Lemma 1.3.1. For $\varepsilon \in (0,1)$, let η_ε be the boundary of the connected component of $\{x \in E : u_1(x) > 1 - \varepsilon\}$ that contains A_1. By Sard's theorem (see, e.g., [Milnor (1965)]), there is a set $\mathcal{N}_0$ having zero Lebesgue measure so that for every $\varepsilon \in (0,1) \setminus \mathcal{N}_0$, η_ε is a C^∞-smooth simple closed curve. Take a decreasing sequence $\{\varepsilon_n, n \geq 1\} \in (0,1) \setminus \mathcal{N}_0$ with $\lim_{n\to\infty} \varepsilon_n = 0$. Since $\{x \in E : u_1(x) > 1 - \varepsilon_n\}$ decreases to A_1, we may assume that each η_{ε_n} is contained inside U_1. As $\psi = 1$ on U_1, we can see by the Green-Gauss formula and Theorem 1.5.2 that the period of v around a_1^* equals

$$\begin{aligned}
\lim_{n\to\infty} \int_{\eta_{\varepsilon_n}} \frac{\partial(\psi v)(\xi)}{\partial \mathbf{n}_\xi} \sigma(d\xi) &= \lim_{n\to\infty} \frac{1}{1-\varepsilon_n} \int_{\eta_{\varepsilon_n}} \frac{\partial G^* f(\xi)}{\partial \mathbf{n}_\xi} u_1(\xi) \sigma(d\xi) \\
&= \lim_{n\to\infty} \frac{1}{1-\varepsilon_n} \int_{D\setminus \mathrm{ins}(\eta_n)} (\nabla u_1 \cdot \nabla G^* f + u_1 \Delta G^* f)\, dx \\
&= \int_D \nabla u_1(x) \cdot \nabla G^* f(x) dx + \int_D u_1(x) \Delta G^* f(x) dx \\
&= 2\mathcal{N}(G^* f)(a_1^*) = 0.
\end{aligned}$$

Here $\mathbf{n}$ denotes the unit inward normal vector field on η_{ε_n} for the interior of η_{ε_n}.

(ii) Conversely, assume that v is a continuous function on D^* that is harmonic in D and has zero period around $a_1^* \in D^*$. Let the relatively compact open subsets $U_1 \subset U_2$ of E, the smooth function ψ and the smooth curves η_{ε_n} be defined as above. Set

$$v_0 = \psi v - v(a_1^*)\varphi \quad \text{and} \quad f(z) = -\frac{1}{2}\mathbb{1}_D(z)\Delta v_0(z) = -\frac{1}{2}\mathbb{1}_D(z)\Delta(\psi v) \text{ for } z \in D.$$

We claim that

$$v_0(z) = G^D f(z) \quad \text{for every } z \in D. \tag{1.6.5}$$

Put $w_0 = v_0 - G^D f$. Just as in the above proof of (i), w_0 is harmonic on D. We have seen that $G^D f$ is continuously extendable to $\overline{D}$ with value 0 on ∂D. By the assumption on v and Lemma 1.1.4, v_0 has the same property and so does w_0. Hence (1.6.5) follows from the maximum principle Lemma 1.1.3.

In view of the proofs of Proposition 1.3.8 and Theorem 1.5.2, we see that $\varphi \in \mathcal{F}_e^* = \mathcal{F}^*$. We have therefore

$$\psi v = G^D f + v(a_1^*)\varphi \in \mathcal{F}^* \quad \text{with} \quad \Delta(\psi v) \in L^2(D; dx).$$

Since v has zero period at a_1^*, we have by the Green-Gauss formula that

$$\begin{aligned}
0 &= \lim_{n\to\infty} \int_{\eta_{\varepsilon_n}} \frac{\partial v(\xi)}{\partial \mathbf{n}_\xi} \sigma(d\xi) = \lim_{n\to\infty} \int_{\eta_{\varepsilon_n}} \frac{\partial (\psi v)(\xi)}{\partial \mathbf{n}_\xi} \sigma(d\xi) \\
&= \lim_{n\to\infty} \frac{1}{1-\varepsilon_n} \int_{\eta_{\varepsilon_n}} \frac{\partial (\psi v)(\xi)}{\partial \mathbf{n}_\xi} u_1(\xi)\sigma(d\xi) \\
&= \lim_{n\to\infty} \frac{1}{1-\varepsilon_n} \int_{D\setminus \mathrm{ins}(\eta_{\varepsilon_n})} (\nabla u_1 \cdot \nabla(\psi v) + u_1 \Delta(\psi v))\, dx \\
&= \int_D \nabla u_1(x) \cdot \nabla(\psi v)(x) dx + \int_D u_1(x)\Delta(\psi v)(x) dx \\
&= 2\mathcal{N}(\psi v)(a_1^*).
\end{aligned}$$

Hence, we conclude by Theorem 1.5.1 that $\psi v \in \mathcal{D}(\mathcal{A}^*)$.

According to Theorem 1.5.2, we have $\psi v = G^* g$ for $g = -\frac{1}{2}\mathbb{1}_D \cdot \Delta(\psi v)$. Since $g = 0$ on U_1, v is $\mathbf{Z}^*$-harmonic in U_1. This together with Theorem 1.6.3 implies that v is $\mathbf{Z}^*$ harmonic in D^* □

1.6.3 *Harmonic conjugate*

The next theorem is a consequence of Theorem 1.6.4. Note that in multiply connected planar domains, classical harmonic functions (i.e. with respect to Brownian motion) in D can only *locally* be realized as the imaginary (or real) part of an analytic function in D. Theorem 1.6.4 shows that BMD is the right tool to study complex analysis in multiply connected domains in $\mathbb{R}^2$.

Theorem 1.6.5. *Suppose that $D := E \setminus K$ is connected. If v is $\mathbf{Z}^*$-harmonic on D^*, then $-v\big|_D$ admits a harmonic conjugate u on D uniquely up to an additive real constant in D so that $f(z) = u(z) + iv(z)$, $z \in D$, is an analytic function in D.*

Proof. Fix some $z_0 \in D$ and the value $u(z_0)$. For any $z \in D$, define

$$u(z) = u(z_0) + \int_\gamma \frac{\partial v(\xi)}{\partial \mathbf{n}_\xi} \sigma(d\xi), \tag{1.6.6}$$

where γ is a C^2-smooth simple curve in D that connects z_0 to z, $\sigma(d\xi)$ is the arc-length measure along γ and $\mathbf{n}$ the unit normal vector field along γ in the counter-clockwise direction (that is, if γ is parameterized by $(x(t), y(t))$, then $\mathbf{n}$ is the unit vector pointing to the same direction as $(-y'(t), x'(t))$). By the zero period property of v, the value of $v(x)$ is independent of the choice of the smooth C^2 simple curve γ that joins z_0 to z and hence well defined. One checks easily that (u, v) satisfies the Cauchy-Riemann equation and hence $f(z) := u(z) + iv(z)$ is an analytic function in D. □

1.7 Green function and Poisson kernel of BMD

Let $D = \mathbb{H} \setminus K$, $K = \bigcup_{i=1}^{N} A_i$, be an $(N+1)$-connected domain where $\{A_i, 1 \le i \le N\}$ are mutually disjoint compact continua contained in $\mathbb{H}$, and $G^0(z,\zeta)$ be the Green function of the ABM $\mathbf{Z}^0 = (Z_t^0, \zeta^0, \mathbb{P}_z^0)$ on D. Consider the BMD $\mathbf{Z}^* = (Z_t^*, \zeta^*, \mathbb{P}_z^*)$ on $D^* = D \cup \{a_1^*, \dots, a_N^*\}$ obtained from the ABM $\mathbf{Z}^{\mathbb{H}}$ on $\mathbb{H}$ by rendering each compact continuum A_i into one point a_i^*. Define functions $\varphi^{(i)}(z)$, $z \in \mathbb{H}$, $1 \le i \le N$, by (1.1.20) with $E = \mathbb{H}$.

As we saw in Proposition 1.3.8, $\mathbf{Z}^*$ is transient. Denote by G^* the 0-order resolvent of $\mathbf{Z}^*$:

$$G^* f(z) = \mathbb{E}_z^* \left[\int_0^\infty f(Z_t^*) dt \right], \quad z \in D^*, \ f \in \mathcal{B}_+(D^*).$$

$G^* f(z)$ is finite m-a.e. $z \in D^*$ for any $f \in L^1_+(D^*; m)$ (cf. [Chen and Fukushima (2012), Proposition 2.1.3]).

Theorem 1.7.1. *For any Borel measurable function $f \ge 0$ on D^* and $z \in D^*$,*

$$G^* f(z) = \int_D G^*(z,\zeta) f(\zeta) m(d\zeta),$$

where

$$G^*(z,\zeta) := G^0(z,\zeta) + 2 \sum_{i,j=1}^{N} \varphi^{(i)}(z) \left(\mathcal{A}^{-1}\right)_{ij} \varphi^{(j)}(\zeta), \quad z, \zeta \in D, \ z \ne \zeta, \tag{1.7.1}$$

$$G^*(a_i^*, \zeta) := 2 \sum_{j=1}^{N} \left(\mathcal{A}^{-1}\right)_{ij} \varphi^{(j)}(\zeta), \quad 1 \le i \le N, \quad \zeta \in D. \tag{1.7.2}$$

Here $\mathcal{A}$ is an $N \times N$-matrix whose (i,j)-entry a_{ij} is the period of $\varphi^{(i)}$ around A_j for $1 \le i, j \le N$.

Proof. For $f \in C_c(D)$, $G^* f \in \mathcal{F}^*$ is $\mathbf{Z}^*$-harmonic in $D^* \setminus \mathrm{supp}[f]$ and

$$G^* f(z) = G^0 f(z) + \sum_{i=1}^{N} \varphi^{(i)}(z) \lambda_i, \quad \text{where } \lambda_i = G^* f(a_i^*), \quad z \in D^*. \tag{1.7.3}$$

Since $G^* f(z)$ is finite m-a.e. on D^* for a non-negative $f \in C_c(D)$ and $G^* f$ is $\mathbf{Z}^*$-excessive, $\lambda_i = G^* f(a_i^*) < \infty$, $1 \le i \le N$, on account of [Chen and Fukushima (2012), Theorem A.2.13 (v)] and Theorem 1.3.2 (ii).

By virtue of Theorem 1.6.4, we have for a smooth simple curve $\gamma_j \subset D \setminus \mathrm{supp}[f]$ surrounding A_j,

$$\frac{1}{2} \sum_{i=1}^{N} a_{ij} \lambda_i = -\frac{1}{2} \int_{\gamma_j} \frac{\partial G^0 f(z)}{\partial \mathbf{n}_z} ds(z), \quad 1 \le i \le N.$$

By taking (1.1.46) into account and using Proposition 1.1.5 (iii) along with the symmetry of $G^0(z,\zeta)$, the righthand side of the above identity can be seen to be equal to

$$-\frac{1}{2}\int_{\mathrm{supp}[f]}\left[\int_{\gamma_j}\frac{\partial}{\partial \mathbf{n}_z}G^0(z,\zeta)ds(z)\right]f(\zeta)m(d\zeta)=\int_D \varphi^{(j)}(\zeta)f(\zeta)m(d\zeta).$$

Thus, for every $f\in C_c(D)$, $\{\lambda_i := G^*f(a_i^*), 1\le i\le N\}$ satisfies

$$\frac{1}{2}\sum_{i=1}^N a_{ij}\lambda_i = (\varphi^{(j)},f)_m \quad \text{for } 1\le j\le N.$$

Since $\{\varphi^{(j)}; 1\le j\le N\}$ are linearly independent as continuous bounded functions on D on account of Lemma 1.1.4, the linear space $\mathcal{V}$ formed by the N-dimensional vectors

$$\left\{((\varphi^{(1)},f)_m,\ldots,(\varphi^{(N)},f)_m);\ f\in C_c(D)\right\}$$

is $\mathbb{R}^N$. Indeed, were the dimension of $\mathcal{V}$ less than N, there would exist a non-zero vector $c=(c_1,\ldots,c_N)\in\mathbb{R}^N$ that is perpendicular to $\mathcal{V}$. This would imply that $\sum_{j=1}^N c_j\varphi^{(j)}=0$ on D, which is impossible. Hence for every $b=(b_1,\ldots,b_N)\in\mathbb{R}^N$, the linear equation system

$$\frac{1}{2}\sum_{i=1}^N a_{ij}\lambda_i = b_j \quad \text{for } 1\le j\le N,$$

has a solution $(\lambda_1,\ldots,\lambda_N)$. Consequently, the matrix $\mathcal{A}$ is invertible. Denote the matrix inverse of $\mathcal{A}$ by $\mathcal{A}^{-1}$. From (1.7.3),

$$G^*f(z)=G^0f(z)+2\sum_{i,j=1}^N \varphi^{(i)}(z)\left(\mathcal{A}^{-1}\right)_{ij}(\varphi^{(j)},f)_m, \quad z\in D,$$

which yields (1.7.1). In view of (1.1.14), properties **(Z.1)**, **(Z.2)** of the planar BM **Z** and Lemma 1.1.4 (i), we also get from the above that

$$G^*f(a_i^*)=\lim_{z\to A_i,\ z\in D}G^*f(z)=2\sum_{i,j=1}^N\left(\mathcal{A}^{-1}\right)_{ij}(\varphi^{(j)},f)_m, \quad 1\le i\le N,$$

yielding (1.7.2). □

We call $G^*(z,\zeta)$, $z\in D^*$, $\zeta\in D$, of Theorem 1.7.1 the *Green function of the BMD* $\mathbf{Z}^*$. One can deduce the symmetry of the matrix $\mathcal{A}$ from (1.7.1) and the symmetry of the Green functions of $\mathbf{Z}^0$ and $\mathbf{Z}^*$.

In view of Lemma 1.1.4 (i) and Proposition 1.1.5 (i), both $G^0(z,\zeta)$ and $\varphi^{(j)}(\zeta)$ are harmonic in $\zeta\in D\setminus\{z\}$ and continuously extendable to $\partial\mathbb{H}$ to be zero there. By the Schwarz reflection, they extend to be harmonic across $\partial\mathbb{H}$. We can thus well define the *Poisson kernel of the BMD* $\mathbf{Z}^*$ by

$$K^*(z,\zeta)=-\frac{1}{2}\frac{\partial}{\partial\mathbf{n}_\zeta}G^*(z,\zeta),\ z\in D^*, \quad \zeta\in\partial\mathbb{H}.$$

Recall that $\mathbf{n}_\zeta$ denotes the unit outward normal vector at $\zeta\in\partial\mathbb{H}$.

For each $\zeta \in \partial\mathbb{H}$, $K^*(z,\zeta)$ is a $\mathbf{Z}^*$-harmonic function of $z \in D^*$ because so is $G^*(z,\zeta)$ in $z \in D^* \setminus \{\zeta\}$ for each $\zeta \in D$. It follows from (1.7.2) that

$$K^*(a_i^*,\zeta) = -\sum_{j=1}^N \left(\mathcal{A}^{-1}\right)_{ij} \frac{\partial}{\partial \mathbf{n}_\zeta}\varphi^{(j)}(\zeta) \geq 0 \quad \text{for } 1 \leq i \leq N \text{ and } \zeta \in \partial\mathbb{H}. \tag{1.7.4}$$

For $z \in D$ and $\zeta \in \partial\mathbb{H}$, we have by (1.7.1),

$$\begin{aligned} K^*(z,\zeta) &= -\frac{1}{2}\frac{\partial}{\partial \mathbf{n}_\zeta} G^0(z,\zeta) - \sum_{i,j=1}^N \varphi^{(i)}(z)\left(\mathcal{A}^{-1}\right)_{ij} \frac{\partial}{\partial \mathbf{n}_\zeta}\varphi^{(j)}(\zeta) \\ &= -\frac{1}{2}\frac{\partial}{\partial \mathbf{n}_\zeta} G^0(z,\zeta) + \sum_{i=1}^N \varphi^{(i)}(z) K^*(a_i^*,\zeta). \end{aligned} \tag{1.7.5}$$

For each $\zeta \in \partial\mathbb{H}$, $K^*(z,\zeta)$ is therefore positive for $z \in D$ in view of Proposition 1.1.5 (ii). As it is $\mathbf{Z}^*$-harmonic in $z \in D^*$, it is also positive for $z \in D^*$.

We now present two potential theoretic lemmas for the sake of later use. First, we express the function $\varphi^{(j)}(z)$, $z \in D$, as the Green potential of a positive measure concentrated on A_j for each fixed $1 \leq j \leq N$. We put $D^j = D \cup A_j (= \mathbb{H} \setminus \bigcup_{k \neq j} A_k)$ and let $\mathbf{Z}^{D^j} = (Z_t^{D_j}, \zeta^{D_j}, \mathbb{P}_z^{D_j})$ be the ABM on D^j. Let $G_\alpha^{D^j}(z,z')$ and $G^{D^j}(z,z')$ be the resolvent density and 0-order resolvent density (the Green function) of $\mathbf{Z}^{D^j}$, respectively, that are defined by (1.1.3) for D^j in place of D. We then have

Lemma 1.7.2. *For each $1 \leq j \leq N$, there exists a unique finite positive measure ν_j concentrated on A_j such that*

$$\varphi^{(j)}(z) = \int_{A_j} G^{D^j}(z,z')\nu_j(dz') \quad \text{for any } z \in D^j. \tag{1.7.6}$$

Proof. Denote by $p_t^{D^j}$ the transition function of $\mathbf{Z}^{D^j}$. Since $\mathbb{C} \setminus D^j$ is non-polar, $\mathbb{P}_z^{D^j}(\zeta^{D^j} < \infty) = 1$, $z \in D^j$, as was observed at the beginning of Section 1.1, so that, for any $z \in D^j$, $p_\ell^{D^j}\mathbb{1}(z) < 1$ for some $\ell \in \mathbb{N}$, which implies the transience of $\mathbf{Z}^{D^j}$ in view of [Fukushima, Oshima and Takeda (2011), Lemma 1.6.5].

The Dirichlet form of $\mathbf{Z}^{D^j}$ on $L^2(D^j)$ is $(\frac{1}{2}\mathbf{D}_{D^j}, H_0^1(D^j))$ where $\mathbf{D}_{D^j}$ denote the Dirichlet integral over D^j ([Fukushima, Oshima and Takeda (2011), Example 4.4.1]). Its extended Dirichlet space $H_{0,e}^1(D^j)$ is a Hilbert space with inner product $\mathbf{D}_{D^j}$ due to the transience of $\mathbf{Z}^{D^j}$. According to [Fukushima, Oshima and Takeda (2011), Theorem 4.3.3], $\varphi^{(j)}$ is a quasi-continuous version of the 0-order equilibrium potential $e_{A_j}^{(0)} \in H_{0,e}^1(D^j)$ of the set A_j.

By the 0-order version of [Fukushima, Oshima and Takeda (2011), Theorem 2.1.5, Lemma 2.2.6], $e_{A_j}^{(0)}$ equals the 0-order potential $U\nu_j \in H_{0,e}^1(D^j)$ of a unique positive Radon measure ν_j concentrated on A_j of finite 0-order energy. Since ν_j is then of finite 1-order energy, it admits a unique α-order potential $U_\alpha\nu_j \in H_0^1(D^j)$.

On the other hand, $G_\alpha^{D^j}\nu_j(z) = \int_{A_j} G_\alpha^{D^j}(z,z')\nu_j(dz')$ is a quasi-continuous version of $U_\alpha\nu_j(z)$ by [Fukushima, Oshima and Takeda (2011), Exercise 4.2.2]. Furthermore, $U_\alpha\nu_j$ is $\mathbf{D}_{D^j}$-convergent to $U\nu_j \in H^1_{0,e}(D^j)$ as $\alpha \downarrow 0$ by [Fukushima, Oshima and Takeda (2011), Lemma 2.2.11]. Since $G_\alpha^{D^j}\nu_j(z) \uparrow G^{D^j}\nu_j(z) := \int_{A_j} G^{D^j}(z,z')\nu_j(dz')$ as $\alpha \downarrow 0$ for each $z \in D^j$, $G^{D^j}\nu_j$ is a quasi-continuous version of $e^{(0)}_{A_j}$ on account of [Chen and Fukushima (2012), Theorem 2.3.4].

Thus, (1.7.6) holds for q.e. $z \in D^j$. As the both hand sides of (1.7.6) are $p_t^{D^j}$-excessive and $p_t^{D^j}(z,\cdot)$ is absolutely continuous with respect to the Lebesgue measure, (1.7.6) holds for every $z \in D^j$. The both sides of (1.7.6) vanish for $z \in K \setminus A_j$. □

We next consider, instead of $\varphi^{(j)}$, the hitting probability $\psi^{(j)}(z) = \mathbb{P}_z^{\mathbb{H}}(\sigma_{A_j} < \infty)$, $z \in \mathbb{H}$, $1 \leq j \leq N$, of the set A_j with respect to the ABM $\mathbf{Z}^{\mathbb{H}} = (Z^{\mathbb{H}}, \mathbb{P}_z^{\mathbb{H}})$ on $\mathbb{H}$. $\mathbf{Z}^{\mathbb{H}}$ is transient and its Dirichlet form on $L^2(\mathbb{H})$ equals $(\frac{1}{2}\mathbf{D}_{\mathbb{H}}, H_0^1(\mathbb{H}))$ where $\mathbf{D}_{\mathbb{H}}$ is the Dirichlet integral over $\mathbb{H}$. Hence $\psi^{(j)}$ is a quasi-continuous version of the 0-order equilibrium potential $e^{(0)}_{A_j} \in H^1_{0,e}(\mathbb{H})$, and we see exactly in the same way as the above lemma that there exists a unique positive finite measure $\mu^{(j)}$ concentrated on A_j satisfying

$$\begin{aligned} \psi^{(j)}(z) &= \int_{A_j} G^{\mathbb{H}}(z,w)\mu^{(j)}(dw) \quad \text{for } z \in \mathbb{H}, \quad \text{and} \\ \mu^{(j)}(A_j) &= \frac{1}{2}\mathbf{D}_{\mathbb{H}}(\psi^{(j)}, \psi^{(j)}) = \mathrm{Cap}_0^{\mathbb{H}}(A_j). \end{aligned} \tag{1.7.7}$$

Here $\mathrm{Cap}_0^{\mathbb{H}}$ is the 0-order capacity relative to $(H^1_{0,e}(\mathbb{H}), \frac{1}{2}\mathbf{D}_{\mathbb{H}})$ evaluated for a compact set $C \subset \mathbb{H}$ (cf. [Fukushima, Oshima and Takeda (2011), Lemma 2.2.7]) by

$$\mathrm{Cap}_0^{\mathbb{H}}(C) = \inf\left\{\frac{1}{2}\mathbf{D}_{\mathbb{H}}(u,u) : u \in C_c^1(\mathbb{H}) \text{ with } u \geq 1 \text{ on } C\right\}.$$

The definition of $\mathrm{Cap}_0^{\mathbb{H}}$ extends to any subset A of $\mathbb{H}$ by

$$\mathrm{Cap}_0^{\mathbb{H}}(A) = \sup\left\{\mathrm{Cap}_0^{\mathbb{H}}(C) : C \text{ compact}, \ C \subset A\right\}.$$

Since $\varphi^{(j)} \leq \psi^{(j)}$, we get from the above explicit expression of $\psi^{(j)}$ and (1.1.12),

$$\frac{\partial\varphi^{(j)}(x+iy)}{\partial y}\bigg|_{y=0} \leq \frac{\partial\psi^{(j)}(x+iy)}{\partial y}\bigg|_{y=0} \leq \frac{2}{\pi}\int_{A_j} \frac{y'}{(x-x')^2+(y')^2}\mu^{(j)}(dw'), \tag{1.7.8}$$

where $w' = x' + iy'$. The following comes directly from (1.7.7) and (1.7.8).

Lemma 1.7.3. *It holds that*

$$\int_{-\infty}^{\infty} \frac{\partial\varphi^{(j)}(x+iy)}{\partial y}\bigg|_{y=0} dx \leq 2\mathrm{Cap}_0^{\mathbb{H}}(A_j), \quad 1 \leq j \leq N. \tag{1.7.9}$$

Proposition 1.7.4.

(i) *The function* $K^*(z,\zeta)$ *of* (1.7.5) *can be continuously extended from* $z \in D$ *to* $z \in \overline{\mathbb{H}} \setminus \{\zeta\}$ *by setting its value at* $z \in \partial\mathbb{H} \setminus \{\zeta\}$ *to be zero and its value at* $z \in A_i$ *to be* $K^*(a_i^*,\zeta)$, $1 \le i \le N$. *The extended function is jointly continuous in* $(z,\zeta) \in (\overline{\mathbb{H}} \setminus J) \times J$ *for any closed interval* $J \subset \partial\mathbb{H}$,

(ii) *For any* $g \in C_b(\partial\mathbb{H})$, *the integral*

$$H^*g(z) = \int_{\partial\mathbb{H}} K^*(z,\zeta)g(\zeta)ds(\zeta), \quad z \in D^*,$$

gives a well defined bounded Z^**-harmonic function on* D^* *and*

$$\lim_{z\to\zeta, z\in D} H^*g(z) = g(\zeta), \quad \zeta \in \partial\mathbb{H}. \tag{1.7.10}$$

(iii) *It holds that*

$$\lim_{z\to\infty} K^*(z,\zeta) = 0 \quad \text{uniformly in } \zeta \text{ on any compact interval of } \partial\mathbb{H}. \tag{1.7.11}$$

(iv) *It holds that*

$$\mathbb{P}_z^*\left(\zeta^* < \infty,\ Z^*_{\zeta^*-} \in \partial\mathbb{H}\right) = 1 \quad \text{for any } z \in D^*. \tag{1.7.12}$$

(v) *It holds for any* $g \in C_b(\partial\mathbb{H})$ *that*

$$\mathbb{E}_z^*\left[g(Z^*_{\zeta^*-})\right] = H^*g(z), \quad z \in D^*. \tag{1.7.13}$$

Proof. (i) The first term of the right-hand side of (1.7.5) admits a continuous extension to $z \in \overline{\mathbb{H}} \setminus \{\zeta\}$ by setting its value at $z \in (\partial\mathbb{H} \setminus \{\zeta\}) \cup K$ to be zero due to its explicit expression (1.1.27) along with a similar consideration to the proof of Lemma 1.1.4 (i). The extended function is easily seen to have the stated joint continuity. The second term can be extended to a joint continuous function in $(z,\zeta) \in \overline{\mathbb{H}} \times \partial\mathbb{H}$ by defining its value at $z \in A_i$ to be $K^*(a_i^*,\zeta)$, $1 \le i \le N$, by virtue of Lemma 1.1.4 (i).

(ii) By Lemma 1.7.3 and (1.1.28),

$$\int_{\partial\mathbb{H}} K^*(a_j^*,\xi)ds(\xi) \le 2\sum_{j=1}^N |b_{ij}|\mathrm{Cap}_0^{\mathbb{H}}(A_j),$$

where b_{ij} is the (i,j)-entry of $\mathcal{A}^{-1}$, and for any $B \in \mathcal{B}(\partial\mathbb{H})$,

$$\int_B K^*(z,\xi)ds(\xi) = \mathbb{P}_z^0\left(Z^0_{\zeta^0-} \in B\right) + \sum_{i=1}^N \varphi^{(i)}(z)\int_B K^*(a_i^*,\xi)ds(\xi). \tag{1.7.14}$$

Thus, $z \mapsto \int_{\partial\mathbb{H}} K^*(z,\xi)ds(\xi)$ is a bounded function on D^*. Since for each $\xi \in \partial\mathbb{H}$, $z \mapsto K^*(z,\xi)$ is $\mathbf{Z}^*$-harmonic on D^*, we see by Fubini theorem that for every bounded $g \in C(\partial\mathbb{H})$, H^*g is a bounded $\mathbf{Z}^*$-harmonic function on D^*. Property (1.7.10) follows from (1.1.28) and Lemma 1.1.4.

(iii) In view of the domination $G^0(z,\zeta) \le G^{\mathbb{H}}(z,\zeta)$ and (1.1.12), we have for $z = x+iy \in D$, $\zeta = \xi + i0 \in \partial\mathbb{H}$,

$$-\frac{1}{2}\frac{\partial}{\partial \mathbf{n}_\zeta}G^0(z,\zeta) \le \frac{1}{\pi}\frac{y}{(x-\xi)^2+y^2}. \tag{1.7.15}$$

Hence, the first term of the expression (1.7.5) of $K^*(z,\zeta)$ converges to 0 as $z\to\infty$ uniformly in ζ on any compact interval of $\partial\mathbb{H}$, and so does its second term on account of (1.1.22) and the continuity of $\frac{\partial\varphi^{(j)}(\zeta)}{\partial\mathbf{n}_\zeta}$ in $\zeta\in\partial\mathbb{H}$ for every $1\le j\le N$. Thus, we get (1.7.11).

(iv) is due to Theorem 1.4.2.

(v) Let

$$h(z)=H^*g(z)-\mathbb{E}_z^*\left[g\left(Z^*_{\zeta^*-}\right)\right],\quad z\in D^*.$$

Since

$$\mathbb{E}_z^*\left[g\left(Z^*_{\zeta^*-}\right)\right]=\mathbb{E}_z^0\left[g(Z^0_{\zeta^0-});Z^0_{\zeta^0-}\in\partial\mathbb{H}\right]+\Sigma_{j=1}^N\varphi^{(j)}(z)\mathbb{E}^*_{a_j^*}\left[g\left(Z^*_{\zeta^*-}\right)\right],\quad z\in D^*,$$

it follows from (1.7.14) that $h(z)=\sum_{j=1}^N\alpha_j\varphi^{(j)}(z)$, $z\in D$, for some constants α_j, $1\le j\le N$. Consider the open set $O_{L,\delta}=\{z\in\mathbb{H}:|z|<L,\ \Im z>\delta\}$ for $L>0$, $\delta>0$ with $K\subset O_{L,\delta}$. According to Lemma 1.1.4 (i), $\lim_{z\to\infty,\,z\in D}\varphi^{(j)}(z)=0$ and $\lim_{\eta\downarrow 0}\varphi^{(j)}(\xi+i\eta)=0$ locally uniformly in $\xi\in\mathbb{R}$ for each $1\le j\le N$. Therefore, for any $\varepsilon>0$, there exist $L>0$, $\delta>0$ such that $|h(z)|<\varepsilon$ for any $z\in\partial O_{L,\delta}$. As h is $\mathbf{Z}^*$-harmonic on D^*,

$$h(z)=\mathbb{E}_z^*\left[h\left(Z^*_{\tau_{O_{L,\delta}}}\right)\right],\quad z\in(D\cap O_{L,\delta})\cup K^*,$$

so $|h(z)|<\varepsilon$ for any $z\in(D\cap O_{L,\delta})\cup K^*$. Thus h vanishes identically on D^*. □

Property (1.7.13) legitimizes the term *Poisson kernel of* $\mathbf{Z}^*$ for $K^*(z,\zeta)$ with $z\in D^*$ and $\zeta\in\partial\mathbb{H}$.

In this section, we have fixed an $(N+1)$-connected domain $D=\mathbb{H}\setminus K$. Denote by diag the diagonal set of $D\times D$. We shall occasionally denote $G^*(z,\zeta)$ defined on $D^*\times D\setminus\text{diag}$ and $K^*(z,\zeta)$ defined on $D^*\times\partial\mathbb{H}$ by $G^*_D(z,\zeta)$ and $K^*_D(z,\zeta)$, respectively, to indicate their relevance to D.

1.8 Conformal invariance of BMD and its Green function

Let ϕ be a conformal map from $\mathbb{H}$ onto $\mathbb{H}$. Let $\{A_j,1\le j\le N\}$ be mutually disjoint continua contained in $\mathbb{H}$. Put $\widetilde{A}_j=\phi(A_j)$, $1\le j\le N$, $D=\mathbb{H}\setminus K$, $K=\bigcup_{j=1}^N A_j$, and $\widetilde{D}=\mathbb{H}\setminus\widetilde{K}$, $\widetilde{K}=\bigcup_{j=1}^N\widetilde{A}_j$. $\phi\big|_D$ is then a conformal map from D onto $\widetilde{D}$.

Just as in the preceding section, we consider a topological space $D^*=D\cup K^*$, $K^*=\{a_1^*,\dots,a_N^*\}$, obtained from $\mathbb{H}$ by rendering each set A_j into a single point a_j^*, $1\le j\le N$, along with the BMD $\mathbf{Z}^*=(Z_t^*,\zeta^*,\mathbb{P}_z^*)$ on D^* and the Green function (0-order resolvent density) $G^*(z,\zeta)$, $z,\zeta\in D^*$, of $\mathbf{Z}^*$, which admits an explicit expression (1.7.1), (1.7.2).

Analogously, we can consider a topological space $\widetilde{D}^*=\widetilde{D}\cup\widetilde{K}^*$, $\widetilde{K}^*=\{\widetilde{a}_1^*,\dots,\widetilde{a}_N^*\}$, the BMD $\widetilde{\mathbf{Z}}^*=(\widetilde{Z}_t^*,\widetilde{\zeta}^*,\widetilde{\mathbb{P}}_z^*)$ on $\widetilde{D}^*$ and the Green function $\widetilde{G}^*(\widetilde{z},\widetilde{\zeta})$, $\widetilde{z},\widetilde{\zeta}\in\widetilde{D}^*$, of $\widetilde{\mathbf{Z}}^*$. Clearly $\phi|_D$ extends to a homeomorphism between D^* and $\widetilde{D}^*$ by setting $\phi(a_i^*)=\widetilde{a}_i^*$, $1\le i\le N$.

The Lebesgue measure m (resp. $\widetilde{m}$) on D (resp. $\widetilde{D}$) is extended to D^* (resp. $\widetilde{D}^*$) by setting $m(K^*) = 0$ (resp. $\widetilde{m}(\widetilde{K}^*) = 0$). Let

$$\mu(B) = \int_B |\phi'(z)|^2 \mathbb{1}_D(z) m(dz), \quad B \in \mathcal{B}(D^*), \tag{1.8.1}$$

$$A_t = \int_0^{t\wedge\zeta^*} |\phi'(Z_s^*)|^2 \mathbb{1}_D(Z_s^*) ds, \quad t \geq 0. \tag{1.8.2}$$

We refer to [Chen and Fukushima (2012), §A.3] for definitions of a smooth measure, a positive continuous additive functional (PCAF), its Revuz measure, its support and PCAF in the strict sense as well.

Theorem 1.8.1 (Conformal invariance of BMD I).

(i) *$\{A_t,\ t \geq 0\}$ is a PCAF in the strict sense of $\mathbf{Z}^*$ whose Revuz measure relative to m equals μ.*
(ii) *A_t is strictly increasing in $t \in [0, \zeta^*)$ $\mathbb{P}_z^*$-a.s. for any $z \in D^*$. The support of $\{A_t,\ t \geq 0\}$ equals D^*.*
(iii) *Let τ_t be the inverse of A_t; that is, $\tau_t := \inf\{s > 0 : A_s > t\}$. Then, for every $w \in \widetilde{D}^*$, the process $\left(\phi(Z_{\tau_t}^*), A_{\zeta^*}, \mathbb{P}_{\phi^{-1}w}^*\right)$ is identical in law with the process $(\widetilde{Z}_t^*, \widetilde{\zeta}^*, \widetilde{\mathbb{P}}_w^*)$.*
(iv) *It holds that*

$$G^*(\phi^{-1}\widetilde{z}, \phi^{-1}\widetilde{\zeta}) = \widetilde{G}^*(\widetilde{z}, \widetilde{\zeta}) \qquad \text{for } \widetilde{z}, \widetilde{\zeta} \in \widetilde{D} \text{ with } \widetilde{z} \neq \widetilde{\zeta}. \tag{1.8.3}$$

Proof. (i) Denote by G_α^* the α-order resolvent of $\mathbf{Z}^*$:

$$G_\alpha^* f(z) = \mathbb{E}_z^*\left[\int_0^\infty e^{-\alpha t} f(Z_t^*) dt\right], \quad z \in Z^*,$$

Consider a sequence $\{U_n\}$ of relatively compact open sets increasing to $\mathbb{H}$ with $K \subset U_1$, $\overline{U}_n \subset U_{n+1}$, $n \geq 1$, and put $D_n^* = (U_n \setminus K) \cup K^*$, $\mu_n = \mathbb{1}_{D_n^*} \cdot \mu$ and $A_t^n = \int_0^{t\wedge\zeta^*} \mathbb{1}_{D_n^*}(Z_s^*) dA_s$ for $t \geq 0$ and $n \geq 1$. Then $D_n^* \uparrow D^*$ as $n \to \infty$, and it follows from (1.3.19) and Proposition 1.7.4 (iv) that

$$\mathbb{P}_z\left(\lim_{n\to\infty} \tau_{D_n^*} = \zeta^* < \infty\right) = 1 \quad \text{for every } z \in D^*. \tag{1.8.4}$$

Since $|\phi'(z)|^2$ is continuous in $z \in \mathbb{H}$, μ and μ_n are smooth measures on D^*, and we have for each $n \geq 1$

$$0 \leq \mathbb{E}_z^*\left[\int_0^\infty e^{-t} dA_t^n\right] = G_1^*(\mathbb{1}_{U_n\setminus K}|\phi'|^2)(z) \leq \sup_{z\in U_n} |\phi'(z)|^2 < \infty \quad \text{for } z \in D^*.$$

Hence A_t^n is a PCAF in the strict sense of $\mathbf{Z}^*$ and so is A_t because of (1.8.4) and $A_t = A_t^n$ for any $t \leq \tau_n$. It further follows from

$$\mathbb{E}_{h\cdot m}^*\left[\int_0^\infty e^{-\alpha t} f(Z_t^*) dA_t\right] = \langle G_\alpha^* h, f \cdot \mu\rangle, \quad f, h \in \mathcal{B}_+(D^*),$$

that μ is the Revuz measure of A in view of [Chen and Fukushima (2012), Theorem 4.1.1].

(ii) By virtue of Theorem 1.7.1, $G^* \mathbb{1}_{\{K^*\}}(z) = 0$, for any $z \in D^*$. Hence the time set $\{t \geq 0 : \mathbf{Z}_t^* \in K^*\}$ is of zero Lebesgue measure $\mathbb{P}_z^*$-a.s. for any $z \in D^*$. As ϕ is conformal, $|\phi'(z)|^2$ is not only continuous but also strictly positive on $\mathbb{H}$. Hence, we obtain the assertions in (ii).

(iii) As the support of $\{A_t, t \geq 0\}$ equals D^*, the time changed process $\mathbf{X} = (Z^*_{\tau_t}, A_{\zeta^*}, \mathbb{P}^*_z)$ of $\mathbf{Z}^*$ by its PCAF in the strict sense A_t is a strong Markov process on D^* in view of [Chen and Fukushima (2012), Theorem A.3.9]. Since τ_t is continuous in $t \geq 0$, a.s., $\mathbf{X}$ is a diffusion on D^*. The transition function of $\mathbf{X}$ will be denoted by $p_t^{\mathbf{X}}$, which is μ-symmetric according to [Chen and Fukushima (2012), Theorem 5.2.1].

We next let $\mathbf{Y} = \left(\phi(\mathbf{Z}^*_{\tau_t}), A_{\zeta^*}, \mathbb{P}^*_{\phi^{-1}(w)}\right)_{w \in \widetilde{D}^*}$, which is the image process of $\mathbf{X}$ under the continuous one-to-one map ϕ from D^* onto $\widetilde{D}^*$. We can readily verify that $\mathbf{Y}$ is a Markov process on $\widetilde{D}^*$ with the transition function $p_t^{\mathbf{Y}}$ given by

$$p_t^{\mathbf{Y}} f(w) = p_t^{\mathbf{X}}(f \circ \phi)(\phi^{-1} w), \quad w \in \widetilde{D}^*, \quad f \in b\mathcal{B}(\widetilde{D}^*). \tag{1.8.5}$$

By combining the relation of the corresponding resolvents

$$G_\alpha^{\mathbf{Y}} f(w) = G_\alpha^{\mathbf{X}}(f \circ \phi)(\phi^{-1} w), \quad w \in \widetilde{D}^*, \quad f \in b\mathcal{B}(\widetilde{D}^*), \quad \alpha > 0, \tag{1.8.6}$$

with [Chen and Fukushima (2012), Theorem A.2.2], we also see that $G_\alpha^{\mathbf{Y}} f(\phi(\mathbf{Z}^*_{\tau_t}))$ is right continuous in $t \geq 0$, $\mathbb{P}^*_{\phi^{-1}(w)}$ a.s., yielding the strong Markov property of $\mathbf{Y}$ according to [Blumenthal and Getoor (1968), Theorem I.8.11]. Thus, $\mathbf{Y}$ is a diffusion on $\widetilde{D}^*$.

Due to the change of variable formula

$$\int_D u(\phi(\zeta)) |\phi'(\zeta)|^2 m(d\zeta) = \int_{\widetilde{D}} u(\widetilde{\zeta}) \widetilde{m}(d\widetilde{\zeta}), \tag{1.8.7}$$

which means that $\widetilde{m}$ is the image measure of μ under ϕ, we get from (1.8.5) $\|p_t^{\mathbf{Y}} f\|_{L^2(\widetilde{D}^*; \widetilde{m})} = \|p_t^{\mathbf{X}}(f \circ \phi)\|_{L^2(D^*;\mu)}$ and the $\widetilde{m}$-symmetry of $\mathbf{Y}$:

$$(p_t^{\mathbf{Y}} f, g)_{L^2(\widetilde{D}^*; \widetilde{m})} = (p_t^{\mathbf{X}}(f \circ \phi), g \circ \phi)_{L^2(D^*;\mu)}.$$

On the other hand, the part process $\mathbf{Z}^0 = (Z_t^0, \zeta^0, \mathbb{P}_z^0)$ of $\mathbf{Z}^*$ on D is the ABM, while, on account of [Chen and Fukushima (2012), Proposition 4.1.10], the part process of $\mathbf{X}$ on D is the time change of $\mathbf{Z}^0$ by its PCAF

$$A_t^0 = \int_0^{t \wedge \tau_D} |\phi'(Z_s^*)|^2 ds = \int_0^{t \wedge \zeta^0} |\phi'(Z_s^0)|^2 ds.$$

Hence, the part process of $\mathbf{Y}$ on $\widetilde{D}$ equals $\left(\phi(Z^0_{\tau_t}), A^0_{\zeta^0}, \mathbb{P}^0_{\phi^{-1}(w)}\right)_{w \in \widetilde{D}}$, which is identical in law with the ABM on $\widetilde{D}$ by virtue of Theorem A.2.1.

Since $\mathbf{Z}^*$ admits no killing on D^*, so does the time changed process $\mathbf{X}$. Hence the image process $\mathbf{Y}$ of $\mathbf{X}$ by ϕ admits no killing on $\widetilde{D}^*$. Due to Theorem 1.3.7 on the uniqueness of BMD, $\mathbf{Y}$ is identical in law with the BMD on $\widetilde{D}^*$.

(iv) Denote by $G^{\mathbf{X}}$ (resp. $G^{\mathbf{Y}}$) the 0-order resolvent of $\mathbf{X}$ (resp. $\mathbf{Y}$). We then get from (1.8.6) that for $f \in \mathcal{B}_+(\widetilde{D}^*)$ and $w \in \widetilde{D}^*$,

$$G^{\mathbf{Y}} f(w) = G^{\mathbf{X}}(f \circ \phi)(\phi^{-1}w),$$

while for $h \in \mathcal{B}_+(D^*)$ and $z \in D^*$,

$$\begin{aligned}
G^{\mathbf{X}} h(z) &= \mathbb{E}_z^*\left[\int_0^{A_{\zeta^*-}} h(Z^*_{\tau_t})dt\right] \\
&= \mathbb{E}_z^*\left[\int_0^{\zeta^*} h(Z_s^*)|\phi'(Z_s^*)|^2 ds\right] = G^*(h|\phi'|^2)(z).
\end{aligned}$$

Accordingly, we have, for $f \in \mathcal{B}_+(\widetilde{D}^*)$ and $w \in \widetilde{D}^*$,

$$\begin{aligned}
\int_{\widetilde{D}^*} \widetilde{G}^*(w,w')f(w')\widetilde{m}(dw') &= G^{\mathbf{Y}} f(w) = G^*(f\circ\phi \cdot |\phi'|^2)(\phi^{-1}w) \\
&= \int_{D^*} G^*(\phi^{-1}w, z')f(\phi(z'))|\phi'|^2(z')m(dz') \\
&= \int_{D^*} G^*(\phi^{-1}w, \phi^{-1}(\phi(z'))f(\phi(z'))\mu(dz'),
\end{aligned}$$

which equals by the change of variable formula (1.8.7) to

$$\int_{\widetilde{D}^*} G^*(\phi^{-1}w, \phi^{-1}w')f(w')\widetilde{m}(dw').$$

Since the Green functions $G^*(z,z')$, $\widetilde{G}^*(w,w')$ admit the explicit expression (1.7.1), the identity obtained in the above along with (1.1.14) and Lemma 1.1.4 (i) yields (1.8.3). □

Theorem 1.8.1 will be applied to a homothetic transformation $\phi(z) = cz$, $c > 0$, and a parallel translation $\phi(z) = z + r$, $r \in \mathbb{R}$, in Section 3.1. It will be also applied to the inversion $\phi(z) = -\frac{1}{z}$ in Section A.8 of the Appendix.

The conformal invariance of BMD remains valid under slightly different settings from Theorem 1.8.1. For instance, consider two continua B and $\widetilde{B}$, and a conformal map ϕ from $D = \mathbb{C} \setminus B$ onto $\widetilde{D} = \mathbb{C} \setminus \widetilde{B}$ sending ∞ to ∞. Let $D^* = D \cup \{b^*\}$ be the topological space obtained from $\mathbb{C}$ by rendering the set B into a singleton b^*, m be the Lebesgue measure on D extended to D^* by setting $m(\{b^*\}) = 0$ and $\mathbf{Z}^* = (Z_t^*, \zeta^*, \mathbb{P}_z^*)$ be the BMD on D^*. Notice that, according to Theorem 1.3.10, $\mathbf{Z}^*$ is conservative in the sense that $\mathbb{P}_z^*(\zeta^* = \infty) = 1$ for every $z \in D^*$.

Let $\widetilde{D}^*, \widetilde{m}$ and $\widetilde{\mathbf{Z}}^*$ be defined analogously with $\widetilde{B}$ in place of B. Finally define a measure μ on D^* and an additive functional $\{A_t\}$ of $\mathbf{Z}^*$ by (1.8.1) and (1.8.2), respectively.

Proposition 1.8.2 (Conformal invariance of BMD II). *Under the above setting, A_t is then a PCAF in the strict sense of $\mathbf{Z}^*$ and strictly increasing in $t \in (0,\infty)$ $\mathbb{P}^*_z$-a.s. for any $z \in D^*$. Let τ_t be its inverse. Then the process $\left(\phi(Z^*_{\tau_t}), \mathbb{P}^*_{\phi^{-1}w}\right)$ is identical in law with $\widetilde{\mathbf{Z}}^* = (\widetilde{Z}^*_t, \widetilde{\mathbb{P}}^*_w)$ for every $w \in \widetilde{D}^*$.*

Proof. It suffices to show that A_t is a PCAF of $\mathbf{Z}^*$ in the strict sense. The rest of the proof is the same as the proof of Theorem 1.8.1.

Take a large $\ell > 0$ so that $B_\ell := \{z \in \mathbb{C}; |z| < \ell\} \supset B$ and put $D^*_\ell = (B_\ell \setminus B) \cup \{b^*\}$. Let $\mathbf{Z}^{\ell,*} = (Z^{\ell,*}_t, \mathbb{P}^{\ell,*}_z)$ be the part of $\mathbf{Z}^*$ on D^*_ℓ. By Theorem 1.4.1, $\mathbf{Z}^{\ell,*}$ is then the BMD on D^*_ℓ. Consequently, we can see in a similar way to the proof of Theorem 1.7.1 that the Green function (0-order resolvent density) $G^{\ell,*}$ of $\mathbf{Z}^{\ell,*}$ admits the expression

$$G^{\ell,*}(z,\zeta) = G^{\ell,0}(z,\zeta) + 2p_\ell^{-1}\varphi_\ell(z)\varphi_\ell(\zeta), \quad z \in D^*_\ell,\ \zeta \in D_\ell \setminus B,$$

where $G^{\ell,0}$ is the Green function of the ABM on $D_\ell \setminus B$, $\varphi_\ell(z) = \mathbb{P}^{\ell,*}_z(\sigma_{\{b^*\}} < \infty)$ for $z \in D^*_\ell$, and p_ℓ is the period of φ_ℓ around B,

In view of (1,1,14), $G^{\ell,0}$ has only a logarithmic singularity on diagonal and $G^{\ell,*}(b^*,\zeta)$ is bounded in $\zeta \in D_\ell \setminus B$. Furthermore, by the assumption that ϕ sends ∞ to ∞, $\phi(D_\ell \setminus B)$ is bounded. Hence according to the formula (1.8.7)

$$\mu(D_\ell \setminus A) = \text{the Lebesgue measure of } \phi(D_\ell \setminus A) < \infty.$$

Therefore,

$$\mathbb{E}^*_z\left[A_{\tau_{D^*_\ell}}\right] = \int_{D_\ell \setminus A} G^{\ell,*}(z,\zeta)\mu(d\zeta) < \infty \quad \text{for every } z \in D^*_\ell.$$

Let $\tau := \lim_{\ell\to\infty} \tau_{D^*_\ell}$. It follows from (1.3.19) that $\tau = \zeta^*$ on $\{\tau < \infty\}$. As BMD $\mathbf{Z}^*$ has infinite lifetime ζ^*, it holds that $\mathbb{P}^*_z(\tau = \infty) = 1$ for every $z \in D^*$. Consequently, A_t is a PCAF of $\mathbf{Z}^*$ in the strict sense. □

Remark 1.8.3 (Excursion reflected Brownian motion). In [Lawler (2006)], G.F. Lawler introduced the notion of an *excursion reflected Brownian motion* (ERBM) quite similar to BMD in order to study Loewner differential equations for multiply connected planar domains. The investigations in this direction were followed up by his Ph.D student S. Drenning [Drenning (2011)].

When $N = 1$ and $E = \mathbb{C}$ so that K consists of only one continuum A, the ERBM coincides with the BMD on the space $(\mathbb{C} \setminus A) \cup \{a^*\}$ which is obtained from $\mathbb{C}$ by rendering the set A into a singleton a^*.

Indeed, in the special case that $A = \mathbb{D}$ the unit disk centered at the origin, denote by $(\mathbb{C} \setminus \mathbb{D}) \cup \{0^*\}$ the space obtained from $\mathbb{C}$ by rendering $\mathbb{D}$ into a singleton 0^*. In [Drenning (2011)], the transition semigroup P_t of the ERBM on $(\mathbb{C} \setminus \mathbb{D}) \cup \{0^*\}$ was specified explicitly in an analogous manner to the semigroup of Walsh's Brownian motion. P_t satisfies a Feller property. By this explicit expression of P_t, one can readily identify the ERBM on $(\mathbb{C} \setminus \mathbb{D}) \cup \{0^*\}$ with the BMD on it. See [Chen and Fukushima (2015), §6] for details.

For a general continuum $A \subset \mathbb{C}$, one can use a conformal map ϕ from $\mathbb{C} \setminus \mathbb{D}$ onto $\mathbb{C} \setminus A$ and Proposition 1.8.2 to see that the BMD on $(\mathbb{C} \setminus A) \cup \{a^*\}$ coincides with the image process by ϕ of the BMD $\mathbf{Z}^*$ on $(\mathbb{C} \setminus \mathbb{D}) \cup \{0^*\}$ being time changed by the PCAF $A_t = \int_0^t |\phi'(Z_s)|^2 ds$. On the other hand, the ERBM on $(\mathbb{C} \setminus A) \cup \{a^*\}$ was defined in [Drenning (2011)] to be the stochastic process obtained from the ERBM on $(\mathbb{C} \setminus \mathbb{D}) \cup \{0^*\}$ via the conformal map ϕ through exactly the same procedure. Thus, the BMD on $(\mathbb{C} \setminus A) \cup \{a^*\}$ must be identical with the ERBM on it.

When $N \geq 2$ however, no characterization of ERBM was given in [Lawler (2006); Drenning (2011)] although several properties that it should satisfy were stated in a descriptive way. However, in view of the localization property of BMD described in Proposition 1.4.3, one can construct ERBM on holes one by one when $N \geq 2$ and then identify it with BMD.

1.9 Complex Poisson kernel $\Psi(z, \zeta)$ of BMD on a standard slit domain

In Section 1.7, we have considered an $(N+1)$-connected domain $D = \mathbb{H} \setminus K$, $K = \bigcup_{i=1}^N A_i$, and the Poisson kernel

$$K^*(z, \zeta), \quad z \in D^* = D \cup \{a_1^*, \dots, a_N^*\}, \quad \zeta \in \partial\mathbb{H},$$

of the BMD $\mathbf{Z}^*$ on D^*. $K^*(z,\zeta)$ has an explicit expression described by (1.7.4) and (1.7.5). In the present section, we assume specifically that D is a *standard slit domain* so that A_i, $1 \leq i \leq N$, are mutually disjoint horizontal line segments contained in $\mathbb{H}$. A_i, a_i^* will be designated by C_i, c_i^*, respectively, for $1 \leq i \leq N$.

For a horizontal line segment C with endpoints z, z^r, denote by $C^{0,+}$ and $C^{0,-}$ distinct two copies of the set $C^0 = C \setminus \{z, z^r\}$ indicating the upper side and lower side of C^0, respectively, and define

$$C^p = C^{0,+} \cup C^{0,-} \cup \{z, \ z^r\}. \tag{1.9.1}$$

This set will be identified with the collection of prime ends of $\mathbb{C}_\infty \setminus C$ in Section 2.1.1.

As $K^*(z,\zeta)$ is a $\mathbf{Z}^*$-harmonic function in $z \in D^*$ for each fixed $\zeta \in \partial\mathbb{H}$, there exists an analytic function $\Psi(z,\zeta)$ in $z \in D$ having $K^*(z,\zeta)$, $z \in D$, as its imaginary part uniquely up to an additive real constant by virtue of Theorem 1.6.5.

Theorem 1.9.1.

(i) *The limit* $\lim_{z\to\infty} \Psi(z,\zeta)$ *exists and is real-valued. Moreover,* $\limsup_{z\to\infty} \sup_{\zeta \in J} |\Psi(z,\zeta)| < \infty$ *for any compact interval* $J \subset \partial\mathbb{H}$.

(ii) $\Psi(z,\zeta)$ *is determined uniquely by the normalization condition*

$$\lim_{z\to\infty} \Psi(z,\zeta) = 0. \tag{1.9.2}$$

$\Psi(z,\zeta)$ *is then jointly continuous in* (z,ζ) *on* $\left(D \cup \left(\cup_{i=1}^N C_i^p\right) \cup (\partial\mathbb{H} \setminus J)\right) \times J$, *for any compact subinterval* J *of* $\partial\mathbb{H}$.

Proof. (i) By (1.7.15), we have

$$-\frac{1}{2}\frac{\partial}{\partial y}\frac{\partial}{\partial \mathbf{n}_\zeta}G^0(z,\zeta)\Big|_{y=0} \le \frac{1}{\pi}\frac{1}{(x-\xi)^2}, \tag{1.9.3}$$

which combined with the expression (1.7.5) of $K^*(z,\zeta)$ and the estimate (1.7.9) implies that for any open interval $I \subset \partial\mathbb{H}$ containing ζ,

$$\int_{\partial\mathbb{H}\setminus I}\left|\frac{\partial K^*(x+iy,\zeta)}{\partial y}\right|_{y=0} dx \text{ is finite and continuous in } \zeta \in I. \tag{1.9.4}$$

Since $z \mapsto \Psi(z,\zeta)$ is an analytic function on D whose imaginary part is continuously extended to $\partial\mathbb{H}\setminus\zeta$ to be zero there by Proposition 1.7.4 (i), it can be extended to an analytic function on $E = \mathbb{C}\setminus(\cup_{k=1}^N C_k)\cup(\cup_{k=1}^N \Pi C_k)\setminus\{\zeta\}$ by the reflection principle, where Π denotes the reflection with respect to $\partial\mathbb{H}$ (cf. [Ahlfors (1979), Theorem 4.24]). Denote the extended analytic function $z \mapsto \Psi(z,\zeta)$ as $u(z)+iv(z)$ with $v(z) = K^*(z,\zeta)$. The real part u can then be evaluated by

$$u(z)-u(z_0) = \int_C v_y dx - v_x dy \left(= -\int_C \frac{\partial v(z')}{\partial \mathbf{n}_{z'}}ds(z')\right), \tag{1.9.5}$$

where C is any rectifiable simple curve connecting z_0 with z in E. We fix an arbitrary compact interval $J \subset \partial\mathbb{H}$ and take $\zeta \in J,\ z_0 = \xi_0 + i0 \in \partial\mathbb{H}\setminus J$ that is located to the right of J. Choose $\ell_0 > 0$ with $J \subset (-\ell_0,\ell_0)$, $\bigcup_{i=1}^N C_i \subset R_{\ell_0} := \{x+iy : -\ell_0 < x < \ell_0,\ 0 < y < \ell_0\}$.

By virtue of Proposition 1.7.4 (iii), $v(z) = K^*(z,\zeta)$ satisfies the properties (1.1.50) uniformly in $\zeta \in J$. Therefore by Lemma 1.1.7,

$$\sup_{\zeta\in J,\ell>2\ell_0}\int_{\Sigma_\ell}\left|\frac{\partial K^*(z,\zeta)}{\partial \mathbf{n}_z}\right| ds(z) < \infty. \tag{1.9.6}$$

For $z \in \mathbb{H}\setminus R_{2\ell_0}$, let $\ell \ge 2\ell_0$ be such that $z \in \Sigma_l$ and C be the curve consisting of the line segment $\{z' \in \partial\mathbb{H} : \xi_0 \le \Re z' \le \ell\}$ and a part of Σ_ℓ that connects z_0 to z through $E \cap \overline{\mathbb{H}}$. For $z \in \partial\mathbb{H}$ with $\xi_0 < \Re z$, we just take $C = \{z' \in \partial\mathbb{H} : \xi_0 \le \Re z' \le \Re z\}$. For $z \in \partial\mathbb{H}$ located to the left of J, we take C analogously.

We can then deduce from (1.9.4), (1.9.5) and (1.9.6) that $u(z)$ is bounded near ∞ of the Riemann sphere $\mathbb{C}\cup\{\infty\}$ uniformly in $\zeta \in J$. This combined with (1.7.11) yields the second assertion of (i). In particular, the analytic function $\Psi(\frac{1}{z})$ is uniformly bounded near the origin $\mathbf{0}$, which is therefore a removable singularity of this analytic function near $\mathbf{0}$, yielding the first assertion of (i).

(ii) Write $\Psi(z,\zeta) = u(z,\zeta) + iK^*(z,\zeta)$. By Proposition 1.7.4 (i), $K^*(z,\zeta)$ is jointly continuous in (z,ζ) on $[\mathbb{C}\setminus\bigcup_{k=1}^N(C_k\cup\Pi C_k)\setminus J]\times J$, and so are $\frac{\partial}{\partial x}K^*(z,\zeta)$ and $\frac{\partial}{\partial y}K(z,\zeta)$ because $K^*(z,\zeta)$ is harmonic in z. Fix some $z_0 = \xi_0 + i0 \in \partial\mathbb{H}\setminus J$ located to the right of J. As $\lim_{x\to\infty} u(x+i0,\zeta) = 0$ by the normalization (1.9.2), we see from (1.9.5) that $u(z,\zeta)$, $z \in D\cup(\partial\mathbb{H}\setminus J)$, $\zeta \in J$, is determined by

$$u(z,\zeta) = \int_C \frac{\partial}{\partial y}K^*(x+iy,\zeta)dx - \frac{\partial}{\partial x}K^*(x+iy,\zeta)dy - \int_{\xi_0}^\infty \frac{\partial}{\partial y}K^*(x+i0,\zeta)dx,$$

for any rectifiable simple curve C connecting z_0 to z in $D \cup (\partial\mathbb{H} \setminus J)$. Consequently, for any $z_1,\ z_2 \in D \cup (\partial\mathbb{H} \setminus J)$,

$$u(z_2, \zeta) - u(z_1, \zeta) = \int_C \frac{\partial}{\partial y} K^*(x + iy, \zeta) dx - \frac{\partial}{\partial x} K^*(x + iy, \zeta) dy \tag{1.9.7}$$

for a rectifiable simple curve C joining z_1 to z_2 through $D \cup (\partial\mathbb{H} \setminus J)$. The joint continuity of $u(x, \zeta)$ on $(D \cup \partial\mathbb{H} \setminus J) \times J$ follows from these two formulae and (1.9.4).

Let us verify the continuous extendability of $\Psi(z, \zeta)$ from $z \in D$ to $z \in C_j^p$ for each $1 \leq j \leq N$, where $C_j^p = C_j^{0,+} \cup C_j^{0,-} \cup \{z_j, z_j^r\}$ is defined by (1.9.1) for $C = C_j$, z_j, z_j^r being the left and right endpoint of C_j. We write $z_j = a + ic,\ z_j^r = b + ic,\ (a < b,\ c > 0)$, denote the open slit $C_j \setminus \{z_j,\ z_j^r\}$ by C_j^0 and consider the rectangles

$$R_+ = \{z = x + iy :\ a < x < b,\ c < y < c + \delta\},$$
$$R_- = \{z = x + iy :\ a < x < b,\ c - \delta < y < c\},$$

and $R = R_+ \cup C_j^0 \cup R_-$, for $\delta > 0$ such that $\left(\bigcup_{k \neq j} C_k\right) \cap R = \emptyset$.

Since, as a function of z, $K^*(z, \zeta)$ takes a constant value on C_j by Proposition 1.7.4 (i), $\Psi(z, \zeta)$ can be extended to be analytic from R_+ (resp. R_-) to R across C_j^0 by the Schwarz reflection (cf. [Ahlfors (1979), Theorem 24 in §4.6.5]). Extend $\Psi(z, \zeta)$ from $z \in D$ to $z \in D \cup C_j^{0,+} \cup C_j^{0,-}$ by setting its value at $z \in C_j^{0,+}$ (resp. $C_j^{0,-}$) $\lim_{z' \in R_+, z' \to z} \Psi(z', \zeta)$ (resp. $\lim_{z' \in R_-, z' \to z} \Psi(z', \zeta)$). Due to the Cauchy integral formula for analytic functions, the extended function is jointly continuous in $(z, \zeta) \in (D \cup C_j^{0,+} \cup C_j^{0,-}) \times \partial\mathbb{H}$.

We next show that $\Psi(z, \zeta)$ can also be continuously extended to the left endpoint z_j of C_j. Take $\varepsilon \in (0, (b-a)/2)$ such that $B(z_j, \varepsilon) \setminus C_j \subset D$. Then $w = \psi(z) = (z - z_j)^{1/2}$ maps $B(z_j, \varepsilon) \setminus C_j \subset D$ conformally onto $B(0, \sqrt{\varepsilon}) \cap \mathbb{H}$. Consequently, $f(w, \zeta) = \Psi(\psi^{-1}(w), \zeta) = \Psi(w^2 + z_j, \zeta)$ is an analytic function in $w \in B(0, \sqrt{\varepsilon}) \cap \mathbb{H}$ that is continuous up to $B(0, \sqrt{\varepsilon}) \cap \partial\mathbb{H} \setminus \{0\}$ and $\Im f(w, \zeta) = K^*(w^2 + z_j, \zeta)$ takes a constant value there. By the Schwarz reflection, $f(w, \zeta)$ extends to an analytic function on $B(0, \sqrt{\varepsilon}) \setminus \{0\}$. Thus, 0 is an isolated singularity of $f(w, \zeta)$. Since $\Im f(w, \zeta)$ is bounded near 0, it has to be a removable singularity (cf. [Ahlfors (1979), Exercise 5 in §4.3.2]).

Hence $f(w, \zeta)$ can be extended to be analytic in $w \in B(0, \sqrt{\varepsilon})$ and the function so extended is jointly continuous in $(z, \zeta) \in B(0, \sqrt{\varepsilon}) \times \partial\mathbb{H}$ due to the Cauchy integral formula. This means that $\lim_{z \in D \cup C_j^{0,+} \cup C_j^{0,-},\ z \to z_j} \Psi(z, \zeta)$ exists and equals $f(0, \zeta)$ and that the function $\Psi(z, \zeta)$ so extended is jointly continuous in $(z, \zeta) \in (D \cup C_j^{0,+} \cup C_j^{0,-} \cup \{z_j\}) \times \partial\mathbb{H}$.

An analogous statement for the right endpoint z_j^r of C_j can be also made. □

We call $\Psi(z, \zeta)$, $z \in D$, $\zeta \in \partial\mathbb{H}$, subjected to the normalization condition (1.9.2) the *complex Poisson kernel* of the BMD $\mathbf{Z}^*$ for the standard slit domain D. We shall occasionally denote it by $\Psi_D(z, \zeta)$ to indicate its relevance to D.

Let

$$K_{\mathbb{H}}(z, \zeta) = \frac{1}{\pi} \frac{\Im z}{|z - \zeta|^2}, \quad \Psi_{\mathbb{H}}(z, \zeta) = -\frac{1}{\pi} \frac{1}{z - \zeta}, \quad z \in \mathbb{H}, \zeta \in \partial\mathbb{H}. \tag{1.9.8}$$

$K_{\mathbb{H}}(z,\zeta) = \Im\Psi_{\mathbb{H}}(z,\zeta)$ and $K_{\mathbb{H}}(z,\zeta)$ is the Poisson kernel to express the harmonic function for the ABM $\mathbf{Z}^{\mathbb{H}}$ on $\mathbb{H}$ so that we may call $\Psi_{\mathbb{H}}(z,\zeta)$ the *complex Poisson kernel* of $\mathbf{Z}^{\mathbb{H}}$. The following lemma concerns the difference of the complex Poisson kernels of the BMD for the standard slit domain $D = \mathbb{H} \setminus \bigcup_{j=1}^N C_j$ and of the ABM on $\mathbb{H}$.

Lemma 1.9.2. *For each $\zeta \in \partial\mathbb{H}$, the function $\mathbf{H}(z,\zeta) = \Psi(z,\zeta) - \Psi_{\mathbb{H}}(z,\zeta)$ of $z \in D$ can be extended from D to an analytic function on $D \cup \Pi D \cup \partial\mathbb{H}$, where Π denotes the mirror reflection relative to $\partial\mathbb{H}$.*

Proof. The function $\mathbf{H}(z,\zeta)$ is analytic in $z \in D$ and it follows from (1.1.27) and (1.7.5) that for $z \in D$ and $\zeta \in \partial\mathbb{H}$,

$$\begin{aligned}\Im\mathbf{H}(z,\zeta) &= K^*(z,\zeta) - K_{\mathbb{H}}(z,\zeta)\\ &= -\mathbb{E}_z^{\mathbb{H}}\left[K_{\mathbb{H}}(Z^{\mathbb{H}}_{\sigma_K},\zeta)\right] - \sum_{i,j=1}^N \varphi^{(i)}(z)\left(\mathcal{A}^{-1}\right)_{ij}\frac{\partial}{\partial \mathbf{n}_\zeta}\varphi^{(j)}(\zeta).\end{aligned}$$

According to Lemma 1.1.4 (i), we have for any $\xi \in \partial\mathbb{H}$, $\lim_{z\to\xi}\varphi^{(i)}(z) = 0$. By the same consideration, we also have $\lim_{z\to\xi}\mathbb{E}_z^{\mathbb{H}}\left[K_{\mathbb{H}}(Z^{\mathbb{H}}_{\sigma_K},\zeta)\right] = 0$. Hence $z \mapsto \mathbf{H}(z,\zeta)$ extends to an analytic function on $D \cup \Pi D \cup \partial\mathbb{H}$ by the Schwarz reflection. □

Chapter 2

Chordal Komatu-Loewner differential equation and BMD

2.1 Komatu-Loewner left differential equation generated by a Jordan arc

2.1.1 $\mathbb{H}$-*hulls and canonical maps for multiply connected domains*

A subset F of $\mathbb{H}$ is called an $\mathbb{H}$-*hull* if $\overline{F}$ is compact, $F = \overline{F} \cap \mathbb{H}$ and $\mathbb{H} \setminus F$ is simply connected. We start with the following Riemann mapping theorem.

Theorem 2.1.1. *For an $\mathbb{H}$-hull F, there exists a unique conformal map g_F^0 from $\mathbb{H} \setminus F$ onto $\mathbb{H}$ such that*

$$\lim_{z\to\infty} \left(g_F^0(z) - z\right) = 0.$$

In this case, g_F^0 satisfies

$$g_F^0(z) = z + \frac{a_F^0}{z} + o(1/|z|) \quad \text{as } z \to \infty, \tag{2.1.1}$$

for some non-negative constant a_F^0.

Proof. First, we claim that every point $w \in \partial(\mathbb{H}\setminus F)$ is regular for $(\mathbb{H}\setminus F)^c$. Indeed, since $\mathbb{H}\setminus F$ is simply connected, $\log(z-w)$ is a well-defined and single-valued analytic function in $\mathbb{H} \setminus F$. Let $h(z) = -\Re\frac{1}{\log(z-w)}$. The function h is harmonic in $\mathbb{H} \setminus F$, continuous on $\overline{\mathbb{H} \setminus F}$ with $h(w) = 0$ and $h > 0$ on $\overline{(\mathbb{H} \setminus F)} \cap B(w, 1/2) \setminus \{w\}$. Hence h is a barrier at w for the bounded domain $(\mathbb{H}\setminus F)\cap B(w, 1/2)$. Thus, w is a regular point for $((\mathbb{H} \setminus F)\cap B(w, 1/2))^c$ (see, e.g., [Karatzas and Shreve (1998), Proposition 2.15 in Chapter 4]), and hence w is a regular point for $(\mathbb{H} \setminus F)^c$.

Define for $z \in \mathbb{H} \setminus F$,

$$v_0(z) = \Im z - h_F(z), \quad \text{where } h_F(z) := \mathbb{E}_z^{\mathbb{H}}\left[\Im Z_{\sigma_F}^{\mathbb{H}}; \sigma_F < \infty\right] \geq 0.$$

Both v_0 and h_F are harmonic functions in $\mathbb{H} \setminus F$. Since every point $w \in \partial(\mathbb{H} \setminus F)$ is regular for $(\mathbb{H} \setminus F)^c$, $v_0(z)$ vanishes continuously on $\partial(\mathbb{H} \setminus F)$. As F is bounded, there is some constant $\rho > 0$ so that $F \subset B_\rho := \{z \in \mathbb{C} : |z| < \rho\}$. Thus there is a constant $\lambda_0 > 0$ so that for every $z \in \mathbb{H} \setminus B_{2\rho}$,

$$h_F(z) \leq \rho \mathbb{P}_z^{\mathbb{H}}(\sigma_F < \infty) \leq \rho \mathbb{P}_z^{\mathbb{H}}(\sigma_{\mathbb{H}\cap\partial B_\rho} < \infty) \leq \lambda_0 \rho/|z|, \tag{2.1.2}$$

where the last inequality is due to Proposition 1.1.2. Consequently,

$$\lim_{z\in\mathbb{H},|z|\to\infty}|v_0(z)-\Im z| = \lim_{z\in\mathbb{H},|z|\to\infty}h_F(z)=0.$$

On the other hand, since $\lim_{z\in\mathbb{H}\setminus F, z\to\zeta}h_F(z)=0$ for $\zeta\in\partial\mathbb{H}\setminus\overline{F}$, h_F extends to a harmonic function on $\mathbb{C}\setminus\overline{F}\cup\Pi F$ by Schwarz reflection $h_F(z)=-h_F(\bar{z})$. Any harmonic function h defined on a neighborhood of the closure of the disk $B_r(z_0)=\{z\in\mathbb{C}:|z-z_0|<r\}$ with $z_0\in\mathbb{C}$ and $r>0$ has the Poisson integral representation

$$h(z)=\frac{1}{2\pi r}\int_{\partial B_r(z_0)}\frac{r^2-|z-z_0|^2}{|w-z|^2}h(w)\sigma(dw)\quad\text{for } z\in B_r(z_0),$$

where σ denotes the Lebesgue arclength measure on $\partial B_r(z_0)$. From this expression, one readily obtains

$$|h_x(z_0)|\le\frac{2}{r}\max_{w\in\partial B_r(z_0)}|h(w)|\quad\text{and}\quad|h_y(z_0)|\le\frac{2}{r}\max_{w\in\partial B_r(z_0)}|h(w)|. \tag{2.1.3}$$

In particular, applying the above to h_F and $r=1$, we get

$$|\nabla h_F(z_0)|\le 4\max_{w\in\partial B_r(z_0)}|h_F(w)|\quad\text{for any } z_0\in\mathbb{C}\setminus B_{2\rho+1},$$

which combined with (2.1.2) yields

$$|\nabla v_0(z)-(0,1)|\le|\nabla h_F(z)|\le 4\lambda_0\rho/|z|\quad\text{for every } z\in\mathbb{H}\setminus B_{2\rho+1}. \tag{2.1.4}$$

Fix some $z_0=x_0+iy_0\in\mathbb{H}\setminus F$ with $y_0>2\rho+1$ and a real value c_1. For any $z\in\mathbb{H}\setminus F$, define

$$u_0(z)=c_1+\int_\gamma\frac{\partial v_0(\xi)}{\partial\mathbf{n}_\xi}ds(\xi),$$

where γ is a piecewise C^2-smooth simple curve in $\mathbb{H}\setminus F$ that connects z_0 to z, $ds(\xi)$ is the arc-length measure along γ and $\mathbf{n}$ the unit normal vector field along γ in the counter-clockwise direction (that is, if γ is parameterized by $(x(t),y(t))$, then $\mathbf{n}$ is the unit vector pointing to the same direction as $(-y'(t),x'(t))$). Since $\mathbb{H}\setminus F$ is simply connected, the value $u_0(z)$ is well defined, independent of the choice of the piecewise smooth curve γ in $\mathbb{H}\setminus F$ connecting z_0 to z. It is easy to verify that (u_0,v_0) satisfies the Cauchy-Riemann equation and so $f_0(z):=u_0(z)+iv_0(z)$ is an analytic function in $\mathbb{H}\setminus F$ (cf. [Conway (1995), §13.3]).

For $z=x+iy\in\mathbb{H}\setminus B_\rho$, let γ the curve consisting of the horizontal line segment γ_1 going from $z_0=x_0+iy_0$ to $x+iy_0$ and the vertical line segment γ_2 going from $x+iy_0$ to z. Then

$$u_0(z)=c_1+\int_{x_0}^{x}\frac{\partial v_0(s,y_0)}{\partial y}ds-\int_{y_0}^{y}\frac{\partial v_0(x,t)}{\partial x}dt. \tag{2.1.5}$$

In view of (2.1.4), there is some constant $c_2>0$ so that

$$|u_0(z)-\Re z|\le c_2\ln|z|\quad\text{and}\quad|f_0(z)-z|\le c_2\ln|z|\quad\text{for } z\in\mathbb{H}\setminus B_{2\rho+1}. \tag{2.1.6}$$

As $\Im f_0(z) = v_0(z)$ is continuously vanishing on $B_\rho^c \cap \partial\mathbb{H}$, by Schwarz reflection principle $f_0(z)$ can be extended to be an analytic function on $\mathbb{C} \setminus \bar{B}_\rho$ by setting $f_0(\bar{z}) = \overline{f_0(z)}$ for $z \in \mathbb{H} \setminus \bar{B}_\rho$. As $f_0(z)$ tends to infinity as $z \to \infty$ by (2.1.6), $g(z) := 1/f_0(1/z)$ is an analytic function in the disk $B_{1/\rho}$ with $g(0) = 0$. It follows from the Taylor expansion of g at 0 and (2.1.6) that

$$f_0(z) = z + a_0 + \frac{a_1}{z} + \frac{a_2}{z^2} + \cdots \quad \text{as } z \to \infty.$$

Since f_0 maps $\partial\mathbb{H} \cap \bar{B}_\rho^c$ to $\partial\mathbb{H}$, all the coefficients a_k are real numbers. By choosing suitable c_1 in (2.1.5), we can make $a_0 = 0$ and thus (2.1.1) holds for f_0 because $a_1 = \lim_{\mathbb{R}\ni y\to\infty} y(y - \Im f_0(iy)) = \lim_{\mathbb{R}\ni y\to\infty} y(y - v_0(iy)) \geq 0$.

In the following, we restrict the definition of f_0 back to $\mathbb{H} \setminus F$. As $\Im f_0(z) = v_0(z) > 0$ in $\mathbb{H}\setminus F$, f_0 maps $\mathbb{H}\setminus F$ into $\mathbb{H}$. Since $\Im f_0(z) = v_0(z)$ vanishes continuously on $\partial(\mathbb{H} \setminus F)$ and $f_0(z) \to \infty$ as $z \to \infty$, for a compact subset $K \subset \mathbb{H}$, its pre-image $f_0^{-1}(K)$ is a compact subset of $\mathbb{H}\setminus F$; in other words, f_0 is a proper map from $\mathbb{H}\setminus F$ to $\mathbb{H}$.

Observe that by (2.1.1), the function g above satisfies $g'(0) = 1$. Consequently, there is $L_0 > 1$ so that $F \subset B_{L_0}$ and f_0 is one-to-one in $\mathbb{H} \setminus B_{L_0}$. Since $\Im f_0(z) = v_0(z) \leq \Im z$ for $z \in \mathbb{H} \setminus F$, for $w \in f_0(\mathbb{H} \setminus F)$ with $\Im w > L_0$, any of its pre-image z has $\Im z > L_0$ and hence is unique. Thus by Lemma A.5.1 in the Appendix, f_0 is a conformal map from $\mathbb{H} \setminus F$ onto $\mathbb{H}$.

We next show that conformal map from $\mathbb{H}\setminus F$ onto $\mathbb{H}$ satisfying (2.1.1) is unique. Suppose f is another such kind of conformal map. Then $\varphi(z) = f_0(f^{-1}(z))$ is a conformal map from $\mathbb{H}$ onto $\mathbb{H}$ and has the property that $|\varphi(z) - z| \to 0$ as $|z| \to \infty$. Since any conformal map from $\mathbb{H}$ onto $\mathbb{H}$ is a Möbius transform, the only one that has the property $\lim_{z\in\mathbb{H},|z|\to\infty} |\varphi(z) - z| = 0$ is the identity map. This proves that $f = f_0$. □

An $(N+1)$-connected domain $D = \mathbb{H}\setminus\bigcup_{j=1}^N A_j$ is called a *standard slit domain* if $\{A_j : 1 \leq j \leq N\}$ are mutually disjoint horizontal line segments in $\mathbb{H}$. Let $D = \mathbb{H}\setminus\bigcup_{j=1}^N A_j$ be an $(N+1)$-connected domain and F be an $\mathbb{H}$-hull with $F \subset D$.

Proposition 2.1.2. *There exists a unique conformal map g_F from $D \setminus F$ onto a standard slit domain such that*

$$g_F(z) = z + \frac{a_F}{z} + o(1/|z|), \qquad z \to \infty, \tag{2.1.7}$$

for some real constant a_F. Moreover, $a_F \geq 0$.

Proof. Let $g_F^0 \colon \mathbb{H} \setminus F \to \mathbb{H}$ be the conformal map of Theorem 2.1.1 and

$$\widehat{D} := g_F^0(D \setminus F) = \mathbb{H} \setminus \cup_{j=1}^N \widehat{A}_j, \quad \text{where } \widehat{A}_j = g_F^0(A_j) \text{ for } 1 \leq j \leq N.$$

Consider the $(2N)$-connected domain $U = \mathbb{C} \setminus \left(\cup_{j=1}^N (\widehat{A}_j \cup \Pi(\widehat{A}_j)\right)$ obtained from $\widehat{D}$ by the mirror reflection $\Pi z = \bar{z}$.

A $(2N)$-connected domain $\mathbb{C} \setminus \bigcup_{j=1}^{2N} C_j$ is called a *parallel slit plane* if each set C_j is a horizontal line segment. Consider the family $\mathcal{F}$ of univalent functions f on $U \subset \mathbb{C}$ satisfying

$$f(z) = z + \frac{a}{z} + o(1/|z|) \quad \text{as } z \to \infty, \tag{2.1.8}$$

for some constant $a \in \mathbb{C}$.

Put $\mu = \sup_{f \in \mathcal{F}} \Re a$. Note that $\mu \geq 0$ as $\mathcal{F}$ contains the identity map z. It is known that there exists then $f_0 \in \mathcal{F}$ satisfying

$$f_0(z) = z + \frac{a_0}{z} + o(1/|z|) \quad \text{as } z \to \infty \quad \text{with} \quad \Re a_0 = \mu,$$

and that such a function f_0 necessarily becomes a conformal map from U onto a parallel slit plane ([Tsuji (1959), Theorem IX.22]). Define $f_1(z) = \overline{f_0(\overline{z})}$, $z \in U$. Then $f_1 \in \mathcal{F}$ and f_1 admits the above expansion near ∞ with $\overline{a}_0$ in place of a_0. As $\Re \overline{a}_0 = \Re a_0 = \mu$, we get $f_1 = f_0$ on U by the uniqueness of a conformal map from U onto a parallel slit plane ([Tsuji (1959), Theorem IX.23]). Hence $f_0(z) \in \partial\mathbb{H}$ only if $z \in \partial\mathbb{H}$. By (2.1.8), $\Im f_0(iy_0) > 0$ for a large $y_0 > 0$. Take any $z \in \widehat{D}$. Then $f_0(z) \in \mathbb{H}$. Indeed, $f_0(z) \notin \partial\mathbb{H}$. If $\Im f_0(z) < 0$, then, for a continuous curve γ connecting iy_0 and z through $\widehat{D}$, $f_0(\gamma)$ must cross $\partial\mathbb{H}$, a contradiction. Accordingly $\widehat{f} = f_0|_{U \cap \mathbb{H}}$ gives a conformal map from $\widehat{D}$ onto a standard slit domain satisfying (2.1.8) for some non-negative a. The composition $g_F = \widehat{f} \circ g_F^0$ then gives a conformal map from $D \setminus F$ onto a standard slit domain satisfying (2.1.7) with $a_F = a_F^0 + a$, which is a non-negative number; see Figure 2.1.

Suppose f is another conformal map from $D \setminus F$ onto a standard slit domain that satisfies (2.1.7). Then $f \circ g_F^{-1}$ is a conformal map from the standard slit domain $g_F(D \setminus F)$ onto a standard slit domain that maps $\partial\mathbb{H}$ onto itself. By the Schwarz reflection, it extends to a conformal map from the parallel slit plane $\widehat{U} := g_F(D \setminus F) \cup \Pi(g_F(D \setminus F)) \cup \partial\mathbb{H}$ onto a parallel slit plane satisfying (2.1.8) for some real a. As the identity map on $\widehat{U}$ is also such a conformal map, by the uniqueness theorem cited above, $f \circ g_F^{-1}$ has to be the identity map on $\widehat{D}$ and so $f = g_F$ on $D \setminus F$. □

The map g_F of Proposition 2.1.2 is called the *canonical map from* $D \setminus F$. The property (2.1.7) is called the *hydrodynamic normalization*, while the non-negative constant a_F in it is called the *half-plane capacity* of the $\mathbb{H}$-hull F *relative to the multiply connected domain* $D = \mathbb{H} \setminus \bigcup_{j=1}^{N} A_j$. In Theorem 2.2.3, we will present an alternative probabilistic proof of the existence of this canonical map g_F.

We also call the map g_F^0 of Theorem 2.1.1 the *canonical map from* $\mathbb{H} \setminus F$. The non-negative constant a_F^0 in (2.1.1) has been simply called the *half-plane capacity* of F and denoted by $\mathrm{hcap}(F)$.

In the rest of this section, we shall consider a special case that the hulls F are generated by a single Jordan arc. Before doing so, we first state some general facts about an ideal boundary of a multiply connected domain called a prime end.

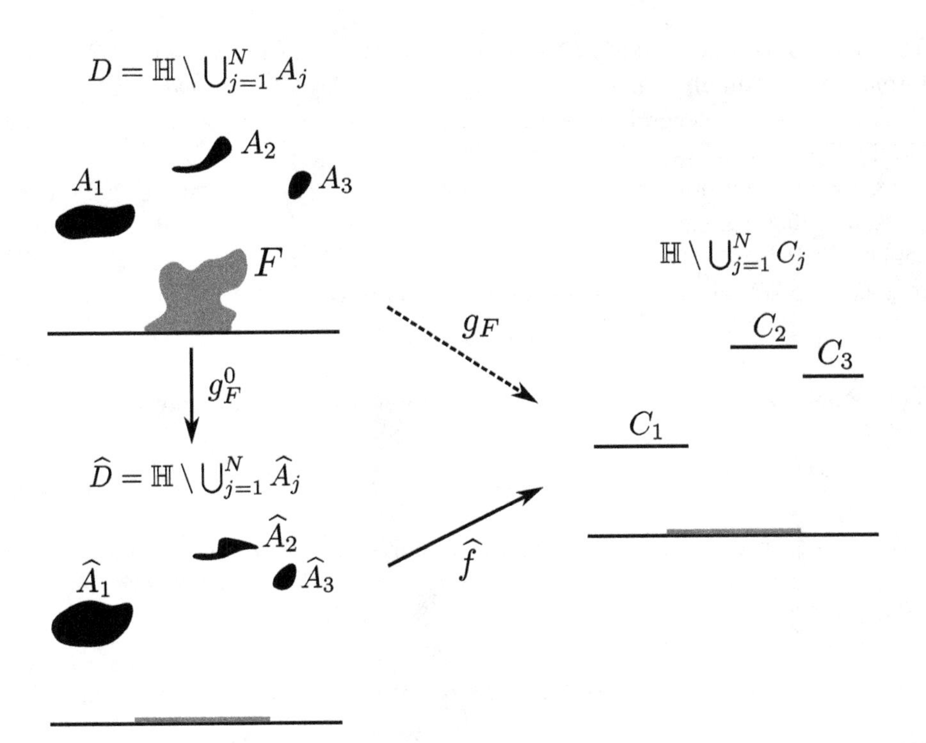

Fig. 2.1 Canonical maps g_F^0, $\widehat{f}$ and $g_F = \widehat{f} \circ g_F^0$.

Consider a general finitely connected domain $D \subset \mathbb{C}$ expressed as (1.1.19). A closed Jordan arc q in $\overline{D}$ is called a *crosscut* if its endpoints lie on a single component of ∂D and other points of q lie inside D. A crosscut obviously separates the domain D into two components.

A sequence $\{q_n\}$ of crosscuts is called a *null-chain* if all q_n are disjoint, there is a component of $D \setminus q_n$ denoted by $\text{ins}(q_n)$ such that $\text{ins}(q_{n+1}) \subset \text{ins}(q_n)$ for all $n \geq 1$ and $\lim_{n\to\infty} \text{diam}(q_n) = 0$. Two null-chains $\{q_n\}$ and $\{q_n'\}$ are said to be *equivalent* if, for every m, there exists n such that $\text{ins}(q_n') \subset \text{ins}(q_m)$, and the same property holds with q_n and q_n' being exchanged. The equivalent class of null chains by this relation is called a *prime end* of D.

Denote by $\mathcal{P}(D)$ the collection of all prime ends of D. We endow a topology on $D \cup \mathcal{P}(D)$ as follows. A subset U of $D \cup \mathcal{P}(D)$ is open if $U \cap D$ is open, and for every prime end $p \in U \cap \mathcal{P}(D)$, there exists a null-chain $\{q_n\} \in p$ such that $\text{ins}(q_n) \subset U \cap D$ for some n. Then by definition a sequence $\{z_k\}$ in D converges to a prime end p if and only if, for some null-chain $\{q_n\} \in p$ and each n, it holds that $z_k \in \text{ins}(q_n)$ for all k sufficiently large.

It is well known as *Carathéodory's theorem* that a conformal map between two finitely connected domains D_1 and D_2 extends to a homeomorphism between $D_1 \cup \mathcal{P}(D_1)$ and $D_2 \cup \mathcal{P}(D_2)$ (cf. [Tsuji (1959), Theorem IX.1], [Pommerenke (1975),

Theorem 9.6], [Conway (1995), Theorem 14.3.4]). Carathéodory's theorem has been formulated for simply connected domains D_1 and D_2. But we can readily prove this fact under the general finite connectivity, for instance, via a proof of [Conway (1995), Theorem 15.3.4].

A standard slit domain D admits a simple description of the space $\mathcal{P}(D)$. For a horizontal line segment C with endpoints z, z^r, denote by $C^{0,+}$ and $C^{0,-}$ distinct copies of $C^0 = C \setminus \{z, z^r\}$ indicating the upper side and the lower side of C^0, respectively. We now recall the definition (1.9.1) of the set C^p:

$$C^p = C^{0,+} \cup C^{0,-} \cup \{z, z^r\}. \tag{2.1.9}$$

For a standard slit domain $D = \mathbb{H} \backslash \cup_{j=1}^N C_j$, it follows immediate from the definition of $\mathcal{P}(D)$ that the identification

$$\mathcal{P}(D) = \partial\mathbb{H} \cup (\cup_{j=1}^n C_j^p), \tag{2.1.10}$$

holds. A sequence $\{z_k\}$ in D converges to a point of $C_j^{0,+}$ (resp. $C_j^{0,-}$) if and only if the sequence converges to the corresponding point of C_j^0 from the above (resp. the below) of C_j^0, $1 \le j \le N$.

2.1.2 *The map $g_{t,s}$ and an expression of $a_t - a_s$ for a Jordan arc*

Fix a standard slit domain $D = \mathbb{H} \setminus \{C_1, \dots, C_N\}$ and a Jordan arc

$$\gamma : [0, t_\gamma) \mapsto \overline{D} \qquad \text{with } \gamma(0) \in \partial\mathbb{H} \text{ and } \gamma(0, t_\gamma) \subset D,\ t_\gamma \in (0, \infty]. \tag{2.1.11}$$

For each $t \in [0, t_\gamma)$, the set $F_t = \gamma(0, t](\subset D)$ is obviously a hull (with $F_0 = \emptyset$) and the family $\{F_t : t \in [0, t_\gamma)\}$ of hulls are *growing* in the sense that $F_s \subsetneq F_t$ for $0 < s < t < t_\gamma$.

For simplicity, the canonical map g_{F_t} from $D \setminus F_t$ onto a standard slit domain $D_t = \mathbb{H} \setminus \bigcup_{j=1}^N C_{t,j}$ will be designated by g_t (see Figure 2.2), while the associated capacity a_{F_t} will be denoted by a_t. It then holds that for each $t \in [0, t_\gamma)$,

$$g_t(z) = z + \frac{a_t}{z} + o(1/|z|) \quad \text{as } z \to \infty. \tag{2.1.12}$$

Note that a_t is a non-negative real-valued function of t.

It is known that, for a bounded simply connected domain D_0 with ∂D_0 being a closed Jordan curve, the space of its prime ends $\mathcal{P}(D_0)$ can be identified with the boundary curve ∂D_0 (cf. [Tsuji (1959), Theorem IX.2]). In the same manner, we can make the identification

$$\mathcal{P}(D \setminus \gamma(0, t]) = \gamma^+[0, t) \cup \gamma^-[0, t) \cup \{\gamma(t)\} \cup (\partial\mathbb{H} \setminus \gamma(0)) \cup (\cup_{j=1}^N C_j^p), \tag{2.1.13}$$

where $\gamma^+[0, t) = \{\gamma(s)^+ : 0 \le s < t\}$ and $\gamma^-[0, t) = \{\gamma(s)^- : 0 \le s < t\}$ are distinct copies of $\gamma[0, t)$ indicating its right and left side, respectively. By

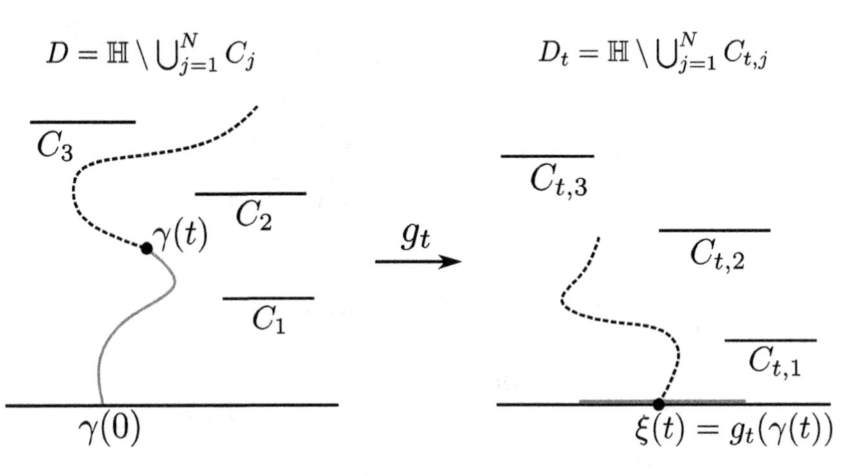

Fig. 2.2 Canonical map g_t.

Carathéodory theorem, g_t extends to a homeomorphism (denoted by g_t again) between $(D \setminus \gamma(0,t]) \cup \mathcal{P}(D \setminus \gamma(0,t])$ and $D_t \cup \mathcal{P}(D_t)$. Under the conformal map g_t, $\gamma^+[0,t) \cup \gamma^-[0,t) \cup \{\gamma(t)\} \cup (\partial\mathbb{H} \setminus \gamma(0))$ (resp. C_j^p) is homeomorphic with $\partial\mathbb{H}$ (resp. $C_{t,j}^p$) for $1 \le j \le N$.

For $0 < s < t < t_\gamma$, define

$$g_{t,s} = g_s \circ g_t^{-1}, \tag{2.1.14}$$

which is a conformal map from D_t onto $D_s \setminus g_s(\gamma[s,t])$. We can then deduce from (2.1.12) that

$$g_{t,s}(z) = z + \frac{a_s - a_t}{z} + o(1/|z|), \qquad z \to \infty. \tag{2.1.15}$$

On account of the above mentioned homeomorphic extension of g_t, one can define

$$\beta_0(t,s) = g_t(\gamma(s)^-) \in \partial\mathbb{H}, \quad \beta_1(t,s) = g_t(\gamma(s)^+) \in \partial\mathbb{H}, \quad \xi(t) = g_t(\gamma(t)) \in \partial\mathbb{H}. \tag{2.1.16}$$

Then it can be verified that

$$\beta_0(t,s) < \xi(t) < \beta_1(t,s), \quad g_{t,s}(\beta_0(t,s)) = g_{t,s}(\beta_1(t,s)) = \xi(s), \tag{2.1.17}$$

and $\Im g_{t,s}(x + i0)$ is continuous in $x \in \mathbb{R}$ with

$$\Im g_{t,s}(x + i0+) \begin{cases} = 0 & \text{for } x \in \partial\mathbb{H} \setminus (\beta_0(t,s), \beta_1(t,s)), \\ > 0 & \text{for } x \in (\beta_0(t,s), \beta_1(t,s)). \end{cases} \tag{2.1.18}$$

See below for Figure 2.3.

We let $\ell_{t,s} = [\beta_0(t,s), \beta_1(t,s)] (\subset \partial\mathbb{H})$. Because of (2.1.18), $g_{t,s}$ extends by the Schwarz reflection to an analytic function on

$$\mathbb{C} \setminus \Gamma_t, \quad \text{where} \quad \Gamma_t = \bigcup_{k=1}^N (C_{t,k} \cup \Pi(C_{t,k})) \cup \ell_{t,s}. \tag{2.1.19}$$

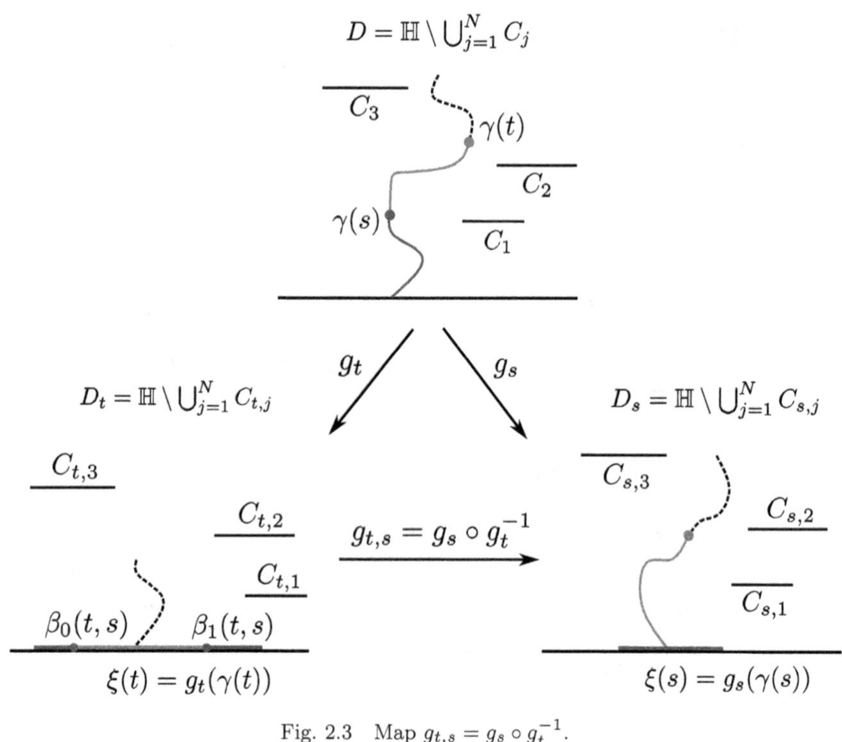

Fig. 2.3 Map $g_{t,s} = g_s \circ g_t^{-1}$.

In what follows, we may and do assume that $\mathbf{0} \notin \ell_{t,s}$.

Let $\widetilde{\Gamma}_t = \bigcup_{k=1}^N (\widetilde{C}_{t,k} \cup \widetilde{\Pi(C_{t,k})}) \cup \widetilde{\ell}_{t,s}$ be the image of Γ_t under the inversion $w = 1/z$. Then $g_{t,s}(1/w)$ is analytic on $\mathbb{C} \setminus \widetilde{\Gamma}_t \setminus \{\mathbf{0}\}$, and so is $f(w) = \left(g_{t,s}(1/w) - \frac{1}{w}\right)/w$. Since (2.1.15) implies

$$\lim_{w \to 0} f(w) = \lim_{z \to \infty} z\{g_{t,s}(z) - z\} = a_s - a_t,$$

$\mathbf{0}$ is a removable singularity of f and f extends to an analytic function on $\mathbb{C} \setminus \widetilde{\Gamma}_t$ with

$$\mathbf{0} \in \mathbb{C} \setminus \widetilde{\Gamma}_t, \quad f(\mathbf{0}) = a_s - a_t. \tag{2.1.20}$$

We note that $\lim_{\zeta \to \infty} f(\zeta) = \lim_{z \to 0} z(g_{t,s}(z) - z) = 0$ and so $\lim_{R \to \infty} \int_{|\zeta|=R} \frac{f(\zeta)}{\zeta} d\zeta = 0$. Therefore we have from (2.1.20) and Cauchy's integral formula

$$a_s - a_t = \frac{1}{2\pi i} \int_{\cup_{k=1}^{2N+1} \widetilde{\gamma}_k} \frac{f(\zeta)}{\zeta} d\zeta, \tag{2.1.21}$$

where $\widetilde{\gamma}_1, \dots, \widetilde{\gamma}_N$ (resp. $\widetilde{\gamma}_{N+1}, \dots, \widetilde{\gamma}_{2N}$) are analytic contours surrounding $\widetilde{C}_{t,1}, \dots, \widetilde{C}_{t,N}$ (resp. $\widetilde{\pi(C_{t,1})}, \dots, \widetilde{\pi(C_{t,N})}$) and $\widetilde{\gamma}_{2N+1}$ is an analytic contour surrounding $\widetilde{\ell_{t,s}}$. They are disjoint and of clockwise orientation, and $\mathbf{0}$ is located outside all of them.

Lemma 2.1.3. *It holds for $0 \le s < t < t_\gamma$ that*

$$a_t - a_s = \frac{1}{\pi}\int_{\beta_0(t,s)}^{\beta_1(t,s)} \Im g_{t,s}(x+i0+)dx. \tag{2.1.22}$$

Proof. It follows from (2.1.21) that

$$a_s - a_t = \frac{1}{2\pi i}\int_{\cup_{k=1}^{2N+1}\widetilde{\gamma}_k} \frac{g_{t,s}(\frac{1}{\zeta}) - \frac{1}{\zeta}}{\zeta^2} d\zeta = -\frac{1}{2\pi i}\int_{\cup_{k=1}^{2N+1}\gamma_k} (g_{t,s}(\eta) - \eta)\, d\eta,$$

where γ_k is the image of $\widetilde{\gamma}_k$ under $\eta = \frac{1}{\zeta}$ for $1 \le k \le 2N+1$. Hence $\gamma_1, \dots, \gamma_N$ (resp. $\gamma_{N+1}, \dots, \gamma_{2N}$) are analytic contours surrounding $C_{t,1}, \dots, C_{t,N}$ (resp. $\Pi(C_{t,1}), \dots, \Pi(C_{t,N})$) and γ_{2N+1} is an analytic contour surrounding $\ell_{t,s}$, all oriented clockwise.

Since $\Im g_{t,s}(\eta)$ is constant on $C_{t,k}$, $g_{t,s}(\eta)$ admits analytic extensions across $C_{t,k}$ from both sides. Hence the integral $\int_{\gamma_k}(g_{t,s}(\eta)-\eta)d\eta$ is equal to $\int_{C_{t,k}}(g_{t,s}(\eta)-\eta)d\eta$ for $1 \le k \le N$ and to $\int_{\Pi(C_{t,k})}(g_{t,s}(\eta)-\eta)d\eta$ for $N+1 \le k \le 2N$. Here the integral on $C_{t,k}$ (and on $\Pi(C_{t,k})$) is understood to be the sum of the integrals along its upper side and lower side with clockwise orientation. Taking γ_{2N+1} to be a rectangle with width 2ε surrounding $\ell_{t,s}$, we then have

$$\begin{aligned} a_s - a_t &= -\frac{1}{2\pi i}\sum_{k=1}^{N}\int_{C_k \cup \Pi(C_k)} (g_{t,s}(\eta) - \eta)\, d\eta \\ &\quad - \lim_{\varepsilon \downarrow 0}\frac{1}{2\pi i}\int_{\beta_0(t,s)}^{\beta_1(t,s)} \Big((g_{t,s}(x+i\varepsilon) - x - i\varepsilon) - (\overline{g_{t,s}(x+i\varepsilon) - x + i\varepsilon})\Big)\, dx. \end{aligned}$$

Since $a_s - a_t$ is real, we conclude that $a_s - a_t$ equals

$$-\frac{1}{2\pi}\sum_{k=1}^{N}\int_{C_k \cup \Pi(C_k)} \Im\,(g_{t,s}(\eta) - \eta)\, d\eta - \frac{1}{\pi}\int_{\beta_0(t,s)}^{\beta_1(t,s)} \Im g_{t,s}(x+i0+)dx.$$

But $\Im(g_{t,s}(\eta) - \eta)$ takes the same constant value from both sides of each C_k and of each $\pi(C_k)$ so that the sum in the above vanishes. This establishes (2.1.22). □

Corollary 2.1.4. *a_t is a strictly increasing left-continuous function in $t > 0$ with $a_0 = 0$.*

Proof. a_t is strictly increasing in view of (2.1.18) and (2.1.22). As $s \uparrow t$, $\gamma(s)^- \to \gamma(t)$ (resp. $\gamma(s)^+ \to \gamma(t)$) so that $\beta_0(t,s) = g_t(\gamma(s)^-) \uparrow g_t(\gamma(t)) = \xi_t$ (resp. $\beta_1(t,s) = g_t(\gamma(s)^+) \downarrow g_t(\gamma(t)) = \xi_t$). a_t is thus left continuous. Since $g_0(z) = z$, $z \in D$, on account of the uniqueness of g_0 and (2.1.12), we have $a_0 = 0$. □

2.1.3 *Chordal Komatu-Loewner left differential equation for a Jordan arc*

Under the setting of the preceding section, we now derive the Komatu-Loewner equation for the canonical map g_t from $D \setminus \gamma[0,t]$ onto D_t which involves its left derivative $\dfrac{\partial^- g_t(z)}{\partial a_t}$ with respect to the strictly increasing left continuous function a_t studied in the preceding section.

Let $\mathbf{Z}^{*,t}$ be the BMD on the standard slit domain D_t, and $\Psi_t(z,\zeta)$ with $(z,\zeta) \in D_t \times \partial\mathbb{H}$ be the complex Poisson kernel of $\mathbf{Z}^{*,t}$ appeared in Theorem 1.9.1, namely, the analytic function on D_t with imaginary part being the Poisson kernel $K_t^*(z,\zeta)$ of $\mathbf{Z}^{*,t}$ satisfying the normalization $\lim_{z\to\infty} \Psi_t(z,\zeta) = 0$.

Theorem 2.1.5. *The family of the conformal maps $\{g_t, t \in [0,t_\gamma)\}$ satisfies the following chordal Komatu-Loewner left differential equation: for $t \in (0,t_\gamma)$ and $z \in D \setminus \gamma[0,t]$, $g_t(z)$ is left differentiable with respect to a_t and*

$$\frac{\partial^- g_t(z)}{\partial a_t} = -\pi \Psi_t(g_t(z), \xi(t)), \quad g_0(z) = z. \tag{2.1.23}$$

Proof. We consider the analytic function

$$F(z) = g_{t,s}(z) - z, \qquad z \in D_t,$$

which satisfies

$$F(z) = \frac{a_s - a_t}{z} + o(1/|z|) \quad \text{as } |z| \to \infty. \tag{2.1.24}$$

Then, $f(z) := \Im F(z)$ is harmonic on D_t that can be extended continuously to each slit $C_{t,i}$ with a constant limit, say f_i, and to $\partial\mathbb{H}$ with the limit function (2.1.18) of compact support $\ell_{t,s}$. As $\lim_{z\to\infty} f(z) = 0$ by (2.1.24), the formula (1.1.43) holds according to Proposition 1.1.6: for $z \in D_t$,

$$f(z) = \sum_{k=1}^{N} f_k \varphi_t^{(k)}(z) - \frac{1}{2}\int_{\partial\mathbb{H}} \frac{\partial G_t^0(z,\zeta)}{\partial \mathbf{n}_\zeta} f(\zeta) ds(\zeta), \tag{2.1.25}$$

where $\{\varphi_t^{(i)}\}$ is the harmonic basis and $G_t^0(z,\zeta)$ is the Green function of the ABM for the standard slit domain $D_t = \mathbb{H} \setminus \cup_{k=1}^N C_{t,k}$.

Since f is the imaginary part of the analytic function F, its period around $C_{t,i}$ vanishes and we have from (1.1.44)

$$0 = -\sum_{k=1}^{N} f_k a_{t,ki} - \int_{\partial\mathbb{H}} \frac{\partial \varphi_t^{(i)}(\zeta)}{\partial \mathbf{n}_\zeta} f(\zeta) ds(\zeta), \tag{2.1.26}$$

for every $1 \le i \le N$, where $a_{t,ki}$ denotes the period of $\varphi_t^{(i)}$ around the slit $C_{t,k}$.

Denote by $\mathcal{A}_t$ the symmetric matrix with (i,j)-entry $a_{t,ij}$ and by $q_{t,ij}$ the (i,j)-entry of $\mathcal{A}_t^{-1}$. Multiply both sides of (2.1.26) by $\sum_{j=1}^N \varphi_t^{(j)}(z) q_{t,ji}$ and sum over in i, and finally add the resulting identity to (2.1.25). We get

$$f(z) = \int_{\partial\mathbb{H}} K_t^*(z,\zeta) f(\zeta) ds(\zeta),$$

in view of the expression (1.7.5) of the Poisson kernel $K_t^*(z,\zeta)$ of $\mathbb{Z}^{*,t}$. Since f vanishes on $\partial\mathbb{H}\setminus[\beta_0(t,s),\beta_1(t,s)]$, we have

$$f(z)=\int_{\beta_0(t,s)}^{\beta_1(t,s)} K_t^*(z,x)f(x)dx,$$

and thus

$$g_{t,s}(z)-z=\int_{\beta_0(t,s)}^{\beta_1(t,s)} \Psi_t(z,x)\Im g_{t,s}(x)dx+c, \tag{2.1.27}$$

for some real constant c. Due to the normalizations (2.1.12) and (1.9.2), we let $z\to\infty$ in (2.1.27) to get $c=0$. We then substitute $z=g_t(w)$, $w\in D\setminus\gamma[0,t]$, to obtain

$$g_s(z)-g_t(z)=\int_{\beta_0(t,s)}^{\beta_1(t,s)} \Psi_t(g_t(z),x)\Im g_{t,s}(x)dx, \quad z\in D\setminus\gamma[0,t]. \tag{2.1.28}$$

By Theorem 1.9.1 (ii), $\Psi_t(z,\zeta)=u_t(z,\zeta)+iK_t^*(z,\zeta)$ is continuous in $\zeta\in\partial\mathbb{H}$ for each $z\in D$, and consequently (2.1.22), (2.1.28) and the mean value theorem of integration imply that, for some $x',x''\in(\beta_0(t,s),\beta_1(t,s))$,

$$\frac{g_s(z)-g_t(z)}{a_s-a_t}=-\pi u_t(g_t(z),x')-i\pi K_t^*(g_t(z),x''), \quad z\in D\setminus\gamma[0,t]. \tag{2.1.29}$$

If we let $s\uparrow t$, then both $\beta_0(t,s)$ and $\beta_1(t,s)$ converge to $\xi(t)$ as was observed in the proof of Corollary 2.1.4 and we arrive at the desired equation (2.1.23). □

In Section 2.5, the left differential equation (2.1.23) will be strengthened into a genuine differential equation. Before doing so, we need to investigate the continuity of various relevant quantities $g_t(z)$, a_t, D_t, $\beta_0(t,s)$, $\beta_1(t,s)$, and $\Psi_t(z,\zeta)$. That will be our tasks in Sections 2.2–2.4.

2.2 Probabilistic representation of canonical map and half-plane capacity

2.2.1 *Expression of $\Im g_F$ by BMD and ABM*

Let $D=\mathbb{H}\setminus K$, where $K=\bigcup_{j=1}^N A_j$, is an $(N+1)$-connected domain, and F be an $\mathbb{H}$-hull with $F\subset D$. In this section, we will present a probabilistic approach to the existence of the canonical map g_F from $D\setminus F$ onto a standard slit domain. More specifically, we will first construct the imaginary part $\Im g_F$ of g_F in terms of the ABM (absorbed Brownian motion) on $\mathbb{H}$ and the BMD (Brownian motion with darning) on $D\cup K^*$ for $K^*=\{a_1^*,\ldots,a_N^*\}$. In the next section, we shall apply this representation to obtain a probabilistic expression of a_F, the capacity of F relative to D as is introduced in Section 2.1.1, in terms of BMD in the case when D is a standard slit domain.

Denote by $\mathbf{Z}^{\mathbb{H}}=(Z_t^{\mathbb{H}},\mathbb{P}_z^{\mathbb{H}})$ the ABM on $\mathbb{H}$ and by $\mathbf{Z}^*=(Z_t^*,\mathbb{P}_z^*)$ the BMD on $D^*=D\cup K^*$ with $K^*=\{a_1^*,\ldots,a_N^*\}$ obtained from $\mathbf{Z}^{\mathbb{H}}$ by regarding each compact continuum A_i as one point a_i^*.

For $r > 0$, let $\Gamma_r = \{z = x + iy : y = r\}$ and

$$v^*(z) := \lim_{r \to \infty} r \cdot \mathbb{P}^*_z(\sigma_{\Gamma_r} < \sigma_F), \quad z \in D^* \setminus F. \tag{2.2.1}$$

Theorem 2.2.1.

(i) *The function v^* on $D^* \setminus F$ is well defined and is Z^*-harmonic on $D^* \setminus F$. Furthermore*

$$v^*(z) = v(z) + \sum_{j=1}^{N} \mathbb{P}^{\mathbb{H}}_z\left(\sigma_K < \sigma_F,\ Z^{\mathbb{H}}_{\sigma_K} \in A_j\right) v^*(a^*_j),\ z \in D \setminus F, \tag{2.2.2}$$

where

$$v(z) = \Im z - \mathbb{E}^{\mathbb{H}}_z\left[\Im Z^{\mathbb{H}}_{\sigma_{F\cup K}};\ \sigma_{F\cup K} < \infty\right]\ (\geq 0), \tag{2.2.3}$$

$$v^*(a^*_i) = \sum_{j=1}^{N} \frac{M_{ij}}{1 - R^*_j} \int_{\eta_j} v(z)\nu_j(dz), \quad 1 \leq i \leq N. \tag{2.2.4}$$

Here $\eta_1, \ldots, \eta_N$ are mutually disjoint smooth Jordan curves surrounding $A_1, \ldots, A_N$, respectively,

$$\nu_i(dz) = \mathbb{P}^*_{a^*_i}\left(Z^*_{\sigma_{\eta_i}} \in dz\right), \quad 1 \leq i \leq N, \tag{2.2.5}$$

$$R^*_i = \int_{\eta_i} \mathbb{P}^{\mathbb{H}}_z\left(\sigma_K < \sigma_F,\ Z^{\mathbb{H}}_{\sigma_K} \in A_i\right) \nu_i(dz), \quad 1 \leq i \leq N, \tag{2.2.6}$$

and M_{ij} is the (i,j)-entry of the matrix $M = \sum_{n=0}^{\infty}(Q^*)^n$ for a matrix Q^* with entries

$$q^*_{ij} = \begin{cases} \mathbb{P}^*_{a^*_i}\left(\sigma_{K^*} < \sigma_F,\ Z^*_{\sigma_{K^*}} = a^*_j\right)/(1 - R^*_i) & \text{if } i \neq j, \\ 0 & \text{if } i = j, \end{cases} \quad 1 \leq i, j \leq N. \tag{2.2.7}$$

(ii) *$v^*\big|_{D\setminus F}$ admits a unique harmonic conjugate u^* such that $f(z) = u^*(z) + iv^*(z)$, $z \in D \setminus F$, is analytic on $D \setminus F$ and*

$$f(z) = z + \frac{a}{z} + o(1/z), \qquad z \to \infty, \tag{2.2.8}$$

for some real constant a.

(iii) *The preceding statements remain valid when $F = \emptyset$ under the interpretation that $\sigma_\emptyset = \infty$.*

A proof of this theorem will be given in Section A.4 of the Appendix by a series of lemmas.

Lemma 2.2.2.

(i) *The function $v(z), z \in D \setminus F$ defined by (2.2.3) is strictly positive on $D \setminus F$.*

(ii) *The function $v^*(z)$, $z \in D^* \setminus F$, defined by* (2.2.1) *is strictly positive on $D^* \setminus F$. Its restriction to $D \setminus F$ can be extended continuously to $\overline{\mathbb{H}}$ by setting its value on $F \cup \partial\mathbb{H}$ to be zero and its value on each continuum A_i to be $v^*(a_i^*)$ for $1 \le i \le N$. Further*

$$\lim_{y \downarrow 0} v^*(x + iy) = 0 \quad \text{uniformly in } x \in \mathbb{R}. \tag{2.2.9}$$

Proof. (i) v is non-negative harmonic on $D \setminus F$ in the classical sense. It is positive when $\Im z$ is large enough. So it is positive everywhere on the connected set $D \setminus F$ by Lemma 1.1.3 (i).

(ii) By Theorem 2.2.1, $v^*(z)$, $z \in D^* \setminus F$, admits the expressions (2.2.2), (2.2.3) and (2.2.4). Hence it is positive on $D^* \setminus F$ by (i).

Using the planar Brownian motion $\mathbf{Z} = \{Z_t, \mathbb{P}_z, z \in \mathbb{C}\}$ and the function $f_1(z) = 1_{F \cup K}(z)\Im z$, we have

$$\mathbb{E}_\zeta^{\mathbb{H}}\left[\Im Z^{\mathbb{H}}_{\sigma_{F \cup K}}; \sigma_{F \cup K} < \infty\right] = \mathbb{E}_\zeta\left[\Im Z_{\sigma_{F \cup K}}; \sigma_{F \cup K} < \sigma_{\partial\mathbb{H}}\right] = \mathbb{E}_\zeta\left[f_1(Z_{\sigma_{F \cup K \cup \partial\mathbb{H}}})\right]$$

for every $\zeta \in \mathbb{H} \setminus F \cup K$, which tends to $\Im z$ as $\zeta \to z \in \partial(F \cup K)$, because any connected component of $F \cup K$ does not reduce to a singleton and $f_1\big|_{\partial(F \cup K)}$ is continuous on a neighborhood of $z \in \partial(F \cup K)$ so that the properties (**Z**.1) and (**Z**.2) of the planar BM **Z** stated in Section 1.1 apply. Furthermore, for $f_2(z) = 1_{A_i}(z)$, $z \in \mathbb{C}$, the function

$$\mathbb{P}_\zeta^{\mathbb{H}}\left(\sigma_K < \sigma_F,\ Z^{\mathbb{H}}_{\sigma_K} \in A_i\right) = \mathbb{E}_\zeta\left[f_2(Z_{\sigma_{F \cup K \cup \partial\mathbb{H}}})\right], \quad \zeta \in \mathbb{H} \setminus (F \cup K),$$

is continuously extendable to $\mathbb{H}$ by setting its value on A_i to be 1 and its value on $F \cup (\cup_{j \neq i} A_j)$ to be zero.

It follows from (2.2.3) that $0 \le v(z) \le \Im z$. We also have, for $b = \inf\{\Im z : z \in K\} > 0$,

$$\mathbb{P}_\zeta^{\mathbb{H}}\left(\sigma_K < \sigma_F, Z^{\mathbb{H}}_{\sigma_F} \in A_i\right) \le \mathbb{P}_\zeta(\sigma_K < \sigma_{\partial\mathbb{H}}) \le \frac{\Im \zeta}{b} \quad \text{whenever } \Im\zeta \in (0, b).$$

Hence (2.2.2) implies (2.2.9). Thus, $v^*\big|_{D \setminus F}$ admits the asserted continuous extension to $\overline{\mathbb{H}}$. □

Theorem 2.2.3. *The analytic function f of Theorem* 2.2.1 (ii) *is the canonical map from $D \setminus F$ onto a standard slit domain. Moreover,*

$$\lim_{z \in D \setminus F,\ z \to w} \Im f(z) = 0 \qquad \text{for any } w \in \partial\mathbb{H} \setminus \overline{F},$$

and f can be extended to $(D \setminus F) \cup \Pi(D \setminus F) \cup (\partial\mathbb{H} \setminus \overline{F})$ by Schwarz reflection as a univalent function.

Proof. Recall $\Pi z = \bar{z}$ is the mirror reflection with respect to $\partial\mathbb{H}$. On account of Lemma 2.2.2, we can extend f by Schwarz reflection to an analytic map from

$$D_1 := \mathbb{C} \setminus (\overline{F} \cup K \cup \Pi(\overline{F} \cup K)),$$

into $\mathbb{C}$, which satisfies

$$f(z) = z + \frac{a}{z} + o(1/|z|), \quad z \to \infty. \tag{2.2.10}$$

We denote by $\overline{\mathbb{C}} = \mathbb{C} \cup \{\infty\}$ the Riemann sphere. By virtue of (2.2.10), $f(\infty) = \infty$ and f is analytic in a neighborhood of ∞ in the sense that $u_f(z) = 1/f(1/z)$ is analytic in a neighborhood of $\mathbf{0}$. Thus, f further extends to an analytic map from

$$D_1 = \overline{\mathbb{C}} \setminus (\overline{F} \cup K \cup \Pi(\overline{F} \cup K)), \tag{2.2.11}$$

into $\overline{\mathbb{C}}$ and the multiplicity of f near ∞ equals 1 because $u_f'(\mathbf{0}) \neq 0$.

Let $D_2 = \overline{\mathbb{C}} \setminus f(\partial D_1)$, where $f(\partial D_1)$ denotes $\bigcap_{C \Subset D_1} \overline{f(D_1 \setminus C)}$, the set of limit points of $f(z)$ as z approaches to ∂D_1 (the intersection is over all compact subsets of D_1). By Lemma 2.2.2 again,

$$f(\partial D_1) \subset \left\{ z = x + iy \in \mathbb{C} : y = \pm v^*(a_j^*) \text{ for } j = 1, \dots, N \text{ or } y = 0 \right\}. \tag{2.2.12}$$

This implies that the complement of D_2 has empty interior and so

$$f(\partial D_1) = \partial D_2. \tag{2.2.13}$$

We next show that ∞ is an interior point of D_2. Since f has multiplicity 1 near infinity, this will then imply by Theorem A.5.2 in the Appendix that f is a conformal mapping from D_1 onto D_2. To this end, it suffices to show that f maps F and each A_j (understood in the sense of the limit points) into a bounded subset in $\mathbb{C}$.

Let ϕ be the conformal map from $\mathbb{H} \setminus F$ onto $\mathbb{H}$ that satisfies the hydrodynamic normalization (2.1.1) at infinity and set $g = f \circ \phi^{-1}$. Clearly, g is well-defined and analytic in $\mathbb{H}$. Let O be an open neighborhood of the compact set $J = \phi(F) \subset \partial\mathbb{H}$ (understood in the sense of limit points). The imaginary part of $g(z)$ tends to 0 as $y = \Im z \to 0$ in $O \cap \mathbb{H}$. Thus g extends analytically across the real line to be an analytic function in the disk O. In particular, g is bounded in every compact subset O. This in particular implies that $f(F)$ is bounded.

Similarly, for each $j = 1, \dots, N$, since A_j is a compact continuum, there is a conformal map ϕ_j of the complement of the interval $[0, 1]$ in $\mathbb{C}$ onto the complement of A_j. Let O_j be a relatively compact open neighborhood of A_j that is disjoint from the other boundary components. Then $\widehat{O}_j = \phi_j^{-1}(O_j)$ is an open neighborhood of $[0, 1]$, and $g_j = f \circ \phi_j$ is analytic in $\widehat{O}_j$. Since $\phi_j(z) \to A_j$ as $z \to [0, 1]$, the imaginary part $\Im g_j(z) = v^*(\phi_j(z)) \to v^*(a_j^*)$ as $z \to [0, 1]$. By the argument similar to that in the previous paragraph, g_j is bounded on every compact subset of O_i and so $f = g_j \circ \phi_j^{-1}$ is bounded near A_j.

In view of the above, we conclude from (2.2.10) that the pre-image of $\infty \in D_2$ under f is ∞ with multiplicity 1. Theorem A.5.2 together with (2.2.12) implies that f is conformal from D_1 onto D_2.

Since f sends $D \setminus F$, $\Pi(D \setminus F)$ and $\partial\mathbb{H} \setminus \overline{F}$ into $\mathbb{H}$, $\Pi\mathbb{H}$ and $\partial\mathbb{H}$, respectively, by Lemma 2.2.2, f is a conformal map from $D \setminus F$ onto $\mathbb{H} \setminus f(\partial D_1)$. Because this map is a topological homeomorphism, $\mathbb{H} \setminus f(\partial D_1)$ has to be $(N+1)$-connected and thus a slit domain with N disjoint horizontal line segments being removed from $\mathbb{H}$.

None of these horizontal line segments degenerates to a single point. In fact, if one of them reduces to a point p, then, due to (2.2.10), p is not a pole nor an essential singularity of the analytic function f^{-1} so p is removable, contradicting the assumption that the boundary components of D_1 are continua. This shows that f is the canonical map from $D \setminus F$ onto a standard slit domain. □

2.2.2 *Representing a_F by BMD and uniform bound of $g_F(z) - z$*

We now assume that the $(N+1)$-connected domain $D = \mathbb{H} \setminus K$ with $K = \bigcup_{j=1}^N A_j$ is a standard slit domain, namely, $\{A_1, \ldots, A_N\}$ are mutually disjoint line segments parallel to the x-axis located in $\mathbb{H}$. Let F be an $\mathbb{H}$-hull with $F \subset D$, g_F be the half-plane canonical map from $D \setminus F$ onto a standard slit domain and a_F be the half-plane capacity of F relative to D. By making use of Theorem 2.2.3, we give an expression of a_F in terms of the BMD $\mathbf{Z}^*$ and then apply it to obtain a uniform bound of $g_F(z) - z$ in terms of a_F, making $o(1/|z|)$ in the hydrodynamic normalization (2.1.7) more precise.

Theorem 2.2.4.

(i) *It holds that, for any $z \in D \setminus F$,*

$$\Im(z - g_F(z)) = \mathbb{E}_z^* \left[\Im Z_{\sigma_F}^* \, ; \, \sigma_F < \infty \right]. \tag{2.2.14}$$

(ii) *For any $\rho > 0$ with $B_\rho \cap \mathbb{H} \supset F \cup K$, it holds that*

$$a_F = \frac{2\rho}{\pi} \int_0^\pi \mathbb{E}_{\rho e^{i\theta}}^* \left[\Im Z_{\sigma_F}^* ; \sigma_F < \infty \right] \sin\theta d\theta. \tag{2.2.15}$$

(iii) *If F is non-empty, then $a_F > 0$.*

(iv) *Take any $\rho > 0$ with $B_\rho \cap \mathbb{H} \supset F \cup K$. It holds then that, for any $z \in \mathbb{H} \setminus \overline{B}_{3\rho}$,*

$$\left| z - g_F(z) - \frac{\Im z}{|z|^2} a_F \right| \le 3(1+\pi) c_{1/2} \, \rho \, \frac{a_F}{|z|^2}, \tag{2.2.16}$$

where $c_{1/2}$ is the constant (1.1.17) for $\kappa = \frac{1}{2}$.

Proof. (i) Let $u(z) = \Im(z - g_F(z))$, $z \in D \setminus F$. By Theorem 2.2.3, $\Im g_F = v^*\big|_{D\setminus F}$. In other words, $\Im g_F$ is the restriction to $D \setminus F$ of the BMD-harmonic function v^* on $D^* \setminus F$. for $D^* = D \cup \{a_1^*, \ldots, a_N^*\}$. As $D = \mathbb{H} \setminus \bigcup_{j=1}^N A_j$ is a standard slit domain, $\Im z$ is continuous on $\mathbb{H}$, constant on each A_j and with zero period around each A_j. Therefore, by virtue of Theorem 1.6.4, u becomes BMD-harmonic on $D^* \setminus F$ if we define its value at each a_j^* by its limit value there (designated by the same notation u).

Choose a decreasing bounded open sets $\{U_k\}$ with

$$F \subset U_k \subset D, \quad k \geq 1, \quad F = \bigcap_{k=1}^{\infty} \overline{U}_k \cap \mathbb{H}.$$

It follows from (2.1.7) that $\lim_{z\to\infty} u(z) = 0$. This combined with Lemma 2.2.2 yields the existence of $R_n \uparrow \infty$ and $\delta_n \downarrow 0$, such that, for $G_n = \{z \in \mathbb{H} : |z| < R_n, \ \Im z > \delta_n\}$,

$$G_1 \supset U_1 \cup K, \quad |u(z)| \leq 1/n \text{ for every } z \in \partial G_n, \quad n \geq 1. \tag{2.2.17}$$

In particular, u can be continuously extended to the one-point compactification $D^* \cup \{\partial\}$ by setting $u(\partial) = 0$.

Since u is BMD-harmonic on $D^* \setminus F$, we have, for any $z \in G_1 \setminus \overline{U}_1 \setminus K$ and any $k \geq 1, \ n \geq 1$,

$$\begin{aligned} u(z) &= \mathbb{E}_z^* \left[u(Z^*_{\tau_{G_n \setminus \overline{U}_k}}) \right] \\ &= \mathbb{E}_z^* \left[u(Z^*_{\sigma_{\partial U_k}}); \sigma_{\partial U_k} < \tau_{G_n} \right] + \mathbb{E}_z^* \left[u(Z^*_{\tau_{G_n}}); \tau_{G_n} \leq \sigma_{\partial U_k} \right]. \end{aligned}$$

We let $n \to \infty$. By (2.2.17), the last term in the above tends to 0. We see by (1.7.12) that $\tau_{G_n} \leq \zeta^* < \infty$ $\mathbb{P}_z^*$-a.s. Let $\tau = \lim_{n\to\infty} \tau_{G_n}$. Then $\tau \leq \zeta^* < \infty$ and the path continuity (1.3.19) implies $Z^*_\tau = \lim_{n\to\infty} Z^*_{\tau_{G_n}} \in \bigcap_n (\mathbb{H} \setminus G_n) = \{\partial\}$ so that $\tau = \zeta^*$ and

$$u(z) = \mathbb{E}_z^* \left[u(Z^*_{\sigma_{\partial U_k}}); \sigma_{\partial U_k} < \zeta^* \right], \quad z \in G_1 \setminus \overline{U}_1 \setminus K.$$

Set $\sigma = \lim_{k\to\infty} \sigma_{\partial U_k} \leq \sigma_F \wedge \zeta^*$. If $\sigma = \zeta^*$, $\lim_{k\to\infty} Z^*_{\sigma_{\partial U_k}} = \partial$ where u vanishes. If $\sigma < \zeta^*$, $Z^*_\sigma = \lim_{k\to\infty} Z^*_{\sigma_{\partial U_k}} \in \bigcap_{k=1}^{\infty} \overline{U}_k \cap \mathbb{H} = F$ and $\sigma = \sigma_F$. As $\Im g_F(z) = v^*(z)$ continuously extends to ∂F to be zero there by Lemma 2.2.2, we get the desired identity (2.2.14). Notice that this identity holds for any $z \in D^* \setminus F$.

(ii) Since the half-plane capacity a_F of the $\mathbb{H}$-hull F relative to D is non-negative, it follows from (2.1.7) that

$$a_F = \lim_{y\to\infty} \Re[iy(g_F(iy) - iy)] = \lim_{y\to\infty} yu(iy),$$

which combined with (2.2.14) gives

$$a_F = \lim_{y\to\infty} y\mathbb{E}^*_{iy} \left[\Im Z^*_{\sigma_F}; \sigma_F < \infty \right]. \tag{2.2.18}$$

On the other hand, the strong Markov property of $\mathbf{Z}^*$ and Proposition 1.1.2 yield that, for $y > \rho$,

$$\begin{aligned} \mathbb{E}^*_{iy} \left[\Im Z^*_{\sigma_F}; \sigma_F < \infty \right] &= \mathbb{E}^*_{iy} \left[\mathbb{E}^*_{Z^{\mathbb{H},*}_{\sigma_{B_\rho \cap \mathbb{H}}}} \left[\Im Z^*_{\sigma_F}; \sigma_F < \infty \right]; \sigma_{B_\rho} < \infty \right] \\ &= \mathbb{E}^{\mathbb{H}}_{iy} \left[\mathbb{E}^*_{Z^{\mathbb{H}}_{\sigma_{B_\rho \cap \mathbb{H}}}} \left[\Im Z^*_{\sigma_F}; \sigma_F < \infty \right]; \sigma_{B_\rho} < \infty \right] \\ &= \frac{2\rho}{\pi} \frac{1}{y} \int_0^\pi (1 + O(1/y)) \mathbb{E}^*_{\rho e^{i\theta}} \left[\Im Z^*_{\sigma_F}; \sigma_F < \infty \right] \sin\theta d\theta. \end{aligned}$$

The desired identity (2.2.15) follows from this and (2.2.18).

(iii) Clearly the function $u(z) = \Im(z - g_F(z))$ is harmonic on $D \setminus F$. As u is non-negative by (2.2.14), it is either positive everywhere on $D \setminus F$ or identically 0 there in view of Lemma 1.1.3.

Suppose u vanishes identically on $D \setminus F$, then so does the function $v(z) = \mathbb{E}_z^D \left[\Im Z^D_{\sigma_F}; \sigma_F < \infty\right], z \in D \setminus F < \infty$, defined in terms of the ABM $\mathbf{Z}^D$ on D becasue $0 \le v \le u$ on D by (2.2.14). Since $\Im Z^D_{\sigma_F} > 0$ on $\{\sigma_F < \infty\}$, we have $p_F(z) := \mathbb{P}_z^D(\sigma_F < \infty) = 0$ for all $z \in D \setminus F$.

On the other hand, as F is non-empty $\mathbb{H}$-hull and no connected component of F reduces to a singlton, F is non-polar relative to the planar BM $\mathbf{Z}$ by **(Z.1)** and so is it relative to $\mathbf{Z}^D$ in view of [Fukushima, Oshima and Takeda (2011), Theorem 4.4.3 (ii)]. In particular $p_F(z) > 0$ for every $z \in D \setminus F$ by Lemma 1.1.3 again, a contradiction. Hence, $u > 0$ everywhere on $D \setminus F$ and $a_F > 0$ in view of the expressions (2.2.14) and (2.2.15).

(iv) Define $h(z) = z - g_F(z) + \frac{a_F}{z}$, $z \in D \setminus F$. Then $\Im h(z) = \Im(z - g_F(z)) - \frac{\Im z}{|z|^2} a_F$. Using the identity (2.2.14), the strong Markov property of $\mathbf{Z}^*$ and Proposition 1.1.2, we have for any $z \in \mathbb{H} \setminus \overline{B_\rho}$

$$\begin{aligned}\Im(z - g_F(z)) &= \mathbb{E}_z^* \left[\mathbb{E}^*_{Z^*_{\sigma_{B_\rho \cap \mathbb{H}}}} \left[\Im Z^*_{\sigma_F}; \sigma_F < \infty\right]; \sigma_{B_\rho} < \infty\right] \\ &= \mathbb{E}_z^{\mathbb{H}} \left[\mathbb{E}^*_{Z^{\mathbb{H}}_{\sigma_{B_\rho \cap \mathbb{H}}}} \left[\Im Z^*_{\sigma_F}; \sigma_F < \infty\right]; \sigma_{B_\rho} < \infty\right] \\ &= \frac{2\rho}{\pi} \frac{\Im z}{|z|^2} \int_0^\pi (1 + c(\theta, \vartheta, \eta) \frac{\rho}{|z|}) \mathbb{E}^*_{\rho e^{i\theta}} \left[\Im Z^*_{\sigma_F}; \sigma_F < \infty\right] \sin\theta d\theta.\end{aligned}$$

Here $c(\theta, \vartheta, \eta)$ is a function satisfying (1.1.17). Therefore, it follows from the expression (2.2.15) of a_F and (1.1.17) with $\kappa = \frac{1}{2}$ that

$$\Im h(z) = \frac{\Im z}{|z|^2} \frac{2\rho}{\pi} \int_0^\pi c(\theta, \vartheta, \eta) \frac{\rho}{|z|} \mathbb{E}^*_{\rho e^{i\theta}} \left[\Im Z^*_{\sigma_F}; \sigma_F < \infty\right] \sin\theta d\theta, \quad |z| > 2\rho,$$

and

$$|\Im h(z)| \le \frac{1}{|z|} c_{1/2} \frac{\rho}{|z|} a_F \le c_{1/2}\, a_F \frac{\rho}{|z|^2}, \quad z \in \mathbb{H} \setminus \overline{B}_{2\rho}.$$

As $\lim_{z \downarrow z_0 \in \partial\mathbb{H} \setminus [-2\rho, 2\rho]} \Im h(z) = 0$, h extends to be an analytic function (denoted by h again) on $\mathbb{C} \setminus \overline{B}_{2\rho}$ by Schwarz reflection. The function $w(\zeta) = \Im h(\zeta)$ is then harmonic on $\mathbb{C} \setminus B(2\rho)$.

Take any $z \in \mathbb{H}$ with $|z| > 3\rho$ and express $w(\zeta)$ on the disk $B_s(z) = \{w \in \mathbb{C} : |w - z| < s\}$ for $s = |z| - 2\rho$ by the Poisson integral. In view of (2.1.3), both $|w_x(z)|$ and $|w_y(z)|$ are then not greater than $\dfrac{2}{s} \max\limits_{\zeta \in \partial B_s(z)} |w(\zeta)|$. Hence, we get from the above bound of $\Im h(z)$ that

$$|h'(z)| \le \frac{2}{|z| - 2\rho} c_{1/2}\, a_F \frac{\rho}{|z|^2} \le \frac{6\rho}{|z|^3} c_{1/2}\, a_F \quad \text{for } |z| > 3\rho. \tag{2.2.19}$$

Since $\lim_{y\uparrow\infty} h(iy) = 0$, it follows that

$$|h(iy)| = \left|\int_y^\infty h'(it)dt\right| \le 3\rho c_{1/2}\frac{1}{y^2}a_F, \quad y > 3\rho.$$

For $z = re^{i\vartheta}$ with $r > 3\rho$, we can use the circular arc C connecting z with ir to get the desired (2.2.16) from (2.2.19) as

$$\begin{aligned}|h(z)| &\le |h(ir)| + \left|\int_C h'(\zeta)d\zeta\right| \le |h(ir)| + \frac{\pi}{2}r\max_{|\zeta|=r}|h'(\zeta)| \\ &\le 3\rho c_{1/2}\frac{1}{r^2}a_F + 3\rho\pi c_{1/2}\frac{1}{r^2}a_F = 3(1+\pi)c_{1/2}\frac{\rho}{|z|^2}a_F.\end{aligned}$$

□

2.2.3 *Representing a_F^0 by ABM and uniform bound of g_F^0*

Theorem 2.2.5. *Let F be an $\mathbb{H}$-hull and g_F^0 be the conformal map from $\mathbb{H}\setminus F$ onto $\mathbb{H}$ satisfying the hydrodynamic normalization (2.1.1) with the $\mathbb{H}$-capacity a_F^0 of F.*

(i) *It holds that*

$$\Im g_F^0(z) = \Im z - \mathbb{E}_z^{\mathbb{H}}\left[\Im Z_{\sigma_F}^{\mathbb{H}}; \sigma_F < \infty\right], \quad z \in \mathbb{H}\setminus F, \tag{2.2.20}$$

(ii) *For any $\rho > 0$ with $B_\rho \cap \mathbb{H} \supset F$,*

$$a_F^0 = \frac{2\rho}{\pi}\int_0^\pi \mathbb{E}_{\rho e^{i\theta}}^{\mathbb{H}}\left[\Im Z_{\sigma_F}^{\mathbb{H}}; \sigma_F < \infty\right]\sin\theta d\theta. \tag{2.2.21}$$

(iii) *$a_F^0 > 0$ if F is non-empty.*

(iv) *Take any $\rho > 0$ with $B_\rho \cap \mathbb{H} \supset F$. It holds that, for any $z \in \mathbb{H}\setminus \overline{B}_{3\rho}$.*

$$\left|z - g_F^0(z) - \frac{\Im z}{|z|^2}a_F^0\right| \le \frac{3(1+\pi)c_{1/2}\rho\, a_F^0}{|z|^2}. \tag{2.2.22}$$

This is a special case of Theorem 2.2.4 that $K = \emptyset$. The same proof as above works. (2.2.20) and (2.2.21) have appeared in [Lawler (2005), (3.5), (3.7)]. To prove (2.2.20), one needs to approximate the set $\mathbb{H}\setminus F$ as in the proof of Theorem 2.2.4. Inequality (2.2.22) has been also shown in [Lawler (2005), Proposition 3.46] by establishing it first for $\rho = 1$ and then using the scaling property

$$g_{rF}^0(z) = rg_F^0(z/r) \quad \text{and} \quad a_{rF}^0 = r^2 a_F^0 \quad \text{for } r > 0.$$

This kind of simple scaling property is not available for g_F and a_F unless $K = \emptyset$; see Lemma 4.1.5 for the case of the standard slit domain.

Various type of chordal Loewner right differential equations for the upper half-plane $\mathbb{H}$ have been obtained based on the estimate (2.2.22) (cf. [Lawler (2005), Proposition 4.4], [Chen, Fukushima and Suzuki (2017), Proposition 2.3]).

2.3 Continuity of various objects generated by Jordan arc

In this section, we return to the setting of Sections 2.1.2 and 2.1.3. We consider a standard slit domain $D = \mathbb{H} \setminus K$, with $K = \bigcup_{j=1}^N C_j$ for mutually disjoint horizontal line segments $C_j \subset \mathbb{H}$, $1 \leq j \leq N$, and a Jordan arc $\gamma = \{\gamma(t); t \in [0, t_\gamma)\}$ satisfying (2.1.11). The Jordan curve γ generates a growing family of $\mathbb{H}$-hulls $\{F_t = \gamma(0, t] : 0 \leq t < t_\gamma\}$ with $F_0 := \emptyset$. For each $t \in [0, t_\gamma)$, let g_t be the canonical map from $\mathbb{H} \setminus F_t$. In particular, $g_t(z)$ satisfies the hydrodynamic normalization (2.1.7) at $z = \infty$, and the associated half-plane capacity a_t is strictly increasing and left-continuous in $t \in (0, t_\gamma]$ with $a_0 = 0$ in view of Corollary 2.1.4.

In what follows, we use the notation $K^p = \bigcup_{j=1}^N C_j^p$, where C_j^p is the set defined by (2.1.9) for the horizontal line segment C_j, $1 \leq j \leq N$.

A primary goal of this chapter is to turn the Komatu-Loewner left-differential equation for $g_t(z)$ obtained in Theorem 2.1.5 into a genuine ordinary differential equation. To this end, we need to establish continuity of various objects associated with the conformal maps $\{g_t(z), t \in [0, t_\gamma)\}$ along with a Lipschitz continuity property of complex Poisson kernels on standard slit domains that will be studied in the next section.

2.3.1 *Joint continuity of $\Im g_t(z)$ and equi-continuity of $\{g_t(z)\}$*

In this section, we first derive the joint continuity of $\Im g_t(z)$ in (t, z) from its probabilistic representation given in Theorem 2.2.1 under a general setting and then we apply it to get continuity properties on $g_t(z)$.

Denote by $\mathbf{Z}^* = (Z_t^*, \zeta^*, \mathbb{P}_z^*)$ the BMD on $D^* = D \cup K^*$ with $K^* = \{c_1^*, \ldots, c_N^*\}$ obtained from the ABM $\mathbf{Z}^{\mathbb{H}} = (Z_t^{\mathbb{H}}, \zeta^{\mathbb{H}}, \mathbb{P}_z^{\mathbb{H}})$ on $\mathbb{H}$ by regarding each slit C_i as a single point c_i^*.

For each $t \in [0, t_\gamma)$, the functions v^*, v and the quantities R_i^*, q_{ij}^*, M_{ij} specified by Theorem 2.2.1 for C_i, c_i^*, $1 \leq i \leq N$, and F_t in place of A_i, a_i^*, $1 \leq i \leq N$, and F will be designated by v_t^*, v_t and $R_i^*(t)$, $q_{ij}^*(t)$, $M_{ij}(t)$ respectively. By virtue of Theorem 2.2.3, it holds that

$$v_t^*(z) = \Im g_t(z) \qquad \text{for } z \in D \setminus F_t. \tag{2.3.1}$$

Define

$$h(t, z) = \mathbb{E}_z^{\mathbb{H}} \left[\Im Z_{\sigma_{F_t \cup K}}^{\mathbb{H}} ; \sigma_{F_t \cup K} < \infty \right], \quad t \in [0, t_\gamma),\ z \in D \setminus F_t;$$

$$k_j(t, z) = \mathbb{P}_z^{\mathbb{H}}(\sigma_K < \sigma_{F_t};\ Z_{\sigma_K}^{\mathbb{H}} \in C_j), \quad t \in [0, t_\gamma),\ z \in D \setminus F_t,\ 1 \leq j \leq N;$$

$$k_j^*(t, c_i^*) = \mathbb{P}_{c_i^*}^*(\sigma_{K^*} < \sigma_{F_t};\ Z_{\sigma_{K^*}}^* \in c_j^*), \quad t \in [0, t_\gamma),\ 1 \leq i, j \leq N.$$

Lemma 2.3.1.

(i) *For each* $T \in (0, t_\gamma)$, $z \in D \setminus F_T$ *and* $1 \le i, j \le N$,

$$\lim_{s \to t} h(s,z) = h(t,z), \ t \in (0,T], \quad \lim_{s \downarrow 0} h(s,z) = \mathbb{E}_z^{\mathbb{H}}\left[\Im Z_{\sigma_K}^{\mathbb{H}}; \sigma_K < \infty\right], \quad (2.3.2)$$

$$\lim_{s \to t} k_j(s,z) = k_j(t,z), \ t \in (0,T], \quad \lim_{s \downarrow 0} k_j(s,z) = \mathbb{P}_z^{\mathbb{H}}\left(\sigma_K < \infty, \ Z_{\sigma_K}^{\mathbb{H}} \in C_j\right), \quad (2.3.3)$$

$$\begin{aligned} &\lim_{s \to t} k_j^*(s, c_i^*) = k_j^*(t, c_i^*) \quad \textit{for } t \in (0,T], \quad \textit{and} \\ &\lim_{s \downarrow 0} k_j^*(s, c_i^*) = \mathbb{P}_{c_i^*}^*\left(\sigma_{K^*} < \infty, \ Z_{\sigma_{K^*}}^* \in c_j^*\right). \end{aligned} \quad (2.3.4)$$

(ii) *Define* $\lambda_k = \Im z$ *for* $z \in C_k$ *with* $1 \le k \le N$. *For any* $1 \le i, j \le N$,

$$\lim_{z \to z_0 \in C_i} h(t,z) = \lambda_i \quad \textit{and} \quad \lim_{z \to z_0 \in \partial\mathbb{H} \setminus \{\gamma(0)\}} h(t,z) = 0 \quad (2.3.5)$$

uniformly in $t \in [0,T]$, *and*

$$\lim_{z \to z_0 \in C_i} k_j(t,z) = \delta_{ij} \quad \textit{and} \quad \lim_{z \to z_0 \in \partial\mathbb{H}} k_j(t,z) = 0 \quad (2.3.6)$$

uniformly in $t \in [0,T]$.

Proof. (i) First notice that the sample path $Z_t^{\mathbb{H}}$ of the ABM $\mathbf{Z}^{\mathbb{H}}$ on $\mathbb{H}$ takes values in the one-point compactification $\mathbb{H} \cup \{\partial\}$ of $\mathbb{H}$ with $Z_t^{\mathbb{H}} = \partial$ for any $t \ge \zeta^{\mathbb{H}}$ and $Z_t^{\mathbb{H}}$ is continuous in $t \in [0, \infty)$.

To show the first identity of (2.3.2), suppose that $s_n \uparrow t \in (0,T]$. Then $(\bigcup_n F_{s_n}) \cup K = \gamma(0,t) \cup K$ and $\sigma_{F_{s_n} \cup K} \downarrow \sigma_{\gamma(0,t) \cup K}$ by [Blumenthal and Getoor (1968), (10.4)(d)]. As $\sigma_{F_t \cup K} = \sigma_{\gamma(0,t) \cup K} \wedge \sigma_{\{\gamma(t)\}}$ by [Blumenthal and Getoor (1968), (10.4)(b)] and the one-point set $\{\gamma(t)\}$ is polar for the ABM $\mathbf{Z}^{\mathbb{H}}$, we have $\sigma_{F_{s_n} \cup K} \downarrow \sigma_{F_t \cup K}$ a.s. so that $h(s_n, z) \to h(t,z)$, $n \to \infty$.

Next suppose that $s_n \downarrow t \in (0,T)$. The lifetime of the ABM $\mathbf{Z}^{\mathbb{H}}$ is finite a.s. because so is the exit time of the planar Brownian motion from $\mathbb{H}$. If $\sigma_{F_{s_n} \cup K} < \infty$ for any n, then $\sigma_{F_{s_n} \cup K} < \zeta^{\mathbb{H}}$ for any n and $\sigma := \lim_{n \to \infty} \sigma_{F_{s_n} \cup K} \le \zeta^{\mathbb{H}} < \infty$. The above mentioned continuity of $Z_t^{\mathbb{H}}$ then implies that

$$Z_\sigma^{\mathbb{H}} = \lim_n Z_{\sigma_{F_{s_n} \cup K}}^{\mathbb{H}} \in (\bigcap_n F_{s_n}) \cup K = F_t \cup K.$$

Hence $\sigma = \sigma_{F_t \cup K}$ and $\bigcap_n \{\sigma_{F_{s_n} \cup K} < \infty\} = \{\sigma_{F_t \cup K} < \infty\}$ so that $h(s_n, z) \to h(t,z)$, $n \to \infty$, completing the proof of the first identity of (2.3.2).

To show the second identity of (2.3.2), assume that $s_n \downarrow 0$. Decompose $h(s_n, z) = \mathrm{I}_n + \mathrm{II}_n$ with

$$\mathrm{I}_n = \mathbb{E}_z^{\mathbb{H}}\left[\Im Z_{\sigma_{F_{s_n}}}; \sigma_{F_{s_n}} < \sigma_K\right] \quad \text{and} \quad \mathrm{II}_n = \mathbb{E}_z^{\mathbb{H}}\left[\Im Z_{\sigma_K}; \sigma_K < \sigma_{F_{s_n}}\right].$$

As $\lim_{n \to \infty} \Im Z_{\sigma_{F_{s_n}}} = 0$ boundedly, $\lim_{n \to \infty} \mathrm{I}_n = 0$. If $\sigma := \lim_n \sigma_{F_{s_n}} < \infty$, then by the above mentioned continuity of $\mathbf{Z}_t^{\mathbb{H}}$ again,

$$Z_\sigma^{\mathbb{H}} = \lim_n Z_{\sigma_{F_{s_n}}}^{\mathbb{H}} \in (\bigcap_n F_{s_n}) \cup \{\partial\} = \{\partial\},$$

which means that $\sigma = \zeta^{\mathbb{H}}$. Hence

$$\lim_n \mathrm{II}_n = \mathbb{E}_z^{\mathbb{H}}\left[\Im Z_{\sigma_K}; \sigma_K < \zeta^{\mathbb{H}}\right] = \mathbb{E}_z^{\mathbb{H}}\left[\Im Z_{\sigma_K}; \sigma_K < \infty\right],$$

yielding the second identity of (2.3.2).

Property (2.3.3) can be proved in the same way as above only with an additional remark that

$$\bigcap_n \{\sigma_K < \sigma_{F_{s_n}}\} = \{\sigma_K \le \sigma_{F_t}\} = \{\sigma_K < \sigma_{F_t}\} \quad \text{when} \quad s_n \uparrow t,$$

As for (2.3.4), recall that the BMD $\mathbf{Z}^* = (Z_t^*, \zeta^*, \mathbb{P}_z^*)$ on D^* has continuous sample paths taking values in D_∂^* by (1.3.19) and finite lifetime ζ^* by (1.4.1). So exactly the same proof as that of (2.3.3) works in getting (2.3.4).

(ii) For each $t \in [0, T]$, we have

$$\begin{aligned}|h(t,z) - \lambda_i| &\le \mathbb{E}_z^{\mathbb{H}}\left[|\Im Z_{\sigma_{F_t}} - \lambda_i|; \sigma_{F_t} < \sigma_K\right] + \mathbb{E}_z^{\mathbb{H}}\left[|\Im Z_{\sigma_K} - \lambda_i|; \sigma_K < \sigma_{F_t}\right] \\ &\le (\sup\{\Im z : z \in F_T\} + \lambda_i) \cdot \mathbb{P}_z^{\mathbb{H}}(\sigma_{F_T} < \sigma_K) + \mathbb{E}_z^{\mathbb{H}}\left[|\Im Z_{\sigma_K} - \lambda_i|; \sigma_K < \infty\right],\end{aligned}$$

which yields the first conclusion in (2.3.5) by making the same consideration as in the proof of Lemma 1.1.4 (i). The second conclusion in (2.3.5) follows from

$$h(t,z) \le \sup\{\Im z : z \in F_T \cup K\} \cdot \mathbb{P}_z^{\mathbb{H}}(\sigma_{F_T \cup K} < \infty),$$

while (2.3.6) follows from $\left|k_j(t,z) - \mathbb{P}_z^{\mathbb{H}}(\sigma_K < \infty,\ Z_{\sigma_K}^{\mathbb{H}} \in C_j)\right| \le \mathbb{P}_z^{\mathbb{H}}(\sigma_{F_T} < \sigma_K)$ and $k_j(t,z) \le \mathbb{P}_z^{\mathbb{H}}(\sigma_K < \infty)$. □

Proposition 2.3.2. *For each $t \in [0, t_\gamma)$, the restriction $v_t^*\big|_{D \setminus F_t}$ of the function v_t^* to $D \setminus F_t$ can be extended continuously to $\overline{\mathbb{H}}$ by setting its value on $F_t \cup \partial\mathbb{H}$ to be 0 and on each slit C_i to be $v_t^*(c_i^*)$. The function so extended is jointly continuous in $(t, z) \in [0, T] \times \left(\overline{\mathbb{H}} \setminus \gamma[0, T]\right)$ for each $T \in (0, t_\gamma)$. Moreover,*

$$0 < \inf_{t \in [0,T],\ 1 \le k \le N} v_t^*(c_k^*) \le \sup_{t \in [0,T],\ 1 \le k \le N} v_t^*(c_k^*) < \infty, \tag{2.3.7}$$

and

$$\lim_{t \downarrow 0} v_t^*(z) = v_0^*(z) = \Im z, \quad z \in \overline{\mathbb{H}}. \tag{2.3.8}$$

Proof. The stated continuous extendability of $v_t^*\big|_{D \setminus F_t}$ to $\overline{\mathbb{H}}$ follows from Lemma 2.2.2.

In view of (2.3.3) and (2.3.4), $R_i^*(t)$ and $q_{ij}^*(t)$ are continuous in $t \in [0, T]$ with the interpretation that $\sigma_{F_0} = \sigma_\emptyset = \infty$. Notice that we have $\sum_{j=1}^N q_{ij}^*(t) < 1$ for any $t \in [0, T]$ and $1 \le i \le N$ because

$$\mathbb{P}_{c_i^*}^{\mathbb{H},*}\left(\sigma_{K^*} < \sigma_{F_t},\ Z_{\sigma_{K^*}}^{\mathbb{H},*} \ne c_i^*\right) = \int_{\eta_i} \mathbb{P}_z^{\mathbb{H}}(\sigma_K < \sigma_{F_t}) \nu_i(dz) - R_i^*(t) < 1 - R_i^*(t).$$

Hence, $a := \max_{1\le i\le N,\, 0\le t\le T} \sum_{j:j\ne i} q_{ij}^*(t) < 1$, which means that $M_{ij}(t)$ is also continuous in $t \in [0, t_\gamma)$ as the sum of a series whose $(n+1)$-th term is continuous and dominated by a^n.

Theorem 2.2.1 now reads

$$\begin{cases} v_t^*(z) = v_t(z) + \sum_{i=1}^N k_i(t,z) v_t^*(c_i^*), \\ v_t(z) = \Im z - h(t,z), \quad t \in (0,T], \quad z \in D \setminus F_T. \\ v_t^*(c_i^*) = \sum_{j=1}^N \frac{M_{ij}(t)}{1-R_j^*(t)} \int_{\gamma_j} v_t(z)\nu_j(dz). \end{cases}$$

It follows from (2.3.2) and (2.3.3) and the continuity of $R_j^*(t)$, $M_{ij}(t)$ verified above that, for each $T \in (0, t_\gamma)$ and $z \in D \setminus F_T$, the functions $v_t(z)$, $v_t^*(c_i^*)$ and $v_t^*(z)$ are continuous in $t \in [0,T]$.

Notice that $v_t^*(z)$ is harmonic in $z \in D \setminus F_T$; for any disk B with $\overline{B} \subset D \setminus F_T$, $v_t^*(z) = \int_{\partial B} p(z,\xi) v_t^*(\xi) ds(\xi)$ for $z \in B$, where $p(z,\xi)$ is the Poisson kernel of the disk B. Consequently, $v_t^*(z)$ is jointly continuous in $(t,z) \in [0,T] \times (D \setminus F_T)$. Owing to Lemma 2.3.1 (ii), this joint continuity can be further extended to $(t,z) \in [0,T] \times (\overline{\mathbb{H}} \setminus \gamma[0,T])$. Properties (2.3.7) is obvious, and so is (2.3.8) because D is now a standard slit domain and $g_0(z) = z$, $z \in D$. □

Theorem 2.3.3. *For each $t \in [0, t_\gamma)$, the conformal map $g_t(z)$ extends continuously to $D \cup K^p \cup \partial\mathbb{H} \setminus \gamma[0,t]$. Moreover, the family $\{g_s(z); s \in [0,t]\}$ is equi-continuous at each $z \in D \cup K^p \cup \partial\mathbb{H} \setminus \gamma[0,t]$ and*

$$\sup_{s\in[0,t]} |g_s(z)| < \infty, \quad \text{for any } z \in D \cup K^p \cup \partial\mathbb{H} \setminus \gamma[0,t]. \tag{2.3.9}$$

Proof. For each $t \in (0, t_\gamma)$, let $\widehat{F}_t = \gamma[0,t] \cup \Pi(\gamma[0,t])$, where, as before, Π denotes the mirror reflection with respect to $\partial\mathbb{H}$. By the Schwarz reflection principle, for each $t > 0$, we can extend $g_t(z)$ to be an analytic function in $\mathbb{C} \setminus (K \cup \Pi(K) \cup \widehat{F}_t)$. By (2.3.1) and Proposition 2.3.2, $v_s^*(z) = \Im g_s(z)$ is continuously extendable in z to $\mathbb{C} \setminus \widehat{F}_t$ to be jointly continuous in $(s,z) \in [0,t] \times (\mathbb{C} \setminus \widehat{F}_t)$.

Since for each $s \ge 0$, $v_s^*(z)$ is harmonic in $z \in \mathbb{C} \setminus (K \cup \Pi(K) \cup \widehat{F}_t)$, it follows from the integral representation for harmonic functions in disks in terms of Poisson kernels that $\nabla_z v_s^*(z)$ is jointly continuous in $(s,z) \in [0,t] \times (\mathbb{C} \setminus (K \cup \Pi(K) \cup \widehat{F}_t))$. For any $z \in D \cup \partial\mathbb{H} \setminus \gamma[0,t]$ and any disk $B_r(z)$ with $\overline{B_r(z)} \subset \mathbb{C} \setminus (K \cup \Pi(K) \cup \widehat{F}_t)$, $|g_s'(w)| = |\nabla_w v_s^*(w)|$, $w \in B_r(z)$, and so $\sup_{w\in\overline{B_r(z)},\, s\in[0,t]} |g_s'(w)| := M$ is finite. Hence

$$|g_s(w) - g_s(z)| = \left| \int_z^w g_s'(\zeta) d\zeta \right| \le M|w-z|, \quad w \in B_r(z), \quad s \in [0,t],$$

the asserted equi-continuous at z.

To verify (2.3.9) for $z \in D \cup \partial\mathbb{H} \setminus \gamma[0,t]$, choose $\rho > 0$ with $\mathbb{H} \cap B_\rho \supset \gamma[0,t] \cup K$ and take any $z_0 \in \mathbb{H} \setminus \overline{B}_{3\rho}$. Theorem 2.2.4 (iv) then implies that $\sup_{s\in[0,t]} |g_s(z_0)| < \infty$ because a_s is positive and increasing by Corollary 2.1.4. We then take a smooth curve C in $\overline{\mathbb{H}} \setminus (K \cup \gamma[0,t])$ connecting z and z_0 and use the identity $u_s^*(z) - u_s^*(z_0) =$

$\int_C(-(v_s^*)_y)dx+(v_s^*)_x dy$ holding for $u_s^* = \Re g_s$ to conclude that $\sup_{s\in[0,t]}|u_s^*(z)| < \infty$ and so $\sup_{s\in[0,t]}|g_s(z)| < \infty$.

We next show that the family $\{g_s(z); s \in [0,t]\}$ is in fact equi-continuous and satisfies (2.3.9) at each $z \in K^p$. Since $\Im g_s(z)$ takes a constant value on each slit C_j continuously, by Schwarz reflection principle, $g_s(z) = u_s^*(z)+iv_s^*(z)$ can be extended to be an analytic function across C_j^+ (resp. C_j^-). As the harmonic function $v_s^*(z)$ is jointly continuous in (s,z), the same argument as above applies and we conclude that the family $\{g_s(z); s \in [0,t]\}$ is equi-continuous and satisfies (2.3.9) at every point in $C_j^+ \cup C_j^-$.

For the left endpoint z_1 of the slit C_j, take $\varepsilon > 0$ small so that it is less than one half of the length of C_j and that $B(z_1,\varepsilon)\setminus C_j \subset D$. Then $\psi(z) = (z-z_1)^{1/2}$ maps $B(z_1,\varepsilon)\setminus C_j$ conformally onto $B(0,\sqrt{\varepsilon})\cap\mathbb{H}$. Consequently, $g_s\circ\psi^{-1}(z) = g_s(z^2+z_1)$ is an analytic function in $z \in B(0,\sqrt{\varepsilon})\cap\mathbb{H}$ that is continuous up to $B(0,\sqrt{\varepsilon})\cap\partial\mathbb{H}\setminus\{0\}$ and $v_s^*(z^2+z_1) = \Im g_s(z^2+z_1)$ takes constant value there. By Schwarz reflection, $g_s(z^2+z_1)$ extends to $B(0,\sqrt{\varepsilon})\setminus\{0\}$. Thus, 0 is an isolated singularity of $g_s(z^2+z_1)$. Since $\Im g_s(z^2+z_1)$ is bounded near the origin, it has to be a removable singularity (cf. [Ahlfors (1979), Exercise 5 in §4.3.2]). It follows that $g_s(z^2+z_1)$ can be extended to be analytic in $z \in B(0,\sqrt{\varepsilon})$ and $\Im g_s(z^2+z_1) = v_s^*(z^2+z_1)$ is jointly continuous in (s,z). Thus, the same argument as above is applicable again in concluding that $\{g_s(z); s\in[0,t]\}$ is equi-continuous and satisfies (2.3.9) at each point $z \in B(z_1,\varepsilon/2)\cap(D\cup K^p)$. The case for the right endpoint of C_j can be dealt with analogously. □

We will show in next section that for every $t \in (0,t_\gamma)$, $g_s(z)$ is in fact jointly continuous in $(s,z) \in [0,t]\times(D\cup K^p\cup\partial\mathbb{H}\setminus\gamma[0,t])$.

2.3.2 *Continuity of $g_{t,s}(z)$, a_t, D_t and $\xi(t)$*

The aim of this section is to show, using Proposition 2.3.2 and Theorem 2.3.3 in the preceding section, the continuity of $g_t(z)$ in t with certain uniformity in z, and thereby derive the continuity of the function a_t in (2.1.12), the standard slit domain D_t and the position $\xi(t)\in\partial\mathbb{H}$ defined in (2.1.16).

Recall the conformal map $g_{t,s} = g_s\circ g_t^{-1}$ defined in Section 2.1.2 for $0 \le s < t < t_\gamma$, which maps $D_t = \mathbb{H}\setminus K_t$ with $K_t = \bigcup_{i=1}^N C_{t,i}$ onto $D_s\setminus g_s(\gamma[s,t])$. The map $g_{t,s}$ extends homeomorphically to $D_t\cup K_t^p\cup\partial\mathbb{H}$. In the proof of the next proposition, we will use the following identity that follows from (2.1.27) with $c=0$:

$$g_{t,s}(z) - z = \int_{\ell_{t,s}} \Psi_t(z,x)\Im g_{t,s}(x)dx \quad \text{for } s<t \text{ and } z\in D_t\cup K_t^p\cup\partial\mathbb{H}. \qquad (2.3.10)$$

Here, $\ell_{t,s}$ denotes the interval $(\beta_0(t,s),\beta_1(t,s))\subset\partial\mathbb{H}$.

Theorem 2.3.4. *For each fixed $t\in(0,t_\gamma)$, $\lim_{s\uparrow t} g_{t,s}(z) = z$ uniformly in z on each compact subset of $D_t\cup K_t^p\cup(\partial\mathbb{H}\setminus\{\xi(t)\})$.*

Proof. We let $M_{\gamma[0,t]} = \sup_{s\in[0,t]} \Im\gamma(s)$. Using the homeomorphic extension of g_t stated in Section 2.1.2, we have by (2.2.2), (2.2.3) and (2.3.7),

$$\begin{aligned}\sup_{0\le s<t}\sup_{x\in\ell_{t,s}} \Im g_{t,s}(x) &= \sup_{0\le s<t}\sup_{s\le s'\le t} [\Im g_s(\gamma(s')^-) + \Im g_s(\gamma(s')^+)] \\ &= \sup_{0\le s'\le t}\sup_{0\le s\le s'} [v_s^*(\gamma(s')^-) + v_s^*(\gamma(s')^+)] \\ &\le 2M_{\gamma[0,t]} + 2\sum_{j=1}^N \sup_{0\le s\le t} v_s^*(c_j^*) =: M_1 < \infty. \quad (2.3.11)\end{aligned}$$

For any compact subset L of $D_t\cup K_t^p\cup(\partial\mathbb{H}\setminus\{\xi(t)\})$, choose $\varepsilon>0$ and $\delta>0$ such that $L\cap B_\varepsilon(\xi(t))=\emptyset$ and $\ell_{t,t-\delta}\subset B_\varepsilon(\xi(t))$. We then see from Theorem 1.9.1 (ii) that $M_2=\sup\{|\Psi_t(z,\zeta)| : z\in L,\ \zeta\in\ell_{t,t-\delta}\}$ is finite. Hence (2.3.10) implies that, for any $s\in(t-\delta,t)$, $\sup_{z\in L}|g_{t,s}(z)-z|\le M_1M_2|\ell_{t,s}|<2M_1M_2\,\varepsilon$. □

The inverse $g_{t,s}^{-1}=g_t\circ g_s^{-1}$ of $g_{t,s}$ is a conformal map from $D_s\setminus g_s(\gamma[s,t])$ onto the standard slit domain D_t satisfying

$$g_{t,s}^{-1}(z) = z + \frac{a_t-a_s}{z} + o(1/|z|) \quad \text{as } |z|\to\infty, \quad 0\le s<t\le t_\gamma. \qquad (2.3.12)$$

Theorem 2.3.5. *For each fixed $s\in[0,t_\gamma)$, $\lim_{t\downarrow s} g_{t,s}^{-1}(z)=z$ uniformly in z on each compact subset of $D_s\cup K_s^p\cup(\partial\mathbb{H}\setminus\{\xi(s)\})$.*

Proof. Without loss of generality, we may assume that $s=0$ and so $g_{t,s}^{-1}=g_t$.

Fix $\delta\in(0,t_\gamma)$. By virtue of Theorem 2.3.3, the family $\{g_t(z)=u_t^*(z)+iv_t^*(z); 0\le t\le\delta\}$ is equi-continuous and satisfies (2.3.9) at each $z\in D\cup K^p\cup\partial\mathbb{H}\setminus\gamma[0,\delta]$. Therefore, Arzela-Ascoli Theorem (cf. [*Conway* (1978), Theorem VII.1.23]) applies in concluding that every sequence $t_n\downarrow 0$ with $t_n<\delta$ admits a subsequence still denoted as t_n such that $\widetilde{g}_{t_n}(z) := g_{t_n}(z)-z$ converges locally uniformly on $D\cup K^p\cup\partial\mathbb{H}\setminus\gamma[0,\delta]$ to a function f, which is analytic on $D\setminus\gamma[0,\delta]$. It follows from (2.3.8) that $\Im f(z)=0$ and so f is a constant on $D\cup K^p\cup\partial\mathbb{H}\setminus\gamma[0,\delta]$.

To see that this constant equals 0, choose $\rho>0$ with $B_\rho\cap\mathbb{H}\supset K\cup\gamma[0,\delta]$ and take any $z\in\mathbb{H}\setminus B_{3\rho}$. Theorem 2.2.4 (iv) then implies that $|\widetilde{g}_{t_n}(z)|\le a_\delta(|z|^{-1}+3(1+\pi)c_{1/2}\rho|z|^{-2})$. It suffices to let $n\to\infty$ first and then $|z|\to\infty$. □

Theorems 2.3.4 and 2.3.5 together with Theorem 2.3.3 immediately yield the following:

Theorem 2.3.6. *For every $0<t<t_\gamma$, $g_s(z)$ is jointly continuous in $(s,z)\in[0,t]\times((D\cup K^p\cup\partial\mathbb{H})\setminus\gamma[0,t])$.*

By the Schwarz reflection, $g_{t,s}^{-1}$, $0\le s<t<t_\gamma$, extends to a conformal map from $\mathbb{C}\setminus\Lambda_s$ onto $\mathbb{C}\setminus\Gamma_t$ still satisfying (2.3.12), where

$$\Lambda_s=\bigcup_{k=1}^N(C_{s,k}\cup\Pi(C_{s,k}))\cup g_s(\gamma[s,t])\cup\Pi(g_s(\gamma[s,t])) \text{ and } \Gamma_t=\bigcup_{k=1}^N(C_{t,k}\cup\Pi(C_{t,k}))\cup\ell_{t,s}.$$

Theorem 2.3.5 in particular implies that, for each fixed $s\in[0,t_\gamma]$,

$$\lim_{t\downarrow s} g_{t,s}^{-1}(z)=z \quad \text{uniformly on each compact subset of } \mathbb{C}\setminus\Lambda_s^0, \qquad (2.3.13)$$

where $\Lambda_s^0 = \bigcup_{k=1}^N (C_{s,k} \cup \Pi(C_{s,k})) \cup \{\xi(s)\}$. This readily leads us to the right continuity of a_t in the following manner.

We may assume that $\xi(s) \neq \mathbf{0}$. Let $\widetilde{\Lambda}_s^0 = \bigcup_{k=1}^N (\widetilde{C_{s,k}} \cup \widetilde{\pi(C_{s,k})}) \cup \widetilde{\{\xi(s)\}}$ be the image of Λ_s^0 under the inversion $w = \frac{1}{z}$. Then $h_{t,s}(w) = \frac{1}{w}\left(g_{t,s}^{-1}(\frac{1}{w}) - \frac{1}{w}\right)$ is analytic on $\mathbb{C} \setminus \widetilde{\Lambda}_s^0 \setminus \{\mathbf{0}\}$. Just as (2.1.21), we can then get from (2.3.12) the integral formula

$$a_t - a_s = \frac{1}{2\pi i} \int_{\bigcup_{k=1}^{2N+1} \widetilde{\gamma}_k} \frac{h_{t,s}(\zeta)}{\zeta} d\zeta, \quad s < t, \tag{2.3.14}$$

where $\widetilde{\gamma}_1, \ldots, \widetilde{\gamma}_N$ (resp. $\widetilde{\gamma}_{N+1}, \ldots, \widetilde{\gamma}_{2N}$) are analytic contours surrounding $\widetilde{C_{s,1}}, \ldots, \widetilde{C_{s,N}}$ (resp. $\widetilde{\pi(C_{s,1})}, \ldots, \widetilde{\pi(C_{s,N})}$) and $\widetilde{\gamma_{2N+1}}$ is an analytic contour containing $\widetilde{\{\xi(s)\}}$ inside such that $\mathbf{0} \notin \cup_{k=1}^{2N+1} \widetilde{\gamma}_k$.

By virtue of (2.3.13), $h_{t,s}(\zeta)$ converges to 0 as $t \downarrow s$ uniformly in ζ on each $\widetilde{\gamma}_k$, $1 \leq k \leq 2N+1$, and consequently we get $\lim_{t \downarrow s} a_t = a_s$ from (2.3.14). This together with Corollary 2.1.4 yields the following result.

Theorem 2.3.7. *a_t is a strictly increasing continuous function in $t \in [0, t_\gamma)$ with $a_0 = 0$.*

We next denote by $\mathcal{D}$ the collection of "labeled (or, ordered)" standard slit domains. For instance, $\mathbb{H} \setminus \{C_1, C_2, C_3, \ldots, C_N\}$ and $\mathbb{H} \setminus \{C_2, C_1, C_3, \ldots, C_N\}$ are considered as different elements of $\mathcal{D}$ in general although they correspond to the same subset $\mathbb{H} \setminus \bigcup_{i=1}^N C_i$ of $\mathbb{H}$. For $D, \widetilde{D} \in \mathcal{D}$, define their distance $d(D, \widetilde{D})$ by

$$d(D, \widetilde{D}) = \max_{1 \leq i \leq N} (|z_i - \widetilde{z}_i| + |z_i^r - \widetilde{z}_i^r|), \tag{2.3.15}$$

where, for $D = \mathbb{H} \setminus \{C_1, C_2, \ldots, C_N\}$, z_i (resp. z_i^r) denotes the left (resp. right) endpoint of C_i, $1 \leq i \leq N$, while $\widetilde{z}_i$ and $\widetilde{z}_i^r$, $1 \leq i \leq N$, are the corresponding points to $\widetilde{D}$. We view $\{D_t : 0 \leq t \leq t_\gamma\}$ as a one parameter subfamily of $\mathcal{D}$.

Theorem 2.3.8. *$\{D_t; 0 \leq t < t_\gamma\}$ is continuous in t in the sense that, for any $t_0 \in [0, t_\gamma)$ and $\varepsilon > 0$, there exists $\delta > 0$ such that $d(D_t, D_{t_0}) < \varepsilon$ for any $t \in [0, t_\gamma)$ with $|t - t_0| < \delta$.*

Proof. We deduce the right continuity from Theorem 2.3.5. The left continuity can be proved by Theorem 2.3.4 in an analogous manner.

For simplicity, we assume that $N = 1$ so that $D_t = \mathbb{H} \setminus C_t$ for a one line segment C_t parallel to $\partial\mathbb{H}$ for each $t \in [0, t_\gamma)$. We denote by z_t, z_t^r the left and right endpoints of C_t and by C_t^p the set of prime ends $\{z_t\} \cup C_t^{0,+} \cup \{z_t^r\} \cup C_t^{0,-}$.

For a fixed $t_0 \in [0, t_\gamma)$, take $\delta_0 > 0$ with $t_0 + \delta_0 < t_\gamma$ and a neighborhood U of C_{t_0} with compact closure $\overline{U} \subset \mathbb{H} \setminus \gamma_{t_0+\delta_0}$ and write $z_{t_0} = a + ci$, $z_{t_0}^r = b + ci$, $a < b$, $c > 0$. Choose $\varepsilon > 0$ such that $\varepsilon < \frac{b-a}{3}$ and $B_\varepsilon(z_{t_0}) \subset U$ and $B_\varepsilon(z_{t_0}^r) \subset U$. Let

$$w_1 = (a + \varepsilon) + ic, \quad w_2 = (b - \varepsilon) + ic, \quad \Lambda = \{x + ic : a + \varepsilon \leq x \leq b - \varepsilon\},$$

and let $w_1^\pm$, $w_2^\pm$ and $\Lambda^\pm$ be the corresponding points and parts of $C_{t_0}^{0,\pm}$.

By virtue of Theorem 2.3.5, there exists $\delta \in (0, \delta_0)$ such that, for any $t \in (t_0, t_0 + \delta)$, the conformal map g_{t,t_0}^{-1} sending $D_{t_0} \setminus g_{t_0}(\gamma[t_0, t])$ onto D_t enjoys the property that

$$|g_{t,t_0}^{-1}(z) - z| < \varepsilon/2 \quad \text{for any } z \in \left(\overline{U} \setminus (B_\varepsilon(z_{t_0}) \cup B_\varepsilon(z_{t_0}^r) \cup \Lambda)\right) \cup \Lambda^+ \cup \Lambda^-. \tag{2.3.16}$$

By fixing $t \in (t_0, t_0 + \delta)$, we write g_{t,t_0}^{-1} as f. As was observed in Section 2.1.2, f extends to a homeomorphism from $C_{t_0}^p$ onto C_t^p. Let $\Gamma = f(\partial B_\varepsilon(z_{t_0}) \setminus \{w_1\})$. (2.3.16) means that $\mathrm{diam}(\Gamma) = \sup_{\zeta, \zeta' \in \Gamma} |\zeta - \zeta'| < 3\varepsilon$.

We note that

$$\Re f(w_1^-) < \frac{a+b}{2} < \Re f(w_2^-), \quad \Re f(w_1^+) < \frac{a+b}{2} < \Re f(w_2^+). \tag{2.3.17}$$

There are three cases below concerning the locations of $f(w_1^\pm)$:

(I) $f(w_1^-) \in C_t^{0,-} \cup \{z_t\}$, $f(w_1^+) \in C_t^{0,+} \cup \{z_t\}$, (II) $f(w_1^-) \in C_t^{0,+}$, (III) $f(w_1^+) \in C_t^{0,-}$.

Since $\Gamma \subset D_t$, we readily see that, in case (I), $\sup_{\zeta \in \Gamma} |z_t - \zeta| \leq \mathrm{diam}(\Gamma)$ and consequently

$$|z_t - z_{t_0}| \leq |z_{t_0} - z_0| + |z_0 - f(z_0)| + |f(z_0) - z_t| < 9\varepsilon/2,$$

where z_0 is an arbitrary chosen point of $\partial B_\varepsilon(z_{t_0}) \setminus \{w_1\}$.

In case (II), we can observe on account of the first bound of (2.3.17) that there must be a point $w_0^- \in \Lambda^-$ with $a + \varepsilon < \Re w_0^- < b - \varepsilon$ and $f(w_0^-) = z_t$. Then

$$a - \frac{\varepsilon}{2} \leq \Re\, w_0^- - \frac{\varepsilon}{2} \leq \Re f(w_0^-) = \Re z_t \leq \Re f(w_1^-) \leq \Re w_1^- + \frac{\varepsilon}{2} = a + \frac{3}{2}\varepsilon,$$

and $|a - \Re z_t| < \frac{3}{2}\varepsilon$. Clearly $|c - \Im z_t| = |\Im w_1^- - \Im f(w_1^-)| \leq \frac{\varepsilon}{2}$. Hence $|z_t - z_{t_0}| < 2\varepsilon$.

We have $|z_t - z_{t_0}| < 2\varepsilon$ in case (III) as well. By thinking of $f(w_2^\pm)$ in place of $f(w_1^\pm)$, we obtain the same estimates for $|z_t^r - z_{t_0}^r|$. $\square$

Theorem 2.3.9. *$\xi(t)$ is continuous in $t \in [0, t_\gamma)$. Moreover*

$$\lim_{t \downarrow s} \beta_0(t, s) = \xi(s) \quad \text{and} \quad \lim_{t \downarrow s} \beta_1(t, s) = \xi(s). \tag{2.3.18}$$

Proof. We first give a proof of the right continuity of $\xi(t)$ and (2.3.18). For any $\varepsilon > 0$ with $B_\varepsilon(\xi(s)) \cap \mathbb{H} \subset D_s$, choose $\delta > 0$ such that $g_s(\gamma[s,t]) \cup \Pi(g_s(\gamma[s,t])) \subset B_\varepsilon(\xi(s))$ for any $t \in (s, s+\delta)$. Let $C = \partial B_\varepsilon(\xi(s))$ and $\Gamma = g_{t,s}^{-1}(C)$. Then $\ell_{t,s} \subset \mathrm{ins}(\Gamma)$ and in particular $\xi(t) \in \mathrm{ins}(\Gamma)$. By Theorem 2.3.5, we have for a sufficiently small $\delta > 0$

$$|g_{t,s}^{-1}(z) - z| < \varepsilon \quad \text{for any } z \in C \text{ and } t \in (s, s+\delta),$$

which particularly means that $\mathrm{diam}(\Gamma) < 4\varepsilon$. By taking any $z \in C$, we then get for any $t \in (s, s+\delta)$

$$|\xi(s) - \xi(t)| \leq |\xi(s) - z| + |z - g_{t,s}^{-1}(z)| + |g_{t,s}^{-1}(z) - \xi(t)| < 6\varepsilon,$$

$$|\xi(s) - \beta_0(t,s)| \le |\xi(s) - z| + |z - g_{t,s}^{-1}(z)| + |g_{t,s}^{-1}(z) - \beta_0(t,s)| < 6\varepsilon,$$

and similarly, $|\xi(s) - \beta_1(t,s)| < 6\varepsilon$.

We next prove the left continuity of $\xi(t)$. For any $\varepsilon > 0$ with $B_\varepsilon(\xi(t)) \cap \mathbb{H} \subset D_t$, there exists $\delta > 0$ such that $\ell(s,t) \subset B_\varepsilon(\xi(t))$ for any $s \in (t-\delta, t)$ in view of the proof of Corollary 2.1.4.

Let $C = \partial B_\varepsilon(\xi(t))$ and $\Gamma = g_{t,s}(C)$ for $s \in (t-\delta, t)$. Then $g_s(\gamma[s,t]) \cup \Pi(g_s(\gamma[s,t])) \subset \mathrm{ins}(\Gamma)$ and in particular $\xi(s) \in \mathrm{ins}(\Gamma)$. By Theorem 2.3.4, we have for a sufficiently small $\delta > 0$

$$|g_{t,s}(z) - z| < \varepsilon \quad \text{for any } z \in C \text{ and } s \in (t-\delta, t),$$

which particularly means that $\mathrm{diam}(\Gamma) < 4\varepsilon$. By taking any $z \in C$, we then get for any $s \in (t-\delta, t)$

$$|\xi(t) - \xi(s)| \le |\xi(t) - z| + |z - g_{t,s}(z)| + |g_{t,s}(z) - \xi(s)| < 6\varepsilon.$$

□

2.4 BMD complex Poisson kernel under perturbation of standard slit domain

2.4.1 *Small perturbation of standard slit domain*

The aim of this section is to establish the following theorem, which combined with theorems obtained in the preceding section will enable us in the next section to derive the Komatu-Loewner differential equation (2.1.23) for the canonical map $g_t(z)$ from $\mathbb{H} \setminus \gamma[0,t]$ but with the left derivative $\frac{\partial^-}{\partial a_t}$ being replaced by the genuine derivative $\frac{\partial}{\partial a_t}$.

Recall the distance d defined by (2.3.15) on the space $\mathcal{D}$ of all 'labeled' standard slit domains. For each $D \in \mathcal{D}$, denote by $\Psi_D(z,\zeta)$, $(z,\zeta) \in D \times \partial\mathbb{H}$, its associated BMD-complex Poisson kernel as defined in Theorem 1.9.1. By Theorem 1.9.1, $\Psi_D(z,\zeta)$ can be extended to be a continuous function on $(D \cup K^p \cup (\partial\mathbb{H} \setminus J)) \times J$ for any compact interval $J \subset \partial\mathbb{H}$.

Theorem 2.4.1. *The correspondence $D \mapsto \Psi_D(z,\zeta)$ is Lipschitz continuous in the following sense: Let U_j, $1 \le j \le N$, be open rectangles in $\mathbb{H}$ with sides parallel to x and y axis, and V_j, $1 \le j \le N$, be bounded open sets in $\mathbb{H}$ satisfying*

$$\overline{U}_j \subset V_j \subset \overline{V}_j \subset \mathbb{H} \quad \text{and} \quad \overline{V}_j \cap \overline{V}_k = \emptyset \quad \text{for } j \neq k. \tag{2.4.1}$$

Let $a > 0$ and $b > 0$ be so that the subcollection $\mathcal{D}_0$ of $\mathcal{D}$ defined by

$$\mathcal{D}_0 = \left\{\mathbb{H} \setminus \cup_{j=1}^N C_j \in \mathcal{D} : C_j \subset U_j,\ |z_j - z_j^r| \ge a,\ \mathrm{dist}(C_j, \partial U_j) \ge b,\ 1 \le j \le N\right\} \tag{2.4.2}$$

is non-empty. There exists $\varepsilon_0 > 0$ such that for any $\varepsilon \in (0, \varepsilon_0)$ and for any $D \in \mathcal{D}_0$ and $\widetilde{D} \in \mathcal{D}$ with $d(D, \widetilde{D}) < \varepsilon$, there exists a diffeomorphism $\widetilde{f}_\varepsilon$ from $\mathbb{H}$ onto $\mathbb{H}$ satisfying

(i) $\widetilde{f}_\varepsilon$ *is sending* D *onto* $\widetilde{D}$, *linear on each* U_j *and the identity map on* $\mathbb{H} \setminus \bigcup_{j=1}^N \overline{V}_j$;

(ii) *for some positive constant* L_1 *independent of* $\varepsilon \in (0, \varepsilon_0)$, $D \in \mathcal{D}_0$ *and* $\widetilde{D} \in \mathcal{D}$,

$$|z - \widetilde{f}_\varepsilon(z)| \leq L_1 \varepsilon \quad \textit{for all } z \in \mathbb{H}; \tag{2.4.3}$$

(iii) *for any compact subset* Q *of* $\overline{\mathbb{H}}$ *containing* $\bigcup_{j=1}^N \overline{V}_j$ *and for any compact subset* J *of* $\partial\mathbb{H}$,

$$\left|\Psi_D(z, \zeta) - \Psi_{\widetilde{D}}(\widetilde{f}_\varepsilon(z), \zeta)\right| \leq L_{Q,J}\, \varepsilon \quad \textit{for every } z \in (Q \setminus K \setminus J) \cup K^p \textit{ and } \zeta \in J, \tag{2.4.4}$$

where $L_{Q,J}$ *is a positive constant independent of* $\varepsilon \in (0, \varepsilon_0)$, $D \in \mathcal{D}_0$ *and* $\widetilde{D} \in \mathcal{D}$.

We notice that the mapping $\widetilde{f}_\varepsilon$ in the above statement can be extended to a homeomorphism between $\overline{\mathbb{H}}$ and $\overline{\mathbb{H}}$ by setting $\widetilde{f}_\varepsilon(z) = z$, $z \in \partial\mathbb{H}$. A construction of a diffeomorphism $\widetilde{f}_\varepsilon$ satisfying Theorem 2.4.1 (i), (ii) will be carried out by the next lemma following the method of interior variations in partial differential equations (cf. [Garabedian (1964)]).

Let U_j and V_j be as in the statement of Theorem 2.4.1. For any $\varepsilon > 0$ and any $D \in \mathcal{D}_0$, take any $\widetilde{D} \in \mathcal{D}$ with $d(D, \widetilde{D}) < \varepsilon$. The quantities associated with $\widetilde{D}$ will be designated with $\widetilde{\ }$. For each $1 \leq j \leq N$, let $\delta_j \in \mathbb{R}$, $b_j \in \mathbb{C}$ be constants that are uniquely determined by

$$\begin{cases} \widetilde{z}_j - z_j = \delta_j z_j + b_j, \\ \widetilde{z}_j^r - z_j^r = \delta_j z_j^r + b_j, \end{cases}$$

where $z_j = x_{j1} + i x_{j2}$ (resp. $z_j^r = x_{j1}^r + i x_{j2}^r$) is the left (resp. right) endpoint of the slit C_j. Since $\delta_j = \frac{(\widetilde{x}_{j1} - x_{j1}) + (x_{j1}^r - \widetilde{x}_{j1}^r)}{x_{j1} - x_{j1}^r}$ and $|b_j| \leq |\widetilde{z}_j - z_j| + |\delta_j||z_j|$, we have

$$\frac{|\delta_j|}{\varepsilon} \leq \frac{1}{a}, \quad \frac{|b_j|}{\varepsilon} \leq 1 + \frac{M_0}{a}, \qquad \text{where } M_0 = \sup_{z \in \cup_{j=1}^N U_j} |z|. \tag{2.4.5}$$

Define a linear map

$$F_{j,\varepsilon}(z) = \frac{1}{\varepsilon}(\delta_j z + b_j), \quad 1 \leq j \leq N, \tag{2.4.6}$$

whose coefficients are bounded uniformly in $\varepsilon > 0$, $D \in \mathcal{D}_0$ and $\widetilde{D} \in \mathcal{D}$ by (2.4.5). Choose a smooth function $q(x_1, x_2)$, $z = x_1 + i x_2 \in \mathbb{H}$, taking value in $[0, 1]$ such that

$$q(x_1, x_2) = \begin{cases} 1 & \text{if } x_1 + i x_2 \in U_j, \quad 1 \leq j \leq N, \\ 0 & \text{if } x_1 + i x_2 \in \mathbb{H} \setminus \cup_{j=1}^N \overline{V}_j, \end{cases}$$

and define a map $\widetilde{f}_\varepsilon$ by

$$\begin{cases} \widetilde{f}_\varepsilon(z) = z + \varepsilon F_\varepsilon(x_1, x_2), \quad \text{where} \\ F_\varepsilon(x_1, x_2) = q(x_1, x_2) \sum_{j=1}^N \mathbf{1}_{V_j}(z) F_{j,\varepsilon}(z), \quad z = x_1 + i x_2. \end{cases} \tag{2.4.7}$$

Lemma 2.4.2. *There exists $\varepsilon_0 > 0$ such that for any $\varepsilon \in (0, \varepsilon_0)$ and for any $D \in \mathcal{D}_0$ and $\widetilde{D} \in \mathcal{D}$ with $d(D, \widetilde{D}) < \varepsilon$, the map $\widetilde{f}_\varepsilon$ defined by* (2.4.7) *is a diffeomorphism from $\mathbb{H}$ onto $\mathbb{H}$ satisfying the properties* (i) *and* (ii) *of Theorem* 2.4.1.

Proof. In view of (2.4.5), (2.4.6) and (2.4.7), $F_\varepsilon(x_1, x_2)$ and its derivatives are bounded on $\mathbb{H}$ uniformly in $\varepsilon > 0$, $D \in \mathcal{D}_0$ and $\widetilde{D} \in \mathcal{D}$. Let $M = \sup_{\varepsilon>0,\, z\in\mathbb{H}} \sum_{k=1}^2 \left|\frac{\partial F_\varepsilon(x_1,x_2)}{\partial x_k}\right| (< \infty)$ and $\varepsilon_1 = 1/(2M)$. Then $|F_\varepsilon(z_i) - F_\varepsilon(z_2)| \le M|z_1 - z_2|$ so that, for any $\varepsilon \in (0, \varepsilon_1)$, $\varepsilon|F_\varepsilon(z_1) - F_\varepsilon(z_2)| \le \frac{1}{2}|z_1 - z_2|$ and $|\widetilde{f}_\varepsilon(z_1) - \widetilde{f}_\varepsilon(z_2)| \ge |z_1 - z_2| - \varepsilon|F_\varepsilon(z_1) - F_\varepsilon(z_2)| > \frac{1}{2}|z_1 - z_2|$. Therefore (2.4.7) defines a continuous injection $\widetilde{f}_\varepsilon$ from $\mathbb{H}$ for any $\varepsilon \in (0, \varepsilon_1)$. $\widetilde{f}_\varepsilon$ is linear on each U_j, sending C_j onto $\widetilde{C}_j$ for $1 \le j \le N$, and an identity map on $\mathbb{H} \setminus \bigcup_{j=1}^N \overline{V}_j$. We can also see that $\widehat{f}_\varepsilon \left(\bigcup_{j=1}^N \overline{V}_j\right) \subset \bigcup_{j=1}^N \overline{V}_j$ because $\widehat{f}_\varepsilon$ can be extended to an injection from $\mathbb{C}$ by setting $\widehat{f}_\varepsilon(z) = z$, $z \in \mathbb{C} \setminus \mathbb{H}$.

In what follows, we write $\widetilde{f}_\varepsilon(z) = \widetilde{z}$, $z = x_1 + ix_2$, $\widetilde{z} = \widetilde{x}_1 + i\widetilde{x}_2$. We then have

$$\frac{\partial(\widetilde{x}_1, \widetilde{x}_2)}{\partial(x_1, x_2)} = \begin{cases} 1 + \varepsilon L(x_1, x_2) & \text{if } x_1 + ix_2 \in \bigcup_{j=1}^N \overline{V}_j, \\ 1 & \text{if } x_1 + ix_2 \in \mathbb{H} \setminus \bigcup_{j=1}^N \overline{V}_j, \end{cases}$$

where $L(x_1, x_2)$ is a uniformly bounded function on $\bigcup_{j=1}^N \overline{V}_j$ in $\varepsilon > 0$, $D \in \mathcal{D}_0$ and $\widetilde{D} \in \mathcal{D}$. Hence $\frac{\partial(\widetilde{x}_1,\widetilde{x}_2)}{\partial(x_1,x_2)} > 0$ for $x_1 + ix_2 \in \mathbb{H}$, for any $\varepsilon \in (0, \varepsilon_2)$ for some ε_2 independent of $D \in \mathcal{D}_0$. Consequently $\widetilde{f}_\varepsilon^{-1}$ is a smooth mapping and $\widetilde{f}_\varepsilon$ is an open map

We let $\varepsilon_0 = \varepsilon_1 \wedge \varepsilon_2$. For $\varepsilon \in (0, \varepsilon_0)$, $\widetilde{U} = \widetilde{f}_\varepsilon(\mathbb{H})$ is a connected open subset of $\mathbb{H}$. On the other hand, note that f_ε is an identity map on the relative closure of $\mathbb{H} \setminus \bigcup_{j=1}^N \overline{V}_j$ in $\mathbb{H}$. Since $\bigcup_{j=1}^N \overline{V}_j$ is a compact subset of $\mathbb{H}$, its image under f_ε is also compact. It follows that

$$\widetilde{U} = \widetilde{f}_\varepsilon(\mathbb{H}) = \left(\mathbb{H} \cap \overline{\mathbb{H} \setminus \cup_{j=1}^N \overline{V}_j}\right) \cup \widetilde{f}_\varepsilon(\cup_{j=1}^N \overline{V}_j),$$

is relatively closed in $\mathbb{H}$. Hence $\widetilde{U} = \mathbb{H}$. □

We remark that F_ε in (2.4.7) depends on $\varepsilon > 0$, D and $\widetilde{D}$, while Garabedian [Garabedian (1964), §15.1] treated a case for a fixed map from D independent of $\varepsilon > 0$.

In order to prove Theorem 2.4.1 (iii) in the next section, we prepare a proposition. We denote by $G(z, w)$ the Green function of the domain $D \in \mathcal{D}_0$ defined by (1.1.14) but with the superscript ^{0} dropped. Let $\varepsilon_0 > 0$ be as in Lemma 2.4.2. The Green function of $\widetilde{D} = \widetilde{f}_\varepsilon(D)$ is denoted by $\widetilde{G}(\widetilde{z}, \widetilde{w})$ and we define

$$g(z, w, \varepsilon) = \widetilde{G}(\widetilde{f}_\varepsilon(z), \widetilde{f}_\varepsilon(w)), \quad z, w \in D, \quad \varepsilon \in (0, \varepsilon_0). \tag{2.4.8}$$

We introduce a second order self-adjoint elliptic differential operator $\mathcal{A}_\varepsilon$ by

$$\begin{cases} (\mathcal{A}_\varepsilon u)(x_1, x_2) = \sum_{k,\ell=1}^{2} \frac{\partial}{\partial x_k}\left(A_{k\ell}^{(\varepsilon)} \frac{\partial u}{\partial x_\ell}\right), \quad \textit{where} \\ A_{k\ell}^{(\varepsilon)} = \frac{1}{2}\frac{\partial(\widetilde{x}_1, \widetilde{x}_2)}{\partial(x_1, x_2)} \sum_{j=1}^{2} \frac{\partial x_k}{\partial \widetilde{x}_j}\frac{\partial x_\ell}{\partial \widetilde{x}_j}, \quad 1 \le k, \ell \le 2. \end{cases} \tag{2.4.9}$$

Proposition 2.4.3.

(i) *$g(z,w,\varepsilon)$ is a fundamental solution of $\mathcal{A}_\varepsilon$ in the sense that*

$$\mathcal{A}_\varepsilon(g_\varepsilon f)(z) = -f(z), \quad z \in D, \tag{2.4.10}$$

for any $f \in C_c(D)$, where $(g_\varepsilon f)(z) = \int_D g(z,w,\varepsilon) f(w) dw_1 dw_2$.

(ii) *$\mathcal{A}_\varepsilon = \frac{1}{2}\Delta + \varepsilon \mathcal{B}^{(\varepsilon)}$, where*

$$\mathcal{B}^{(\varepsilon)} = \sum_{k,\ell=1}^{2} b_{k\ell}^{(\varepsilon)} \frac{\partial^2}{\partial x_k \partial x_\ell} + \sum_{k,\ell=1}^{2} \frac{\partial b_{k\ell}^{(\varepsilon)}}{\partial x_k} \frac{\partial}{\partial x_\ell}. \tag{2.4.11}$$

Here $b_{k\ell}^{(\varepsilon)}$, $1 \le k,\ell \le 2$, are smooth functions on $\mathbb{H}$ with $b_{k\ell}^{(\varepsilon)} = b_{\ell k}^{(\varepsilon)}$ vanishing on $(\mathbb{H} \setminus \bigcup_{i=1}^N \overline{V}_i) \cup (\bigcup_{i=1}^N U_i)$ that together with their derivatives are bounded on $\mathbb{H}$ uniformly in $\varepsilon \in (0,\varepsilon_0)$, $D \in \mathcal{D}_0$ and $\widetilde{D} \in \mathcal{D}$.

(iii) *Put $\Lambda = \bigcup_{i=1}^N (\overline{V}_i \setminus U_i)$. Then for any $\varepsilon \in (0,\varepsilon_0)$, $\zeta \in \overline{\mathbb{H}} \setminus \Lambda$ and $w \in \overline{\mathbb{H}}$,*

$$g(\zeta,w,\varepsilon) - G(\zeta,w) = \varepsilon \int_{\{z=x_1+ix_2 \in \Lambda\}} \mathcal{B}_z^{(\varepsilon)} G(z,\zeta) g(z,w,\varepsilon) dx_1 dx_2. \tag{2.4.12}$$

(iv) *There exists $\widetilde{\varepsilon}_0 \in (0,\varepsilon_0]$ independent of $D \in \mathcal{D}_0$ such that for any $\varepsilon \in (0,\widetilde{\varepsilon}_0)$, $\zeta \in \overline{\mathbb{H}} \setminus \Lambda$ and $w \in \overline{\mathbb{H}}$*

$$g(\zeta,w,\varepsilon) - G(\zeta,w) = \varepsilon \int_\Lambda \mathcal{B}_z^{(\varepsilon)} G(z,\zeta)(G(z,w) + \varepsilon \eta^{(\varepsilon)}(z,w)) dx_1 dx_2, \tag{2.4.13}$$

where $z = x + iy$ and $\eta^{(\varepsilon)}$ is a continuous function on $\overline{\mathbb{H}} \times \overline{\mathbb{H}}$ that is bounded uniformly in $\varepsilon \in (0,\widetilde{\varepsilon}_0)$, $D \in \mathcal{D}_0$ and $\widetilde{D} \in \mathcal{D}$.

(v) *For each compact $J \subset \partial\mathbb{H}$, the function $\frac{\partial}{\partial \mathbf{n}_\zeta} \mathcal{B}_z^{(\varepsilon)} G(z,\zeta)$ is bounded in $(z,\zeta) \in \Lambda \times J$ that is uniform in $\varepsilon \in (0,\varepsilon_0)$, $D \in \mathcal{D}_0$ and $\widetilde{D} \in \mathcal{D}$.*

(vi) *For each $1 \le i \le N$, $\mathcal{B}_z^{(\varepsilon)} \varphi^{(i)}(z)$ is bounded in z on Λ that is uniform in $\varepsilon \in (0,\varepsilon_0)$, $D \in \mathcal{D}_0$ and $\widetilde{D} \in \mathcal{D}$. Here $\varphi^{(i)}(z) = \mathbb{P}_z^{\mathbb{H}}(\sigma_K < \infty,\ Z_{\sigma_K}^{\mathbb{H}} \in C_i)$, $z \in D$, and $K = \cup_{j=1}^N C_j$.*

(vii) *For each compact set $J \subset \partial\mathbb{H}$ and for $k = 1,2$,*

$$\begin{aligned} &\sup_{x_1 \in \mathbb{R}, \zeta \in J} \int_0^\infty \mathbb{1}_\Lambda(z) \left| \frac{\partial^2}{\partial x_k \partial \zeta_2} G(\zeta,z) \right| dx_2, \quad \text{and} \\ &\sup_{x_2 > 0, \zeta \in J} \int_{-\infty}^\infty \mathbb{1}_\Lambda(z) \left| \frac{\partial^2}{\partial x_k \partial \zeta_2} G(\zeta,z) \right| dx_1 \end{aligned} \tag{2.4.14}$$

are bounded in $D \in \mathcal{D}_0$.

(viii) *Fix $1 \le j \le N$. It holds for $k = 1,2$ that*

$$\sup_{x_1 \in \mathbb{R}} \int_0^\infty \mathbb{1}_\Lambda(z) \left| \frac{\partial}{\partial x_k} \varphi^{(j)}(z) \right| dx_2 \quad \text{and} \quad \sup_{x_2 > 0} \int_{-\infty}^\infty \mathbb{1}_\Lambda(z) \left| \frac{\partial}{\partial x_k} \varphi^{(j)}(z) \right| dx_1, \tag{2.4.15}$$

are bounded in $D \in \mathcal{D}_0$.

In what follows, the constant $\widetilde{\varepsilon}_0$ in the statement of (iv) above will simply be denoted as ε_0. A proof of this proposition will be given in Section A.6 of the Appendix through a series of lemmas.

Observe that BMD-complex Poisson kernels can be obtained from Green functions by taking normal derivatives at points of $\partial\mathbb{H}$, periods around slits and line integrals of normal derivatives along smooth curves; see (1.7.5), Proposition 1.1.5 (iii) and (1.9.5). In the next section, we shall prove Theorem 2.4.1 (iii) for $\widetilde{f}_\varepsilon$ defined by (2.4.7), by using the perturbation formulae (2.4.12) and (2.4.13) for Green functions and by estimating the three operations just mentioned.

We let $V = \cup_{j=1}^N V_j$, $U = \cup_{j=1}^N U_j$ so that $\Lambda = \overline{V} \setminus U$. For $b > 0$ in (2.4.2), define

$$b_0 = b \wedge \operatorname{dist}(\overline{V}, \partial\mathbb{H}). \tag{2.4.16}$$

2.4.2 *Lipschitz continuity of $\Psi_D(z,\xi)$*

In this section, we give a proof of Theorem 2.4.1 (iii).

First, we notice the following Green's first formula for the self-adjoint differential operator (2.4.9) with smooth coefficients considered on a bounded domain $U \subset D \cup \partial\mathbb{H}$ with smooth boundary ∂U:

$$\sum_{i,j=1}^{2} \int_U A_{ij}^{(\varepsilon)} \frac{\partial v}{\partial x_i} \frac{\partial u}{\partial x_j} dx_1 dx_2 + \int_U v\mathcal{A}_\varepsilon u \, dx_1 dx_2 = \int_{\partial U} v \sum_{i,j=1}^{2} A_{ij}^{(\varepsilon)} \frac{\partial u}{\partial x_i} \frac{\partial x_j}{\partial \mathbf{n}} ds, \tag{2.4.17}$$

for smooth functions u, v on $\overline{U}$. This follows from the Gauss divergence theorem

$$\int_U \left(\frac{\partial F_1}{\partial x_1} + \frac{\partial F_2}{\partial x_2} \right) dx_1 dx_2 = \int_{\partial U} F_1 dx_2 - \int_{\partial U} F_2 dx_1,$$

by the substitution

$$F_i = \sum_{j=1}^{2} a_{ij} \left(v \frac{\partial u}{\partial x_j} \right), \quad i = 1, 2.$$

For a smooth function u on $D \cup \partial\mathbb{H}$ and a smooth simple curve $\gamma \subset D \cup \partial\mathbb{H}$, define

$$I(u;\gamma) = \int_\gamma \frac{\partial u(w)}{\partial \mathbf{n}_w} ds(w). \tag{2.4.18}$$

The corresponding quantity for $\widetilde{D}$ is denoted by $\widetilde{I}(\widetilde{u};\widetilde{\gamma})$.

Lemma 2.4.4. *For a smooth simple curve γ in $D \cup \partial\mathbb{H}$ of finite length and a smooth function $\widetilde{u}$ on $\widetilde{D} \cup \partial\mathbb{H}$, define $\widetilde{\gamma} = \widetilde{f}_\varepsilon(\gamma)$. Then*

$$\widetilde{I}(\widetilde{u};\, \widetilde{\gamma}) = 2 \int_\gamma \sum_{k,\ell=1}^{2} A_{k\ell}^{(\varepsilon)}(w) \frac{\partial \widetilde{u}(\widetilde{f}_\varepsilon(w))}{\partial w_k} \frac{\partial w_\ell}{\partial \mathbf{n}} ds(w). \tag{2.4.19}$$

Proof. Extend γ to a smooth simple closed curve in $D \cup \partial\mathbb{H}$ and denote its enclosed interior by U. Define $\widetilde{U} = \widetilde{f}_\varepsilon(U)$. For any smooth functions $\widetilde{u}$, $\widetilde{v}$ on $\widetilde{U}$, substitute $u(z) = \widetilde{u}(\widetilde{f}_\varepsilon(z))$, $v(z) = \widetilde{v}(\widetilde{f}_\varepsilon(z))$ in (2.4.17). By change of variables, we have

$$\frac{1}{2}\int_{\widetilde{U}} \sum_{j=1}^{2} \frac{\partial \widetilde{v}}{\partial \widetilde{x}_j}\frac{\partial \widetilde{u}}{\partial \widetilde{x}_j} d\widetilde{x}_1 d\widetilde{x}_2 = \sum_{k,\ell=1}^{2} \int_U A_{k.\ell}^{(\varepsilon)} \frac{\partial \widetilde{v}}{\partial x_k}\frac{\partial \widetilde{u}}{\partial x_\ell} dx_1 dx_2,$$

which implies

$$\frac{1}{2}\int_{\widetilde{U}} \widetilde{v}\widetilde{\Delta}\widetilde{u} d\widetilde{x}_1 \widetilde{x}_2 = \int_U \widetilde{v}\mathcal{A}_\varepsilon \widetilde{u} dx_1 dx_2,$$

if $\widetilde{v}$ has a compact support in $\widetilde{U}$. The last identity actually holds for any smooth function $\widetilde{v}$ on $\widetilde{U}$ by approximating it with smooth functions of compact support in $\widetilde{U}$. Hence (2.4.17) for $\widetilde{u} \circ f_\varepsilon, \widetilde{v} \circ f_\varepsilon$ in place of u, v along with Green's first formula yields

$$\int_{\partial \widetilde{U}} \widetilde{v} \frac{\partial \widetilde{u}}{\partial \widetilde{\mathbf{n}}} d\widetilde{s} = 2\int_{\partial U} \widetilde{v} \sum_{i,j=1}^{2} A_{ij}^{(\varepsilon)} \frac{\partial \widetilde{u}}{\partial x_i}\frac{\partial x_i}{\partial \mathbf{n}} ds,$$

holding for any smooth function $\widetilde{v}$ on $\partial\widetilde{U}$. Approximating the indicator function of $\widetilde{\gamma}$ by uniformly bounded smooth functions $\widetilde{v}$ on $\partial\widetilde{U}$, we arrive at (2.4.19). $\square$

For a function $\phi(z, \zeta)$ on $((\overline{\mathbb{H}} \setminus K) \cup K^p) \times \partial\mathbb{H} \setminus d$ where $d = \{(z, \zeta) : z = \zeta \in \partial\mathbb{H}\}$, and a function $\psi(z, \zeta, \varepsilon)$ on $[((\overline{\mathbb{H}} \setminus K) \cup K^p) \times \partial\mathbb{H} \setminus d] \times (0, \varepsilon_0)$ for some $\varepsilon_0 > 0$, we write

$$\phi(z, \zeta) \sim \psi(z, \zeta, \varepsilon), \quad z \in (\overline{\mathbb{H}} \setminus K) \cup K^p, \quad \zeta \in \partial\mathbb{H}, \quad \varepsilon \in (0, \varepsilon_0),$$

if, for any compact set $Q \subset \overline{\mathbb{H}}$ containing $\bigcup_{j=1}^N U_j$ and any compact set $J \subset \partial\mathbb{H}$, there exists a positive constant $L_{Q,J}$ independent of $\varepsilon \in (0, \varepsilon_0)$, $D \in \mathcal{D}_0$ and $\widetilde{D} = \widetilde{f}_\varepsilon(D) \in \mathcal{D}$ such that

$$|\phi(z, \zeta) - \psi(z, \zeta, \varepsilon)| \le L_{Q,J} \cdot \varepsilon, \quad z \in (Q \setminus K \setminus J) \cup K^p, \ \zeta \in J, \ \varepsilon \in (0, \varepsilon_0).$$

Using this notation, the third assertion of Theorem 2.4.1 can be simply expressed as

$$\Psi_D(z, \zeta) \sim \Psi_{\widetilde{D}}(\widetilde{f}_\varepsilon(z), \zeta), \quad z \in (\overline{\mathbb{H}} \setminus K) \cup K^p, \quad \zeta \in \partial\mathbb{H}, \quad \varepsilon \in (0, \varepsilon_0).$$

We also use the analogous notations:

$u(z) \sim v(z, \varepsilon)$, $z \in (\overline{\mathbb{H}} \setminus K) \cup K^p$, $\varepsilon \in (0, \varepsilon_0)$ for functions $u(z)$ on $(\overline{\mathbb{H}} \setminus K) \cup K^p$ and $v(z, \varepsilon)$ on $(\overline{\mathbb{H}} \setminus K) \cup \partial_p K \times (0, \varepsilon_0)$ if, for any compact set $Q \subset \overline{\mathbb{H}}$ containing $\bigcup_{j=1}^N U_j$, there exists a positive constant L_Q independent of $\varepsilon \in (0, \varepsilon_0)$, $D \in \mathcal{D}_0$ and $\widetilde{D} = \widetilde{f}_\varepsilon(D) \in \mathcal{D}$ such that $|u(z) - v(z, \varepsilon)| \le L_Q \cdot \varepsilon$, $z \in (Q \setminus K) \cup K^p$, $\varepsilon \in (0, \varepsilon_0)$;

$f(\zeta) \sim g(\zeta, \varepsilon)$, $\zeta \in \partial\mathbb{H}$, $\varepsilon \in (0, \varepsilon_0)$, for functions $f(\zeta)$ on $\partial\mathbb{H}$ and $g(\zeta, \varepsilon)$ on $\partial\mathbb{H} \times (0, \varepsilon_0)$, if, for any compact set $J \subset \partial\mathbb{H}$, there exists a positive constant L_2 independent of $\varepsilon \in (0, \varepsilon_0)$, $D \in \mathcal{D}_0$ and $\widetilde{D} = \widetilde{f}_\varepsilon(D) \in \mathcal{D}$ such that

$|f(\zeta)-g(\zeta,\varepsilon)| \le L_2\cdot\varepsilon,\ \zeta\in J,\ \varepsilon\in(0,\varepsilon_0);\ c\sim c(\varepsilon),\quad \varepsilon\in(0,\varepsilon_0),$ for a constant c and a function $c(\varepsilon)$ of $\varepsilon\in(0,\varepsilon_0)$, if there exists a positive constant L_3 independent of $\varepsilon\in(0,\varepsilon_0)$, $D\in\mathcal{D}_0$ and $\widetilde{D}=\widetilde{f}_\varepsilon(D)\in\mathcal{D}$ such that $|c-c(\varepsilon)|\le L_3,\ \varepsilon\in(0,\varepsilon_0)$.

Recall the explicit formula (1.7.4)–(1.7.5) of the BMD-Poisson kernel $K^*(z,\zeta)$ for D:

$$K^*(z,\zeta)=-\frac{1}{2}\frac{\partial}{\partial\mathbf{n}_\zeta}G(z,\zeta)-\Phi(z)\mathcal{A}^{-1}\frac{\partial}{\partial\mathbf{n}_\zeta}\Phi(\zeta)^{\rm tr},\quad z\in D,\ \zeta\in\partial\mathbb{H}. \tag{2.4.20}$$

where $\Phi(z)$ is the vector with entries $\varphi^{(i)}(z)$, $1\le i\le N$, being the harmonic basis for D and $\mathcal{A}$ is an $N\times N$-matrix whose (i,j)-component a_{ij} is the period of $\varphi^{(j)}$ around the slit C_i, $1\le i,j\le N$. The corresponding notions for $\widetilde{D}=\widetilde{f}_\varepsilon(D)$ will be designated with $\widetilde{\ }$. Let $\varepsilon_0>0$ be as in the paragraph below the statement of Proposition 2.4.3.

Proposition 2.4.5. *There exists $\widehat{\varepsilon}_0\in(0,\varepsilon_0]$ such that*

$$K^*(z,\zeta)\sim\widetilde{K}^*(\widetilde{f}_\varepsilon(z),\zeta),\quad z\in(\overline{\mathbb{H}}\setminus K)\cup K^p,\quad \zeta\in\partial\mathbb{H},\quad \varepsilon\in(0,\widehat{\varepsilon}_0). \tag{2.4.21}$$

To prove this proposition, we prepare three lemmas.

Lemma 2.4.6. *It holds that*

$$\frac{\partial}{\partial\mathbf{n}_\zeta}G(w,\zeta)\sim\frac{\partial}{\partial\mathbf{n}_\zeta}\widetilde{G}(\widetilde{f}_\varepsilon(w),\zeta),\quad w\in(\overline{\mathbb{H}}\setminus K)\cup K^p,\quad \zeta\in\partial\mathbb{H},\quad \varepsilon\in(0,\varepsilon_0). \tag{2.4.22}$$

$$\varphi^{(i)}(w)\sim\widetilde{\varphi}^{(i)}(\widetilde{f}_\varepsilon(w))\quad w\in(\overline{\mathbb{H}}\setminus K)\cup K^p,\quad 1\le i\le N,\quad \varepsilon\in(0,\varepsilon_0). \tag{2.4.23}$$

$$\frac{\partial}{\partial\mathbf{n}_\zeta}\varphi^{(i)}(\zeta)\sim\frac{\partial}{\partial\mathbf{n}_\zeta}\widetilde{\varphi}^{(i)}(\zeta),\quad \zeta\in\partial\mathbb{H},\quad 1\le i\le N,\quad \varepsilon\in(0,\varepsilon_0). \tag{2.4.24}$$

Furthermore, there exists a function $\eta_{ij}(\varepsilon)$ bounded in $\varepsilon\in(0,\varepsilon_0)$ uniformly in $D\in\mathcal{D}_0$ and $\widetilde{D}=\widetilde{f}_\varepsilon(D)$ such that

$$a_{ij}-\widetilde{a}_{ij}=\varepsilon\eta_{ij}(\varepsilon),\quad 1\le i,j\le N. \tag{2.4.25}$$

Proof. Due to the expression (1.1.12) of the Green function $G^{\mathbb{H}}$ of the ABM on $\mathbb{H}$, we have $\lim_{w\in\mathbb{H},w\to\infty}G^{\mathbb{H}}\mathbb{1}_\Lambda(w)=0$ and so $G^{\mathbb{H}}\mathbb{1}_\Lambda(w)$ is bounded in $w\in\mathbb{H}$. By taking the partial derivative in ζ_2 at $\zeta\in\partial\mathbb{H}$ on both sides of (2.4.13) and using the symmetry of G, Proposition 2.4.3 (v) as well as the domination $G\le G^{\mathbb{H}}$, we get (2.4.22).

Let $\gamma_i\subset U_i$ be a smooth Jordan curve surrounding C_i for each $1\le i\le N$. By Proposition 1.1.5 (iii), $\varphi^{(i)}(w)=-\frac{1}{2}I(G(\cdot,w);\gamma_i)$. On the other hand, since $A^{(\varepsilon)}=\frac{1}{2}\Delta$ on U by Proposition 2.4.3 (ii), (2.4.19) applied to $\widetilde{u}(\cdot)=\widetilde{G}(\cdot,\widetilde{f}_\varepsilon(w))$ reads

$$\widetilde{\varphi}^{(i)}(\widetilde{f}_\varepsilon(w))=-\frac{1}{2}\widetilde{I}(\widetilde{u};\widetilde{\gamma}_i)=-\frac{1}{2}I(g(\cdot,w,\varepsilon);\gamma_i). \tag{2.4.26}$$

By taking the period around C_i with respect to ζ on both sides of (2.4.13), we obtain (2.4.23) by (2.4.26), Proposition 2.4.3 (vi) and the domination $G\le G^{\mathbb{H}}$.

We now use (2.4.12) instead of (2.4.13). In both sides of (2.4.12), we take the period around C_i with respect to w and use (2.4.26) to obtain

$$\varphi^{(i)}(\zeta) - \widetilde{\varphi}^{(i)}(\widetilde{f}_\varepsilon(\zeta)) = -\varepsilon \int_\Lambda \mathcal{B}_z^{(\varepsilon)} G(z,\zeta)\widetilde{\varphi}^{(i)}(\widetilde{f}_\varepsilon(z))dx_1 dx_2. \tag{2.4.27}$$

Keeping in mind that $0 \le \widetilde{\varphi}^{(i)}(\widetilde{f}_\varepsilon(z)) \le 1$, we perform two operations in (2.4.27) with respect to ζ. Firstly, we take the normal derivative in ζ_2 at $\zeta \in \partial\mathbb{H}$ and use Proposition 2.4.3 (v) to have (2.4.24). Secondly we take the period for the curve $\gamma_j \subset U_j$ surrounding C_j and use (2.4.19) as well as Proposition 2.4.3 (vi) again to get (2.4.25). □

The space $\mathcal{D}$ is the collection of all labeled standard slits domains equipped with metric d of (2.3.14). Let us define an open subset $\mathcal{S}$ of the Euclidean space $\mathbb{R}^{3N}$ by

$$\mathcal{S} = \Big\{ \mathbf{s} := (\mathbf{y}, \mathbf{x}, \mathbf{x}^r) \in \mathbb{R}^{3N} : \mathbf{y},\ \mathbf{x},\ \mathbf{x}^r \in \mathbb{R}^N,\ \mathbf{y} > \mathbf{0},\ \mathbf{x} < \mathbf{x}^r,$$
$$\text{either } x_j^r < x_k \text{ or } x_k^r < x_j \text{ whenever } y_j = y_k,\ j \neq k \Big\}. \tag{2.4.28}$$

The space $\mathcal{D}$ can be identified with $\mathcal{S}$ as a topological space. To $\mathbf{s} = (\mathbf{y}, \mathbf{x}, \mathbf{x}^r) \in \mathcal{S}$, we correspond $D \in \mathcal{D}$ having $x_i + iy_i$ and $x_i^r + iy_i$ as the left and right endpoints of the slit C_i, respectively, $1 \le i \le N$. With this identification, the subspace $\mathcal{D}_0$ of $\mathcal{D}$ defined by (2.4.2) can be seen to be a bounded closed subset and hence a compact subset of $\mathcal{D}$.

Lemma 2.4.7. *Let b_{ij} (resp. $\widetilde{b}_{ij}$) be the (i,j)-component of $\mathcal{A}^{-1}$ (resp. $\widetilde{\mathcal{A}}^{-1}$), $1 \le i,j \le N$. There exists then $\widehat{\varepsilon}_0 \in (0, \varepsilon_0]$ such that*

$$\widetilde{b}_{ij} \sim b_{ij}, \quad \varepsilon \in (0, \widehat{\varepsilon}_0), \qquad 1 \le i,j \le N, \tag{2.4.29}$$

Proof. The symmetric matrix with components a_{ij} (resp. $\widetilde{a}_{ij}$) will be denoted by $\mathcal{A}(D)$ (resp. $\mathcal{A}(\widetilde{D})$) to indicate it to be a function of $\mathcal{D}$ (resp. $\widetilde{D}$). The symmetric matrix with components $\eta_{ij}(\varepsilon)$ in (2.4.25) is denoted by $C(\varepsilon, D, \widetilde{D})$ to indicate its dependence on D and $\widetilde{D}$. Then (2.4.25) reads

$$\mathcal{A}(\widetilde{D}) = \mathcal{A}(D) - \varepsilon C(\varepsilon, D, \widetilde{D}). \tag{2.4.30}$$

For $D \in \mathcal{D}$ and $\delta > 0$, let $\mathcal{N}_\delta(D) = \{\widehat{D} \in \mathcal{D} : \text{dist}(\widehat{D}, D) < \delta\}$. We first show that $\mathcal{A}(D)^{-1}$ is uniformly bounded in $D \in \mathcal{D}_0$.

For each $D \in \mathcal{D}_0$, denote by $\lambda_i(\varepsilon, D, \widetilde{D})$, $1 \le i \le N$, the eigenvalues of $\mathcal{A}(D)^{-1}C(\varepsilon, D, \widetilde{D})$. Since $\sup_{\varepsilon \in (0,\varepsilon_0)} \sup_{\widetilde{D} \in \mathcal{N}_\varepsilon(D)} |\eta_{ij}(\varepsilon, D, \widetilde{D})| < \infty$, $1 \le i,j \le N$, by Lemma 2.4.6, we have

$$\max_{1 \le i \le N} \sup_{\varepsilon \in (0,\varepsilon_0)} \sup_{\widetilde{D} \in \mathcal{N}_\varepsilon(D)} |\lambda_i(\varepsilon, D, \widetilde{D})| =: \kappa(D) < \infty.$$

For each $D \in \mathcal{D}_0$, choose $\varepsilon(D) \in (0, \varepsilon_0)$ with $\varepsilon(D)\kappa(D) < 1/2$. As $\mathcal{D}_0$ is compact subset of $\mathcal{D}$, we can select finite number of domains $D_1, \cdots, D_\ell \in \mathcal{D}_0$ such that

$$\mathcal{D}_0 \subset \bigcup_{k=1}^{\ell} \mathcal{N}_{\varepsilon_k}(D_k) \quad \text{for } \varepsilon_k = \varepsilon(D_k).$$

Now, for any $D \in \mathcal{D}_0$, pick $D_k \in \mathcal{D}_0$ with $D \in \mathcal{N}_{\varepsilon_k}(D_k)$. By (2.4.30),

$$\mathcal{A}(D) = \mathcal{A}(D_k) - \varepsilon_k C(\varepsilon_k, D_k, D) = \mathcal{A}(D_k)\left[I - \varepsilon_k \mathcal{A}(D_k)^{-1} C(\varepsilon_k, D_k, D)\right].$$

Let $\lambda_{ki} = \lambda_{ki}(\varepsilon, D_k, D), 1 \leq i \leq N$, be the eigenvalues of $M_k = \mathcal{A}(D_k)^{-1}C(\varepsilon_k, D_k, D)$. Then $\varepsilon_k|\lambda_{ki}| < 1/2, 1 \leq i \leq N$, and, for the orthogonal matrix O_k diagonalizing M_k,

$$\mathcal{A}(D)^{-1} = (I - \varepsilon_k M_k)^{-1}\mathcal{A}(D_k)^{-1} = [I + \cdot O_k E_k O_k^{-1}] \cdot \mathcal{A}(D_k)^{-1}, \tag{2.4.31}$$

where E_k is a diagonal matrix with the diagonal entry $(\varepsilon_k\lambda_k)/(1-\varepsilon_k\lambda_k)$, whose absolute value is less than 1. Hence $\mathcal{A}(D)^{-1}$ is uniformly bounded in $D \in \mathcal{D}_0$.

It follows from (2.4.30) that

$$\mathcal{A}(\widetilde{D}) = \mathcal{A}(D)[I - \varepsilon \mathcal{A}(D)^{-1} C(\varepsilon, D, \widetilde{D})].$$

We see now that $\mathcal{A}(D)^{-1}C(\varepsilon, D, \widetilde{D})$ is bounded in $\varepsilon \in (0, \varepsilon_0)$ uniformly in $D \in \mathcal{D}_0$ and $\widetilde{D} \in \mathcal{N}_\varepsilon(D)$, and consequently we can use its eigenvalues as in the above to find $\widehat{\varepsilon}_0 \in (0, \varepsilon_0)$ such that, for any $\varepsilon \in (0, \widehat{\varepsilon}_0)$,

$$\mathcal{A}(\widetilde{D})^{-1} = (I + \varepsilon \widehat{C}(\varepsilon, D, \widehat{D}))\mathcal{A}(D)^{-1} = \mathcal{A}(D)^{-1} + \varepsilon \widehat{C}(\varepsilon)\mathcal{A}(D)^{-1},$$

where $\widehat{C}(\varepsilon)$ is matrix whose entries are bounded uniformly in $\varepsilon \in (0, \widehat{\varepsilon}_0)$, $D \in \mathcal{D}_0$ and $\widetilde{D} = \widetilde{f}_\varepsilon(D)$. □

Recall the 0-order capacity $\mathrm{Cap}_0^{\mathbb{H}}$ of the subsets of $\mathbb{H}$ relative to the transient extended Sobolev space $(H^1_{0,e}(\mathbb{H}), \frac{1}{2}\mathbf{D})$ that has appeared in Lemma 1.7.3. Just as this lemma was obtained from (1.7.7) and (1.7.8), we immediately obtain the following lemma from them.

Lemma 2.4.8. *It holds that*

$$0 \leq -\frac{\partial \varphi^{(j)}(\zeta)}{\partial \mathbf{n}_\zeta} \leq \frac{2}{\pi b_0}\mathrm{Cap}_0^{\mathbb{H}}(C_j), \quad \zeta \in \partial\mathbb{H}, \quad 1 \leq j \leq N, \tag{2.4.32}$$

where b_0 *is a positive number defined by* (2.4.16).

Proof of Proposition 2.4.5. Let $\widehat{\varepsilon}_0 > 0$ be as in Lemma 2.4.7. For each $1 \leq i, j \leq N$, b_{ij} is designated as $b_{ij}(D)$ to indicate it to be a function of $D \in \mathcal{D}$. As has been observed, $\mathcal{D}_0$ is a compact subspace of $\mathcal{D}$, and so there exist finite number of domains $D_1, \cdots, D_m \in \mathcal{D}_0$ such that $\mathcal{D}_0 \subset \bigcup_{k=1}^m \mathcal{N}_{\widehat{\varepsilon}_0/2}(D_k)$. Let $L_5 = \max_{1\leq k\leq m}|b_{ij}(D_k)|$. Then, by (2.4.29),

$$|b_{ij}(D)| \leq L_5 + (\widehat{\varepsilon}_0/2)L_4 =: L_6 < \infty \quad \text{for any } D \in \mathcal{D}_0. \tag{2.4.33}$$

This combined with (2.4.32) yields

$$\begin{aligned}\left|b_{ij}\frac{\partial\varphi^{(j)}(\zeta)}{\partial\mathbf{n}_\zeta} - \widetilde{b}_{ij}\frac{\partial\widetilde{\varphi}^{(j)}(\zeta)}{\partial\mathbf{n}_\zeta}\right| &\leq |b_{ij}|\left|\frac{\partial\varphi^{(j)}(\zeta)}{\partial\mathbf{n}_\zeta} - \frac{\partial\widetilde{\varphi}^{(j)}(\zeta)}{\partial\mathbf{n}_\zeta}\right| + \left|\frac{\partial\widetilde{\varphi}^{(j)}(\zeta)}{\partial\mathbf{n}_\zeta}\right||b_{ij} - \widetilde{b}_{ij}| \\ &\leq L_6\left|\frac{\partial\varphi^{(j)}(\zeta)}{\partial\mathbf{n}_\zeta} - \frac{\partial\widetilde{\varphi}^{(j)}(\zeta)}{\partial\mathbf{n}_\zeta}\right| \\ &\quad + \frac{2}{\pi b_0}\mathrm{Cap}_0^{\mathbb{H}}(C_j)|b_{ij} - \widetilde{b}_{ij}|, \quad \zeta \in \partial\mathbb{H}.\end{aligned}$$

It then follows from (2.4.24) and (2.4.29) that

$$b_{ij}\frac{\partial\varphi^{(j)}(\zeta)}{\partial\mathbf{n}_\zeta}\sim\widetilde{b}_{ij}\frac{\partial\widetilde{\varphi}^{(j)}(\zeta)}{\partial\mathbf{n}_\zeta},\quad \varepsilon\in(0,\widehat{\varepsilon}_0). \tag{2.4.34}$$

As $\varphi^{(i)}(w)$ is uniformly bounded by 1, (2.4.23) also implies

$$\varphi^{(i)}(w)b_{ij}\frac{\partial\varphi^{(j)}(\zeta)}{\partial\mathbf{n}_\zeta}\sim\widetilde{\varphi}^{(i)}(\widetilde{f}_\varepsilon(w))\widetilde{b}_{ij}\frac{\partial\widetilde{\varphi}^{(j)}(\zeta)}{\partial\mathbf{n}_\zeta},\quad \varepsilon\in(0,\widehat{\varepsilon}_0).$$

The proof of (2.4.21) is now complete in view of (2.4.20) and (2.4.22). □

Recall that $K^*(z,\zeta)$ is the imaginary part of the complex Poisson kernel $\Psi_D(z,\zeta)$. In order to establish an analogous assertion to Proposition 2.4.5 for the real part of $\Psi_D(z,\zeta)$, we need to prepare two lemmas and a proposition. The first lemma concerns an estimate of the density function of the exit distribution of the ABM on a slit domain.

For $R>0$, consider a slit domain $\Delta_R=\mathbb{C}\setminus A_R$ where $A_R=\{z=x+i0:|x|\le R\}$. The boundary of Δ_R (the space of its prime ends as was defined in Section 2.1.2) is $A_R^p=A_R^{0,+}\cup A_R^{0,-}\cup\{-R+i0,R+i0\}$, where $A_R^{0,+}$ (resp. $A_R^{0,-}$) denotes the upper (resp. lower) side of the line segment $\{z=x+i0:|x|<R\}$.

For $\delta>0$, let $E_{R,\delta}$ be the ellipse surrounding A_R^p defined by $x^2/(R+\delta)^2+y^2/(2\delta)^2=1,\ z=x+iy\in E_{R,\delta}$.

Lemma 2.4.9. *Let $\mathbf{Z}^{\Delta_R}=(Z_t^{\Delta_R},\zeta^{\Delta_R},\mathbb{P}_z^{\Delta_R})$ be the ABM on Δ_R. It holds for any $\delta>0$ that*

$$\mathbb{E}_z^{\Delta_R}\left[f\left(Z^{\Delta_R}_{\zeta^{\Delta_R}-}\right)\right]\le\frac{1}{2\pi}\left(1+\frac{R}{\delta}\right)\int_{A_R^p}f(\xi)\frac{1}{\sqrt{R^2-\xi^2}}d\xi, \tag{2.4.35}$$

for any non-negative Borel function f on A_R^p, whenever $z\in\mathbb{C}$ is located outside the ellipse $E_{R,\delta}$.

Proof. We consider the map $\psi(z')=z'+(R^2/4z')$ that is a conformal map from $\mathbb{C}_{R/2}=\mathbb{C}\setminus B_{R/2}$ onto Δ_R, sending $\frac{R}{2}e^{i\theta}\in\partial B_{R/2}$ to $R\cos\theta\in A_R^p$, $0\le\theta<2\pi$. For $S=\frac{R}{2}+\delta$, $\delta>0$, the image $\psi(\partial B_S)$ of the circle ∂B_S is the ellipse $x^2/[S+(R^2/(4S)]^2+y^2/[S-(R^2/(4S)]^2=1,\quad z=x+iy$, which is located inside of the ellipse $E_{R,\delta}$. Hence, if a point $z\in\mathbb{C}$ is located outside $E_{R,\delta}$, then its preimage $z'=\psi^{-1}(z)$ satisfies $|z'|>S=\frac{R}{2}+\delta$.

As in the proof of Proposition 1.1.2, we make use of the conformal invariance of the ABM $\mathbf{Z}^{\mathbb{C}_{R/2}}:=(Z_t^{\mathbb{C}_{R/2}},\zeta^{\mathbb{C}_{R/2}},\mathbb{P}_{z'}^{\mathbb{C}_{R/2}})$ on $\mathbb{C}_{R/2}$ according to Theorem A.2.1 in the Appendix as follows. Let $\check{\mathbf{Z}}^{\mathbb{C}_{R/2}}:=(\check{Z}_t,\check{\zeta},\mathbb{P}_{z'}^{\mathbb{C}_{R/2}})_{z'\in\mathbb{C}_{R/2}}$ be the time change of $\mathbf{Z}^{\mathbb{C}_{R/2}}$ by means of its positive continuous additive functional $A_t=\int_0^t|\psi'(Z_s^{\mathbb{C}_{R/2}})|^2ds$, namely $\check{Z}_t=Z^{\mathbb{C}_{R/2}}_{A_t^{-1}}$, $\check{\zeta}=A_{\zeta^{\mathbb{C}_{R/2}}}$. Then, for any $z'\in\mathbb{C}_{R/2}$,

$$\left(\psi(\check{Z}_t),\check{\zeta},\mathbb{P}_{z'}^{\mathbb{C}_{R/2}}\right)\text{ is identical in law with }(Z_t^{\Delta_R},\zeta^{\Delta_R},\mathbb{P}^{\Delta_R}_{\psi(z')}).$$

If we define $p_{(\alpha,\beta)}(z) = \mathbb{P}_z^{\Delta_R}\left(Z^{\Delta_R}_{\zeta^{\Delta_R}-} \in (\alpha,\beta)\right)$ for $z \in \Delta_R$ and $\alpha, \beta \in A_R^{0,+}, \alpha < \beta$, we get from the above and the exterior Poisson integral formula for a bounded harmonic function on $\mathbb{C}_{R/2}$ (cf. [Helms (2009), Lemma 5.2.2])

$$
\begin{aligned}
p_{(\alpha,\beta)}(\psi(z')) &= \mathbb{P}_{z'}^{\mathbb{C}_{R/2}}\left(\check{Z}_{\check{\zeta}-} = (R/2)e^{i\theta},\ \cos^{-1}(\beta/R) < \theta < \cos^{-1}(\alpha/R)\right) \\
&= \frac{1}{2\pi}\int_{\cos^{-1}(\beta/R)}^{\cos^{-1}(\alpha/R)} \frac{|z'|^2 - R^2/4}{|z' - (R/2)e^{i\theta}|^2} d\theta \\
&\le \frac{1}{2\pi}\left[1 + \frac{2R}{2|z'| - R}\right]\int_\alpha^\beta \frac{d\xi}{\sqrt{R^2 - \xi^2}}, \quad z' \in \mathbb{C}_{R/2}.
\end{aligned}
$$

which yields (2.4.35) because, if $z = \psi(z')$ is located outside the ellipse $E_{R,\delta}$, then $|z'| > \frac{R}{2} + \delta$ as has been observed. □

Lemma 2.4.10. *Define for $R > 0$*

$$
I_R(w) = \int_{-R}^{R}\left(\log_+ \frac{1}{|w - x|}\right)\frac{1}{\sqrt{R^2 - x^2}} dx, \quad w \in \mathbb{R}.
$$

Here $\log_+ a = 0 \vee (\log a)$ for $a > 0$. Then

$$
I_R(w) \le c_1 + \frac{c_2}{\sqrt{R}} + c_3 \log_+ \frac{1}{R} + c_4 \log_+ R, \quad w \in \mathbb{R}
$$

for some positive constants c_k, $1 \le k \le 4$, independent of $R > 0$ and $w \in \mathbb{R}$, while $I_R(w) = 0$ for $|w| \ge R + 1$.

Proof. $I_R(w) \le \frac{1}{\sqrt{R}}(I_+(w) + I_+(-w))$ for $I_+(w) = \int_{(0,R)}\left(\log_+ \frac{1}{|w-x|}\right)\frac{1}{\sqrt{R-x}}dx$. When $w \in (0,R)$, let $(0,R) = J_1 + J_2 + J_3$, where $J_1 := (0,w)$, $J_2 := (w, (w+R)/2)$ and $J_3 = ((w+R)/2, R)$. The contributions to the integral $I_+(w)$ from J_1, J_2, J_3 are bounded by 4, 3/2 and $1 + 2\sqrt{R}\log_+ R/2$, respectively. Note that $I_+(w)$ vanishes when $w > R+1$ or $w < -1$, and $I_+(w) \le 4$ when $w \in (R, R+1)$. When $-1 < w < 0$, $I_+(w) \le d_1 + d_2\sqrt{R} + d_3\sqrt{R}\log_+ \frac{1}{R}$ for positive constants d_1, d_2, d_3 independent of $R > 0$ and $w \in \mathbb{R}$. Thus we obtain the stated bound. □

We write $\Psi_D(w,\zeta) = u(w,\zeta) + iK^*(w,\zeta)$, $w \in D$, $\zeta \in \partial\mathbb{H}$. Fix a compact interval $J \subset \partial\mathbb{H}$, take a point $w_0 = \xi_0 + i0 \in \partial\mathbb{H}$ to the right of J and consider the half line $\Gamma = \{w \in \partial\mathbb{H} : \xi_0 \le \Re w\} \subset \partial\mathbb{H}$. For each $w \in D$, let $\gamma(w)$ be any smooth simple curve in D joining w to w_0. As we saw in the proof of Theorem 1.9.1, $u(w,\zeta)$, $w \in D$, $\zeta \in J$, can be evaluated by

$$
u(w,\zeta) = I(K^*(\cdot,\zeta); \gamma(w)) + I(K^*(\cdot,\zeta); \Gamma), \tag{2.4.36}
$$

which is independent of the choice of the curve $\gamma(w)$.

We make a special choice of $w_0 = \xi_0 + i0$ and $\gamma(w)$ for $w = w_1 + iw_2 \in D$ as follows. Let $a_0 = \sup_{w \in \overline{V}} \Re w$ and take ξ_0 with $a_0 < \xi_0$. Denote by ℓ_j the imaginary part of point of C_j, $1 \le i \le N$. $\gamma(w)$ is then a curve consisting of a vertical line segment γ_0 starting at w_0 ending at (ξ_0, η), a horizontal line segment γ_1 through D starting at (ξ_0, η) ending at (w_1, η) and a vertical line γ_2 starting at (w_1, η) ending at w.

η is chosen to satisfy

$$\left(\min_{1\leq j\leq N} |\eta - \ell_j|\right) \wedge \eta > \delta_0, \quad \text{for some positive } \delta_0 \text{ independent of } w \in D. \tag{2.4.37}$$

We occasionally write $\gamma_0, \gamma_1, \gamma_2$ as $\gamma_0(\xi_0, \eta)$, $\gamma_1(w_1, \eta)$, $\gamma_2(w, \eta)$, respectively, to indicate their dependence on the endpoints.

Proposition 2.4.11. *There exists a positive $\varepsilon_0' \leq \varepsilon_0$ independent of $D \in \mathcal{D}_0$ such that, for any positive $\varepsilon < \varepsilon_0'$ and for any relatively compact open set Q with $\overline{V} \subset Q \subset \mathbb{H}$, the inequality*

$$|\widetilde{I}(\widetilde{G}(\cdot, \widetilde{f}_\varepsilon(z)); \widetilde{\gamma}(\widetilde{f}_\varepsilon(w))| \leq C_0 + C_1 \log_+ \frac{1}{|x_1 - w_1|} + C_2 \log_+ \frac{1}{|x_2 - \eta|}, \tag{2.4.38}$$

holds for any $z = x_1 + ix_2 \in \Lambda$, $w = w_1 + iw_2 \in Q \setminus K$, where C_0, C_1, C_2 are positive constants independent of $\varepsilon \in (0, \varepsilon_0')$, $D \in \mathcal{D}_0$ and $\widetilde{f}_\varepsilon(D)$.

Proof. Let $Q = (-\alpha, \alpha) \times (0, \beta)$ for $\alpha > 0, \beta > 0$ with $Q \supset \overline{V}$. We assume for η that $\eta < \beta$ besides (2.4.37). We also assume for ξ_0 that $a_0 < \xi_0 < \alpha$. By (2.4.19) and the definition (2.4.8) of $g(\cdot, \cdot, \varepsilon)$, we have for $w \in D$ and $z \in \Lambda$

$$\widetilde{I}(\widetilde{G}(\cdot, \widetilde{f}_\varepsilon(z)), \widetilde{\gamma}(\widetilde{f}_\varepsilon(w)) = 2 \int_{\gamma(w)} \sum_{k,\ell=1}^{2} A_{k\ell}^{(\varepsilon)} \frac{\partial g(z, w', \varepsilon)}{\partial w_k'} \frac{\partial w_\ell'}{\partial \mathbf{n}} ds(w'), \tag{2.4.39}$$

which equals $\widetilde{I}_0 + \widetilde{I}_1 + \widetilde{I}_2$ where

$$\begin{cases} \widetilde{I}_0 = 2\sum_{k=1}^2 \int_{\gamma_0(\xi_0,\eta)} A_{k1}^{(\varepsilon)}(w') g_{w_k'}(z, w', \varepsilon) dw_2', \quad j = 1, 2, \\ \widetilde{I}_1 = 2\sum_{k=1}^2 \int_{\gamma_1(w_1,\eta)} A_{k2}^{(\varepsilon)}(w') g_{w_k'}(z, w', \varepsilon) dw_1', \\ \widetilde{I}_2 = 2\sum_{k=1}^2 \int_{\gamma_2(w,\eta)} A_{k1}^{(\varepsilon)}(w') g_{w_k'}(z, w', \varepsilon) dw_2', \quad j = 1, 2. \end{cases}$$

Let $M_1 > 0$ be a uniform bound of $|A_{k\ell}^{(\varepsilon)}(w)|$, $w \in D, 1 \leq k, \ell \leq 2$. Then $|\widetilde{I}_i| \leq 2M_1 \widetilde{J}_i$, $i = 0, 1, 2$, for

$$\begin{cases} \widetilde{J}_0 = \sum_{k=1}^2 \int_{\gamma_0} \left|\frac{\partial}{\partial w_k'} g(z, w', \varepsilon)\right| dw_2', \quad (\gamma_0 = \gamma_0(\xi_0, \eta)), \\ \widetilde{J}_i = \sum_{k=1}^2 \int_{\gamma_i} \left|\frac{\partial}{\partial w_k'} g(z, w', \varepsilon)\right| dw_i', \quad i = 1, 2, \quad (\gamma_1 = \gamma_1(w_1, \eta), \ \gamma_2 = \gamma_2(w, \eta)). \end{cases} \tag{2.4.40}$$

In view of (1.1.14),

$$g(w', z, \varepsilon) = -\frac{1}{2\pi} \log |\widetilde{f}_\varepsilon(w') - \widetilde{f}_\varepsilon(z)|^2 + \frac{1}{2\pi} \int_{\partial \widetilde{D}} \widetilde{\Pi}(\widetilde{f}_\varepsilon(z), d\widetilde{\zeta}) \log |\widetilde{\zeta} - \widetilde{f}_\varepsilon(w')|^2, \tag{2.4.41}$$

where $\widetilde{\Pi}(\widetilde{f}_\varepsilon(z), \cdot)$ is the exit distribution from the standard slit domain $\widetilde{D}$ of the planar Brownian motion starting at $\widetilde{f}_\varepsilon(z)$ for $z \in \Lambda$. The first and second term of the right hand side of (2.4.41) will be designated by $L_1(z, w', \varepsilon)$ and $L_2(z, w', \varepsilon)$, respectively.

First of all, we examine the contributions of $L_1(z, w', \varepsilon)$ to $\widetilde{J}_i$'s. Denote by $F_1(x_1, x_2)$ and $F_2(x_1, x_2)$ the real and imaginary parts of $F_\varepsilon(x_1, x_2)$ defined by (2.4.7) for $z = x_1 + ix_2$. We take $M_2 > 0$ to be a uniform bound of $|\frac{\partial}{\partial x_k} F_i(x_1, x_2)|$ for $1 \leq i, k \leq 2$, and we let $\varepsilon_1 = \frac{1}{8M_2} \wedge \varepsilon_0$. We then have, for any $\varepsilon \in (0, \varepsilon_1)$ and for any $z, w' \in \mathbb{H}$, $z \neq w'$

$$\frac{1}{2\pi}\left|\frac{\partial}{\partial w_1'} \log |\widetilde{f}_\varepsilon(w') - \widetilde{f}_\varepsilon(z)|^2\right| \leq \frac{41}{16\pi} \frac{|w_1' - x_1| + |w_2' - x_2|}{(w_1' - x_1)^2 + (w_2' - x_2)^2}. \tag{2.4.42}$$

To verify this, put $\widetilde{f}_\varepsilon(z) = f_1(x_1, x_2) + if_2(x_1, x_2)$, $z = x_1 + ix_2$, and write

$$\begin{cases} f_1(w') - f_1(z) = (w_1' - x_1) + \varepsilon\{F_1(w') - F_1(z)\} =: A_1 + \varepsilon B_1 \\ f_2(w') - f_2(z) = (w_2' - x_2) + \varepsilon\{F_2(w') - F_2(z)\} =: A_2 + \varepsilon B_2. \end{cases}$$

Then

$$\left|\frac{\partial}{\partial w_1'} \log |\widetilde{f}_\varepsilon(w') - \widetilde{f}_\varepsilon(z)|^2\right| = 2\frac{|(A_1 + \varepsilon B_1)(1 + \varepsilon F_{1w_1}) + (A_2 + \varepsilon B_2) \cdot \varepsilon F_{2w_1}|}{(A_1 + \varepsilon B_1)^2 + (A_2 + \varepsilon B_2)^2}.$$

Since $|B_k| \leq M_2(|A_1|+|A_2|)$, $k = 1, 2$, the denominator of the above ratio dominates $(A_1^2 + A_2^2) + 2\varepsilon(A_1B_1 + A_2B_2) \geq (1 - 4\varepsilon M_2)(A_1^2 + A_2^2) \geq \frac{1}{2}(A_1^2 + A_2^2)$ for $\varepsilon \in (0, \varepsilon_1)$. For $\varepsilon \in (0, \varepsilon_1)$, the numerator is dominated by $(9/8)|A_1 + \varepsilon B_1| + (1/8)|A_2 + \varepsilon B_2| \leq (41/32)[|A_1| + |A_2|]$. Hence we get (2.4.42), which remains valid under the replacement of $\frac{\partial}{\partial w_1'}$ by $\frac{\partial}{\partial w_2'}$

For $\varepsilon \in (0, \varepsilon_1)$, let $h_0(z, \xi_0, \eta)$, $h_1(z, w_1, \eta)$ and $h_2(z, w, \eta)$ be the contributions of $L_1(z, w', \varepsilon)$ to $\widetilde{J}_0, \widetilde{J}_1$ and $\widetilde{J}_2$, respectively. It follows from (2.4.40) and (2.4.42) that, for $z = x_1 + ix_2 \in \mathbb{H}$ with $x_2 \leq \beta$, $w \in \overline{Q}$ and $\varepsilon \in (0, \varepsilon_1)$,

$$\begin{cases} \frac{8\pi}{41}|h_0(z, \xi_0, \eta)| \leq & 2\log_+ \frac{1}{\xi_0 - a_0} + \pi + \log_+[(\xi_0 - x_1)^2 + \beta^2], \\ \frac{8\pi}{41}|h_1(z, w_1, \eta)| \leq & \pi + 2\log_+ \frac{1}{|\eta - x_2|} + \frac{1}{2}\log_+[(w_1 - x_1)^2 + \beta^2] \\ & +\frac{1}{2}\log_+[(\xi_0 - x_1)^2 + \beta^2], \\ \frac{8\pi}{41}|h_2(z, w_1, \eta)| \leq & \pi + 2\log_+ \frac{1}{|w_1 - x_1|} + \log_+[(w_1 - x_1)^2 + \beta^2]. \end{cases} \tag{2.4.43}$$

Hence, we have, for $z \in \mathbb{H}$ with $\Im z < \beta$ and $w \in Q$,

$$\begin{cases} \frac{8\pi}{41}|h_0(z, \xi_0, \eta)| \leq \log_+(4\alpha^2 + \beta^2) + \pi + 2\log_+(1/(\xi_0 - a_0)), \\ \frac{8\pi}{41}|h_1(z, w_1, \eta)| \leq \log_+(4\alpha^2 + \beta^2) + \pi + 2\log_+(1/|\eta - x_2|), \\ \frac{8\pi}{41}|h_2(z, w, \eta)| \leq \log_+(4\alpha^2 + \beta^2) + \pi + 2\log_+(1/|w_1 - x_1|), \end{cases} \tag{2.4.44}$$

so that the contribution of $L_1(z, w', \varepsilon)$ to the left-hand side of (2.4.38) is dominated by the right-hand side of (2.4.38).

Next, let $K_0(z, \xi_0, \eta, \varepsilon)$, $K_1(z, w_1, \eta, \varepsilon)$ and $K_2(z, w, \eta, \varepsilon)$ be the contributions of $L_2(z, w', \varepsilon)$ to $\widetilde{J}_0, \widetilde{J}_1$ and $\widetilde{J}_2$, respectively. It follows from (2.4.40) and (2.4.41) that, $K_0(z, \xi_0, \eta, \varepsilon) \leq K_{01} + K_{02}$ with

$$K_{01} = \int_{\partial\mathbb{H}} \Pi(\widetilde{f}_\varepsilon(z), d\zeta)h_0(\zeta, \xi_0, \eta), \quad K_{02} = \sum_{j=1}^{N} \int_{C_j^p} \Pi(\widetilde{f}_\varepsilon(z), d\widetilde{\zeta})h_0(\zeta, \xi_0, \eta),$$

where $\widetilde{\zeta} = \widetilde{f}_\varepsilon(\zeta)$. Define K_{11} and K_{12} (resp. K_{21} and K_{22}) as above for $h_1(\zeta, w_1, \eta)$ (resp. $h_2(\zeta, w, \eta)$) in place of $h_0(\zeta, \xi_0, \eta)$. Then, we also have the bounds

$$K_1(z, w_1, \eta, \varepsilon) \leq K_{11} + K_{12}, \quad K_2(z, w_1, \eta, \varepsilon) \leq K_{21} + K_{22}.$$

We will show that actually each K_{ij}, $0 \leq i, j \leq 2$, is uniformly bounded by a constant, yielding the desired estimate (2.4.38).

To estimate K_{01}, K_{11} and K_{21}, let us examine the integral

$$I(z, \kappa, \varepsilon) = \int_{\partial \mathbb{H}} \widetilde{\Pi}(\widetilde{f}_\varepsilon(z), dx) \log_+[(x-\kappa)^2 + \beta^2], \quad z \in \Lambda, \quad \kappa \in (-\alpha, \alpha). \quad (2.4.45)$$

By Lemma 2.4.2, the map $\widetilde{f}_\varepsilon$ has the property (2.4.3) and consequently, there exists $\varepsilon_2 \in (0, \varepsilon_1]$ such that, for any $\varepsilon \in (0, \varepsilon_2)$,

$$\widetilde{f}_\varepsilon(\Lambda) \subset Q, \quad \mathrm{dist}(\widetilde{f}_\varepsilon(\Lambda), \partial \mathbb{H}) \geq d_1,$$

for some constant $d_1 > 0$. Hence for any $z \in \Lambda$, $\widetilde{f}_\varepsilon(z) = \xi_1 + i\xi_2$ has the property that $-\alpha < \xi_1 < \alpha$ and $d_1 < \xi_2 < \beta$.

Since $\Pi(\widetilde{f}_\varepsilon(z), \cdot)$ is dominated by the Poisson kernel density on $\partial \mathbb{H}$,

$$\begin{aligned} I(z, \kappa, \varepsilon) &\leq \frac{1}{\pi} \int_{-\infty}^{\infty} \frac{\xi_2 \log_+[(x-\kappa)^2 + \beta^2]}{(x-\xi_1)^2 + \xi_2^2} dx \\ &= \frac{1}{\pi} \int_{-\infty}^{\infty} \frac{\log_+[\xi_2^2 (x-\kappa_0)^2 + \beta^2]}{x^2+1} dx \\ &\leq \frac{1}{\pi} \int_{-\infty}^{\infty} p(x, \kappa_0) dx + 2 \log_+ \beta, \end{aligned}$$

where

$$p(x, \kappa_0) = \frac{\log[(x-\kappa_0)^2 + 1]}{x^2+1}, \quad \kappa_0 = \frac{\kappa - \xi_1}{\xi_2}.$$

Put $\alpha_0 = \frac{2\alpha}{d_1}$. As $-\alpha_0 < \kappa_0 < \alpha_0$,

$$\begin{cases} 0 < x - \kappa_0 < x + \alpha_0, & \text{if} \quad x > \alpha_0, \\ x - \alpha_0 < x - \kappa_0 < 0, & \text{if} \quad x < -\alpha_0, \\ -2\alpha_0 < x - \kappa_0 < 2\alpha_0, & \text{if} \quad -\alpha_0 \leq x \leq \alpha_0, \end{cases}$$

so that

$$\begin{aligned} \int_{\mathbb{R}} p(x, \kappa_0) dx \leq & \int_{-\infty}^{\alpha_0} \frac{\log[(x-\alpha_0)^2 + 1]}{x^2+1} dx + \int_{\alpha_0}^{\infty} \frac{\log[(x+\alpha_0)^2+1]}{x^2+1} dx \\ & + 2\alpha_0 \log(4\alpha_0^2 + 1). \end{aligned}$$

Accordingly $I(z, \kappa, \varepsilon)$ is bounded by a constant independent of $z \in \Lambda, \kappa \in (-\alpha, \alpha)$ and $\varepsilon \in (0, \varepsilon_2)$. In view of (2.4.43) holding for $z = x_1 + i0 \in \partial \mathbb{H}$, we thus conclude that $K_{01} + K_{11} + K_{21}$ is bounded by a constant independent of $z \in \Lambda$, $w \in \overline{Q}$, η and $\varepsilon \in (0, \varepsilon_1)$.

As for K_{02} and K_{12}, we see from the estimate (2.4.43) for $z \in \bigcup_{j=1}^N C_j$ that K_{02} is bounded by a constant independent of $z \in \Lambda, w \in Q, \varepsilon \in (0, \varepsilon_1)$, and so is K_{12} due to the condition (2.4.37) imposed on η.

K_{22} requires to examine the boundedness of the integral

$$\int_{C_j^p} \widetilde{\Pi}(\widetilde{f}_\varepsilon(z), d\widetilde{\zeta}) \log_+ \frac{1}{|\zeta_1 - w_1|}, \quad \widetilde{\zeta} = \widetilde{f}_\varepsilon(\zeta),$$

in $z \in \Lambda, w \in Q, \varepsilon > 0$, for each $1 \le i \le N$. For positive constants a, b, L_1 appearing in Theorem 2.4.1, define $b_0 > 0$ by (2.4.16) and denote $\varepsilon_2 \wedge a \wedge \frac{a \wedge b_0}{4L_1}$ by ε_0'.

When $\zeta \in C_j$ and $w \in Q \setminus U_j$, $|\zeta - w| \ge b$ by (2.4.2). When $\zeta \in C_j$ and $w \in U_j$, $\widetilde{\zeta}_1 - \widetilde{w}_1 = (1 + \delta_j)(\zeta_1 - w_1)$ for $\widetilde{\zeta} = \widetilde{f}_\varepsilon(\zeta), \widetilde{w} = \widetilde{f}_\varepsilon(w)$ so that $\log_+(1/|\zeta_1 - w_1|) \le \log 2 + \log_+(1/|\widetilde{\zeta}_1 - \widetilde{w}_1|)$ for $\varepsilon \in (0, \varepsilon_0')$ by (2.4.5). So we are left with estimating

$$\widetilde{q}(z, w, \varepsilon) = \int_{\widetilde{C}_j^p} \widetilde{\Pi}(\widetilde{f}_\varepsilon(z), d\widetilde{\zeta}) \log_+(1/|\widetilde{\zeta}_1 - \widetilde{w}_1|) \quad \text{for } z \in \Lambda, \ w \in U_j,$$

which in turn is dominated by the expectation $\mathbb{E}^{\Delta_j, 0}_{\widetilde{f}_\varepsilon(z)} \log_+ \left(1/|Z^0_{\zeta^0 -} - \widetilde{w}_1|\right)$ with respect to the ABM $\mathbb{Z}^{\Delta_j} = \left(Z^0_t, \zeta^0, \mathbb{P}^{\Delta_j, 0}_z\right)$ on the slit domain $\Delta_j = \mathbb{C} \setminus \widetilde{C}_j$.

Let $\zeta_0 = \zeta_{01} + i\zeta_{02}$ and R be the center and a half of the width of the slit $\widetilde{C}_j$, respectively. If $\varepsilon \in (0, \varepsilon_0')$, then, for any $z \in \Lambda$ and $\zeta \in C_j$,

$$|\widetilde{f}_\varepsilon(z) - \widetilde{f}_\varepsilon(\zeta)| \ge |z - \zeta| - |\widetilde{f}_\varepsilon(z) - z| - |\widetilde{f}_\varepsilon(\zeta) - \zeta| > b/2,$$

by (2.4.3) so that $\mathrm{dist}(\widetilde{f}_\varepsilon(\Lambda), \widetilde{C}_j) > b/2$ and the ellipse $E_{R, b/4}(\zeta_0)$ defined by

$$(x - \zeta_{01})^2/(R + \frac{b}{4})^2 + (y - \zeta_{02})^2/(\frac{b}{2})^2 = 1, \quad x + iy \in E_{R, b/2}(\zeta_0),$$

does not intersect with Λ. By the translation $T_{\zeta_0}(z) = z - \zeta_0$, $z \in \mathbb{C}$, the slit domain $\Delta_j = \mathbb{C} \setminus \widetilde{C}_j$, the ellipse $E_{R, b/4}(\zeta_0)$, and the ABM $\mathbb{Z}^{\Delta_j}$ on Δ_j are sent to Δ_R, $E_{b/4}$, and $\mathbb{Z}^{\Delta_R}$, respectively. Therefore, we get from Lemma 2.4.9 that, for any $z \in \Lambda$ and $w \in U_j$,

$$\widetilde{q}(z, w, \varepsilon) \le \frac{1}{2\pi} \left(1 + \frac{4R}{b}\right) \mathbb{1}_{\widetilde{U}_j}(\widetilde{w}) \int_{\partial_p \widetilde{C}_R} \log_+ \frac{1}{|\widetilde{w}_1 - \xi_{01} - \xi_1|} \frac{1}{\sqrt{R^2 - \xi^2}} d\xi. \tag{2.4.46}$$

On account of (2.4.2) and the stated choices of $\varepsilon_0' > 0$ and Q, the width of $\widetilde{C}_j$ is in $(a/2, 2\alpha)$ and consequently, $R \in (a/4, \alpha)$ for any $\varepsilon \in (0, \varepsilon_0')$. Hence, it follows from (2.4.46) and Lemma 2.4.10 that, for $z \in \Lambda$, $w \in U_j$, the left hand side of (2.4.46) is bounded by a constant which is independent of $\varepsilon \in (0, \varepsilon_0')$, $D \in \mathcal{D}$ and $\widetilde{f}_\varepsilon(D)$. □

The constant ε_0' in Proposition 2.4.11 will be designated by ε_0 again.

Proposition 2.4.12. *It holds that*

$$u(z, \zeta) \sim \widetilde{u}(\widetilde{f}_\varepsilon(z), \zeta), \quad z \in (\overline{\mathbb{H}} \setminus K) \cup K^p, \ \zeta \in \partial\mathbb{H}. \tag{2.4.47}$$

Proof. 1°. Let $\mathcal{L}$ be a linear operator sending a smooth function $h(w)$ on D to

$$(\mathcal{L}h)(w) = 2\int_{\gamma(w)} \sum_{k,\ell=1}^{2} A_{kl}^{(\varepsilon)}(w') \frac{\partial h(w')}{\partial w'_k} \frac{\partial w'_\ell}{\partial \mathbf{n}} ds(w'), \quad w \in D,$$

for the curve $\gamma(w) = \gamma_0(\xi_0,\eta) + \gamma_1(w_1,\eta) + \gamma_2(w,\eta)$ specified in the paragraph above Proposition 2.4.11. On both sides of (2.4.12), we take the normal derivative at $\zeta \in J$ and apply the linear operator $\mathcal{L}$ in w. On account of (2.4.19), (2.4.39) and Proposition 2.4.3 (ii), we then have for $w \in D$, $\zeta \in J$

$$\begin{aligned}
&\widetilde{I}(\widetilde{G}_{\zeta_2}(\zeta,\cdot);\widetilde{\gamma}(\widetilde{f}_\varepsilon(w))) - I(G_{\zeta_2}(\zeta,\cdot);\gamma(w)) \\
&= \varepsilon \int_\Lambda B_z^{(\varepsilon)} G(z,\zeta)_{\zeta_2} \widetilde{I}(\widetilde{G}(\widetilde{f}_\varepsilon(z),\cdot);\widetilde{\gamma}(\widetilde{f}_\varepsilon(w)))dx_1 dx_2 \\
&\quad + 2\varepsilon \sum_{k=1}^{2} \int_{\gamma_0\cup\gamma_2} \mathbb{1}_\Lambda(w') b_{k1}^{(\varepsilon)}(w') G_{\zeta_2 w'_k}(\zeta,w')dw'_2 \\
&\quad + 2\varepsilon \sum_{k=1}^{2} \int_{\gamma_1} \mathbb{1}_\Lambda(w') b_{k2}^{(\varepsilon)}(w') G_{\zeta_2 w'_k}(\zeta,w')dw'_1.
\end{aligned}$$

This combined with Proposition 2.4.11 and Proposition 2.4.3 (ii), (v), (vii), yields

$$I(G_{\zeta_2}(\zeta,\cdot);\gamma(w)) \sim \widetilde{I}(\widetilde{G}_{\zeta_2}(\zeta,\cdot);\widetilde{\gamma}(\widetilde{f}_\varepsilon(w))), \quad w \in (\overline{\mathbb{H}} \setminus K) \cup K^p,\ \zeta \in \partial\mathbb{H},\ \varepsilon \in (0,\varepsilon_0). \tag{2.4.48}$$

2°. As $g(z,w,\varepsilon) = \widetilde{G}(\widetilde{f}_\varepsilon(z),w) \le G^{\mathbb{H}}(\widetilde{f}_\varepsilon(z),w)$ for $z \in \mathbb{H}$ and $w \in \mathbb{H} \setminus \bigcup_{j=1}^N \overline{V}_j$, we have from (1.1.10)

$$I(g(z,\cdot,\varepsilon);\Gamma) \le \int_{-\infty}^{\infty} G_{w_2}^{\mathbb{H}}(\widetilde{f}_\varepsilon(z),w)\big|_{w_2=0} dw_1 = 2.$$

So, we can use (2.4.12) and Proposition 2.4.3 (v) again to get

$$I(G_{\zeta_2}(\zeta,\cdot);\Gamma)) \sim I(g_{\zeta_2}(\zeta,\cdot,\varepsilon);\Gamma), \quad \zeta \in \partial\mathbb{H}.$$

Since $\widetilde{f}_\varepsilon(z) = z$ near $\partial\mathbb{H}$, we have $\widetilde{I}(\widetilde{G}_{\zeta_2}(\zeta,\cdot);\Gamma) = I(g_{\zeta_2}(\zeta,\cdot,\varepsilon);\Gamma)$ for $\zeta \in \partial\mathbb{H}$ and

$$I(G_{\zeta_2}(\zeta,\cdot);\Gamma) \sim \widetilde{I}(\widetilde{G}_{\zeta_2}(\zeta,\cdot);\Gamma), \quad \zeta \in \partial\mathbb{H}, \quad \varepsilon \in (0,\varepsilon_0). \tag{2.4.49}$$

3°. For each $1 \le i \le N$, we take a smooth simple closed curve $\gamma_i \subset U_i$ surrounding C_i. We then make the same procedure as in 1°, but we replace the operation of taking the partial derivative with respect to ζ_2 at $\zeta \in \partial\mathbb{H}$ by the operation of taking a period around C_i in $\zeta \in \gamma_i$ and use (2.4.26). We then utilize Proposition 2.4.3 (ii), (vi), (viii) as well as Proposition 2.4.11. We can also make the same replacement in the procedure of 2° and use Proposition 2.4.3 (vi). We thus arrive at

$$I(\varphi^{(i)};\gamma(w)) \sim \widetilde{I}(\widetilde{\varphi}^{(i)};\widetilde{\gamma}(\widetilde{f}_\varepsilon(w))), \quad w \in D \cup K^p \cup \partial\mathbb{H}, \quad \varepsilon \in (0,\varepsilon_0). \tag{2.4.50}$$

$$I(\varphi^{(i)};\Gamma) \sim \widetilde{I}(\widetilde{\varphi}^{(i)};\Gamma), \quad \varepsilon \in (0,\varepsilon_0). \tag{2.4.51}$$

4°. $I(\varphi^{(i)},\gamma(w))$ will be denoted by $I_D(\varphi^{(i)},\gamma(w))$ to indicate its dependence on $D \in \mathcal{D}$. Let $D_1,\ldots,D_m \in \mathcal{D}_0$ be as in the proof of Proposition 2.4.5 and put, for any compact set $Q \subset \mathbb{H}$ containing U, $\max_{1\le k\le m} \sup_{w\in Q} |I_{D_k}(\varphi^{(i)}, \gamma(w)| = L_7(Q)$. Then, by (2.4.50), $|I_D(\varphi^{(i)}, \gamma(w)| \le L_8(Q) < \infty$ for any $D \in \mathcal{D}_0$ and $w \in Q$. This combined with (2.4.32), (2.4.33), (2.4.34) and (2.4.50) implies

$$I(\varphi^{(i)}, \gamma(w)) b_{ij} \frac{\partial \varphi^{(j)}(\zeta)}{\partial \mathbf{n}_\zeta} \sim \widetilde{I}(\widetilde{\varphi}^{(i)}, \widetilde{\gamma}(\widetilde{f}_\varepsilon(w)) \widetilde{b}_{ij} \frac{\partial \widetilde{\varphi}^{(j)}(\zeta)}{\partial \mathbf{n}_\zeta},$$

for $w \in (\overline{\mathbb{H}} \setminus K) \cup K^p$, $\zeta \in \partial\mathbb{H}$ and $\varepsilon \in (0, \varepsilon_0)$. Similarly it follows from (2.4.32), (2.4.33) and (2.4.50) that

$$I(\varphi^{(i)}, \Gamma) b_{ij} \frac{\partial \varphi^{(j)}(\zeta)}{\partial \mathbf{n}_\zeta} \sim \widetilde{I}(\widetilde{\varphi}^{(i)}, \Gamma) \widetilde{b}_{ij} \frac{\partial \widetilde{\varphi}^{(j)}(\zeta)}{\partial \mathbf{n}_\zeta}, \quad \varepsilon \in (0, \varepsilon_0).$$

The proof of Proposition 2.4.12 is now complete in view of (2.4.48), (2.4.49), the above two relations as well as (2.4.20) and (2.4.36). □

Propositions 2.4.5 and 2.4.12 yield Theorem 2.4.1 (iii).

2.5 Joint continuity of $\Psi_t(z, \zeta)$ and chordal Komatu-Loewner differential equation

We now return to the setting of Sections 2.1.2 and 2.1.3. Recall that $D = \mathbb{H} \setminus K$ is a fixed standard slit domain with $K = \bigcup_{j=1}^N C_i$, γ is a Jordan arc satisfying (2.1.11), for each $t \in [0, t_\gamma)$, g_t is a conformal map from $D \setminus \gamma[0, t]$ onto a standard slit domain $D_t = \mathbb{H} \setminus K_t$ with $K_t = \bigcup_{i=1}^N C_{t,i}$ satisfying (2.1.12), and $\Psi_t(z, \zeta)$ is the complex Poisson kernel of BMD on D_t. For $t = 0$, g_0 is the identity map on D, $D_0 = D$ and $\Psi_0(z, \zeta)$ is the BMD-complex Poisson kernel for D.

By combining Theorem 2.4.1 with Theorem 1.9.1 and Theorem 2.3.8, we are led to

Theorem 2.5.1. *$\Psi_t(z, \zeta)$ is jointly continuous in*

$$(t, z, \zeta) \in \bigcup_{t\in[0,t_\gamma)} \{t\} \times ((\mathbb{H} \setminus K_t) \cup K_t^p \cup (\partial\mathbb{H} \setminus J)) \times J,$$

for each compact set $J \subset \partial\mathbb{H}$.

Proof. Fix $t^* \in [0, t_\gamma)$ and a compact interval $J \subset \partial\mathbb{H}$. We shall apply Theorem 2.4.1 to the single fixed $D_{t^*} = \mathbb{H} \setminus \cup_{j=1}^N C_{t^*,j} \in \mathcal{D}_0$ by choosing U_j, V_j as $C_{t^*,j} \subset U_j \Subset V_j \Subset \mathbb{H}$, and $a > 0$ and $b > 0$ less than the minimum of the width of $C_{t^*,j}$ and the minimum of $\mathrm{dist}(C_{t^*,j}, \partial U_j)$, respectively. Let $\varepsilon_0 > 0$ be as in Theorem 2.4.1. Take any $\varepsilon \in (0, \varepsilon_0)$ and any relatively compact open subset G_1 of $\overline{\mathbb{H}}$ such that, $K_{t^*} \subset G_1$ and, if we define $G_2 = U_\varepsilon(G_1) \cap \overline{\mathbb{H}}$, $G_3 = U_\varepsilon(G_2) \cap \overline{\mathbb{H}}$, then $\overline{G}_3 \cap J = \emptyset$. Here $U_\varepsilon(G_i)$ denotes the ε-neighborhood of G_i, $i = 1, 2$. We write $G_{i,t} = G_i \setminus K_t$, $t \in [0, t_\gamma)$, $1 \le i \le 3$. By virtue of Theorem 1.9.1, there exists $\delta > 0$ such that

$$|\Psi_{t^*}(z_1^*, \zeta_1) - \Psi_{t^*}(z_2^*, \zeta_2)| < \varepsilon, \tag{2.5.1}$$

for any z_1^*, $z_2^* \in G_{3,t^*} \cup K_{t^*}^p$ with $|z_1^* - z_2^*| < \delta$ and any ζ_1, $\zeta_2 \in J$ with $|\zeta_1 - \zeta_2| < \delta$.

Let $\varepsilon' = \dfrac{\varepsilon \wedge \delta}{1+2L_1} < \varepsilon \wedge \delta$. By virtue of Theorem 2.3.8, there exists $\delta_0 > 0$ such that

$$t \in (t^* - \delta_0, t^* + \delta_0) \cap [0, t_\gamma) \implies d(D_t,\ D_{t^*}) < \varepsilon', \tag{2.5.2}$$

which particularly implies that $K_t \subset G_2$ whenever $|t - t^*| < \delta_0$.

Now take any $t_1,\ t_2 \in (t^* - \delta_0, t^* + \delta_0) \cap [0, t_\gamma)$, any $z_1 \in G_{2,t_1} \cup K^p_{t_1},\ z_2 \in G_{2,t_2} \cup K^p_{t_2}$ with $|z_1 - z_2| < \varepsilon'$ and any $\zeta_1, \zeta_2 \in J$ with $|\zeta_1 - \zeta_2| < \delta$. Denote by $f^i_{\varepsilon'}$ the diffeomorphism sending D_{t^*} onto D_{t_i} that appears in Theorem 2.4.1, $i = 1, 2$. There exist $z_i^* \in D_{t^*} \cup K^p_{t^*}$ such that $f^i_{\varepsilon'}(z_i^*) = z_i$, $i = 1, 2$. By (2.4.3), $|z_i - z_i^*| \le L_1\varepsilon' < \varepsilon$ so that $z_i^* \in G_{3,t^*} \cup K^p_{t^*}$, $i = 1, 2$. We further have $|z_1^* - z_2^*| \le |z_1 - z_1^*| + |z_2 - z_2^*| + |z_1 - z_2| < 2L_1\varepsilon' + \varepsilon' = \varepsilon \wedge \delta$. Therefore we obtain from (2.4.4) and (2.5.1) the following desired estimate

$$\begin{aligned}|\Psi_{t_1}(z_1, \zeta_1) - \Psi_{t_2}(z_2, \zeta_2)| &\le |\Psi_{t_1}(z_1, \zeta_1) - \Psi_{t^*}(z_1^*, \zeta_1)| + |\Psi_{t_2}(z_2, \zeta_2) - \Psi_{t^*}(z_2^*, \zeta_2)| \\ &\quad + |\Psi_{t^*}(z_1^*, \zeta_1) - \Psi_{t^*}(z_2^*, \zeta_2)| \\ &< (2L^* + 1)\varepsilon,\end{aligned}$$

where $L^* = L_{\overline{G_3}, J}$. □

Recall the half-plane capacity a_t of $\gamma[0, t]$, $t \in [0, t_\gamma)$, associated with the canonical map g_t from $D \setminus \gamma[0, t]$ as introduced in Section 2.1.2. By virtue of Theorem 2.3.7, a_t is a strictly increasing continuous function of $t \in [0, t_\gamma)$ with $a_0 = 0$.

Theorem 2.5.2. *For each $z \in (D \cup K^p) \setminus \gamma[0, t_\gamma)$ and $t \in (0, t_\gamma)$, $g_t(z)$ is differentiable with respect to a_t and satisfies the differential equation*

$$\frac{\partial g_t(z)}{\partial a_t} = -\pi\Psi_t(g_t(z), \xi(t)), \quad g_0(z) = z, \quad 0 < t \le t_\gamma. \tag{2.5.3}$$

Furthermore, $g_t(z)$ is right differentiable with respect to a_t at $t = 0$ with

$$\left.\frac{\partial^+ g_t(z)}{\partial a_t}\right|_{t=0} = -\pi\Psi_0(z, \gamma(0)), \quad z \in D \cup \partial_p K. \tag{2.5.4}$$

Proof. On account of Theorems 1.9.1 and 2.3.6, the identity (2.1.28) derived in Section 2.1.3 remains valid for $z \in (D \cup K^p) \setminus \gamma[0, t_\gamma)$. This along with (2.1.22) implies, as in the previous derivation of (2.1.29), that there exist $x', x'' \in (\beta_0(t, s), \beta_1(t, s)) \subset \partial\mathbb{H}$ such that

$$\frac{g_s(z) - g_t(z)}{a_s - a_t} = -\pi u_t(g_t(z), x') - i\pi K_t^*(g_t(z), x''), \quad z \in (D \cup K^p) \setminus \gamma[0, t_\gamma). \tag{2.5.5}$$

By letting $s \uparrow t$, we get from (2.5.5) the equation (2.5.3) in the left derivative sense just as in the proof of Theorem 2.1.5. We next let $t \downarrow s$ in (2.5.5). By virtue of Theorem 2.3.9, both x' and x'' converge to $\xi(s)$. This combined with Theorems 2.3.6 and 2.5.1 leads us to the equation (2.5.3) in the right derivative sense (for s in place of t).

Notice that (2.1.17) and (2.1.23) hold also for $s = 0$ with $g_{t,0} = g_t^{-1}$, $a_0 = 0$, $g_0(z) = z$, and accordingly (2.5.5) is also valid for $s = 0$. Further (2.3.18) holds for $s = 0$ with $\xi(0) = \gamma(0)$. Therefore, we obtain (2.5.4) in the same way as above. □

Since the half-plane capacity a_t of $\gamma[0,t]$ is strictly increasing and continuous in $t \in [0, t_\gamma)$, one can reparametrize the Jordan arc $\{\gamma(t),\ t \in [0,t_\gamma)\}$ by $\widetilde{\gamma}(t) := \gamma(\widetilde{a}_{2t})$ for $t \in [0, t_{\widetilde{\gamma}})$, where $\widetilde{a}_t$ is the inverse function of $t \mapsto a_t$ and $t_{\widetilde{\gamma}} := a_{t_\gamma/2}$. Then the capacity $\widetilde{a}_t$ associated with $\widetilde{\gamma}$ equals $2t$ and the equation (2.5.3) becomes the ordinary differential equation called the *Komatu-Loewner equation*

$$\frac{\partial g_t(z)}{\partial t} = -2\pi \Psi_t(g_t(z), \xi(t)), \quad g_0(z) = z, \quad 0 < t \le t_{\widetilde{\gamma}}, \tag{2.5.6}$$

holding for any $z \in (D \cup K^p) \setminus \widetilde{\gamma}[0, t_{\widetilde{\gamma}})$. This reparametrization of γ will be called the *half-plane capacity reparametrization.* The equation (2.5.3) may be called the *intrinsic chordal Komatu-Loewner equation.* In the following, we will use this parametrization of the curve γ and drop the tilde from $\widetilde{\gamma}$ and $t_{\widetilde{\gamma}}$.

2.6 Chordal Komatu-Loewner differential equation for slit motions

We continue to work under the setting of the previous section. We make the half-plane capacity reparametrization of the Jordan arc γ to obtain the ODE (2.5.6) for the canonical map g_t from $D \setminus \gamma[0,t]$ onto the standard slit domain $D_t = \mathbb{H} \setminus K_t$ with $K_t = \bigcup_{i=1}^N C_{t,i}$. Thus, $\{g_t\}$ induces the motion of the slits $\{\bigcup_{i=1}^N C_{t,i}\}$ in t. In this section, we derive a differential equation for this slit motion from the trace of the chordal Komatu-Loewner differential equation (2.5.6) on K^p.

For simplicity, we assume throughout this section that $\overline{\gamma(0,t_\gamma)} \cap (\bigcup_{i=1}^N C_i) = \emptyset$ to derive the desired equation in Theorem 2.6.4. This extra assumption is harmless; it is always fulfilled if t_γ is replaced by any $T \in (0, t_\gamma)$. Therefore, Theorem 2.6.4 proved under this assumption automatically implies the general form of the theorem.

For $t \in [0, t_\gamma)$, the conformal map g_t from $D \setminus \gamma[0,t]$ onto D_t can be extended analytically to K^p in the following manner. In what follows, fix $j \in \{1, \dots, N\}$ and denote the left and right endpoints of C_j by $z_j = a + ic$ and $z_j^r = b + ic$, respectively. Denote the open slit $C_j \setminus \{z_j,\ z_j^r\}$ by C_j^0. Consider the open rectangles

$$R_+ := \{z : a < x < b,\ c < y < c + \delta\}, \qquad R_- := \{z : a < x < b,\ c - \delta < y < c\},$$

and $R := R_+ \cup C_j^0 \cup R_-$, where $\delta > 0$ is sufficiently small so that $R_+ \cup R_- \subset D \setminus \gamma[0, t_\gamma)$. Since $\Im g_t(z)$ takes a constant value on the slit C_j, g_t can be extended to an analytic function g_t^+ (resp. g_t^-) from R_+ (resp. R_-) to R across C_j^0 by the Schwarz reflection.

We next take $\varepsilon > 0$ with $\varepsilon < (b-a)/2$ so that $B(z_j, \varepsilon) \setminus C_j \subset D \setminus \gamma[0, t_\gamma)$. Then $\psi(z) = (z - z_j)^{1/2}$ maps $B(z_j, \varepsilon) \setminus C_j$ conformally onto $B(0, \sqrt{\varepsilon}) \cap \mathbb{H}$. As in the proof of Theorem 2.3.3,

$$f_t^\ell(z) = g_t \circ \psi^{-1}(z) = g_t(z^2 + z_j),$$

can be extended analytically to $B(0, \sqrt{\varepsilon})$ by the Schwarz reflection and by noting that the origin 0 is a removable singularity for f_t^ℓ. Similarly, we can induce an analytic function f_t^r on $B(0, \sqrt{\varepsilon})$ from g_t on $B(z_j^r, \varepsilon) \setminus C_j$.

For an analytic function $u(z)$, its derivatives in z will be denoted by $u'(z)$, $u''(z)$ and so on.

Lemma 2.6.1.

(i) $(g_t^\pm)'(z)$ *and* $(g_t^\pm)''(z)$ *are continuous in* $(t, z) \in [0, t_\gamma) \times R$.

(ii) $\eta_t(z, \zeta) := \Psi_t(g_t(z), \zeta)$ *can be extended to an analytic function* $\eta_t^+(z, \zeta)$ *(resp.* $\eta_t^-(z, \zeta)$*) from* R_+ *(resp.* R_-*) to* R *by the Schwarz reflection, and*

$$(\eta_t^\pm(z, \zeta))' \text{ are continuous in } (t, z, \zeta) \in [0, t_\gamma) \times R \times \partial\mathbb{H}. \tag{2.6.1}$$

(iii) $(g_t^\pm)'(z)$ *are differentiable in* $t \in (0, t_\gamma)$ *and*

$$\partial_t (g_t^\pm)'(z) \text{ are continuous in } (t, z) \in [0, t_\gamma) \times R. \tag{2.6.2}$$

(iv) $(f_t^\ell)'(z)$ *and* $(f_t^\ell)''(z)$ *are continuous in* $(t, z) \in [0, t_\gamma) \times B(\mathbf{0}, \sqrt{\varepsilon})$.

(v) $\widetilde{\eta}_t(z, \zeta) := \Psi_t(f_t^\ell(z), \zeta) = \Psi_t(g_t(\psi^{-1}(z)), \zeta)$ *can be extended to an analytic function from* $B(\mathbf{0}, \sqrt{\varepsilon}) \cap \mathbb{H}$ *to* $B(\mathbf{0}, \sqrt{\varepsilon})$ *and*

$$\widetilde{\eta}_t(z, \zeta) \text{ and } (\widetilde{\eta}_t(z, \zeta))' \text{ are continuous in } (t, z, \zeta) \in [0, t_\gamma) \times B(\mathbf{0}, \sqrt{\varepsilon}) \times \partial\mathbb{H}. \tag{2.6.3}$$

(vi) $(f_t^\ell)'(z)$ *is differentiable in* $t \in (0, t_\gamma)$ *and*

$$\partial_t (f_t^\ell)'(z) \text{ is continuous in } (t, z) \in [0, t_\gamma) \times B(\mathbf{0}, \sqrt{\varepsilon}). \tag{2.6.4}$$

(vii) *The statements* (iv), (v), (vi) *in the above remain valid with* f_t^r *in place of* f_t^ℓ.

Proof. (i) This follows from the Cauchy integral formulae of derivatives of $g_t^\pm$ combined with Theorem 2.3.6.

(ii) This can be proved in the same way as (i) using Theorem 2.3.6 and Theorem 2.5.1.

(iii) We only give a proof for g_t^+. Notice that, if $F(z)$ is an analytic function on R_+ continuously extendable to C_j^0 with $\Im F(z) = i\ell$, $z \in C_j^0$, for a constant $\ell > 0$, then the analytic extension $F^+(z)$ of $F(z)$ to R by Schwarz reflection is given by

$$F^+(z) = \begin{cases} F(z), & z \in R_+ \cup C_j^0 \\ \overline{F}(\overline{z} + 2ic) + 2i\ell, & z \in R_-. \end{cases}$$

Therefore, the identity (2.1.28) extends in z from $R_+ \cup C_j^0$ to R_- with some additional imaginary constant, and consequently we obtain

$$(g_s^+)'(z) - (g_t^+)'(z) = \int_{\beta_0(t,s)}^{\beta_1(t,s)} (\eta_t^+(z, x))' \Im g_{t,s}(x) dx, \quad z \in R, \quad 0 \le s < t < t_\gamma.$$

By the half-plane capacity reparametrization, we have from (2.1.22) that

$$2(s-t) = \frac{1}{\pi}\int_{\beta_0(t,s)}^{\beta_1(t,s)} \Im g_{t,s}(x+i0+)dx, \quad 0 \le s < t < t_\gamma. \tag{2.6.5}$$

Taking quotient of the last two displays and using (2.6.1), we see just as in the proof of Theorem 2.5.2 that $(g_t^{\pm})'(z)$ is differentiable in t and

$$\partial_t (g_t^+)'(z) = -2\pi(\eta_t^+(z,\xi(t)))',$$

whose right-hand side is continuous in (t,z) by (2.6.1) and Theorem 2.3.9.

(iv) Let $\widetilde{\gamma}$ be a closed smooth Jordan curve in $B(\mathbf{0},\sqrt{\varepsilon})$. By Cauchy's integral formula

$$(f_t^\ell)'(z) = \frac{1}{2\pi i}\int_{\widetilde{\gamma}} \frac{f_t^\ell(\zeta)}{(\zeta-z)^2}d\zeta, \quad z \in \text{ins }\widetilde{\gamma}.$$

Since $f_t^\ell(\zeta) = g_t(\zeta^2 + z_j)$ is continuous in t uniformly in $\zeta \in \widetilde{\gamma}$ by Theorem 2.3.6, we get the desired continuity. The same is true for $(f_t^\ell)''(z)$.

(v) Since $\Im\widetilde{\eta}_t(z,\zeta)$ is constant in z on $B(\mathbf{0},\sqrt{\varepsilon}) \cap \partial\mathbb{H} \setminus \{\mathbf{0}\}$, it extends analytically to $B(\mathbf{0},\sqrt{\varepsilon}) \setminus \{\mathbf{0}\}$ by the Schwarz reflection. Note that $\mathbf{0}$ is a removable singularity because $\Im\eta_t(z,\zeta)$ is bounded near $\{\mathbf{0}\}$. The second assertion can be shown as the proof of (ii) using Theorem 2.3.6 and Theorem 2.5.1.

(vi) Taking z to be $\psi^{-1}(z)$ in (2.1.23), we have

$$f_s^\ell(z) - f_t^\ell(z) = \int_{\beta_0(t,s)}^{\beta_1(t,s)} \widetilde{\eta}_t(z,x)\Im g_{t,s}(x)dx \quad \text{for } z \in B(\mathbf{0},\sqrt{\varepsilon}) \cap \mathbb{H} \text{ and } s < t. \tag{2.6.6}$$

By making a similar consideration to the proof of (iii),

$$(f_s^\ell)'(z) - (f_t^\ell)'(z) = \int_{\beta_0(t,s)}^{\beta_1(t,s)} (\widetilde{\eta}_t(z,x))'\Im g_{t,s}(x)dx \quad \text{for } z \in B(\mathbf{0},\sqrt{\varepsilon}) \text{ and } s < t.$$

Taking quotient of the above with (2.6.5) and using (2.6.3), $\partial_t^-(f_t^\ell)'(z) = 2\pi(\widetilde{\eta}_t(z,\xi(t)))'$. Since the right hand side is continuous in t by (2.6.3) and Theorem 2.3.9, we arrive at the conclusion (vi). □

Denote by $z_j(t)$ and $z_j^r(t)$ the left and right endpoints of the slit $C_{t,j}$ for $D_t \in \mathcal{D}$ and $t \in [0,t_\gamma)$. Since g_t is a homeomorphism between C_j^p and $C_{t,j}^p$ for each $t \in [0,t_\gamma)$ as was seen in Section 2.1.1, there exist unique

$$\widetilde{z}_j(t) = \widetilde{x}_j(t) + iy_j \in C_j^p \quad \text{and} \quad \widetilde{z}_j^r(t) = \widetilde{x}_j^r(t) + iy_j \in C_j^p$$

such that $g_t(\widetilde{z}_j(t)) = z_j(t)$ and $g_t(\widetilde{z}_j^r(t)) = z_j^r(t)$; see Figure 2.4 with $N=2$.

Lemma 2.6.2.

(i) *If* $\widetilde{z}_j(t) \in C_j^{0,+}$, *then*

$$(g_t^+)'(\widetilde{z}_j(t)) = 0, \qquad (g_t^+)''(\widetilde{z}_j(t)) \neq 0. \tag{2.6.7}$$

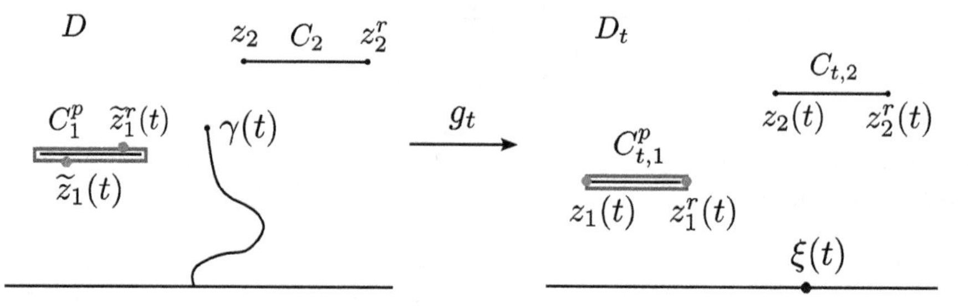

Fig. 2.4 The endpoints $z_j(t)$ and $z_j^r(t)$ of the slit $C_j(t)$ and the pre-image $\widetilde{z}_1(t) = g_t^{-1}(z_1(t))$.

(ii) *If* $\widetilde{z}_j(t) \in C_j^{0,-}$, *then* (2.6.7) *holds with* g_t^- *in place of* g_t^+.

(iii) *If* $\widetilde{z}_j(t) \in C_j^p \cap B(z_j, \varepsilon)$, *then, for* $\psi(z) = (z - z_j)^{1/2}$,

$$(f_t^\ell)'(\psi(\widetilde{z}_j(t))) = 0, \qquad (f_t^\ell)''(\psi(\widetilde{z}_j(t))) \neq 0. \tag{2.6.8}$$

(iv) *If* $\widetilde{z}_j(t) \in C_j^p \cap B(z_j^r, \varepsilon)$, *then, for* $\psi(z) = (z - z_j^r)^{1/2}$,

$$(f_t^r)'(\psi(\widetilde{z}_j(t))) = 0, \qquad (f_t^r)''(\psi(\widetilde{z}_j(t))) \neq 0. \tag{2.6.9}$$

(v) *The above four statements also hold with* $\widetilde{z}_j^r(t)$ *in place of* $\widetilde{z}_j(t)$.

Proof. It suffices to prove (i) and (iii).
(i) g_t^+ is analytic on R and $\widetilde{z}_j(t) \in R$. Suppose $g_t^+(z) - z_j(t)$ has a zero of order m at $\widetilde{z}_j(t)$: for some analytic function h with $h(\widetilde{z}_j(t)) \neq 0$,

$$g_t^+(z) - z_j(t) = g_t^+(z) - g_t^+(\widetilde{z}_j(t)) = (z - \widetilde{z}_j(t))^m h(z).$$

Then, in view of [Ahlfors (1979), Theorem 11 in §4.3.3] (see also [*Conway* (1978), Theorem 7.4 in Chapter IV]), there exist $\varepsilon_0 > 0$ with $B(\widetilde{z}_j(t), \varepsilon_0) \subset R$ and $\delta_0 > 0$, such that, for any $w \in B(z_j(t), \delta_0) \setminus \{z_j(t)\}$, the equation $g_t^+(z) - w = 0$ has m distinct simple roots in $B(\widetilde{z}_j(t), \varepsilon_0)$. Since $z_j(t)$ is an endpoint of $C_{t,j}$ and g_t is homeomorphic between C_j^p and $C_{t,j}^p$, one can find $\delta_{00} \in (0, \delta_0)$ such that, for any $w \in B(z_j(t), \delta_{00}) \cap C_{t,j}$ with $w \neq z_j(t)$, w corresponds to two distinct points $w_+ \in C_{t,j}^{0,+}$, $w_- \in C_{t,j}^{0,-}$ so that there are two distinct points $\widetilde{w}_+, \widetilde{w}_- \in C_j^{0,+} \cap B(\widetilde{z}_j(t), \varepsilon_0)$ with $g_t^+(\widetilde{w}_\pm) = w$. Hence $m = 2$.

(iii) Except for the last part, the following proof is similar to that of (i).

f_t^ℓ is analytic on $B(\mathbf{0}, \sqrt{\varepsilon})$, $\psi(\widetilde{z}_j(t)) \in B(\mathbf{0}, \sqrt{\varepsilon})$ and $f_t^\ell(\psi(\widetilde{z}_j(t)) = z_j(t)$. Suppose $f_t^\ell(z) - z_j(t)$ has a zero of order m at $\psi(\widetilde{z}_j(t))$: for some analytic function h with $h(\psi(\widetilde{z}_j(t))) \neq 0$,

$$f_t^\ell(z) - z_j(t) = f_t^\ell(z) - f_t^\ell(\psi(\widetilde{z}_j(t))) = (z - \psi(\widetilde{z}_j(t)))^m h(z), \ z \in B(\mathbf{0}, \sqrt{\varepsilon}).$$

Then, as in the proof of (i), there exist $\varepsilon_0 > 0$ with $B(\psi(\widetilde{z}_j(t)), \varepsilon_0) \subset B(\mathbf{0}, \sqrt{\varepsilon})$ and $\delta_0 > 0$, such that, for any $w \in B(z_j(t), \delta_0)$, the equation $f_t^\ell(z) - w = 0$ has m distinct simple roots in $B(\psi(\widetilde{z}_j(t)), \varepsilon_0)$.

Just as the proof of (i), one can find $\delta_{00} \in (0, \delta_0)$ such that, for any $w \in B(z_j(t), \delta_{00}) \cap C_{t,j}$ with $w \neq z_j(t)$, there are two distinct points $\widetilde{w}_+, \widetilde{w}_- \in C_j^p \cap B(z_j, \varepsilon)$ with $\psi(\widetilde{w}_+), \psi(\widetilde{w}_-) \in B(\psi(\widetilde{z}_j(t)), \varepsilon_0)$ and $f_t^\ell(\psi(\widetilde{w}_\pm)) = w$. Therefore $m = 2$. □

Let $h^\pm(t, z) = (g_t^\pm)'(z)$, which is a C^1-function in $(t, z) \in (0, t_\gamma) \times R$ by (i) and (iii) of Lemma 2.6.1. We further let $k(t, w) = (f_t^\ell)'(w)$, which is a C^1-function in $(t, w) \in (0, t_\gamma) \times B(0, \sqrt{\varepsilon})$ by (iv) and (vi) of Lemma 2.6.1.

Lemma 2.6.3.

(i) *Assume that* $\widetilde{z}_j(t_0) \in C_j^{0,\pm}$. *Then there exists* $\delta_1 > 0$ *such that, for* $|t - t_0| < \delta_1$, $\widetilde{z}_j(t)$ *is continuously differentiable in* t *and*

$$h^\pm(t, \widetilde{z}_j(t)) = 0, \quad (h^\pm)'(t, \widetilde{z}_j(t)) \neq 0, \tag{2.6.10}$$

$$\frac{d}{dt}\widetilde{z}_j(t) = -\frac{\partial_t h^\pm(t, z)}{(h^\pm)'(t, z)}\Big|_{z=\widetilde{z}_j(t)}. \tag{2.6.11}$$

(ii) *Assume that* $\widetilde{z}_j(t_0) \in C_j^p \cap B(z_j, \varepsilon)$. *Then there exists* $\delta_1 > 0$ *such that, for* $|t - t_0| < \delta_1$, $\psi(\widetilde{z}_j(t))$ *is continuously differentiable in* t *and*

$$k(t, \psi(\widetilde{z}_j(t))) = 0, \quad k'(t, \psi(\widetilde{z}_j(t))) \neq 0, \tag{2.6.12}$$

$$\frac{d}{dt}\psi(\widetilde{z}_j(t)) = -\frac{\partial_t k(t, w)}{k'(t, w)}\Big|_{w=\psi(\widetilde{z}_j(t))}. \tag{2.6.13}$$

Proof. (i) We only give a proof for the case that $\widetilde{z}_j(t_0) \in C_j^{0,+}$.

Note that $h^+(t, x+ic)$ is a C^1 real-valued function in $(t, x) \in (0, t_\gamma) \times (a, b)$, where we have set $C_j^0 = \{x + ic : a < x < b\}$. Hence, by (2.6.7) and the implicit function theorem, there exists $\delta > 0$ and a unique real C^1-function $\widehat{x}_j(t)$, $t \in (t_0 - \delta, t_0 + \delta)$, such that

$$\widehat{x}_j(t_0) = \Re\widetilde{z}_j(t_0), \;\; h^+(t, \widehat{x}_j(t)+ic) = 0 \;\text{ and }\; (h^+)'(t, \widehat{x}_j(t)+ic) \neq 0 \quad \text{for } |t-t_0| < \delta, \tag{2.6.14}$$

with

$$\frac{d}{dt}\widehat{x}_j(t) = -\frac{\partial_t h^+(t, z)}{(h^+)'(t, z)}\Big|_{z=\widehat{x}_j(t)+ic} \quad \text{for } t \in (t_0 - \delta, t_0 + \delta). \tag{2.6.15}$$

We will show that there is some $\delta_1 \in (0, \delta)$ so that

$$\widetilde{z}_j(t) = \widehat{z}_j(t) := \widehat{x}_j(t) + ic \quad \text{for } t \in (t_0 - \delta_1, t_0 + \delta_1), \tag{2.6.16}$$

which establishes (2.6.10) and (2.6.11).

In fact, if $w_t = g_t^+(\widehat{z}_j(t)) \in C_{t,j} \setminus \{z_j(t), z_j^r(t)\}$, then, by the same reasoning as in the proof of Lemma 2.6.2 (i), $\widehat{z}_j(t)$ is a zero of $g_t^+(z) - w_t$ of order 1, namely,

$$g_t^+(z) - w_t = (z - \widehat{z}_j(t))v(z) \quad \text{with } v(\widehat{z}_j(t)) \neq 0$$

so that $h^+(t, \widehat{z}_j(t)) \neq 0$, a contradiction. Hence, $\widehat{z}_j(t)$ equals either $\widetilde{z}_j(t)$ or $\widetilde{z}_j^r(t)$. But, $\widehat{z}_j(t_0) = \widetilde{z}_j(t_0)$ and we can obtain (2.6.16) as follows:

By Theorem 2.3.8 on the continuity of the image domains $\{D_t\}$, there exist $\varepsilon_0 > 0$ and $\delta_0 > 0$ such that

$$|t - t_0| < \delta_0 \quad \Longrightarrow \quad |z_j^r(t) - z_j(t_0)| > \varepsilon_0. \tag{2.6.17}$$

By Theorem 2.3.6 on the joint continuity of $g_t(z)$ and the continuity of $\widehat{z}_j(t)$, there exists $\delta_1 \in (0, \delta_0)$ with

$$|t - t_0| < \delta_1 \quad \Longrightarrow \quad |g_t(\widehat{z}_j(t)) - z_j(t_0)| = |g_t(\widehat{z}_j(t)) - g_{t_0}(\widetilde{z}_j(t_0))| < \varepsilon_0.$$

Hence $g_t(\widehat{z}_j(t)) \neq z_j^r(t)$ so that $g_t(\widehat{z}_j(t)) = z_j(t)$, namely, $\widehat{z}_j(t) = \widetilde{z}_j(t)$.

(ii) As $k(t, x + i0) := (f_t^\ell)'(x + i0)$ is a C^1-smooth real-valued function in $(t, x) \in (0, t_\gamma) \times (-\sqrt{\varepsilon}, \sqrt{\varepsilon})$ we see by (2.6.8) and the implicit function theorem that there exists $\delta > 0$ and a unique C^1-function $\eta(t)$, $t \in (t_0 - \delta, t_0 + \delta)$, such that

$$\eta(t_0) = \Re\psi(\widetilde{z}_j(t_0)), \quad k(t, \eta(t) + i0) = 0 \quad \text{and} \quad k'(t, \eta(t) + i0) \neq 0 \quad \text{for } |t - t_0| < \delta, \tag{2.6.18}$$

with

$$\frac{d}{dt}\eta(t) = -\frac{\partial_t k(t, z)}{k'(t, z)}\Big|_{z=\eta(t)+i0}, \quad t \in (t_0 - \delta, t_0 + \delta). \tag{2.6.19}$$

Define $\widehat{z}_j(t) = \psi^{-1}(\eta(t) + i0) \in C_j^p$. Then (2.6.12) and (2.6.13) hold for $\psi(\widehat{z}_j(t))$ in place of $\psi(\widetilde{z}_j(t))$. We show that

$$\widehat{z}_j(t) = \widetilde{z}_j(t), \qquad t \in (t_0 - \delta_1, t_0 + \delta_1) \text{ for some } \delta_1 \in (0, \delta). \tag{2.6.20}$$

In fact, if $w_t = f_t^\ell(\psi(\widehat{z}_j(t)) = g_t(\widehat{z}_j(t)) \in C_{t,j} \setminus \{z_j(t), z_j^r(t)\}$, then, in the same way as above, $\psi(\widehat{z}_j(t))$ is a zero of $f_t^\ell(w) - w_t$ of order 1, and so

$$f_t^\ell(w) - w_t = (w - \psi(\widehat{z}_j(t)))v(w) \quad \text{with } v(\psi(\widehat{z}_j(t))) \neq 0.$$

Hence, $k(t, \psi(\widehat{z}_j(t))) = v(\psi(\widehat{z}_j(t))) \neq 0$, a contradiction.

We just saw that $\widehat{z}_j(t)$ equals either $\widetilde{z}_j(t)$ or $\widetilde{z}_j^r(t)$. Since $f_t^\ell(w) = g_t(w^2 + z_j)$ is jointly continuous by Theorem 2.3.6, $\psi(\widehat{z}_j(t))$ is continuous and $\psi(\widehat{z}_j(t_0)) = \psi(\widetilde{z}_j(t_0))$, there exists $\delta_1 \in (0, \delta)$ with

$$|t - t_0| < \delta_1 \quad \Longrightarrow \quad |g_t(\widehat{z}_j(t)) - z_j(t_0)| = |f_t^\ell(\psi(\widehat{z}_j(t))) - f_{t_0}^\ell(\psi(\widetilde{z}(t_0)))| < \varepsilon_0,$$

which compared with (2.6.17) yields (2.6.20).

An analogous assertion holds for $\psi(\widetilde{z}_j^r(t))$ for $\psi(z) = (z - z_j^r)^{1/2}$ in a neighborhood of t_0 when $\widetilde{z}(t_0) \in C_j^p \cap B(z_j^r, \varepsilon)$. □

Theorem 2.6.4. *For each $1 \leq j \leq N$, the endpoints $z_j(t) = x_j(t) + iy_j(t)$ and $z_j^r(t) = x_j^r(t) + iy_j(t)$ of $C_{t,j}$ satisfy the following equations.*

$$\frac{d}{dt}y_j(t) = -2\pi\Im\Psi_t(z_j(t), \xi(t)), \tag{2.6.21}$$

$$\frac{d}{dt}x_j(t) = -2\pi\Re\Psi_t(z_j(t), \xi(t)), \tag{2.6.22}$$

$$\frac{d}{dt}x_j^r(t) = -2\pi\Re\Psi_t(z_j^r(t), \xi(t)). \tag{2.6.23}$$

Proof. It suffices to prove (2.6.21) and (2.6.22). By virtue of Theorem 2.5.2, we have

$$\partial_t g_t^{\pm}(z) = -2\pi\Psi_t(g_t^{\pm}(z), \xi(t)) \quad \text{for } z \in C_j^{0,\pm}. \tag{2.6.24}$$

The identity (2.6.6) remains valid for $z \in B(\mathbf{0}, \sqrt{\varepsilon}) \cap \overline{\mathbb{H}}$ by Theorem 2.5.1. Hence, by using (2.6.3), (2.6.5) and Theorem 2.5.1, we can get

$$\partial_t f_t^{\ell}(\psi(z)) = -2\pi\Psi_t(f_t^{\ell}(\psi(z)), \xi(t)) \quad \text{for } z \in C_j^p \cap B(z_j, \varepsilon). \tag{2.6.25}$$

When $\widetilde{z}_j(t) \in C_j^{0,\pm}$, $z_j(t) = g_t^{\pm}(\widetilde{z}_j(t))$ and $\widetilde{z}_j(t)$ is a C^1-function by Lemma 2.6.3(i). Hence, it follows from (2.6.24), (2.6.10) and (2.6.11) that

$$\frac{d}{dt}z_j(t) = \frac{d}{dt}(g_t^{\pm}(\widetilde{z}_j(t)) = \partial_t g_t^{\pm}(\widetilde{z}_j(t)) + (g_t^{\pm})'(\widetilde{z}_j(t))\frac{d}{dt}\widetilde{z}_j(t) = -2\pi\Psi_t(z_j(t), \xi(t)).$$

When $\widetilde{z}_j(t) \in C_j^p \cap B(z_j, \varepsilon)$, $z_j(t) = f_t^{\ell}(\psi(\widetilde{z}_j(t)))$ and $\psi(\widetilde{z}_j(t))$ is a C^1-function by Lemma 2.6.3 (ii). Hence, it follows from (2.6.25), (2.6.12), (2.6.13) that

$$\begin{aligned}\frac{d}{dt}z_j(t) = \frac{d}{dt}f_t^{\ell}(\psi(\widetilde{z}_j(t))) &= \partial_t f_t^{\ell}(\psi(\widetilde{z}_j(t))) + (f_t^{\ell})'(\psi(\widetilde{z}_j(t)))\frac{d}{dt}\psi(\widetilde{z}_j(t)) \\ &= -2\pi\Psi_t(z_j(t), \xi(t)).\end{aligned}$$

□

Remark 2.6.5. The equation (2.6.21)–(2.6.23) was first derived in Bauer-Friedrich [Bauer and Friedrich (2008)] by assuming that $\widetilde{z}_j(t) \in C_j^p \setminus \{z_j, z_j^r\}$ and also by taking for granted the smoothness of "$\frac{d}{dz}g_t(z)$" in (t, z), which is now established by Lemma 2.6.1. If

$$g_t(z_j) = z_j(t) \quad \text{and} \quad g_t(z_j^r) = z_j^r(t) \quad \text{for } t \in (0, t_\gamma) \text{ and } 1 \le j \le N, \tag{2.6.26}$$

then Theorem 2.6.4 is merely a special case of the Komatu-Loewner equation (2.5.6) with $z = z_j$ and $z = z_j^r$, $1 \le j \le N$, respectively. But in general (2.6.26) is not true. □

We call (2.6.21)–(2.6.23) the *Komatu-Loewner differential equation for the slits.*

Chapter 3

Komatu-Loewner evolution (KLE)

3.1 Slit motion $\mathbf{s}(t)$ induced by a motion $\xi(t)$ on $\partial\mathbb{H}$

Recall that $\mathcal{D}$ is the collection of all labelled standard slit domains introduced in the second half of Section 2.3.2, which is a metric space with equipped with metric d defined by (2.3.15).

It is convenient to consider an open subset $\mathcal{S}$ of $\mathbb{R}^{3N}$ defined by

$$\mathcal{S} = \Big\{ \mathbf{s} := (\mathbf{y}, \mathbf{x}, \mathbf{x}^r) \in \mathbb{R}^{3N} : \ \mathbf{y},\, \mathbf{x},\, \mathbf{x}^r \in \mathbb{R}^N,\, \mathbf{y} > \mathbf{0},\, \mathbf{x} < \mathbf{x}^r, \text{ either } x_j^r < x_k \text{ or } x_k^r < x_j \text{ whenever } y_j = y_k \text{ for some } j \neq k \Big\}. \tag{3.1.1}$$

The space $\mathcal{D}$ can be identified with $\mathcal{S}$ as a topological space. The element of $\mathcal{S}$ corresponding to $D \in \mathcal{D}$ will be denoted by $\mathbf{s}(D)$, while the element of $\mathcal{D}$ corresponding to $\mathbf{s} \in \mathcal{S}$ will be denoted by $D(\mathbf{s})$. In particular, for $\mathbf{s} = (\mathbf{y}, \mathbf{x}, \mathbf{x}^r) \in \mathcal{S}$, $z_j = x_j + iy_j$, $z_j^r = x_j^r + iy_j$ correspond to the left and right endpoints of the j-th slit C_j of $D(\mathbf{s}) \in \mathcal{D}$, $1 \le j \le N$. By this correspondence, we easily see the following bi-Lipschitz equivalence of $\mathcal{D}$ and $\mathcal{S}$ as metric spaces: for $\mathbf{s}, \widetilde{\mathbf{s}} \in \mathcal{S}$,

$$\frac{1}{2}\, d(D(\mathbf{s}), D(\widetilde{\mathbf{s}})) \le \|\mathbf{s} - \widetilde{\mathbf{s}}\| \le \sqrt{N}\, d(D(\mathbf{s}), D(\widetilde{\mathbf{s}})), \tag{3.1.2}$$

where $\|\mathbf{s} - \widetilde{\mathbf{s}}\|$ denotes the Euclidean norm of $\mathbf{s} - \widetilde{\mathbf{s}} \in \mathbb{R}^{3N}$.

For $\mathbf{s} = (y_1, \ldots, y_N, x_1, \ldots, x_N, x_1^r, \ldots, x_N^r) \in \mathcal{S}$, we denote by $\Psi_{\mathbf{s}}(z, \xi)$ the BMD complex Poisson kernel of $D(\mathbf{s})$. We can then rewrite the Komatu-Loewner equation (2.6.21)–(2.6.23) for slits as follows: for $t \in [0, t_\gamma]$,

$$\mathbf{s}_j(t) - \mathbf{s}_j(0) = \int_0^t b_j(\xi(s), \mathbf{s}(s))ds, \quad 1 \le j \le 3N, \tag{3.1.3}$$

where $\mathbf{s}_j(t)$ is is the j-th entry of $\mathbf{s}(t)$ for $1 \le j \le 3N$, and, for $\xi \in \mathbb{R}$, $\mathbf{s} \in \mathcal{S}$,

$$b_j(\xi, \mathbf{s}) = \begin{cases} -2\pi \Im \Psi_{\mathbf{s}}(z_j, \xi), & 1 \le j \le N, \\ -2\pi \Re \Psi_{\mathbf{s}}(z_{j-N}, \xi), & N+1 \le j \le 2N, \\ -2\pi \Re \Psi_{\mathbf{s}}(z_{j-2N}^r, \xi), & 2N+1 \le j \le 3N. \end{cases} \tag{3.1.4}$$

In Chapter 2, both the continuous function $\xi(t) \in \partial\mathbb{H}$ and the slit motion $\mathbf{s}(t)$ were produced by a given Jordan arc γ with $\gamma(0, t_\gamma) \subset \mathbb{H}$ and $\gamma(0) \in \partial\mathbb{H}$ through the canonical map g_t from $D \setminus \gamma[0,t]$ and they were shown to satisfy the equation (3.1.3). Now, we instead start with the ODE (3.1.3) and consider the problem of solving it for an arbitrarily given continuous function $\xi(t) \in \partial\mathbb{H}$. Notice that, for each $1 \le j \le 3N$, $b_j(\xi, \mathbf{s})$ is continuous in $(\xi, \mathbf{s}) \in \partial\mathbb{H} \times \mathcal{S}$. This can be shown in a similar manner to the proof of Theorem 2.5.1 by making use of Theorems 2.4.1 and 1.9.1. Consequently, $b_j(\xi(t), \mathbf{s})$ is continuous in $(t, \mathbf{s}) \in [0, \infty) \times \mathcal{S}$ for a continuous function $\xi : [0, \infty) \mapsto \partial\mathbb{H}$.

Lemma 3.1.1.

(i) *For each $1 \le j \le 3N$, the function $b_j(\xi, \mathbf{s})$ defined by* (3.1.4) *is locally Lipschitz continuous in the sense that, for any $\mathbf{s}_0 \in \mathcal{S}$ and any finite open interval $J \subset \partial\mathbb{H}$, there exists a neighborhood $U(\mathbf{s}_0)$ of $\mathbf{s}_0$ and a constant $L_J > 0$ such that*

$$|b_j(\xi, \mathbf{s}_1) - b_j(\xi, \mathbf{s}_2)| \le L_J \|\mathbf{s}_1 - \mathbf{s}_2\| \quad \text{for } \mathbf{s}_1, \mathbf{s}_2 \in U(\mathbf{s}_0) \text{ and } \xi \in J. \tag{3.1.5}$$

(ii) *Given any real continuous function $\xi(t)$ on $[0, \infty)$ and any $\mathbf{s}_0 \in \mathcal{S}$, the Komatu-Loewner equation* (3.1.3) *for slits with initial condition $\mathbf{s}(0) = \mathbf{s}_0$ admits a unique solution $\mathbf{s}(t)$ for $t \in [0, \zeta(\mathbf{s}_0))$, where $[0, \zeta(\mathbf{s}_0))$ is the maximal interval of existence.*

Proof. We need only to show (i), which can be obtained from Theorem 2.4.1 as follows. Take $\mathbf{s}_0 \in S$ with $D(\mathbf{s}_0) = \mathbb{H} \setminus \bigcup_{j=1}^N C_{0,j} \in \mathcal{D}$ and a finite open interval $J \subset \mathbb{R}$. Choose open sets U_j, V_j and positive numbers a and b so that

$$C_{0,j} \subset U_j \Subset V_j \Subset \mathbb{H}, \qquad \min_{1 \le j \le N} |C_{0,j}| > a \quad \text{and} \quad \min_{1 \le j \le N} \operatorname{dist}(C_{0,j}, \partial U_j) > b,$$

where $|C_{0,j}|$ denotes the length of the slit $C_{0,j}$. Define the neighborhood $\widehat{\mathcal{D}}_0$ of $D(\mathbf{s}_0)$ in $\mathcal{D}$ by

$$\widehat{\mathcal{D}}_0 = \{\mathbb{H} \setminus \cup_{j=1}^N C_j \in \mathcal{D} : C_j \subset U_j,\ |z_j - z_j^r| > a,\ \operatorname{dist}(C_j, \partial U_j) > b,\ 1 \le j \le N\}.$$

For $\varepsilon_0 > 0$ in the statement of Theorem 2.4.1, let $\widehat{\mathcal{D}}_{00} = \{D \in \widehat{\mathcal{D}}_0 : d(D(\mathbf{s}_0), D) < \frac{\varepsilon_0}{2}\}$. Any two members D and $\widetilde{D}$ from $\widehat{\mathcal{D}}_{00}$ satisfy $d(D, \widetilde{D}) < \varepsilon_0$ and we deduce from (2.4.4) by taking $\varepsilon = d(D, \widetilde{D})$ there that

$$\begin{cases} |\Psi(z_j, \xi) - \widetilde{\Psi}(\tilde{z}_j, \xi)| \le L \cdot d(D, \widetilde{D}) \\ |\Psi(z_j^r, \xi) - \widetilde{\Psi}(\tilde{z}_j^r, \xi)| \le L \cdot d(D, \widetilde{D}), \quad 1 \le j \le N,\ \xi \in J, \end{cases}$$

where $L > 0$ is a constant independent of $\xi \in J$, D and $\widetilde{D}$, and $\Psi(z, \xi)$ and $\widetilde{\Psi}(z, \xi)$ are the complex Poisson kernels of BMD on D and $\widetilde{D}$, respectively. This along with (3.1.2) yields (3.1.5). □

The maximal time $\zeta(\mathbf{s}_0)$ for the existence of the solution $\mathbf{s}(t)$ in Lemma 3.1.1 (ii) will be called the *explosion time* of the solution of the equation (3.1.3) with initial value $\mathbf{s}_0$.

In Section 4.2, we shall investigate a stochastic differential equation for $(\xi(t), \mathbf{s}(t)) \in \mathbb{R} \times \mathcal{S}$ whose solution path $(\xi(t), \mathbf{s}(t))$ satisfies (3.1.3) as a part of a SDE system. As a preparation for it, we study in the rest of this section specific homogeneities that the BMD complex Poisson kernel $\Psi_{\mathbf{s}}(z,\xi)$ enjoys.

A real-valued function $u(\xi,\mathbf{s})$ on $\mathbb{R}\times\mathcal{S}$ is called *homogeneous with degree* 0 (resp. -1) if

$$u(c\xi,c\mathbf{s})=u(\xi,\mathbf{s})\quad(\text{resp. } u(c\xi,c\mathbf{s})=c^{-1}u(\xi,\mathbf{s}))\quad\text{for any } c>0 \text{ and } (\xi,\mathbf{s})\in\mathbb{R}\times\mathcal{S}.$$

The same definition of the homogeneity is in force for a real-valued function $u(\mathbf{s})$ on $\mathcal{S}$.

For $r\in\mathbb{R}$, denote by $\widehat{r}$ the vector in $\mathbb{R}^{3N}$ whose first N entries are 0 and the last $2N$ entries are all r. Note that $\mathbf{s}(D+r)=\mathbf{s}(D)+\widehat{r}$ for $D\in\mathcal{D}$ and $r\in\mathbb{R}$.

A real-valued function $u(\xi,\mathbf{s})$ on $\mathbb{R}\times\mathcal{S}$ is called *homogeneous in horizontal direction* if

$$u(\xi+r,\mathbf{s}+\widehat{r})=u(\xi,\mathbf{s})\quad\text{for any } \xi\in\mathbb{R},\ \mathbf{s}\in\mathcal{S} \text{ and } r\in\mathbb{R}.$$

In this case, $u(\xi,\mathbf{s})=\widetilde{u}(\mathbf{s}-\widehat{\xi})$ for every $\xi\in\mathbb{R}$ and $\mathbf{s}\in\mathcal{S}$, where $\widetilde{u}(\mathbf{s}):=u(0,\mathbf{s})$.

Proposition 3.1.2. *For each $1\le j\le 3N$, the function $b_j(\xi,\mathbf{s})$ defined by* (3.1.4) *is a homogeneous function of degree -1 and homogeneous in horizontal direction. In particular*

$$b_j(\xi,\mathbf{s})=\widetilde{b}_j(\mathbf{s}-\widehat{\xi})\quad\textit{for } \xi\in\mathbb{R} \textit{ and } \mathbf{s}\in\mathcal{S},\quad\textit{where } \widetilde{b}_j(\mathbf{s}):=b_j(0,\mathbf{s}).\tag{3.1.6}$$

Proof. The stated properties of $b_j(\xi,\mathbf{s})$ follow directly from the corresponding homogeneities of the BMD complex Poisson kernel $\Psi(z,\zeta)$, which are in turn consequences of the conformal invariance of the BMD Green function shown by Theorem 1.8.1.

For $c>0$, let $\phi(z):=cz$ be the homothetic transformation on $\mathbb{H}$. For each $\mathbf{s}\in S$, $\phi|_{D(\mathbf{s})}$ is a conformal map from $D(\mathbf{s})$ onto $D(c\mathbf{s})$. According to Theorem 1.8.1, the BMD Green functions $G^*_{\mathbf{s}}(z,\zeta)$ and $G^*_{c\mathbf{s}}(\widetilde{z},\widetilde{\zeta})$ on $D(\mathbf{s})^*$ and $D(c\mathbf{s})^*$, respectively, are related by

$$G^*_{\mathbf{s}}(z,\zeta)=G^*_{c\mathbf{s}}(cz,c\zeta),\quad z,\zeta\in D(\mathbf{s}).$$

Hence, the corresponding BMD Poisson kernels are related for $\xi\in\partial\mathbb{H}$ by

$$K^*_{\mathbf{s}}(z,\zeta)=\lim_{\delta\downarrow 0}\frac{1}{2\delta}G^*_{\mathbf{s}}(z,\xi+i\delta)=\lim_{\delta\downarrow 0}\frac{1}{2\delta}G^*_{c\mathbf{s}}(cz,c\xi+ic\delta)=cK^*_{c\mathbf{s}}(cz,c\zeta).$$

Since the complex Poisson kernel $\Psi_{\mathbf{s}}(z,\xi)$ with $z\in D(\mathbf{s})$ and $\xi\in\partial\mathbb{H}$ is the unique analytic function in z with the imaginary part $K^*_{\mathbf{s}}(z,\xi)$ satisfying $\lim_{z\to\infty}\Psi_{\mathbf{s}}(z,\xi)=0$, we obtain

$$\Psi_{c\mathbf{s}}(cz,c\xi)=c^{-1}\Psi_{\mathbf{s}}(z,\xi),\quad z\in D(\mathbf{s}).\tag{3.1.7}$$

We then get the homogeneity of $b_j(\xi,\mathbf{s})$ of degree -1 for $1\le j\le 3N$ by letting $z\in D(\mathbf{s})$ in (3.1.7) approach to endpoints of slits of $D(\mathbf{s})$ and by taking the continuity of Ψ in Theorem 1.9.1 (ii) into account.

We next take a parallel translation $\phi(z) = z + (r + i0)$ for $r \in \mathbb{R}$. For each $\mathbf{s} \in S$, $\phi\big|_{D(\mathbf{s})}$ is a conformal map from $D(\mathbf{s})$ onto $D(\mathbf{s} + \widehat{r})$. By Theorem 1.8.1,

$$G^*_{\mathbf{s}}(z, \zeta) = G^*_{\mathbf{s}+\widehat{r}}(z + r, \zeta + r), \quad z, \zeta \in D(\mathbf{s}),$$

so that the corresponding BMD Poisson kernels and BMD complex Poisson kernels are related by

$$K^*_{\mathbf{s}}(z, \xi) = K^*_{\mathbf{s}+\widehat{r}}(z + r, \xi + r) \quad \text{and} \quad \Psi_{\mathbf{s}}(z, \xi) = \Psi_{\mathbf{s}+\widehat{r}}(z + r, \xi + r) \tag{3.1.8}$$

for $z \in D(\mathbf{s})$ and $\xi \in \partial\mathbb{H}$. We then let $z \in D(\mathbf{s})$ approach to endpoints of slits of $D(\mathbf{s})$ as above to conclude that $b_j(\xi, \mathbf{s})$, $1 \le j \le 3N$, are homogeneous in horizontal direction. □

3.2 KLE driven by $\xi(t)$ and its basic properties

Throughout this and the next sections, we fix a pair of functions $(\xi(t), \mathbf{s}(t)) \in \mathbb{R} \times \mathcal{S}$ of $t \in [0, \zeta)$ for $\zeta > 0$, satisfying the following two properties **(I)** and **(II)**:

(I) $\xi(t)$ is a real-valued continuous function of $t \in [0, \zeta)$.
(II) $(\xi(t), \mathbf{s}(t))$, $t \in [0, \zeta)$, satisfies the equation (3.1.3) with b_j, $1 \le j \le 3N$, given by (3.1.4).

We have freedom of choices of such a pair in two ways.

The first way is to take any deterministic real continuous function $\xi(t)$, $t \in [0, \infty)$, substitute it into the right hand side of (3.1.3) and get the unique solution $\mathbf{s}(t)$ on a maximal time interval $[0, \zeta)$ of the resulting ordinary differential equation by using Lemma 3.1.1. Here we identify $\xi(t) \in \mathbb{R}$ with $\xi(t) + i0 \in \partial\mathbb{H}$.

The second way is to choose any solution path $(\xi(t), \mathbf{s}(t))$, $t \in [0, \zeta)$, of the system of the stochastic differential equation that will be formulated in the next chapter. This system of SDE includes the equation (3.1.3) as a part of the system.

We write $D_t = D(\mathbf{s}(t)) \in \mathcal{D}$, $t \in [0, \zeta)$, and define

$$\mathcal{X} = \bigcup_{t \in [0,\zeta)} \{t\} \times D_t,$$

$$\widehat{\mathcal{X}} = \bigcup_{t \in [0,\zeta)} \{t\} \times (D_t \cup K(t)^p \cup (\partial\mathbb{H} \setminus \{\xi(t)\})),$$

where $D_t = \mathbb{H} \setminus K(t)$, $K(t) = \cup_{i=1}^N C_j(t)$ and $K(t)^p = \cup_{j=1}^N C_j(t)^p$. See (1.9.1). The endpoints of the slit $C_j(t)$ are denoted by $z_j(t)$ and $z_j^r(t)$, and we set $C_j^0(t) = C_j(t) \setminus \{z_j(t), z_j^r(t)\}$. Note that $\mathcal{X}$ is a domain of $[0, \zeta) \times \mathbb{H}$ in $\mathbb{R}^3$ because $t \mapsto D_t = D(\mathbf{s}(t))$ is continuous.

Given a pair $(\mathbf{s}(t), \xi(t))$ satisfying **(I)**, **(II)**, we substitute it into the right hand side of the equation

$$\frac{d}{dt} z(t) = -2\pi \Psi_{\mathbf{s}(t)}(z(t), \xi(t)), \tag{3.2.1}$$

and we first study the unique existence of local solutions $z(t)$ of the resulting equation (3.2.1) with initial condition

$$z(\tau) = z_0 \in D_\tau \cup K(\tau)^p \cup (\partial\mathbb{H} \setminus \xi(\tau)), \tag{3.2.2}$$

for $\tau \in [0, \zeta)$.

Proposition 3.2.1.

(i) *$\Psi_{\mathbf{s}(t)}(z, \xi(t))$ is jointly continuous in $(t, z) \in \widehat{\mathcal{X}}$.*

(ii) *For every finite time interval $I \subset [0, \zeta)$,*

$$\limsup_{z\to\infty} \sup_{t\in I} |z\Psi_{\mathbf{s}(t)}(z, \xi(t))| < \infty. \tag{3.2.3}$$

Furthermore the limit $\alpha(t) := \lim_{z\to\infty} z\Psi_{\mathbf{s}(t)}(z, \xi(t))$ exists for each $t \in I$ and $\alpha(t)$ is a real valued bounded continuous function of $t \in I$.

(iii) *$\Psi_{\mathbf{s}(t)}(z, \xi(t))$ is locally Lipschitz continuous in z in the following sense: for any $(\tau, z_0) \in \mathcal{X}$, there exist $t_0 > 0$, $\rho > 0$ and $L > 0$ such that*

$$V = [(\tau - t_0)^+, \tau + t_0] \times \{z : |z - z_0| \le \rho\} \subset \mathcal{X},$$

and

$$|\Psi_{\mathbf{s}(t)}(z_1, \xi(t)) - \Psi_{\mathbf{s}(t)}(z_2, \xi(t))| \le L\,|z_1 - z_2|, \tag{3.2.4}$$

for any (t, z_1), $(t, z_2) \in V$.

(iv) *Fix $1 \le j \le N$. For any $\tau \in [0, \zeta)$ and $z_0 \in C_j^0(\tau)$, take a small open rectangle R centered at z_0 with sides parallel to the axes, $z_j(t) \notin \overline{R}$. $z_j^r(t) \notin \overline{R}$ and $\overline{R} \setminus C_j^0(\tau) \subset D$. There exists then $t_0 > 0$ such that the set $R \setminus C_j^0(t)$ is disconnected consisting of upper and lower open rectangles for every $t \in [(\tau - t_0)^+, \tau + t_0]$. Denote by $\Psi^+_{\mathbf{s}(t)}(z, \xi(t))$ (respectively, $\Psi^-_{\mathbf{s}(t)}(z, \xi(t))$) the extension of $\Psi_{\mathbf{s}(t)}(z, \xi(t))$ from the upper (respectively, lower) rectangle of $R \setminus C_j^0(t)$ to R. Then $\Psi^{\pm}_{\mathbf{s}(t)}(z, \xi(t))$ satisfies the Lipschitz continuity (3.2.4) for any $(t, z_1), (r, z_2) \in V_j$ where $V_j = [(\tau - t_0)^+, \tau + t_0] \times R$.*

(v) *For any $\tau \in [0, \zeta)$ and $z_0 \in \partial\mathbb{H} \setminus \{\xi(\tau)\}$, there exist $t_0 > 0$, $\rho > 0$ and $L > 0$ such that*

$$\begin{aligned} V_0 &= [(\tau - t_0)^+, \tau + t_0] \times \{z \in \overline{\mathbb{H}} : |z - z_0| \le \rho\} \\ &\subset \bigcup_{t\in[(\tau-t_0)^+, \tau+t_0]} \{t\} \times (D_t \cup (\partial\mathbb{H} \setminus \{\xi(t)\}), \end{aligned}$$

and (3.2.4) holds for any (t, z_1), $(t, z_2) \in V_0$.

(vi) *For every $\tau \in [0, \zeta)$ and $z_0 \in D_\tau \cup (\partial\mathbb{H} \setminus \xi(\tau))$, there exists a unique local solution $\{z(t); t \in (\tau - \widetilde{t}_0, \tau + \widetilde{t}_0) \cap [0, \zeta)\}$, $\widetilde{t}_0 > 0$, of (3.2.1) and (3.2.2) satisfying $z(\tau) = z_0$.*

(vii) *Fix $1 \le j \le N$. For each initial time $\tau \in [0, \zeta)$ and initial position $z_0 \in C_j^{0,+}(\tau)$, there exists a unique local solution $\{z(t); t \in (\tau - \widetilde{t}_0, \tau + \widetilde{t}_0) \cap [0, \infty)\}$, $\widetilde{t}_0 > 0$, of the equation (3.2.1) with $\Psi^+_{\mathbf{s}(t)}(z, \xi(t))$ in place of $\Psi_{\mathbf{s}(t)}(z, \xi(t))$ and $z(\tau) = z_0$. An analogous statement holds for $z_0 \in C_j^{0,-}(\tau)$.*

Proof. (i) This can be shown in the same way as the proof of Theorem 2.5.1 on account of the continuity of $\xi(t)$ and using the continuity of $t \mapsto D_t = D(\mathbf{s}(t))$.

(ii) Take $R > 0$ sufficiently large so that the closure of the set $\cup_{t\in I}(\cup_{j=1}^N C_j(t)) \cup \xi(t)$ is contained in $B(\mathbf{0}, R) = \{z \in \mathbb{C} : |z| < R\}$. Extend the analytic function $h(z,t) = \Psi_{\mathbf{s}(t)}(z, \xi(t))$ from $\mathbb{H} \setminus B(\mathbf{0}, R)$ to $\mathbb{C} \setminus B(\mathbf{0}, R)$ by the Schwarz reflection. By (i), $M = \sup_{z\in\partial B(\mathbf{0},R), t\in I} |h(z,t)|$ is finite. Define $\widehat{h}(z,t) = h(1/z, t)$. Since $h(z,t)$ tends to zero as $z \to \infty$, $\widehat{h}(z,t)$ is analytic on $B(\mathbf{0}, 1/R)$ with $\widehat{h}(0,t) = 0$ and, by [Ahlfors (1979), (28)–(29) in Chapter 4], we have

$$\frac{1}{z}\widehat{h}(z,t) = \frac{1}{2\pi i}\int_{|\zeta|=1/R} \frac{\widehat{h}(\zeta,t)}{\zeta(\zeta - z)} d\zeta = \frac{R}{2\pi}\int_0^{2\pi} \frac{h(Re^{-i\theta}, t)}{(e^{i\theta} - Rz)} d\theta, \quad |z| < 1/R.$$

Consequently,

$$\sup_{t\in I} |z\Psi_{\mathbf{s}(t)}(z, \xi(t))| \le 2RM \quad \text{if} \quad |z| \ge 2R, \tag{3.2.5}$$

and

$$\alpha(t) = \lim_{z\to 0} \frac{R}{2\pi}\int_0^{2\pi} \frac{h(Re^{-i\theta}, t)}{(e^{i\theta} - Rz)} d\theta = \frac{1}{2\pi}\int_{\partial B(\mathbf{0},R)} h(\zeta, t) d\zeta,$$

which is real valued and bounded continuous in $t \in I$ by (i).

(iii) Since $\mathcal{X}$ is a domain, there exist, for any $(\tau, z_0) \in \mathcal{X}$, $t_0 > 0$ and $\rho > 0$ such that

$$[(\tau - t_0)^+, \tau + t_0] \times \{z : |z - z_0| \le \rho\} \subset \mathcal{X}.$$

For each $t \in [(\tau - t_0)^+, \tau + t_0]$, $f_t(z) = \Psi_{\mathbf{s}(t)}(z, \xi(t))$ is analytic in $z \in D_t$. Therefore in a similar way as the identity (28)-(29) for $n = 1$ is derived in [Ahlfors (1979), Chapter 4], we can obtain for $C = \{z : |z - z_0| = \rho\}$ the identity,

$$f_t(z_1) - f_t(z_2) = \frac{z_1 - z_2}{2\pi i}\int_C \frac{f_t(\zeta)}{(\zeta - z_1)(\zeta - z_2)} d\zeta \quad \text{for } z_i \in B(z_0, \rho/2) \text{ with } i = 1, 2.$$

It follows from (i) that

$$M = \sup_{t\in[(\tau - t_0)^+, \tau + t_0], \zeta\in C} |f_t(\zeta)| < \infty.$$

Hence, we have (3.2.4) with $L = (4M)/\rho$.

(iv) and (v) can be shown similarly. For (v), we extend $\Psi_{\mathbf{s}(t)}(z, \xi(t))$ by Schwarz reflection.

(vi) follows from (iii) and (v), while (vii) does from (iv), by noting that, for a finite interval $I \subset [0, \zeta)$, $M_I = \sup_{(t,z)\in\widehat{\mathcal{X}}, t\in I}\ 2\pi\Psi_{\mathbf{s}(t)}(z, \xi(t))|$ is finite on account of (i) and (ii). □

Lemma 3.2.2.

(i) *Fix* $1 \leq j \leq N$. *For any* $\tau \in [0, \zeta)$ *and* $z_0 = x_0 + iy_0 \in C_j^{0,+}(\tau)$, *there exists a unique solution* $z(t)$, $t \in [(\tau - t_0)^+, \tau + t_0]$, *of* (3.2.1) *and* (3.2.2) *for some* $t_0 > 0$ *such that*

$$z(\tau) = z_0, \quad z(t) \in C_j^{0,+}(t) \text{ for every } t \in [(\tau - t_0)^+, \tau + t_0]. \tag{3.2.6}$$

An analogous statement holds for $z_0 \in C_j^{0,-}(\tau)$.

(ii) *For any* $\tau \in [0, \zeta)$ *and* $z_0 \in \partial\mathbb{H} \setminus \{\xi(\tau)\}$, *there exists a unique solution* $z(t)$, $t \in [(\tau - t_0)^+, \tau + t_0]$, *of* (3.2.1) *and* (3.2.2) *for some* $t_0 > 0$ *such that*

$$z(\tau) = z_0, \quad z(t) \in \partial\mathbb{H} \setminus \{\xi(t)\} \text{ for every } t \in [(\tau - t_0)^+, \tau + t_0]. \tag{3.2.7}$$

Proof. (i) When $z \in C_j(t)^p$, $\Im\Psi_{\mathbf{s}(t)}(z, \xi(t)) = K^*_{\mathbf{s}(t)}(z, \xi(t))$ is a continuous function $\eta_j(t)$ of t independent of z in view of Proposition 1.7.4 (i) and Proposition 3.2.1 (i). Writing the solution $\mathbf{s}(t) \in \mathcal{S}$ of the slit motion equation (3.1.3) as

$$\mathbf{s}(t) = (y_1(t), \ldots, y_N(t), x_1(t), \ldots, x_N(t), x_1(t)^r, \ldots, x_N(t)^r),$$

the equation (3.1.3) for $y_j(t)$ is reduced to $\dfrac{d}{dt} y_j(t) = -2\pi\eta_j(t)$. This equation under the condition $y_j(\tau) = y_0$ is uniquely solved by $\widetilde{y}_j(t) := -2\pi \int_\tau^t \eta_j(s)ds + y_0$.

Consider the equation for a real valued function $x(t)$:

$$\frac{d}{dt} x(t) = -2\pi\Re\Psi_{\mathbf{s}(t)}(x(t) + i\widetilde{y}_j(t), \xi(t)). \tag{3.2.8}$$

For $z_0 = x_0 + iy_0 \in C_j^{0,+}(\tau)$, the equation (3.2.8) admits a unique local solution $x(t)$ under the condition $x(\tau) = x_0$ by virtue of Proposition 3.2.1 (iv). Since $x_j(\tau) < x(\tau) < x_j(\tau)^r$ and $x_j(t)$, $x_j(t)^r$ are continuous in t as solutions of (3.1.3), we have $x_j(t) < x(t) < x_j(t)^r$, namely, $z(t) := x(t) + i\widetilde{y}_j(t) \in C_j^{0,+}(t)$ for any t with $|t - \tau| < \delta$ by choosing a suficiently small $\delta > 0$.

Accordingly, if $|t - \tau| < \delta$, then

$$-2\pi\Im\Psi_{\mathbf{s}(t)}(z(t), \xi(t)) = -2\pi\eta_j(t) = \frac{d}{dt}\widetilde{y}_j(t) = \frac{d}{dt}\Im z(t),$$

which along with (3.2.8) says that $z(t)$ is the unique solution of (3.2.1) with $z(\tau) = z_0 \in C_j^{0,+}(\tau)$ and the stated property.

(ii) Take $z_0 = x_0 + i0 \in \partial\mathbb{H} \setminus \{\xi(\tau)\}$. By Proposition 3.2.1 (v), the equation

$$\frac{d}{dt} x(t) = -2\pi\Re\Psi_{\mathbf{s}(t)}(x(t), \xi(t)), \quad x(\tau) = x_0, \tag{3.2.9}$$

admits a unique real-valued solution $x(t)$, $t \in I := ((\tau - t_0)^+, \tau + t_0)$, for some $t_0 > 0$ such that $x(t) \neq \xi(t)$ for any $t \in I$. Since $\Im\Psi_{\mathbf{s}(t)}(z, \xi(t)) = 0$ for any $z \in \partial\mathbb{H} \setminus \{\xi(t)\}$ by Proposition 1.7.4 (v), $z(t) = x(t) + i0$ is the unique solution of (3.2.1) with $z(\tau) = z_0$ and the stated property. □

Denote by $z_j(t)$ and $z_j^r(t)$ the left and right endpoints of the jth slit $C_j(t)$ of $\mathbf{s}(t)$. We know from (3.1.3) and (3.1.4)

$$\frac{dz_j(t)}{dt} = -2\pi\Psi_{\mathbf{s}(t)}(z_j(t), \xi(t)), \quad t \in [0, \zeta). \tag{3.2.10}$$

$z_j^r(t)$ satisfies the same equation.

A solution $\{z(t),\ t \in I\}$ of the equation (3.2.1) for a time interval $I \subset [0, \zeta)$ is said to *pass through* $\mathcal{X}$ (resp. $\widehat{\mathcal{X}}$) if $(t, z(t)) \in \mathcal{X}$ (resp. $(t, z(t)) \in \widehat{\mathcal{X}}$) for every $t \in I$.

According to Proposition 3.2.1 (vii), the local solution $z(t)$ of (3.2.1) passing through $\widehat{\mathcal{X}}$ with initial condition $z(\tau) = z_0 \in C_j^{0,\pm}(\tau)$ is unique. Lemma 3.2.2 (i) implies that such a solution $z(t)$ must satisfy $z(t) \in C_j^{0,+}(t)$ (resp. $z(t) \in C_j^{0,-}(t)$) for any t in some open time interval containing τ when $z_0 \in C_j^{0,+}(\tau)$ (resp. $z_0 \in C_j^{0,-}(\tau)$). An analogous property holds for a local solution $z(t)$ passing through $\widehat{\mathcal{X}}$ with condition $z(\tau) \in \partial\mathbb{H} \setminus \xi(\tau)$ according to Proposition 3.2.1 (vi) and Lemma 3.2.2 (ii).

Lemma 3.2.3. *Fix $1 \le j \le N$ and let $[\alpha, \beta]$ be a finite subinterval of $[0, \zeta)$ with $0 < \alpha < \beta < \zeta$.*

(i) *Suppose that $\{z(t); t \in [\alpha, \beta]\}$ is a solution of (3.2.1) passing through $\widehat{\mathcal{X}}$ with $z(\beta) = z_j(\beta)$. Then there exists $t_0 \in (0, \beta - \alpha)$ so that $z(t) \in C_j(t)^p$ for all $t \in [\beta - t_0, \beta]$. The same conclusion holds if $z_j(\beta)$ and $z_j(t)$ are replaced by $z_j^r(\beta)$ and $z_j^r(t)$.*

(ii) *Suppose that $\{z(t); t \in [\alpha, \beta]\}$ is a solution of (3.2.1) passing through $\widehat{\mathcal{X}}$ with $z(\alpha) = z_j(\alpha)$. Then there exists $t_0 \in (0, \beta - \alpha)$ so that $z(t) \in C_j(t)^p$ for all $t \in [\alpha, \alpha + t_0)$. The same conclusion holds if $z_j(\alpha)$ and $z_j(t)$ are replaced by $z_j^r(\alpha)$ and $z_j^r(t)$.*

Proof. We only prove (i) as the proof for (ii) is analogous. For $\zeta \in \mathbb{C}$ and $\varepsilon > 0$, we use $B(\zeta, \varepsilon)$ to denote the ball $\{z \in \mathbb{C} : |z - \zeta| < \varepsilon\}$ centered at ζ with radius ε.

Suppose that $\{z(t),\ t \in [\alpha, \beta]\}$ is a solution of (3.2.1) passing through $\widehat{\mathcal{X}}$ and that $z(\beta) = z_j(\beta)$. As $z_j(t), z_j^r(t)$ and $z(t)$ are continuous in t, wa can find $t_0 \in (0, \beta - \alpha)$ and $\varepsilon > 0$

$$\begin{cases} B(z_j(t), \varepsilon) \subset \mathbb{H} \quad \text{and} \quad z_j^r(t) \notin B(z_j(t), \varepsilon) \\ z(t) \in B(z_j(t), \varepsilon/2) \cap (D_t \cup C_j(t)^p) \quad \text{for every } t \in [\beta - t_0, \beta]. \end{cases} \tag{3.2.11}$$

For each $t \in (\beta - t_0, \beta]$, let

$$\psi_t(z) = \sqrt{z - z_j(t)} \ :\ B(z_j(t), \varepsilon) \setminus C_j(t) \to B(\mathbf{0}, \sqrt{\varepsilon}) \cap \mathbb{H},$$

and

$$f_t(z) = \Psi_{\mathbf{s}(t)}(\psi_t^{-1}(z), \xi(t)) = \Psi_{\mathbf{s}(t)}(z^2 + z_j(t), \xi(t)) \ :\ B(\mathbf{0}, \sqrt{\varepsilon}) \cap \mathbb{H} \to \mathbb{C}.$$

Then f_t is an analytic function on $B(\mathbf{0}, \sqrt{\varepsilon}) \cap \mathbb{H}$, which can be extended to be an analytic function on $B(\mathbf{0}, \sqrt{\varepsilon})\}$ by the Schwarz reflection because $\Im f_t(z)$ admits a constant limit on $B(\mathbf{0}, \sqrt{\varepsilon}) \cap \partial\mathbb{H}$. On account of [Ahlfors (1979), (28) and (29) in Chapter 4], it holds for every $a \in (\sqrt{\varepsilon}/2, \sqrt{\varepsilon})$ and $z \in B(\mathbf{0}, a)$ that

$$f_t(z) - f_t(\mathbf{0}) = z\, h_t(z) \quad \text{with } h_t(z) := \frac{1}{2\pi i} \int_{\partial B(\mathbf{0},a)} \frac{f_t(\zeta)}{\zeta(\zeta - z)} d\zeta. \tag{3.2.12}$$

In particular, $|h_t'(z)|$ is uniformly bounded in $(z,t) \in B(\mathbf{0}, \sqrt{\varepsilon}/2) \times I$ for any finite open interval I with $[\alpha, \beta] \subset I \subset [0, \zeta)$ by virtue of Proposition 3.2.1 (i). Accordingly $h_t(z)$, is Lipschitz continuous on $B(\mathbf{0}, \sqrt{\varepsilon}/2)$ uniformly in $t \in I$:

$$|h_t(z_1) - h_t(z_2)| \le L|z_1 - z_2|, \quad z_1, z_2 \in B(\mathbf{0}, \sqrt{\varepsilon}/2), \tag{3.2.13}$$

for a constant $L > 0$ independent of $t \in I$.

We now let $\widehat{z}(t) = \psi_t(z(t)) = \sqrt{z(t) - z_j(t)}$ for $t \in [\beta - t_0, \beta]$. On account of (3.2.11), $\widehat{z}(\beta) = 0$,

$$\widehat{z}(t) \in B(\mathbf{0}, \sqrt{\varepsilon}/2) \cap \overline{\mathbb{H}} \quad \text{for every } t \in [\beta - t_0, \beta], \tag{3.2.14}$$

and

$$\frac{dz(t)}{dt} = -2\pi \Psi_{\mathbf{s}(t)}(z(t), \xi(t)) = -2\pi f_t(\widehat{z}(t)), \quad t \in [\beta - t_0, \beta]. \tag{3.2.15}$$

We further let $\mathcal{T} = \{t \in (\beta - t_0, \beta] : \widehat{z}(t) = 0\}$. As $\widehat{z}(t)$ is continuous in t, $[\beta - t_0, \beta] \setminus \mathcal{T}$ is a countable union of disjoint open intervals. Denote one of them by $(\widehat{\alpha}, \widehat{\beta})$. Then $\widehat{z}(t) \neq 0$ for any $t \in (\widehat{\alpha}, \widehat{\beta})$ and $\widehat{z}(\widehat{\beta}) = 0$.

Since $\frac{dz_j(t)}{dt} = -2\pi f_t(\mathbf{0})$ by (3.2.10), we have by (3.2.12), (3.2.14) and (3.2.15) that for any $t \in (\widehat{\alpha}, \widehat{\beta})$

$$\frac{d\widehat{z}(t)}{dt} = \frac{1}{2\widehat{z}(t)} \left(\frac{dz(t)}{dt} - \frac{dz_j(t)}{dt} \right) = -\frac{\pi}{\widehat{z}(t)} \left(f_t(\widehat{z}(t)) - f_t(\mathbf{0}) \right) = -\pi\, h_t(\widehat{z}(t)).$$

Since $h_t(z)$ is Lipschitz continuous on $B(\mathbf{0}, \sqrt{\varepsilon}/2)$ uniformly in $t \in I$ by (3.2.13), $\widehat{z}(t)$, $t \in (\widehat{\alpha}, \widehat{\beta})$, is the unique solution of the equation

$$\frac{d\widehat{z}(t)}{dt} = -\pi h_t(\widehat{z}(t)), \quad \widehat{\alpha} < t < \widehat{\beta}, \quad \widehat{z}(\widehat{\beta}) = 0. \tag{3.2.16}$$

On the other hand, $\Im(f_t(z) - f_t(\mathbf{0})) = 0$ on $B(\mathbf{0}, \sqrt{\varepsilon}) \cap \partial\mathbb{H}$ and we get from (3.2.12)

$$\Im h_t(z) = 0 \qquad \text{on } B(\mathbf{0}, \sqrt{\varepsilon}) \cap \partial\mathbb{H}. \tag{3.2.17}$$

Consequently, a real valued solution to (3.2.16) exists. By the uniqueness, $\widehat{z}(t)$ is real valued, namely, $\widehat{z}(t) \in B(\mathbf{0}, \sqrt{\varepsilon}) \cap \partial\mathbb{H}$, which in turn implies $z(t) \in C_j(t)^p$ for any $t \in (\widehat{\alpha}, \widehat{\beta})$. Hence $z(t) \in C_j(t)^p$ for any $t \in (\beta - t_0, \beta]$.

An analogous argument shows that the second part of (i) holds as well. □

Due to (i) and (iii) of Proposition 3.2.1, and a general theorem in ODE (see, e.g., [Hartman (1964)]), there exists, for each $(\tau, z_0) \in \mathcal{X}$, a unique solution $z(t)$ of the equation (3.2.1) satisfying the initial condition $z(\tau) = z_0$ and passing through $\mathcal{X}$ with a maximal time interval $I_{\tau,z_0} (\subset [0, \zeta))$ of existence. Such a solution of

(3.2.1) will be designated by $\varphi(t;\tau,z_0)$, $t\in I_{\tau,z_0}$. Let α and β be the left and right endpoints of I_{τ,z_0}, respectively, both depending on (τ,z_0). Then $(t,\varphi(t;\tau,z_0))\in\mathcal{X}$ for any $t\in I_{\tau,z_0}\setminus\{\alpha,\beta\}$.

Proposition 3.2.4. *For any $(\tau,z_0)\in\mathcal{X}$, the maximal time interval I_{τ,z_0} of existence of the unique solution $\varphi(t;\tau,z_0)$ of* (3.2.1) *with $\varphi(\tau;\tau,z_0)=z_0$ passing through $\mathcal{X}$ is $[0,\beta)$ for some $\beta>\tau$ and*

$$\lim_{t\uparrow\beta}\Im\varphi(t;\tau,z_0)=0,\quad \lim_{t\uparrow\beta}|\varphi(t;\tau,z_0)-\xi(\beta)|=0 \quad \textit{whenever } \beta<\zeta. \tag{3.2.18}$$

Proof. Fix $\beta_0\in(0,\zeta)$ and $z_0\in D_{\beta_0}$. Let (α,β) be the largest subinterval of $(0,\zeta)$ so that the equation (3.2.1) has a unique solution $z(t)=\varphi(t;\beta_0,z_0)$ in $t\in(\alpha,\beta)$ satisfying $z(\beta_0)=z_0$ and passing through $\mathcal{X}$. By (i) and (iii) of Proposition 3.2.1, such an interval (α,β) exists with $0\le\alpha<\beta_0<\beta\le\zeta$. For simplicity, we write $\varphi(t;\tau,z_0)$ as $\varphi(t)$. We claim that

$$\alpha=0 \quad\text{and}\quad \varphi(0+):=\lim_{t\downarrow 0}\varphi(t)\in D. \tag{3.2.19}$$

Since the imaginary part of the right hand side of the equation (3.2.1) is negative, $\Im\varphi(t)$ is decreasing in t. By (i) and (ii) of Proposition 3.2.1, $\varphi(\alpha+):=\lim_{t\downarrow\alpha}\varphi(t)$ exists with $\Im\varphi(\alpha+)>0$. Set $\varphi(\alpha)=\varphi(\alpha+)$, which takes value in $D_\alpha\cup\cup_{j=1}^N C_j(\alpha)^p$. By Proposition 3.2.1 (vii), Lemma 3.2.2 (i) and Lemma 3.2.3, $\varphi(\alpha)\notin\cup_{j=1}^N C_j(\alpha)^p$ as $\varphi(t)\in D_t$ for $t\in(\alpha,\beta_0)$. Thus, $\varphi(\alpha)\in D_\alpha$. If $\alpha>0$, then the solution $\varphi(t)$ of (3.2.1) can be extended to $(\alpha-\varepsilon,\beta_0]$ for some $\varepsilon\in(0,\alpha)$. This contradicts to the maximality of (α,β). Thus $\alpha=0$ and the claim (3.2.19) is proved.

Since $\Im\varphi(t)$ is decreasing in t, $\lim_{t\uparrow\beta}\Im\varphi(t)$ exists. Assume $\beta<\zeta$. Were $\lim_{t\uparrow\beta}\Im\varphi(t)>0$, it follows from (i) and (ii) of Proposition 3.2.1 that $\varphi(\beta-):=\lim_{t\uparrow\beta}\varphi(t)$ exists and takes value in $D_\beta\cup\cup_{j=1}^N C_j(\beta)^p$. By Proposition 3.2.1 (vii), Lemma 3.2.2 (i) and Lemma 3.2.3 again, $\varphi(\beta-)\notin\cup_{j=1}^N C_j(\beta)^p$ as $\varphi(t)\in D_t$ for $t\in(\beta_0,\beta)$. Hence $\varphi(\beta-)\in D_\beta$ and thus, the solution $\varphi(t)$ of (3.2.1) can be extended to $[\beta_0,\beta+\varepsilon)$ for some $\varepsilon\in(0,\zeta-\beta)$. This contradicts to the maximality of (α,β) and so $\lim_{t\uparrow\beta}\Im\varphi(t)=0$.

We now proceed to proving the second claim in (3.2.18). Suppose $\limsup_{t\uparrow\beta}|\varphi(t)-\xi(\beta)|>0$. Then by the continuity of ξ, $\limsup_{t\uparrow\beta}|\varphi(t)-\xi(t)|>0$. Thus, there are an $\varepsilon>0$ and a sequence $\{t_n;n\ge1\}\subset(\beta-\varepsilon,\beta)$ increasing to β such that $\inf_{s\in[\beta-\varepsilon,\beta]}|\varphi(t_n)-\xi(s)|>\varepsilon$ for every $n\ge1$. By (i) and (ii) of Proposition 3.2.1, $\Psi_{\mathbf{s}(t)}(z,\xi(t))$ is bounded on

$$\widehat{\mathcal{X}}_0:=\Big\{(s,z)\in\widehat{\mathcal{X}}: s\in[\beta-\varepsilon,\beta],\ \inf_{s\in[\beta-\varepsilon,\beta]}|z-\xi(s)|\ge\varepsilon/2\Big\},$$

say, by $M>0$. So as long as $(t,\varphi(t))\in\widehat{\mathcal{X}}_0$, $|\frac{d}{dt}\varphi(t)|\le 2\pi M$. Let $\delta=\varepsilon/(4\pi M)$. This observation implies that $|\varphi(t_n)-\varphi(t)|\le 2\pi M(t-t_n)\le\varepsilon/2$ for every $t\in[t_n,t_n+\delta]\cap[t_0,\beta)$. Consequently, $\varphi(\beta-)=\lim_{t\uparrow\beta}\varphi(t)$ exists and takes value in $\partial\mathbb{H}\setminus\{\xi(\beta)\}$. But this contradicts to Proposition 3.2.1 (vi) and Lemma 3.2.2 (ii) as $\varphi(t)\in D_t$ for $t\in[t_1,\beta)$. This implies that $\lim_{t\uparrow\beta}|\varphi(t)-\xi(\beta)|=0$. □

We write $D_0 = D(\mathbf{s}(0)) \in \mathcal{D}$ as D.

Theorem 3.2.5.

(i) *For each $z \in D$, there exists a unique solution $g_t(z)$, $t \in [0, t_z)$, of the equation*

$$\partial_t g_t(z) = -2\pi \Psi_{\mathbf{s}(t)}(g_t(z), \xi(t)) \quad \text{with } g_0(z) = z \in D, \tag{3.2.20}$$

passing through $\mathcal{X}$, where $[0, t_z)$, $t_z > 0$, is the maximal time interval of its existence. It further holds that

$$\lim_{t \uparrow t_z} \Im g_t(z) = 0, \quad \lim_{t \uparrow t_z} |g_t(z) - \xi(t_z)| = 0 \text{ whenever } t_z < \zeta. \tag{3.2.21}$$

(ii) *Define*

$$F_t = \{z \in D : t_z \leq t\}, \quad t > 0. \tag{3.2.22}$$

Then $D \setminus F_t$ is open and g_t is a conformal map from $D \setminus F_t$ onto D_t for each $t > 0$.

Proof. (i) This just follows from Proposition 3.2.4 with $(\tau, z_0) = (0, z)$ and $\varphi(t, 0, z) = g_t(z)$.

(ii) According to [Hartman (1964), Theorem V.2.1], t_z is lower semi-continuous in $z \in D$, namely, if $t^0 < t_z$, then there exists $\varepsilon > 0$ such that $t^0 < t_{\widetilde{z}}$ for any $\widetilde{z} \in B(z, \varepsilon)$, Therefore $D \setminus F_t = \{z \in D : t < t_z\}$ is open.

As $\Psi_{\mathbf{s}(t)}(z, \xi(t))$ is analytic in z, so is $\varphi(t; 0, z)$ on $D \setminus F_t$ (cf. [Coddington and Levinson (1955)]). By virtue of Proposition 3.2.4, the map $z \in D \backslash F_t \mapsto \varphi(t; 0, z) \in D_t$ is one-to-one and onto; in fact, the map $w \in D_t \mapsto \varphi(0; t, w) \in D$ is its unique inverse. □

Note that the complex Poisson kernel of the absorbed Brownian motion (ABM) in $\mathbb{H}$ is

$$\Psi_{\mathbb{H}}(z, \xi) = -\frac{1}{\pi} \frac{1}{z - \xi}, \quad z \in \mathbb{H}, \ \xi \in \partial \mathbb{H}, \tag{3.2.23}$$

whose imaginary part $P(z, \xi) := \Im \Psi_{\mathbb{H}}(z, \xi) = \frac{1}{\pi} \frac{y}{(x-\xi)^2 + y^2}$ is the Poisson kernel of ABM in $\mathbb{H}$.

Let I be a finite subinterval of $[0, \zeta)$, and R, M be the positive constants in the proof of Proposition 3.2.1 (ii).

Lemma 3.2.6. (i) *Let $\widetilde{M} := \sup_{t \in I} |\xi(t)|$. Then*

$$\sup_{t \in I} \left| \frac{\Psi_{\mathbf{s}(t)}(z, \xi(t))}{\Psi_{\mathbb{H}}(z, \xi(t))} \right| \leq 4\pi R M \quad \text{for } |z| \geq (2R) \vee \widetilde{M}.$$

(ii) *For any $R_1 \geq R$,*

$$\sup_{t \in I} \sup_{z \in D_t, \, |z| \leq R_1} |\Psi_{\mathbf{s}(t)}(z, \xi(t)) - \Psi_{\mathbb{H}}(z, \xi(t))| < \infty.$$

Proof. (i) This follows from (3.2.5) as

$$\left|\frac{\Psi_{\mathbf{s}(t)}(z,\xi(t))}{\Psi^{\mathbb{H}}(z,\xi(t))}\right| = \pi|z-\xi(t)|\,|\Psi_{\mathbf{s}(t)}(z,\xi(t))| \leq 2\pi|z\Psi_{\mathbf{s}(t)}(z,\xi(t)| \quad \text{for } t\in I \text{ and } |z|\geq \widetilde{M}.$$

(ii) For $z\in D_t = D(\mathbf{s}(t))$ and $\xi\in\partial\mathbb{H}$, let

$$\mathbf{H}_t(z,\xi) = \Psi_{\mathbf{s}(t)}(z,\xi) - \Psi_{\mathbb{H}}(z,\xi).$$

By virtue of Lemma 1.9.2, $\mathbf{H}_t(z,\xi)$ is, for each $\xi\in\partial\mathbb{H}$, extended in z from D_t to an analytic function on $D_t\cup\Pi D_t\cup\partial\mathbb{H}$.

Choose $\varepsilon>0$ and $\ell>0$ so that the set $\Lambda=\{w=u+iv : |u|<\ell,\ 0\leq v<\varepsilon\}$ contains $J=\overline{\{\xi(t):t\in I\}}$ but does not intersect with the slits of D_t for any $t\in I$. On account of Proposition 3.2.1 (i), we see that, for any $R_1>\ell$, $\sup_{t\in I}\sup_{z\in D_t\setminus\Lambda,\ |z|\leq R_1}|\mathbf{H}_t(z,\xi(t))| = M_1<\infty$. Due to the maximum principle for an analytic function, $\mathbf{H}_t(z,\xi(t))$ has the same bound for $z\in\Lambda$. □

Recall that ζ is the lifetime for the real-valued driving function $\{x(s); s\in[0,\zeta)\}$. For each fixed $T\in(0,\zeta)$, by Lemma 3.2.6,

$$M_T := \sup_{t\in[0,T]}\sup_{z\in D_t} 2\pi|z-\xi(t)||\Psi_{\mathbf{s}(t)}(z,\xi(t))| < \infty. \tag{3.2.24}$$

The next lemma extends [Lawler (2005), Lemma 4.13] from the simply connected domain $\mathbb{H}$ to multiply connected domains.

Lemma 3.2.7. *Let $T\in(0,\zeta)$, and define $R_t := \sup_{0\leq s\leq t}|\xi(s)-\xi(0)|\vee\sqrt{tM_T/2}$ for $t\in[0,T]$. Then for each $t\in[0,T]$, $F_t\subset B(\xi(0),4R_t)$ and*

$$|z-g_s(z)|\leq R_t \quad \textit{for any } z\in D\cap B(\xi(0),4R_t)^c \textit{ and } s\in[0,t]. \tag{3.2.25}$$

Proof. Fix $t\in I$. For $z\in D$ with $|z-\xi(0)|\geq 4R_t$, define $\sigma=\inf\{s:|g_s(z)-z|\geq R_t\}$. If $s\leq t\wedge\sigma$, then $|g_s(z)-z|<R_t$ and

$$|\xi(s)-g_s(z)|\geq|z-\xi(0)|-|\xi(s)-\xi(0)|-|g_s(z)-z|>4R_t-2R_t=2R_t.$$

Hence, we have by (3.2.24)

$$|\partial_s g_s(z)| = \left|2\pi\Psi_{\mathbf{s}(s)}(g_s(z),\xi(s))\right| \leq \frac{M_1}{|g_s(z)-\xi(s)|}\leq\frac{M_1}{2R_t}.$$

Consequently, $|z-g_s(z)| = |\int_0^s\partial_r g_r(z)dr| \leq \frac{M_1}{2R_t}s$ for $s\in[0,t\wedge\sigma]$. We claim that $\sigma\geq t$. Suppose otherwise, then by the definition of σ, we would have $R_t = |z-g_\sigma(z)|\leq\frac{M_1}{2R_t}\sigma$ and so $\sigma\geq\frac{2}{M_1}R_t^2\geq t$. This contradiction establishes that $\sigma\geq t$. So for all $s\in[0,t]$, we have $|g_s(z)-z|\leq R_t$ and $|\xi(s)-g_s(z)|\geq 2R_t$. Thus we have by (3.2.21) that $t<t_z$ and $z\in\mathbb{H}\setminus F_t$. □

Theorem 3.2.8.

(i) *The conformal map* $g_t(z)$ *in* Theorem 3.2.5 *satisfies the hydrodynamic normalization at infinity; for some real constant* a_t,

$$g_t(z) = z + \frac{a_t}{z} + o(1/|z|) \quad \text{as } z \to \infty \text{ for every } t \in [0, \zeta). \tag{3.2.26}$$

Moreover,

$$a_t = -2\pi \int_0^t \alpha(s)ds \quad \text{for every } t \in (0, \zeta).$$

where $\alpha(s) := \lim_{z\to\infty} z\Psi_{\mathbf{s}(s)}(z, \xi(s))$ *as in* Proposition 3.2.1 (ii) *is a continuous function of* s.

(ii) *The set* F_t *defined by* (3.2.22) *is an* $\mathbb{H}$*-hull; that is,* F_t *is relatively closed in* $\mathbb{H}$ *and bounded, and moreover* $\mathbb{H} \setminus F_t$ *is simply connected.*

(iii) $\{F_t\}$ *is strictly increasing in* t. *It has the property*

$$\bigcap_{\delta>0} \overline{g_t(F_{t+\delta} \setminus F_t)} = \{\xi(t)\} \quad \text{for } t \in [0, \zeta). \tag{3.2.27}$$

Proof. (i) Due to the continuity of the solution of (3.2.1) with respect to the initial position (cf. [Hartman (1964), Theorem V.2.1]), it follows from Proposition 3.2.1 (vi), Lemma 3.2.2 (ii) and Lemma 3.2.7 that, by taking a sufficiently large $R > 0$, $\lim_{z\in\mathbb{H},\, z\to w} \Im g_t(z) = 0$ for any $w \in \partial\mathbb{H} \setminus B_R(\mathbf{0})$. Since $\lim_{z\to\infty} g_t(z) = \infty$ by Lemma 3.2.7, we get the expression (3.2.26) with a complex constant a_t by making the Schwarz reflection and by thinking about the function $g_t(1/z)^{-1}$, $|z| > R$. On the other hand, we have from (3.2.20),

$$g_t(z) - z = -2\pi \int_0^t \Psi_{\mathbf{s}(r)}(g_r(z), \xi(r))dr, \quad t \in [0, t_z). \tag{3.2.28}$$

Fix $T \in (0, \zeta)$, by Lemma 3.2.7 and Proposition 3.2.1 (ii), there is a constant $R \geq 1$ so that for every $z \in D$ with $|z| > R$, we have $t_z > T$ and

$$\sup_{s\in[0,T]} |z - g_s(z)| \leq R \quad \text{and} \quad \sup_{|z|>R,\, s\in[0,T]} |z\Psi_{\mathbf{s}(t)}(z, \xi(s))| =: M < \infty.$$

Thus, for every $s \in [0, T]$ and every $z \in D$ with $|z| > 2R$,

$$|g_s(z)| \geq |z| - |z - g_s(z)| > |z|/2 > R,$$

so that

$$|(z - g_s(z))\Psi_{\mathbf{s}(s)}(g_s(z), \xi(t))| \leq \frac{R}{|z|}\frac{|z|}{|g_s(z)|}|g_s(z)\Psi_{\mathbf{s}(s)}(g_s(z), \xi(s))| \leq 2\frac{R}{|z|}M.$$

These together with (3.2.28), Proposition 3.2.1 (ii) and the bounded convergence theorem gives that for every $t \in [0, T]$,

$$\begin{aligned}
a_t &= \lim_{z\to\infty} z(g_s(z) - z) = -2\pi \lim_{z\to\infty} \int_0^t z\Psi_{\mathbf{s}(r)}(g_r(z), \xi(r))dr \\
&= -2\pi \lim_{z\to\infty} \int_0^t g_r(z)\Psi_{\mathbf{s}(r)}(g_r(z), \xi(r))dr \\
&\quad -2\pi \lim_{z\to\infty} \int_0^t (z - g_r(z))\Psi_{\mathbf{s}(r)}(g_r(z), \xi(r))dr \\
&= -2\pi \int_0^t \alpha(s)ds.
\end{aligned}$$

(ii) It follows from Theorem 3.2.5 and Lemma 3.2.7 that F_t is relatively closed and bounded. Were $\mathbb{H} \setminus F_t$ not simply connected, $D \setminus F_t$ would be multiply connected of degree at least $N+2$, which is absurd as the conformal image of $D \setminus F_t$ under g_t is the $(N+1)$-ply connected slit domain D_t.

(iii) Suppose $F_t = F_{t'}$ for some $t' > t \geq 0$. Then both g_t and $g_{t'}$ are conformal maps from $D \setminus F_t$ onto standard slit domains satisfying the hydrodynamic normalization. By the uniqueness, we get $g_t(z) = g_{t'}(z)$, $z \in D \setminus F_t$, which is absurd because $\Im g_t(z)$ is strictly decreasing as t increases.

By Lemma 3.2.7 and the fact that $\lim_{t \to 0} R_t = 0$, we have $\cap_{\delta > 0} \overline{F}_\delta = \{\xi(0)\}$. So (3.2.27) holds for $t = 0$. For every $t_0 \in (0, \zeta)$, $\{\widehat{F}_{t_0} := g_t(F_{t_0+t} \setminus F_{t_0}); t \in [0, \zeta - t_0)\}$ is the family of increasing closed sets associated with the KL-equation (3.2.20) in Theorem 3.2.5 but with $\mathbf{s}(t)$, $\xi(t)$ and D being replaced by $\widehat{\mathbf{s}}(t) := \mathbf{s}(t_0 + t)$, $\widehat{\xi}(t) := \xi(t_0 + t)$ and $\widehat{D} := D(t_0)$, respectively. Thus the same argument for $t = 0$ above applied to $\{\widehat{F}_\delta; \delta > 0\}$ yields that (3.2.27) holds for $t = t_0$. □

Given $(\mathbf{s}(t), \xi(t))$ satisfying **(I)** and **(II)** stated in the beginning of this section, the unique solution $g_t(z)$ of the Komatu-Loewner differential equation (3.2.20) is the canonical map from $D \setminus F_t$ onto $D_t = D(\mathbf{s}(t))$ in the sense of Section 2.1.1, where F_t is the $\mathbb{H}$-hull defined by (3.2.22). We call $\{F_t\}$ the *Komatu-Loewner evolution driven by* $\xi(t)$.

We call the specific property (3.2.27) of the $\mathbb{H}$-hull $\{F_t\}$ its *right continuity at* t *with limit* $\xi(t)$.

The real number a_t appearing in (3.2.26) is called the *half-plane capacity* for the Komatu-Loewner evolution.

3.3 Half-plane capacity for Komatu-Loewner evolution

We keep the setting of Section 3.2 and consider the Komatu-Loewner evolution $\{g_t(z)\}$, $\{F_t\}$ driven by $\xi(t)$. Let a_t be the associated half-plane capacity appearing in (3.2.26).

Theorem 3.3.1. *It holds that* $a_t = 2t$ *for every* $t \geq 0$.

This theorem is an immediate consequence of the following proposition, which compared with the equation (3.2.20) implies that the right derivative $\frac{d^+ a_t}{dt}$ equals 2. Since a_t is continuously differentiable by Theorem 3.2.8, we have $a_t = 2t$.

Proposition 3.3.2. *The half-plane capacity function* a_t *is strictly increasing and continuously differentiable in* $t \in [0, \zeta)$. *The conformal map* $g_t(z)$ *is right differentiable in* a_t *and*

$$\frac{\partial^+ g_t(z)}{d a_t} = -\pi \Psi_{\mathbf{s}(t)}(g_t(z), \xi(t)), \quad g_0(z) = z \in D, \quad t \in [0, t_z). \tag{3.3.1}$$

Here $\frac{\partial^+ g_t(z)}{d a_t}$ *is the right derivative of* $g_t(z)$ *with respect to* a_t.

In order to prove this proposition, we need the following variant of Proposition 1.1.6.

Lemma 3.3.3. *Let $D = \mathbb{H} \setminus K$, $K = \bigcup_{j=1}^N C_j$, be a standard slit domain and f be a bounded harmonic function on D. Assume further that $\lim_{z\to\infty} f(z) = 0$ and f takes a constant value f_j continuously on each C_j for $1 \le j \le N$. Then the vertical limit*

$$f(\xi) = \lim_{y\downarrow 0} f(x+iy) \quad \text{where} \quad \xi = x + i0, \tag{3.3.2}$$

exists for a.e. $\xi \in \partial\mathbb{H}$ *and the identities* (1.1.43) *and* (1.1.44) *in* Proposition 1.1.6 *hold.*

Proof. We let

$$v(z) = f(z) - \sum_{j=1}^N f_j \varphi^{(j)}(z). \tag{3.3.3}$$

which is a bounded harmonic function in D vanishing continuously on each C_j and goes to zero as $z \to \infty$ on account of Lemma 1.1.4 (i). Let $b > 0$ be so that all the slits are above the line $y = 2b$. Denote by $\mathbf{Z}^0 = \{Z_t^0, t \ge 0; \mathbb{P}_z^0, z \in D\}$ the ABM in $\mathbb{C} \setminus K$ with lifetime ζ^0. By the Fatou theorem for bounded classical harmonic functions in the stripe $S := \{z \in \mathbb{H} : 0 < \Im z < 2b\}$, we know that v has non-tangential limit, denoted as $v(\xi)$, for a.e. $\xi \in \partial\mathbb{H}$, and that for every $R > 0$,

$$v(z) = \mathbb{E}_z[v(Z^0_{\tau_{S_R}})] \quad \text{for every } z \in S_R := \{z \in S : |\Re z| \le R\},$$

where $\tau_{S_R} = \inf\{t > 0 : Z_t^0 \notin S_R\}$ is the first exit from S_R by $\mathbf{Z}^0$. Taking $R \to \infty$, we get by the bounded convergence theorem and the fact that $\lim_{z\to\infty} v(z) = 0$,

$$v(z) = \mathbb{E}_z[v(Z^0_{T_0})] \quad \text{for every } z \in S, \tag{3.3.4}$$

where $T_0 := \sigma_{\partial\mathbb{H}} \wedge \sigma_{\ell_{2b}}$ As each $\varphi^{(j)}$ vanishes continuously on $\partial\mathbb{H}$ by Lemma 1.1.4, f has the same non-tangential limit as $v(\xi)$, for a.e. $\xi \in \partial\mathbb{H}$. In particular,

$$f(\xi) = \lim_{y\downarrow 0} f(x+iy) = v(\xi), \tag{3.3.5}$$

exists for a.e. $\xi = x + i0 \in \mathbb{H}$.

Take $L > 0$ so large that all the slits are below the horizontal line $\ell_L := \{z \in \mathbb{C} : \Im z = L\}$. We claim that

$$v(z) = \mathbb{E}_z\left[v(Z^0_{\sigma_{\partial\mathbb{H}}\wedge\sigma_{\ell_L}})\right] \quad \text{for every } z \in D \text{ with } \Im z < L. \tag{3.3.6}$$

For any $z \in D$ with $\Im z < L$, we can choose $b > 0$ so that $b < \Im z$ and that all the slits are above the line $y = 2b$. Let $R_0 > 0$ so that all the slits are contained in the vertical stripe $\{z \in \mathbb{C} : |\Re z| < R_0\}$. Let $C_{R,b,L} := \{z \in D : |\Re z| \le R \text{ and } b \le \Im z \le L\}$. Since v is harmonic in $\mathbb{H} \setminus K$ and vanishes continuously on K, for every $R > R_0$,

$$v(z) = \mathbb{E}_z\left[v(Z^0_{\tau_{C_{R,b,L}}})\right] \quad \text{for every } z \in C_{R,b,L}.$$

Sending $R \to \infty$, we have by the bounded convergence theorem and the fact that $\lim_{z\to\infty} v(z) = 0$,

$$v(z) = \mathbb{E}_z \left[v(Z^0_{\sigma_{\ell_b}\wedge\sigma_{\ell_L}}) \right] \quad \text{for every } z \in D \text{ with } b \le \Im z \le L. \tag{3.3.7}$$

Define $T_1 = \sigma_{\ell_b} \wedge \sigma_{\ell_L} \wedge \zeta^0$, and for $k \ge 1$,

$$T_{2k} = \begin{cases} T_{2k-1} + (\sigma_{\partial\mathbb{H}} \wedge \sigma_{\ell_{2b}}) \circ \theta_{T_{2k-1}} & \text{if } T_{2k-1} < \sigma_{\partial\mathbb{H}} \wedge \sigma_{\ell_L} \wedge \zeta^0, \\ \sigma_{\partial\mathbb{H}} \wedge \sigma_{\ell_L} \wedge \zeta^0 & \text{if } T_{2k-1} = \sigma_{\partial\mathbb{H}} \wedge \sigma_{\ell_L} \wedge \zeta^0, \end{cases}$$

$$T_{2k+1} = \begin{cases} T_{2k} + (\sigma_{\ell_b} \wedge \sigma_{\ell_L}) \circ \theta_{T_{2k}} & \text{if } T_{2k} < \sigma_{\partial\mathbb{H}} \wedge \sigma_{\ell_L} \wedge \zeta^0, \\ \sigma_{\partial\mathbb{H}} \wedge \sigma_{\ell_L} \wedge \zeta^0 & \text{if } T_{2k} = \sigma_{\partial\mathbb{H}} \wedge \sigma_{\ell_L} \wedge \zeta^0. \end{cases}$$

Each T_k is a stopping time not exceeding $\sigma_{\partial\mathbb{H}}\wedge\sigma_{\ell_L}\wedge\zeta^0$, and T_k takes value $\sigma_{\ell_L}\wedge\sigma_{\partial\mathbb{H}}\wedge\zeta^0$ for sufficiently large k. Clearly for any $z \in D$ with $b < \Im z < L$, $v(x) = \mathbb{E}_z[v(Z^0_{T_1})]$ by (3.3.7). By the strong Markov property of Z^0 and (3.3.4),

$$\begin{aligned} v(z) &= \mathbb{E}_z \left[v(Z^0_{\sigma_{\ell_L}}); \sigma_{\ell_L} < \sigma_{\ell_b} \right] + \mathbb{E}_z \left[v(Z^0_{\sigma_{\ell_b}}); \sigma_{\ell_b} < \sigma_{\ell_L} \right] \\ &= \mathbb{E}_z \left[v(Z^0_{\sigma_{\ell_L}}); \sigma_{\ell_L} < \sigma_{\ell_b} \right] + \mathbb{E}_z \left[\mathbb{E}_{Z^0_{\sigma_{\ell_b}}} \left[v(Z^0_{\sigma_{\partial\mathbb{H}}\wedge\sigma_{\ell_{2b}}}) \right]; \sigma_{\ell_b} < \sigma_{\ell_L} \right] \\ &= \mathbb{E}_z \left[v(Z^0_{T_2}); \sigma_{\ell_L} < \sigma_{\ell_b} \right] + \mathbb{E}_z \left[v(Z^0_{T_2}); \sigma_{\ell_b} < \sigma_{\ell_L} \right] \\ &= \mathbb{E}_z \left[v(Z^0_{T_2}) \right]. \end{aligned}$$

Continuing this process, we get by mathematical induction that $v(z) = \mathbb{E}_z[v(Z^0_{T_k})]$ for every $k \ge 2$. Passing $k \to \infty$ yields the desired identity as claimed in (3.3.6). Now passing $L \to \infty$, we get

$$v(z) = \mathbb{E}_z \left[v(Z^0_{\sigma_{\partial\mathbb{H}}}) \right] \quad \text{for } z \in D.$$

This together with (3.3.5) and (1.1.28) of Proposition 1.1.5 gives the identity (1.1.43). The identity (1.1.44) is obtained by taking the period of the both hand sides of (1.1.43) as in the proof of Proposition 1.1.6. □

Fix $T \in (0, \zeta)$ and, for $0 \le s < t \le T$, set $g_{t,s} = g_s \circ g_t^{-1}$, which is a conformal map from D_t onto $D_s \setminus g_s(F_t \setminus F_s)$ satisfying (2.1.15). Its inverse map $g_{t,s}^{-1}$ is a canonical map from $D_s \setminus g_s(F_t \setminus F_s)$ onto the standard slit domain D_t. Therefore, we can draw from the proof of Theorem 2.2.3 the following conclusion.

Let $\ell_{t,s}$ be the set of all limiting points of $g_{t,s}^{-1} \circ g_s(z) = g_t(z)$ as z approaches to $F_t \setminus F_s$. In view of the proof of Theorem 2.2.3, $\ell_{t,s}$ is then a compact subset of $\partial\mathbb{H}$ and $\lim_{z\to\zeta} \Im g_{t,s}(z) = 0$ for any $\zeta \in \partial\mathbb{H} \setminus \ell_{t,s}$.

Set $F(z) = g_{t,s}(z) - z$ and $f(z) = \Im F(z)$ for $z \in D$. Then f is harmonic on D and $\lim_{z\to\infty} f(z) = 0$ by (2.1.15). Hence, f is bounded on D and satisfies all properties required by Lemma 3.3.3. Accordingly,

$$\Im g_{t,s}(\zeta) = \lim_{y\downarrow 0} \Im g_{t,s}(x + iy) = \lim_{y\downarrow 0} f(x + iy), \quad \text{where } \zeta = x + i0, \tag{3.3.8}$$

exists for a.e. $\zeta \in \partial\mathbb{H}$, and we have the identity

$$\Im(g_{t,s}(z) - z) = \sum_{j=1}^{N} f_j \varphi_t^{(j)}(z) - \frac{1}{2}\int_{\ell_{t,s}} \frac{\partial G_t^0(z,\zeta)}{\partial \mathbf{n}_\zeta} \Im g_{t,s}(\zeta) ds(\zeta)$$

as well as the identity (1.1.44) with $\Im g_{t,s}(\zeta)$ in place of $f(\zeta)$.

Consequently, in exactly the same way as (2.1.22) and (2.1.28) were derived in Section 2.1, we can get the following.

Lemma 3.3.4. *For $0 \le s < t \le t_0$, $a_t - a_s = \pi^{-1}\int_{\ell_{t,s}} \Im g_{t,s}(x+i0)dx$ and*

$$g_s(z) - g_t(z) = \int_{\ell_{t,s}} \Psi_{\mathbf{s}(t)}(g_t(z), x) \Im g_{t,s}(x+i0)dx, \quad z \in D \setminus F_t.$$

By the Schwarz reflection, we can extend $g_{t,s}^{-1}$ to a conformal map on

$$D_s \cup \Pi D_s \cup \partial\mathbb{H} \setminus (\overline{g_s(F_t \setminus F_s)} \cup \Pi g_s(F_t \setminus F_s)).$$

Lemma 3.3.5. *For any compact subset V of $D_s \cup \partial\mathbb{H} \setminus \{\xi(s)\}$, $\lim_{t\downarrow s} g_{t,s}^{-1}(z) = z$ uniformly in $z \in V \cup \Pi V$.*

Proof. Without loss of generality, we may assume $s = 0$ and so $g_{t,s}^{-1} = g_t$. Let V be any relatively compact open subset of $D \cup (\partial\mathbb{H} \setminus \{\xi(0)\})$. In Theorem 3.2.5, we considered the family of solution curves $\{(g_t(z),\ 0 \le t < t_z) : z \in D\}$ of (3.2.20) parametrized by the initial position $z = g_0(z) \in D$. We add to this family the solution curve $(g_t(z),\ 0 \le t < t_z)$ of (3.2.20) with initial position $z = g_0(z) \in \partial\mathbb{H} \setminus \{\xi(0)\}$ satisfying $g_t(z) \in \partial\mathbb{H}$, $0 \le t < t_z$, where

$$t_z = \sup\left\{t \in [0,\zeta) : \inf_{s\in[0,t]} |g_s(z) - \xi(s)| > 0\right\}.$$

By Proposition 3.2.1 (vi) and Lemma 3.2.2 (ii), such a solution exists uniquely and takes values in $\partial\mathbb{H}$. Define $F_t(\partial\mathbb{H}) = \{z \in \partial\mathbb{H} \setminus \{\xi(0)\} : t_z \le t\}$, $t > 0$.

Furthermore, by using the mirror reflection $\Pi z = \overline{z}$, the equation (3.2.20) and its solution are extended for $z \in D \cup \Pi D \cup (\partial\mathbb{H} \setminus \{\zeta\})$ as follows. Define

$$\widehat{\Psi}_{\mathbf{s}(t)}(z,\zeta) = \begin{cases} \Psi_{\mathbf{s}(t)}(z,\zeta), & z \in D \cup \partial\mathbb{H} \setminus \{\zeta\},\ \zeta \in \partial\mathbb{H}, \\ \overline{\Psi}_{\mathbf{s}(t)}(\overline{z},\zeta), & z \in \Pi D, \end{cases}$$

$$\widetilde{g}_t(z) = \begin{cases} g_t(z), & z \in D \cup \partial\mathbb{H} \setminus \{\zeta\}, \\ \overline{g}_t(\overline{z}), & z \in \Pi D. \end{cases}$$

It then follows from (3.2.20) that, for $z \in D \cup \Pi D \cup \partial\mathbb{H} \setminus \{\zeta(0)\}$,

$$\partial_t \widehat{g}_t(z) = -2\pi \widehat{\Psi}_{\mathbf{s}(t)}(\widehat{g}_t(z), \xi(t)), \quad \text{with} \quad \widetilde{g}_0(z) = z; \tag{3.3.9}$$

namely, $\widehat{g}_t(z)$ solves the equation (3.3.9) and it is the unique solution even if $z \in \Pi D$ in view of Proposition 3.2.1 (iii).

According to [Hartman (1964), Theorem V.2.1], $\widehat{g}_t(z)$ is jointly continuous on $\mathcal{G} = \{(t,z) : z \in D \cup \Pi D \cup (\partial\mathbb{H} \setminus \{\xi(0)\}),\ t \in [0,t_z)\}$. For the set V as above, Theorem 3.2.8 (iii) implies that there exists $\delta > 0$ and a relatively compact open set U such that

$$V \cup \Pi V \subset U \subset \overline{U} \subset D \cup \Pi D \cup \partial\mathbb{H} \setminus (\overline{F}_\delta \cup \Pi F_\delta \cup F_\delta(\partial\mathbb{H})).$$

So $[0,\delta] \times \overline{U}$ is a compact subset of $\mathcal{G}$. Hence $\sup_{t\in[0,\delta], z\in\overline{U}} |\widehat{g}_t(z)|$ is finite by the continuity of $\widehat{g}_t(z)$ mentioned above, and accordingly $\{\widehat{g}_t(z) : 0 \le t \le \delta\}$ is a normal family of analytic functions on U. This implies that $\lim_{t\downarrow 0} \widehat{g}_t(z) = z$ uniformly in $z \in V \cup \Pi V$. □

Proof of Proposition 3.3.2. By Theorem 3.2.8 (i), a_t is continuously differentiable in $t \in [0,\zeta)$. For $t > s \ge 0$, $a_t - a_s$ is the half-plane capacity associated with the canonical map $g_{t,s}$ from $D_s \setminus (F_t \setminus F_s)$. As $F_t \setminus F_s \neq \emptyset$ in view of Theorem 3.2.8 (iii), we have by Theorem 2.2.4 (iii) that $a_t - a_s > 0$.

For any $\varepsilon_0 > 0$ with $B(\xi(s), \varepsilon_0) \cap \mathbb{H} \subset D_s$, there exists $\delta_1 > 0$ so that

$$g_s(F_t \setminus F_s) \cup \Pi g_s(F_t \setminus F_s) \subset B(\xi(s), \varepsilon_0) \quad \text{for any } t \in (s, s+\delta_1)$$

by virtue of Theorem 3.2.8 (iii). In particular, $\ell_{t,s}$ is in the interior of the region bounded by the Jordan curve $g_{t,s}^{-1}(\partial B(\xi(s), \varepsilon_0))$. By Lemma 3.3.5, there exists $\delta_2 \in (0, \delta_1)$ such that

$$|g_{t,s}^{-1}(z) - z| < \varepsilon_0 \text{ for any } z \in \partial B(\xi(s), \varepsilon_0) \text{ and for any } t \in (s, s+\delta_2).$$

In particular, the diameter of $g_{t,s}^{-1}(\partial B(\xi(s), \varepsilon_0))$ is less than $4\varepsilon_0$. Therefore, we get for any $x \in \ell_{t,s}$

$$|\xi(s) - x| \le |\xi(s) - z| + |z - g_{t,s}^{-1}(z)| + |g_{t,s}^{-1}(z) - x| < 6\varepsilon_0 \quad \text{for } t \in (s, s+\delta_2), \tag{3.3.10}$$

by taking any $z \in \partial B(\xi(s), \varepsilon_0)$.

On the other hand, using the continuity of $\mathbf{s}(t)$, we can show that $\Psi_{\mathbf{s}(t)}(z,x)$ is jointly continuous in (t,z,x) as in the proof of Theorem 2.5.1. Fix $z \in D$. Since $g_t(z)$ is continuous in t, $\Psi_{\mathbf{s}(t)}(g_t(z), x)$ is continuous in $t > 0$ and $x \in \partial\mathbb{H}$. Therefore, for any $\varepsilon > 0$, there exist $\delta_3 > 0$ and $\varepsilon_0 > 0$ such that

$$|\Psi_{\mathbf{s}(t)}(g_t(z), x) - \Psi_{\mathbf{s}(s)}(g_s(z), \xi(s))| < \varepsilon, \tag{3.3.11}$$

for any $t \in (s, s+\delta_3)$ and for any $x \in \partial\mathbb{H}$ with $|x - \xi(s)| < 6\varepsilon_0$. For this ε_0, determine $\delta_2 > 0$ as above and set $\delta_0 = \delta_2 \wedge \delta_3$. It then follows from Lemma 3.3.4, (3.3.10) and (3.3.11) that, for any $t \in (s, s+\delta_0)$,

$$\left| \frac{g_t(z) - g_s(z)}{a_t - a_s} + \pi \Psi_{\mathbf{s}(s)}(g_s(z), \xi(s)) \right| < \varepsilon.$$

This proves the Proposition. □

Chapter 4

Stochastic Komatu-Loewner evolution (SKLE)

4.1 Random Jordan arc and induced process $(\xi(t), \mathbf{s}(t))$

Recall the collection $\mathcal{D}$ of all labelled standard slit domains introduced in the second half of Section 2.3.2, which is a metric space with distance (2.3.15). The space $\mathcal{D}$ can be identified with the space $\mathcal{S}$ defined by (3.1.1) as a topological space. We write $\mathbf{s}(D)$ (resp. $D(\mathbf{s})$) the element in $\mathcal{S}$ (resp. $\mathcal{D}$) corresponding to $D \in \mathcal{D}$ (resp. $\mathbf{s} \in \mathcal{S}$). In particular, for $\mathbf{s} = (\mathbf{y}, \mathbf{x}, \mathbf{x}^r) \in \mathcal{S}$, $z_j = x_j + iy_j$, $z_j^r = x_j^r + iy_j$ are left and right endpoints of the j-th slit C_j of $D(\mathbf{s}) \in \mathcal{D}$.

Given $D \in \mathcal{D}$ and a Jordan arc γ satisfying (2.1.11), the motion $\xi(t) \in \partial\mathbb{H}$ and the slit motion $\mathbf{s}(t) \in \mathcal{S}$ were induced in Sections 2.1.2 and 2.6, respectively, via the canonical map $g_t(z)$ from $D \setminus \gamma[0,t]$, namely, the unique conformal map from $D \setminus \gamma[0,t]$ onto some $D_t \in \mathcal{D}$ satisfying (2.1.7). The pair $(\xi(t), \mathbf{s}(t))$ satisfies the equation (3.1.3). In this section, we shall randomize the Jordan arc γ in a way that the probability law of γ has a domain Markov property and the invariance under the linear conformal map. We shall then describe the implied probabilistic laws of the random process $\mathbf{W}_t = (\xi(t), \mathbf{s}(t))$. We occasionally write g_t as $g_{D\setminus\gamma[0,t]}$ to indicate its dependence on $D \in \mathcal{D}$ and γ. Note that g_t sends $\gamma(t)$ to $\xi(t) \in \partial\mathbb{H}$ and the slits in D to $\mathbf{s}(t) = \mathbf{s}(D_t)$.

4.1.1 *Domain Markov property and conformal invariance*

Fix $D \in \mathcal{D}$ and set

$$\Omega(D) = \big\{\gamma = \{\gamma(t) : 0 \le t < t_\gamma\} : \\ \text{Jordan arc}, \ \gamma(0, t_\gamma) \subset D, \ \gamma(0) \in \partial\mathbb{H}, \ 0 < t_\gamma \le \infty\big\}.$$

Two curves γ, $\widetilde{\gamma} \in \Omega(D)$ are regarded equivalent if $\widetilde{\gamma}$ can be obtained from γ by a reparametrization. Denote by $\dot{\Omega}(D)$ the totality of the equivalence classes of $\Omega(D)$. As was observed in the last paragraph of Section 2.5, each $\gamma \in \Omega(D)$ admits the half-plane capacity reparametrization. Throughout this section, each $\dot{\gamma} \in \dot{\Omega}(D)$ will be represented by a curve (denoted by $\dot{\gamma}$ again) belonging to this class with the *half-plane capacity parametrization* in the following sense: the half-plane capacity a_t for the canonical map $g_t(z)$ from $D\setminus\dot{\gamma}[0,t]$ determined by (2.1.7) equals $2t$ for all

$t \in (0, t_{\dot\gamma})$. We conventionally adjoin a cemetery point δ to $D \cup \partial\mathbb{H}$ and set $\dot\gamma(t) = \delta$ for $t \geq t_{\dot\gamma}$ so that $\dot\gamma$ is regarded as a map from $[0, \infty]$ to the space $D \cup \partial\mathbb{H} \cup \{\delta\}$ equipped with the σ-algebra $\mathcal{B}(D \cup \partial\mathbb{H} \cup \{\delta\})$ generated by $\mathcal{B}(D \cup \partial\mathbb{H})$.

We introduce σ-algebras $\mathcal{G}_t^-(D)$, $t \geq 0$, and $\mathcal{G}(D)$ on $\dot\Omega(D)$ by

$$\mathcal{G}_t^-(D) := \bigcap_{t'>t} \mathcal{G}_{t'}^{0,-}(D) \quad \text{for} \quad \mathcal{G}_t^{0,-}(D) = \sigma\{\dot\gamma(s) : 0 \leq s \leq t,\ \dot\gamma \in \dot\Omega(D)\},$$

$$\mathcal{G}(D) := \sigma\{\dot\gamma(s) : s \geq 0,\ \dot\gamma \in \dot\Omega(D)\}.$$

It follows from the right continuity of $\{\mathcal{G}_t^-\}$ that $t_{\dot\gamma}$ is a $\{\mathcal{G}_t^-\}$-stopping time and $\{t < t_{\dot\gamma}\} \in \mathcal{G}_t^-(D)$ for any $t \geq 0$ (cf. [Blumenthal and Getoor (1968), Example on p. 31]).

For each $t \geq 0$, define the shift operator θ_t sending $\dot\gamma \in \dot\Omega(D)$ to a curve contained in $(D \setminus \dot\gamma[0, t)) \cup \{\delta\}$ by $\theta_t\dot\gamma(\cdot) = \dot\gamma(t + \cdot)$, and set

$$\Theta_t\dot\Omega(D) = \{\theta_t\dot\gamma : \dot\gamma \in \dot\Omega(D) \text{ with } t_{\dot\gamma} > t\}.$$

We equip $\Theta_t\dot\Omega(D)$ with the σ-algebra

$$\mathcal{G}_t^+(D) = \{\Lambda \subset \Theta_t\dot\Omega(D) : \theta_t^{-1}\Lambda \in \mathcal{G}(D)\}. \tag{4.1.1}$$

Given $D \in \mathcal{D}$, $\dot\gamma \in \dot\Omega(D)$ and $t \geq 0$ with $t < t_{\dot\gamma}$, g_t denotes the canonical map from $D \setminus \dot\gamma[0, t]$ onto $D_t = g_t(D \setminus \dot\gamma[0, t])$. Below, we write g_t as $g_t^{\dot\gamma}$ to emphasize its dependence on $\dot\gamma \in \dot\Omega(D)$.

For $\dot\gamma \in \dot\Omega(D)$ and $t < t_{\dot\gamma}$, we define $g_t^{\dot\gamma}\Lambda$ for $\Lambda \in \mathcal{G}_t^+(D)$ by

$$g_t^{\dot\gamma}\Lambda = \left\{ g_t^{\dot\gamma}\widetilde\gamma : \widetilde\gamma \in \Lambda,\ (\theta_t^{-1}\widetilde\gamma)\big|_{[0,t]} = \dot\gamma\big|_{[0,t]} \right\}. \tag{4.1.2}$$

For each $D \in \mathcal{D}$, we consider a family of probability measures $\{\mathbb{P}_{D,z};\ z \in \partial\mathbb{H}\}$ on $(\dot\Omega(D), \mathcal{G}(D))$ that satisfies the property

$$\mathbb{P}_{D,z}(\{\dot\gamma(0) = z\}) = 1 \quad \text{for } z \in \partial\mathbb{H}, \tag{4.1.3}$$

as well as the following two properties **(DMP)** and **(IL)**.

(DMP) (domain Markov property): For any $D \in \mathcal{D}$, $z \in \partial\mathbb{H}$, $t \geq 0$ and $\Lambda \in \mathcal{G}_t^+(D)$,

$$\mathbb{P}_{D,z}\left(\theta_t^{-1}\Lambda,\ t < t_{\dot\gamma}\ \middle|\ \mathcal{G}_t^-(D)\right) = \mathbb{1}_{\{t<t_{\dot\gamma}\}}\, \mathbb{P}_{g_t^{\dot\gamma}(D\setminus\dot\gamma[0,t]),\, g_t^{\dot\gamma}(\dot\gamma(t))}(g_t^{\dot\gamma}\Lambda) \tag{4.1.4}$$

holding for $\mathbb{P}_{D,z}$-a.e. $\dot\gamma \in \dot\Omega(D)$; see Figure 4.1.

Note that $g_t^{\dot\gamma}(\Lambda) \subset \Omega(D_t)$ and the right-hand side of (4.1.4) is well defined because, for the cylinder set

$$\Lambda = \left\{\widetilde\gamma \in \Theta_t\dot\Omega(D) : \widetilde\gamma(s_k) \in B_k,\ 1 \leq k \leq n\right\} \in \mathcal{G}_t^+(D),$$
$$0 \leq s_1 < \cdots < s_n,\ B_k \in \mathcal{B}(D),\ 1 \leq k \leq n,$$

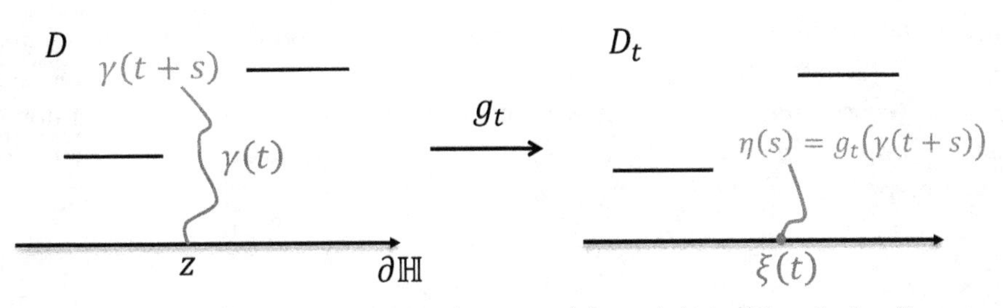

Fig. 4.1 The domain Markov property: the curve $\eta(s) = g_t(\gamma(t+s))$ has the law $\mathbb{P}_{D_t,\xi(t)}$.

we have

$$
\begin{aligned}
g_t^{\dot\gamma}(\Lambda) = \Big\{ &\dot\eta \in \dot\Omega(D_t) : \dot\eta(0) = g_t^{\dot\gamma}(\dot\gamma(t)), \\
&\dot\eta(s_k) \in g_t^{\dot\gamma}(B_k \cap (D \setminus \dot\gamma[0,t])),\ 1 \le k \le n \Big\} \\
\in\ & \mathcal{G}(D_t).
\end{aligned}
$$

Here, we have used the fact that any curve $\dot\eta \in g_t^{\dot\gamma}\Theta_t\dot\Omega(D)$ still obeys the half-plane capacity parametrization on account of the identity $g_{t+s} = g_{D_t \setminus \dot\eta[0,s]} \circ g_t$. From the above remark, it is easy to verify that

$$
g_t^{\dot\gamma}(\Theta_t\dot\Omega(D)) = \Big\{ \dot\eta \in \dot\Omega(D_t) : \dot\eta(0) = g_t^{\dot\gamma}(\dot\gamma(t)) \Big\},
$$

and so

$$
\mathbb{P}_{g_t^{\dot\gamma}(D\setminus\dot\gamma[0,t]),\, g_t^{\dot\gamma}(\dot\gamma(t))}\left(g_t^{\dot\gamma}(\Theta_t\dot\Omega(D)) \right) = 1,
$$

as each $\mathbb{P}_{D_t,w}$ with $w \in \partial\mathbb{H}$ is a probability measure on $\dot\Omega(D_t)$ satisfying $\mathbb{P}_{D_t,w}(\{\dot\eta(0) = w\}) = 1$. The formulation (4.1.4) of **(DMP)** is consistent with this fact when one takes $\Lambda = \Theta_t\dot\Omega(D)$ there.

(IL) (invariance under linear conformal map): For any $D \in \mathcal{D}$ and any linear map f from D onto $f(D) \in \mathcal{D}$,

$$
\mathbb{P}_{f(D),f(z)} = \mathbb{P}_{D,z} \circ f^{-1} \qquad \text{for every } z \in \partial\mathbb{H}. \tag{4.1.5}
$$

Notice that such a map must be of the form $f(z) = cz + r$ for constants $c \ge 0, r \in \mathbb{R}$, which sends $\partial\mathbb{H}$ to $\partial\mathbb{H}$. f maps $\dot\gamma \in \dot\Omega(D)$ to the equivalence class in $\dot\Omega(f(D))$ that $f(\dot\gamma)$ belongs to. The right-hand side of (4.1.5) denotes the image measure of $\mathbb{P}_{D,z}$ by this map.

Remark 4.1.1. If we define the probability measure $\mathbb{P}_{D\setminus\dot\gamma[0,t],\dot\gamma(t)}$ on $(\Theta_t\dot\Omega(D), \mathcal{G}_t^+(D))$ by

$$
\mathbb{P}_{D\setminus\dot\gamma[0,t],\dot\gamma(t)}(\Lambda) = \mathbb{P}_{D_t,g_t(\dot\gamma(t))}(g_t(\Lambda)),
$$

in formal analogue to **(IL)** with $f = g_t$, then (4.1.4) can be restated as

$$
\mathbb{P}_{D,z}\left(\theta_t^{-1}\Lambda \middle| \mathcal{G}_t^-(D)\right) = \mathbb{P}_{D\setminus\dot\gamma[0,t],\dot\gamma(t)}(\Lambda) \quad \text{for every } \Lambda \in \mathcal{G}_t^+(D). \tag{4.1.6}
$$

This explains why we call (4.1.4) the domain Markov property. In fact, in many models of stochastic physics, there are natural candidates of probability measures $\mathbb{P}_{D,z}$ for general multiply connected domains D. Property (4.1.6) is called domain Markov property in SLE literature, while conformal invariance means for any conformal map $f : D \to f(D)$,

$$\mathbb{P}_{D,z}(\Lambda) = \mathbb{P}_{f(D),f(z)}(f(\Lambda)) \quad \text{for any } z \in \partial D; \tag{4.1.7}$$

see [Schramm (2000)]. Taking $f = g_t : D \setminus \gamma[0,t] \to D_t$ in the above gives $\mathbb{P}_{D\setminus\dot\gamma[0,t],\dot\gamma(t)}(\Lambda) = \mathbb{P}_{D_t,g_t(\dot\gamma(t))}(g_t(\Lambda))$. In our formulation above, we consider the family of probability measures $\mathbb{P}_{D,z}$ only for standard slit domains $D \in \mathcal{D}$ and $z \in \partial\mathbb{H}$. Thus we formulated **(DMP)** by combining (4.1.6) with the conformal invariance assumption (4.1.7) under map g_t. □

4.1.2 *Markov property of* $\mathbf{W}_t = (\xi(t), \mathbf{s}(t))$

For each $D \in \mathcal{D}$, $\dot\gamma \in \dot\Omega(D)$ and $t \in [0, t_{\dot\gamma})$, $\dot\gamma$ induces the canonical map from $D\setminus\dot\gamma[0,t]$, namely, the conformal map g_t from $D\setminus\dot\gamma[0,t]$ onto $D_t = g_t(D\setminus\dot\gamma[0,t]) \in \mathcal{D}$ satisfying (2.1.7). The conformal map $g_t(z)$ can be extended to a continuous map from $D \cup K^p \cup \gamma[0,t]^p \cup \partial\mathbb{H}$ onto $\overline{\mathbb{H}}$.

Let $\{\mathbf{s}(t) = \mathbf{s}(D_t),\ t \in [0, t_\gamma)\}$ be the induced slit motion with $D_0 := D$. We will consider the joint process

$$\mathbf{W}_t = \begin{cases} (\xi(t), \mathbf{s}(t)) \in \mathbb{R} \times \mathcal{S} \subset \mathbb{R}^{3N+1}, & 0 \le t < t_{\dot\gamma}, \\ \widehat{\delta}, & t \ge t_{\dot\gamma}. \end{cases}$$

Here the real part of $\xi(t) = g_t(\dot\gamma(t)) \in \partial\mathbb{H}$ is designated by $\xi(t)$ again and $\widehat{\delta}$ is an extra point conventionally adjoined to $\mathbb{R} \times \mathcal{S}$. We shall occasionally write $\mathbf{s}(t)$ as $g_t^D(\mathbf{s})$ with $\mathbf{s} = \mathbf{s}(D)$.

To establish the Markov property of $\mathbf{W}_t$, we need the following measurability results.

Lemma 4.1.2. *Fix $D \in \mathcal{D}$ and $t \ge 0$.*

(i) *For each $z \in D$, $\Im g_t(z)\mathbb{1}_{\{t<t_{\dot\gamma},\, z\notin\dot\gamma(0,t]\}}$ is a $[0,\infty)$-valued $\mathcal{G}_t^-(D)$-measurable function on $\dot\Omega(D)$.*
(ii) *$g_t(z)\mathbb{1}_{\{t<t_{\dot\gamma}, z\notin\dot\gamma(0,t]\}}$ is an $\overline{\mathbb{H}}$-valued $\mathcal{B}(D)\times\mathcal{G}_t^-(D)$-measurable function on $D \times \dot\Omega(D)$.*
(iii) *$\mathbf{W}_t$ is an $\mathbb{R}^{3N+1}$-valued $\mathcal{G}_t^-(D)$-measurable function on $\dot\Omega(D)$.*

Proof. (i) We make use of the probabilistic representation (2.2.1) of $\Im g_t(z)$ for $F = \dot\gamma[0,t]$.

Let $Z^* = (Z_t^*, \zeta^*, \mathbb{P}_z^*)$ be the BMD on $D^* = D \cup \{c_1^*, \dots, c_N^*\}$ obtained from the absorbed Brownian motion on $\mathbb{H}$ by rendering each hole C_k into a single point c_k^*, with life time ζ^*. Take $r > 0$ large enough so that the set $\mathbb{H}_r = \{z \in \mathbb{H} : \Im z < r\}$ contains K.

For $t > 0$, set

$$f_{t,z,r}(\dot\gamma) = \mathbb{P}^*_z\left(\sigma_r < \sigma_{\dot\gamma(0,t]}\right) \mathbb{1}_{\{t<t_{\dot\gamma},\, z\notin\dot\gamma[0,t]\}} \mathbb{1}_{\{\dot\gamma[0,t]\subset\mathbb{H}_r\}},$$

where σ_r is the hitting time of the set $\Gamma_r = \{z \in \mathbb{H} : \Im z = r\}$ by Z^*. It suffices to prove that $f_{t,z,r}(\dot\gamma)$ is a $\mathcal{G}^-_t(D)$-measurable function on $\dot\Omega(D)$ for each fixed $z \in D$ and any $r > 0$, because

$$\lim_{r\to\infty} r f_{t,z,r}(\dot\gamma) = \Im g_t(z)\mathbb{1}_{\{t<t_{\dot\gamma},\, z\notin\dot\gamma[0,t]\}}.$$

Note that

$$f_{t,z,r}(\dot\gamma) = \mathbb{P}^*_z\left(\dot\gamma(0,t] \cap Z^*_{[0,\sigma_r]} = \emptyset, \sigma_r < \infty\right) \mathbb{1}_{\{t<t_{\dot\gamma},\, z\notin\dot\gamma[0,t]\}} \mathbb{1}_{\{\dot\gamma[0,t]\subset\mathbb{H}_r\}}. \quad (4.1.8)$$

Without loss of generality, we take the canonical path space

$$\Xi := \{\omega \in C([0,\infty) \to D^* \cup \{\partial\}) : \omega(t) = \partial \text{ for } t \geq \zeta^*(\omega)\}$$

equipped with Borel σ-field $\mathcal{B}(\Xi) = \sigma\{\omega(t), t \geq 0\}$ as the sample space of BMD Z^* with $Z^*_t(\omega) := \omega(t)$. We consider the direct product $\dot\Omega(D) \times \Xi$ of the measurable space $(\dot\Omega(D), \mathcal{G}^-_t(D))$ and $(\Xi, \mathcal{B}(\Xi))$. The set

$$\Lambda = \Big\{(\dot\gamma,\omega) \in \dot\Omega(D) \times \Xi : \dot\gamma(0,t] \cap \omega([0,\sigma_r]) = \emptyset,\ \sigma_r < \infty,$$
$$t < t_{\dot\gamma},\ z \notin \dot\gamma[0,t],\ \dot\gamma[0,t] \subset \mathbb{H}_r\Big\}$$

is $\mathcal{G}^-_t(D) \times \mathcal{B}(\Xi)$-measurable because

$$\Lambda = \bigcup_{n=1}^{\infty} \bigcup_{s\in\mathbb{Q}_+} \Bigg[\{(\dot\gamma,\omega) : \sigma_r(\omega) \leq s,\ t < t_{\dot\gamma},\ z \notin \dot\gamma[0,t] \subset \mathbb{H}_r\}$$
$$\cap \bigcap_{\substack{u\in[0,t]\cap\mathbb{Q}_+ \\ v\in[0,s]\cap\mathbb{Q}_+}} \Big(\{(\dot\gamma,\omega) : \omega(v) \in \{c^*_1,\ldots,c^*_N\}\}$$
$$\cup \{(\dot\gamma,\omega) : |\dot\gamma(u) - \omega(v)| > 1/n,\ \omega(v) \in D\}\Big)\Bigg],$$

where $\mathbb{Q}_+$ denotes the set of positive rational numbers.

In view of (4.1.8), $f_{t,z,r}(\dot\gamma) = \mathbb{P}^*_z(\Lambda_{\dot\gamma})$ for the $\dot\gamma$-section $\Lambda_{\dot\gamma} = \{\omega \in \Xi : (\dot\gamma,\omega) \in \Lambda\}$ of Λ and so $f_{t,z,r}(\dot\gamma)$ is $\mathcal{G}^-_t(D)$-measurable by the Fubini Theorem, as was to be proved.

(ii) We set

$$u_t(z) = u_t(z,\dot\gamma) := \Re g_t(z)\mathbb{1}_{\{t<t_{\dot\gamma}, z\notin\dot\gamma(0,t]\}} - \infty\mathbb{1}_{\{t\geq t_{\dot\gamma}\}\cup\{z\in\dot\gamma(0,t]\}}$$

and $v_t(z) = v_t(z,\dot\gamma) := \Im g_t(z)\mathbb{1}_{\{t<t_{\dot\gamma},\, z\notin\dot\gamma[0,t]\}}$. Here, the term $-\infty\mathbb{1}$ is put just for a notational convenience in the proof of (iii).

It suffices to prove $u_t(z)$ is $\mathcal{G}^-_t(D)$-measurable. To this end, we construct $u_t(z)$ from $v_t(z)$ by the Cauchy–Riemann relation and line integral as we have done in several places of this book. Here, we have to be careful about path of integration. As it should pass through $D \setminus \dot\gamma(0,t]$, one may image that the path of integration depends on the random curve $\dot{\gamma} \in \dot{\Omega}(D)$. This dependence makes our measurability issue quite complicated.

To choose path of integration in a nice way, we use a localization method. We decompose $\mathbb{H}$ into small squares. Accordingly, we decompose $\dot{\Omega}(D)$ into the sets $\dot{\Omega}(D;t,\mathbf{Q})$ of curves $\dot{\gamma}$ whose traces up to t do not intersect a family $\mathbf{Q}$ of squares specified in advance. (Precise definition of these symbols are given immediately below.) On each $(\cup\mathbf{Q}) \times \dot{\Omega}(D;t,\mathbf{Q}) \subset \mathbb{H} \times \dot{\Omega}(D)$ we can choose deterministic paths of integration that pass through squares of $\mathbf{Q}$, and so the measurability can be addressed in a standard way.

The actual proof goes as follows. We define a set of closed squares of side length 2^{-n} in $\mathbb{R}^2 \simeq \mathbb{C}$ with dyadic vertices:

$$\mathcal{Q}_n := \left\{ [k2^{-n}, (k+1)2^{-n}] \times [l2^{-n}, (l+1)2^{-n}] : k \in \mathbb{Z},\ l \in \mathbb{Z}_+ \right\}.$$

Here $\mathbb{Z}_+$ denotes the set of non-negative integers. Let $\mathcal{X}_{n,D}$ be the set of all finite sequences $\mathbf{Q} = (Q_l)_{l=1}^{L_\mathbf{Q}}$ of $\mathcal{Q}_n$ such that Q_l's are mutually distinct, $Q_l \subset D$ for all l, $Q_l \cap Q_{l+1} \neq \emptyset$ for $1 \leq l \leq L_\mathbf{Q} - 1$, and vertical shifts $Q_1 + iy$ of the initial square Q_1 are in D for all $y > 0$. A sequence $\mathbf{Q} \in \mathcal{X}_{n,D}$ is conventionally extended to non-positive indices by $Q_l := Q_1 + i(1-l)2^{-n}$ with $-l \in \mathbb{Z}_+$. The union $\bigcup_{l=-\infty}^{L_\mathbf{Q}} Q_l$ of squares in $\mathbf{Q}$ is designated as $\cup\mathbf{Q}$. For each $\mathbf{Q} \in \mathcal{X}_{n,D}$, we choose an arbitrary base point $z_\mathbf{Q} \in Q_1$ and then, for each $z \in \cup\mathbf{Q}$, choose a simple smooth curve $C_{\mathbf{Q},z}$ connecting $z_\mathbf{Q}$ and z in $\cup\mathbf{Q}$.

For each $\mathbf{Q} \in \mathcal{X}_{n,D}$, let

$$\dot{\Omega}(D;t,\mathbf{Q}) := \left\{ \dot{\gamma} \in \dot{\Omega}(D) : t < t_{\dot{\gamma}},\ \dot{\gamma}[0,t] \cap (\cup\mathbf{Q}) = \emptyset \right\},$$

which is $\mathcal{G}_t^-(D)$-measurable. We consider the restriction of $v_t(z,\dot{\gamma})$ to $(\cup\mathbf{Q}) \times \dot{\Omega}(D;t,\mathbf{Q})$. This restriction is a $\mathcal{G}_t^-(D) \cap \dot{\Omega}(D;t,\mathbf{Q})$-measurable function of $\dot{\gamma}$ for each $z \in \cup\mathbf{Q}$ and the restriction to $\cup\mathbf{Q}$ of the imaginary part of an analytic function of z for each $\dot{\gamma}$. Hence, we can show that $v_t(z,\dot{\gamma})$ is $\mathcal{B}(\cup\mathbf{Q}) \times (\mathcal{G}_t^-(D) \cap \dot{\Omega}(D;t,\mathbf{Q}))$-measurable and so are its partial derivatives $\partial_x v_t(z)$ and $\partial_y v_t(z)$ in a way similar to the usual proof of the progressive measurability of right-continuous adapted processes (see for instance [Rogers and Williams (1979), II, (73.10)]).

For each $\mathbf{Q} \in \mathcal{X}_{n,D}$, the line integral

$$\widetilde{u}_{t,n}^{\mathbf{Q}}(z,\dot{\gamma}) := \int_{C_{\mathbf{Q},z}} \frac{\partial}{\partial y'} v_t(z',\dot{\gamma})\,dx' - \frac{\partial}{\partial x'} v_t(z',\dot{\gamma})\,dy',$$
$$(z,\dot{\gamma}) \in (\cup\mathbf{Q}) \times \dot{\Omega}(D;t,\mathbf{Q}),$$

defines a $\mathcal{B}(\cup\mathbf{Q}) \times (\mathcal{G}_t^-(D) \cap \dot{\Omega}(D;t,\mathbf{Q}))$-measurable function, because the line integral is a limit of Riemann sums. Notice that $\widetilde{u}_{t,n}^{\mathbf{Q}}(z,\dot{\gamma}) = u_t(z,\dot{\gamma}) - u_t(z_\mathbf{Q},\dot{\gamma})$ and $\lim_{z\to\infty}(g_t(z) - z) = 0$. We set

$$c_{t,n}^{\mathbf{Q}}(\dot{\gamma}) := \lim_{\substack{z\to\infty \\ z\in\cup\mathbf{Q}}} (\widetilde{u}_{t,n}^{\mathbf{Q}}(z,\dot{\gamma}) + iv_t(z,\dot{\gamma}) - z),$$

which is $\mathcal{G}_t^-(D)$-measurable and identical with $-u_t(z_{\mathbf{Q}}, \dot{\gamma})$. Hence $u_{t,n}^{\mathbf{Q}}(z, \dot{\gamma}) := \widetilde{u}_{t,n}^{\mathbf{Q}}(z, \dot{\gamma}) - c_{t,n}^{\mathbf{Q}}(\dot{\gamma})$ is $\mathcal{B}(\cup\mathbf{Q}) \otimes (\mathcal{G}_t^-(D) \cap \dot{\Omega}(D; t, \mathbf{Q}))$-measurable and coincides with the restriction of $u_t(z, \dot{\gamma})$ to $\cup\mathbf{Q}$.

Finally, we set

$$u_{t,n}(z, \dot{\gamma}) := \begin{cases} u_{t,n}^{\mathbf{Q}}(z, \dot{\gamma}) & \text{if } (z, \dot{\gamma}) \in (\cup\mathbf{Q}) \times \dot{\Omega}(D; t, \mathbf{Q}) \text{ for some } \mathbf{Q} \in \mathcal{X}_{n,D} \\ -\infty & \text{otherwise.} \end{cases}$$

Although, the sets $(\cup\mathbf{Q}) \times \dot{\Omega}(D; t, \mathbf{Q})$ indexed by $\mathbf{Q}$ are not mutually disjoint, this definition clearly determines a well-defined $\mathcal{B}(D) \times \mathcal{G}_t^-(D)$-measurable function. Moreover, for every pair $(z, \dot{\gamma}) \in D \times \dot{\Omega}(D)$ with $t < t_{\dot{\gamma}}$ and $z \notin \dot{\gamma}[0, t]$, we can always find $\mathbf{Q} \in \mathcal{X}_{n,D}$ such that $(z, \dot{\gamma}) \in (\cup\mathbf{Q}) \times \dot{\Omega}(D; t, \mathbf{Q})$ if n is sufficiently large. Hence $u_{t,n}$ converges pointwise to u_t as $n \to \infty$, which completes the proof of (ii).

(iii) We begin with the $\mathcal{G}_t^-(D)$-measurability of $\xi(t)\mathbb{1}_{\{t<t_{\dot{\gamma}}\}}$. This measurability is not immediate from (ii), because, in the relation

$$\xi(t) = \lim_{\substack{z \to \dot{\gamma}(t) \\ z \in D \setminus \dot{\gamma}(0,t]}} g_t(z) = \lim_{\substack{z \to \dot{\gamma}(t) \\ z \in D \setminus \dot{\gamma}(0,t]}} u_t(z) (\in \mathbb{R}), \qquad t < t_{\dot{\gamma}},$$

how z approaches $\dot{\gamma}(t)$ affects the measurability of the limits here. Instead of specifying a certain sequence converging to $\dot{\gamma}(t)$, we use the decomposition of the plane such as in (ii) again.

For $n \geq 1$, let $\mathcal{Q}_n'$ be the set of all squares of the form

$$Q = (k2^{-n}, (k+1)2^{-n}] \times (l2^{-n}, (l+1)2^{-1}], \quad k \in \mathbb{Z}, \ l \in \mathbb{Z}_+.$$

To each $Q \in \mathcal{Q}_n'$, we associate an $\mathbb{R} \cup \{-\infty\}$-valued $\mathcal{G}_t^-(D)$-measurable function

$$\widehat{u}_t(\dot{\gamma}; Q) := \sup\{ u_t(z, \dot{\gamma}) : z \in Q \cap \mathbb{Q}^2 \}.$$

Here, we recall that $u_t(z, \dot{\gamma})$ was set to be $-\infty$ if $z \in \dot{\gamma}(0, t]$ in the proof of (ii). We also set

$$E(t, Q) := \{ \dot{\gamma} \in \dot{\Omega}(D) : \dot{\gamma}(t) \in Q \} \in \mathcal{G}_t^-(D).$$

Since

$$\bigcup_{Q \in \mathcal{Q}_n} E(t, Q) = \{ \dot{\gamma} \in \dot{\Omega}(D) \, ; t < t_{\dot{\gamma}} \} \quad \text{(disjoint union)},$$

it follows that

$$\xi(t)\mathbb{1}_{\{t<t_{\dot{\gamma}}\}} = \lim_{n \to \infty} \sum_{Q \in \mathcal{Q}_n} \widehat{u}_t(\dot{\gamma}; Q) \mathbf{1}_{E(t,Q)}(\dot{\gamma}), \quad t < t_{\dot{\gamma}}.$$

This proves the $\mathcal{G}_t^-(D)$-measurability of $\xi(t)\mathbb{1}_{\{t<t_{\dot{\gamma}}\}}$.

The function $z \mapsto g_t(z)\mathbb{1}_{\{t<t_{\dot{\gamma}}\}}$ is continuously extendable from $D \setminus \dot{\gamma}(0, t]$ to $(D \setminus \dot{\gamma}(0, t]) \cup K^p$. In particular, for each $1 \leq j \leq N$, the endpoints of the j-th slit $C_j(t)$ of $D(\mathbf{s}(t))$ are expressed as

$$y_j(t)\mathbb{1}_{\{t<t_{\dot{\gamma}}\}} = \Im g_t(z_j)\mathbb{1}_{\{t<t_{\dot{\gamma}}\}}, \quad x_j(t)\mathbb{1}_{\{t<t_{\dot{\gamma}}\}} = \min_{z \in C_j^p} \Re g_t(z)\mathbb{1}_{\{t<t_{\dot{\gamma}}\}},$$

$$\text{and} \quad x_j^r(t)\mathbb{1}_{\{t<t_{\dot{\gamma}}\}} = \max_{z \in C_j^p} \Re g_t(z)\mathbb{1}_{\{t<t_{\dot{\gamma}}\}}$$

and hence $g_t^-(D)$-measurable, where z_j is the left endpoint of the initial slit C_j. Therefore $\mathbf{s}(t)\mathbb{1}_{\{t<t_{\dot\gamma}\}}$ is $\mathcal{G}_t^-(D)$-measurable. □

For $\xi \in \mathbb{R}$ and $\mathbf{s} \in \mathcal{S}$, we denote the probability measure $\mathbb{P}_{D(\mathbf{s}),\xi+i0}$ on $(\dot\Omega(D(\mathbf{s})), \mathcal{G}(D(\mathbf{s})))$ by $\mathbb{P}_{(\xi,\mathbf{s})}$.

Theorem 4.1.3. (Time homogeneous Markov property) *The process* $\{\mathbf{W}_t,\ t \geq 0;\ \mathbb{P}_{(\xi,\mathbf{s})},\ \xi \in \mathbb{R},\ \mathbf{s} \in \mathcal{S}\}$ *is* $\{\mathcal{G}_t^-(D(\mathbf{s}(0))); t \geq 0\}$*-adapted, and*

$$\mathbb{P}_{(\xi,\mathbf{s})}(\mathbf{W}_0 = (\xi, \mathbf{s})) = 1, \tag{4.1.9}$$

$$\mathbb{P}_{(\xi,\mathbf{s})}\left(\mathbf{W}_{t+s} \in B \mid \mathcal{G}_t^-(D(\mathbf{s}))\right) = \mathbb{P}_{\mathbf{W}_t}(\mathbf{W}_s \in B) \quad \textit{for } t, s \geq 0,\ B \in \mathcal{B}(\mathbb{R} \times \mathcal{S}). \tag{4.1.10}$$

Proof. $\mathbf{W}_t$ is $\mathcal{G}_t(D(\mathbf{s}(0)))$-measurable by Lemma 4.1.2. (4.1.9) follows from (4.1.3).

For $D = D(\mathbf{s}) \in \mathcal{D}$, $\dot\gamma \in \dot\Omega(D)$ and $t \in [0, t_{\dot\gamma})$, consider the canonical map g_t from $D \setminus \dot\gamma[0,t]$ onto $D_t = g_t(D \setminus \dot\gamma[0,t]) \in \mathcal{D}$ sending $\dot\gamma(t)$ to $\xi(t) \in \partial\mathbb{H}$. So, the domain Markov property (4.1.4) simply reads, for $\Lambda \in \mathcal{G}_t^+(D)$ and $\xi \in \partial\mathbb{H}$,

$$\mathbb{P}_{D,\xi}(\theta_t^{-1}\Lambda,\ t < t_{\dot\gamma} \mid \mathcal{G}_t^-(D)) = \mathbb{P}_{D_t,\xi(t)}(g_t\Lambda)\mathbb{1}_{\{t<t_{\dot\gamma}\}}. \tag{4.1.11}$$

Note however that $D_t, \xi(t)$ and $g_t\Lambda$ on the righthand side all depend on $\dot\gamma \in \dot\Omega(D)$ and (4.1.11) is required to hold for $\mathbb{P}_{D,\xi}$-a.e. $\dot\gamma \in \dot\Omega(D)$.

Set, for $t, s > 0$ and $B \in \mathcal{B}(\mathbb{R} \times \mathcal{S})$,

$$\Lambda_{t,s} = \left\{\widetilde\gamma = \theta_t\dot\gamma : \dot\gamma \in \dot\Omega(D),\ t + s < t_{\dot\gamma},\ (\widetilde\xi(s), \widetilde{\mathbf{s}}(s)) \in B\right\}. \tag{4.1.12}$$

Here, by means of the canonical map $g_{t,s} := g_{D\setminus(\dot\gamma[0,t]\cup\widetilde\gamma[0,s])}$ from $D\setminus(\dot\gamma[0,t]\cup\widetilde\gamma[0,s])$ onto $\widetilde D_s \in \mathcal{D}$, we define $\widetilde\xi(s) = g_{t,s}(\widetilde\gamma(s))$ and $\widetilde{\mathbf{s}}(s) = \mathbf{s}(\widetilde D_s)$. Then

$$\theta_t^{-1}\Lambda_{t,s} = \{\dot\gamma \in \dot\Omega(D) : t + s < t_{\dot\gamma},\ \mathbf{W}_{t+s} \in B\} \in \mathcal{G}(D),$$

and consequently, $\Lambda_{t,s} \in \mathcal{G}_t^+(D)$ in view of (4.1.1).

On the other hand, as $g_{t,s}(z) = g_{D_t\setminus g_t\widetilde\gamma[0,s]} \circ g_t(z)$, we have

$$\Lambda_{t,s} = \Big\{\widetilde\gamma = \theta_t\dot\gamma :\ \dot\gamma \in \dot\Omega(D),\ t + s < t_{\dot\gamma},$$
$$(g_{D_t\setminus g_t\widetilde\gamma[0,s]}(g_t(\widetilde\gamma(s))),\ \mathbf{s}(g_{D_t\setminus g_t\widetilde\gamma[0,s]}(D_t))) \in B\Big\}.$$

Hence, for each $\dot\gamma \in \dot\Omega(D)$ with $t < t_{\dot\gamma}$,

$$g_t\Lambda_{t,s} = \{\dot\eta \in g_t\Theta_t\dot\Omega(D) : \left(g_{D_t\setminus\dot\eta[0,s]}(\dot\eta(s)), \mathbf{s}(g_{D_t\setminus\dot\eta[0,s]}(D_t))\right) \in B\},$$

which coincides with $\{\dot\eta \in \dot\Omega(D_t) : \dot\eta(0) = \xi(t),\ \mathbf{W}_s \in B\}$ on account of the remark made right above **(IL)**; see also Figure 4.2.

The conclusion of the theorem now follows from (4.1.11). □

We write $\mathbf{w} = (\xi, \mathbf{s})$ and put

$$P_t(\mathbf{w}, B) = \mathbb{P}_{\mathbf{w}}(\mathbf{W}_t \in B), \quad t \geq 0,\ \mathbf{w} \in \mathbb{R} \times \mathcal{S},\ B \in \mathcal{B}(\mathbb{R} \times \mathcal{S}).$$

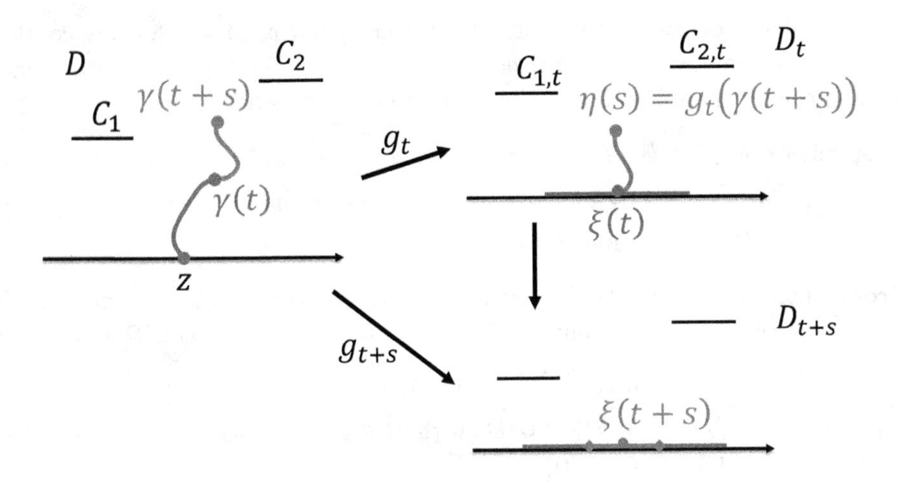

Fig. 4.2 Illustration for the proof of Theorem 4.1.3.

We assume that

(B) $P_t(\mathbf{w}, B)$ is $\mathcal{B}(\mathbb{R} \times \mathcal{S})$-measurable in $\mathbf{w}$ for each $t \geq 0$, $B \in \mathcal{B}(\mathbb{R} \times \mathcal{S})$.

We call it the *transition function* of the time homogeneous Markov process $(\mathbf{W}_t, \mathbb{P}_{\mathbf{w}})$ on $\mathbb{R}\times\mathcal{S}$. It follows from (4.1.10) that, for any $0 \leq t_1 < t_2 < \cdots < t_n$, $f_1, f_2, \ldots, f_n \in \mathcal{B}_b(\mathbb{R} \times \mathcal{S})$, and $\mathbf{w} \in \mathbb{R} \times \mathcal{S}$,

$$\mathbb{E}_{\mathbf{w}}\left[\prod_{k=1}^{n} f_k(\mathbf{W}_{t_k})\right] = \int_{(\mathbb{R}\times\mathcal{S})^n} \prod_{k=1}^{n} f_k(\mathbf{w}_k) P_{t_k - t_{k-1}}(\mathbf{w}_{k-1}, d\mathbf{w}_k)$$

with $t_0 := 0$ and $\mathbf{w}_0 := \mathbf{w}$, the integral being taken first on $\mathbf{w}_n$, then on $\mathbf{w}_{n-1}$, and so on.

4.1.3 *Scaling property and homogeneity in horizontal direction of* W

Lemma 4.1.4. *For $D \in \mathcal{D}, \gamma \in \Omega(D)$, let $a_t = a_t(\gamma, D)$ be the associated half-plane capacity. Then for any $c > 0$*

$$a_t(c\gamma, cD) = c^2 a_t(\gamma, D), \quad t \in [0, t_\gamma). \tag{4.1.13}$$

In particular, if γ is of the half-plane capacity parametrization, then

$$(\dot{c\gamma})(t) = c\,\dot{\gamma}(c^{-2}t), \quad 0 \leq t < c^2 t_{\dot{\gamma}} =: t_{(\dot{c\gamma})}. \tag{4.1.14}$$

is the half-plane capacity parametrization of the curve $c\gamma$ in cD.

Proof. Let $g_t(z)$ be the canonical map from $D \setminus \gamma[0,t]$. Then $g_t^c(z) = c g_t(z/c)$ is the canonical map from $cD \setminus c\gamma[0,t]$. (4.1.13) follows from (2.1.7) and

$$z(g_t^c(z) - z) = c^2 \frac{z}{c}\left(g_t(\frac{z}{c}) - \frac{z}{c}\right).$$

(4.1.14) follows from $a_t(\dot{\gamma}, D) = 2t$ and (4.1.13). □

We make a convention that for the cemetery point δ, $c\delta = \delta$ for any constant $c > 0$. Then the identity (4.1.14) holds for any $t \geq 0$; for $t \geq c^2 t_{\dot\gamma}$, both sides of (4.1.14) equal δ. Keeping this in mind, we show the following:

Proposition 4.1.5. *For $D \in \mathcal{D}$, $z \in \partial\mathbb{H}$ and any $c > 0$,*

$$\begin{aligned}&\{c^{-1}\dot\gamma(c^2t),\ t \geq 0\} \text{ under } \mathbb{P}_{cD,cz} \text{ has the same distribution}\\ &\text{as } \{\dot\gamma(t),\ t \geq 0\} \text{ under } \mathbb{P}_{D,z}.\end{aligned} \tag{4.1.15}$$

Proof. For a fixed $c > 0$, $f(z) = cz$ is a conformal map from D onto $cD \in \mathcal{D}$. By the invariance under linear conformal map (4.1.5), we have for $D \in \mathcal{D}$ and $z \in \partial\mathbb{H}$,

$$\mathbb{P}_{D,z}(\Lambda) = \mathbb{P}_{cD,cz}(f(\Lambda)), \quad \Lambda \in \mathcal{G}(D). \tag{4.1.16}$$

For $\Lambda = \{\dot\gamma \in \dot\Omega(D) : \dot\gamma \in B\} \in \mathcal{G}(D)$ with $B \in \mathcal{B}\left((\overline{\mathbb{H}} \cup \{\delta\})^{[0,\infty)}\right)$, $f(\Lambda) = \{\dot\gamma \in \dot\Omega(cD) : (\dot{\gamma/c}) \in B\}$. By (4.1.14),

$$(\dot{\gamma/c})(t) = c^{-1}\dot\gamma(c^2t), \quad t \geq 0, \tag{4.1.17}$$

and so (4.1.15) follows from (4.1.16). □

Theorem 4.1.6. (A scaling property of $\mathbf{W}$) *For $\xi \in \mathbb{R}$, $\mathbf{s} \in \mathcal{S}$ and $c > 0$,*

$$\begin{aligned}&\{c^{-1}\mathbf{W}_{c^2t},\ t \geq 0\} \text{ under } \mathbb{P}_{(c\xi,c\mathbf{s})} \text{ has the same distribution}\\ &\text{as } \{\mathbf{W}_t,\ t \geq 0\} \text{ under } \mathbb{P}_{(\xi,\mathbf{s})}.\end{aligned} \tag{4.1.18}$$

Proof. For fixed $D \in \mathcal{D}$ and $c > 0$, consider the canonical map g_t^{cD} associated with cD and a curve $\{\dot\gamma(t),\ t \geq 0\} \in \dot\Omega(cD)$. The induced process $\mathbf{W}_t = (\xi(t), \mathbf{s}(t))$, $t \in [0, t_{\dot\gamma})$, is given by $\mathbf{s}(t) = g_t^{cD}(c\mathbf{s})$ for $\mathbf{s} = \mathbf{s}(D)$ and $\xi(t) = g_t^{cD}(\dot\gamma(t))$.

Now, the curve on the left-hand side of (4.1.15) defined for $\dot\gamma \in \dot\Omega(cD)$ belongs to $\dot\Omega(D)$ in view of (4.1.17) and the associated canonical map $\widetilde{g}_t^D$ from D is given by

$$\widetilde{g}_t^D(z) = c^{-1}g_{c^2t}^{cD}(cz), \quad z \in D, \quad \text{for } t \in [0, t_{\dot\gamma/c^2}), \tag{4.1.19}$$

which induces the motion $\{c^{-1}\mathbf{W}_{c^2t} : t \in [0, t_{\dot\gamma/c^2})\}$, because $\widetilde{g}_t^D(\mathbf{s}) = c^{-1}\mathbf{s}(c^2t)$ and $\widetilde{g}_t^D(c^{-1}\dot\gamma(c^2t)) = c^{-1}\xi(c^2t)$ for $t \in [0, t_{\dot\gamma/c^2})$.

Let $\{\mathbf{W}_t,\ t \geq 0\}$ be the $(\mathbb{R} \times S)$-valued motion produced by $D \in \mathcal{D}$ and $\dot\gamma \in \dot\Omega(D)$. Then, for $0 \leq t_1 < t_2 < \cdots < t_n$, $(\mathbf{W}_{t_1}, \mathbf{W}_{t_2}, \ldots, \mathbf{W}_{t_n})$ equals an $(\mathbb{R} \times S)^n$-valued $\mathcal{G}(D)$-measurable function $F(\dot\gamma)$ of $\dot\gamma \in \dot\Omega(D)$ by virtue of Lemma 4.1.2. Therefore, we can conclude from (4.1.15) and the above observation that (4.1.18) holds. □

Lemma 4.1.7. *For $D \in \mathcal{D}, \gamma \in \Omega(D)$, let $a_t = a_t(\gamma, D)$ be the associated half-plane capacity. Then for any $r \in \mathbb{R}$,*

$$a_t(\gamma + r, D + r) = a_t(\gamma, D), \quad t \in [0, t_\gamma). \tag{4.1.20}$$

In particular, the half-plane capacity parametrization of the curve $\gamma+r$ in $D+r$ is given by $\dot\gamma+r$; in other words,

$$(\dot{\overline{\gamma+r}})(t)=\dot\gamma(t)+r, \quad 0\le t<t_{\dot\gamma}. \tag{4.1.21}$$

Proof. Let $g_t(z)$ be the canonical map associated with (γ, D). Then $g_t^r(z) = g_t(z-r)+r$ is the canonical map associated with $(\gamma+r, D+r)$. (4.1.20) follows from (2.1.7) and

$$z(g_t^r(z)-z)=\frac{z}{z-r}\cdot(z-r)\left(g_t(z-r)-(z-r)\right).$$

(4.1.21) follows from $a_t(\dot\gamma, D)=2t$ and (4.1.20). □

The identity (4.1.21) holds for any $t\ge 0$ because both sides of (4.1.21) equal δ when $t\ge t_{\dot\gamma}$.

Proposition 4.1.8. *For $D\in\mathcal{D}$, $z\in\partial\mathbb{H}$ and any $r\in\mathbb{R}$,*

$$\begin{aligned}&\{\dot\gamma(t)-r,\ t\ge 0\} \text{ under } \mathbb{P}_{D+r,z+r} \text{ has the same distribution}\\ &\text{as } \{\dot\gamma(t),\ t\ge 0\} \text{ under } \mathbb{P}_{D,z}.\end{aligned} \tag{4.1.22}$$

Proof. For a fixed $r\in\mathbb{R}$, consider the shift $f(z)=z+r$, $z\in D$. By the invariance under linear conformal map (4.1.5), we have for $D\in\mathcal{D}$, $z\in\partial\mathbb{H}$,

$$\mathbb{P}_{D,z}(\Lambda)=\mathbb{P}_{D+r,z+r}(f(\Lambda)), \quad \Lambda\in\mathcal{G}(D). \tag{4.1.23}$$

$\Lambda\in\mathcal{G}(D)$ can be expressed as $\Lambda=\{\dot\gamma\in\dot\Omega(D):\dot\gamma\in B\}$ for $B\in\mathcal{B}\left((\overline{\mathbb{H}}\cup\{\delta\})^{[0,\infty)}\right)$. Then

$$f(\Lambda)=\{\dot\gamma\in\dot\Omega(D+r):(\dot{\overline{\gamma-r}})\in B\}.$$

This combined with (4.1.21) and (4.1.23) leads us to (4.1.22). □

For $r\in\mathbb{R}$, denote by $\widehat{r}$ the vector in $\mathbb{R}^{3N}$ whose first N entries are 0 and the last $2N$ entries are r. Note that $\mathbf{s}(D+r)=\mathbf{s}(D)+\widehat{r}$ for $D\in\mathcal{D}$, $r\in\mathbb{R}$.

Theorem 4.1.9. (Homogeneity of $(\mathbf{W}_t,\ \mathbb{P}_{(\xi,\mathbf{s})})$ in horizontal direction) *For $\xi\in\mathbb{R}$, $\mathbf{s}\in\mathcal{S}$ and $r\in\mathbb{R}$, $\{(\xi(t)-r,\mathbf{s}(t)-\widehat{r}),\ t\ge 0\}$ under $\mathbb{P}_{(\xi+r,\mathbf{s}+\widehat{r})}$ has the same distribution as $\{(\xi(t),\mathbf{s}(t)),\ t\ge 0\}$ under $\mathbb{P}_{(\xi,\mathbf{s})}$.*

Proof. Fix $D\in\mathcal{D}$, $z=\xi+i0\in\partial\mathbb{H}$, $r\in\mathbb{R}$ and put $\mathbf{s}=\mathbf{s}(D)$. Consider the canonical map g_t^{D+r} associated with $D+r$ and a curve $\{\dot\gamma(t),\ t\ge 0\}\in\dot\Omega(D+r)$. The process $\mathbf{W}_t=(\mathbf{s}(t),\xi(t))$, $t\in[0,t_{\dot\gamma})$, being considered under $\mathbb{P}_{D+r,z+r}=\mathbb{P}_{(\xi+r,\mathbf{s}+\widehat{r})}$ is induced from g_t^{D+r} by

$$\xi(t)=g_t^{D+r}(\dot\gamma(t)), \quad \mathbf{s}(t)=g_t^{D+r}(\mathbf{s}+\widehat{r}).$$

Now, the curve on the left-hand side of (4.1.22) belongs to $\dot\Omega(D)$ in view of (4.1.21) and the associated canonical map $\widetilde{g}_t^D$ is given by

$$\widetilde{g}_t^D(z)=g_t^{D+r}(z+r)-r, \quad z\in D, \quad \text{for } t\in[0,t_{\dot\gamma}).$$

The induced motion is

$$\begin{cases} \widetilde{g}_t^D(\dot\gamma(t) - r) = g_t^{D+r}(\dot\gamma(t)) - r = \xi(t) - r, \\ \widetilde{g}_t^D(\mathbf{s}) = g_t^{D+r}(\mathbf{s} + \widehat{r}) - \widehat{r} = \mathbf{s}(t) - \widehat{r}. \end{cases}$$

The theorem now follows from (4.1.22) by the same reason as in the last paragraph of the proof of Theorem 4.1.6. □

4.2 Stochastic differential equation for $(\xi(t), \mathbf{s}(t))$

We keep the setting in the preceding section. We write $\mathbf{w} = (\xi, \mathbf{s}) \in \mathbb{R} \times \mathcal{S}$. We know from Theorem 4.1.3 that $\mathbf{W} = (\mathbf{W}_t, \mathbb{P}_\mathbf{w})$ is a time homogeneous Markov process taking values in $\mathbb{R} \times \mathcal{S} \subset \mathbb{R}^{3N+1}$. The sample path $\mathbf{W}_t = (\xi(t), \mathbf{s}(t))$ of $\mathbf{W}$ is continuous up to its lifetime $t_{\dot\gamma} \leq \infty$ owing to Theorem 2.3.9, Theorem 2.3.8 and (3.1.2). Let P_t be its transition semigroup defined by

$$P_t f(\mathbf{w}) = \mathbb{E}_\mathbf{w}[f(\mathbf{W}_t)], \qquad t \geq 0, \ \mathbf{w} \in \mathbb{R} \times \mathcal{S}.$$

Denote by $C_\infty(\mathbb{R} \times \mathcal{S})$ the space of all continuous functions on $\mathbb{R} \times \mathcal{S}$ vanishing at infinity.

In this section, we assume that the Markov process $\mathbf{W}$ satisfies additional properties **(C.1)** and **(C.2)** stated below.

(C.1) $P_t(C_\infty(\mathbb{R} \times \mathcal{S})) \subset C_\infty(\mathbb{R} \times \mathcal{S}),\ t > 0, \quad C_c^\infty(\mathbb{R} \times \mathcal{S}) \subset \mathcal{D}(L),$

where L is the infinitesimal generator of $\{P_t,\ t > 0\}$ defined by

$$\begin{aligned} Lf(\mathbf{w}) &= \lim_{t \downarrow 0} \frac{1}{t}(P_t f(\mathbf{w}) - f(\mathbf{w})), \quad \mathbf{w} \in \mathbb{R} \times \mathcal{S}, \\ \mathcal{D}(L) &= \{f \in C_\infty(\mathbb{R} \times \mathcal{S}) : \text{ the right-hand side above} \\ &\qquad \text{converges uniformly in } \mathbf{w} \in \mathbb{R} \times \mathcal{S}\}. \end{aligned} \tag{4.2.1}$$

Under condition **(C.1)**, $\mathbf{W} = \{\mathbf{W}_t, \mathbb{P}_\mathbf{w}\}$ is a *Feller-Dynkin diffusion* in the sense of [Rogers and Williams (1979)]. In view of [Rogers and Williams (1979), III, (13.3)], the restriction $\mathcal{L}$ of L to $C_c^\infty(\mathbb{R} \times \mathcal{S})$ is a second order elliptic partial differential operator expressed as

$$\mathcal{L}f(\mathbf{w}) = \frac{1}{2} \sum_{i,j=0}^{3N} a_{ij}(\mathbf{w}) f_{w_i w_j}(\mathbf{w}) + \sum_{i=0}^{3N} b_i(\mathbf{w}) f_{w_i}(\mathbf{w}) + c(\mathbf{w}) f(\mathbf{w}), \quad \mathbf{w} \in \mathbb{R} \times \mathcal{S}, \tag{4.2.2}$$

where a is a non-negative definite symmetric matrix-valued continuous function, b is a vector-valued continuous function and c is a non-positive continuous function.

Our second assumption on $\mathbf{W}$ is

(C.2) $c(\mathbf{w}) = 0 \quad$ for every $\mathbf{w} \in \mathbb{R} \times \mathcal{S}$.

This condition is clearly satisfied if $\mathbf{W}$ is conservative: $\mathbb{P}_{D,z}(t_{\dot\gamma} = \infty) = 1$ for any $D \in \mathcal{D}$ and $z \in \partial\mathbb{H}$, or, equivalently,

$$P_t 1(\mathbf{w}) = 1 \quad \text{for any } t \geq 0 \text{ and } \mathbf{w} \in \mathbb{R} \times \mathcal{S}. \tag{4.2.3}$$

In fact, $c(\mathbf{w})$ can be evaluated as

$$c(\mathbf{w}) = \lim_{t \downarrow 0} \frac{1}{t}(P_t 1(\mathbf{w}) - 1), \quad \mathbf{w} \in \mathbb{R} \times \mathcal{S},$$

according to Theorem 5.8 and its Remark in [Dynkin (1965), Chapter V]. Hence (4.2.3) implies **(C.2)**. Condition **(C.2)** means that $\mathbf{W}$ admits no killing inside $\mathbb{R}\times\mathcal{S}$, and so it is much weaker than the conservativeness of $\mathbf{W}$.

From (4.2.1), we get for any $f \in C_c^\infty(\mathbb{R} \times \mathcal{S})$,

$$P_t f(\mathbf{w}) - f(\mathbf{w}) = \int_0^t P_s(\mathcal{L}f)(\mathbf{w})ds, \ t \geq 0, \ \mathbf{w} \in \mathbb{R} \times \mathcal{S}.$$

By the Markov property of $\mathbf{W}$, the above implies that for any $f \in C_c^\infty(\mathbb{R} \times \mathcal{S})$,

$$M_t^f := f(\mathbf{W}_t) - f(\mathbf{W}_0) - \int_0^t \mathcal{L}f(\mathbf{W}_s)ds, \quad t \geq 0, \tag{4.2.4}$$

is a martingale with the convention that $\mathbf{W}_t = \widehat{\delta}$ for $t \geq t_{\hat{\gamma}}$ and any function g is extended to $\widehat{\delta}$ by taking value 0. We denote by $W_t^{(j)}$ the j-th coordinate of the process $\mathbf{W}_t$ so that

$$W_t^{(0)} = \xi(t), \quad (W_t^{(1)}, \ldots, W_t^{(3N)}) = \mathbf{s}(t).$$

We choose relatively compact open sets G_n, $n \in \mathbb{N}$, of $\mathbb{R} \times \mathcal{S}$ so that $\overline{G_n} \subset G_{n+1}$ for all n and $G_n \uparrow \mathbb{R} \times \mathcal{S}$ as $n \to \infty$. Let σ_n be the exit time of $\mathbf{W}_t$ from G_n:

$$\sigma_n := \inf\{t > 0 : \mathbf{W}_t \in (\mathbb{R} \times \mathcal{S}) \setminus G_n\}, \quad n \in \mathbb{N}.$$

Note that the lifetime $t_{\hat{\gamma}}$ of $\mathbf{W}$ is the increasing limit of σ_n as $n \to \infty$. For each $n \geq 1$ and $0 \leq j \leq 3N$, take some $f_{n,j} \in C_c^\infty(\mathbb{R} \times \mathcal{S})$ so that $f_{n,j}(\mathbf{w}) = w_j$ for $\mathbf{w} \in \overline{G_n}$ and $f_{n,j}(\mathbf{w}) = 0$ for $\mathbf{w} \in (\mathbb{R}\times\mathcal{S})\setminus G_{n+1}$. By the proof of (i) $\Rightarrow$ (ii) of [Revuz and Yor (1999), Proposition VII.2.4] applied to functions $f_{n,j}$ and restricted to the time interval $[0, \sigma_n]$, (4.2.2), condition **(C.2)** and (4.2.4) imply that the processes

$$M_{t\wedge\sigma_n}^j := W_{t\wedge\sigma_n}^{(j)} - W_0^{(j)} - \int_0^{t\wedge\sigma_n} b_j(\mathbf{W}_s)ds, \quad t \geq 0, \quad 0 \leq j \leq 3N,$$

are martingales with

$$\langle M^j, M^k \rangle_{t\wedge\sigma_n} = \int_0^{t\wedge\sigma_n} a_{jk}(\mathbf{W}_s)ds, \quad t \geq 0, \ 0 \leq j,k \leq 3N. \tag{4.2.5}$$

On the other hand, we have seen in Section 3.1 that $\mathbf{s}(t) = (\mathbf{s}_1(t), \ldots, \mathbf{s}_{3N}(t))$ satisfies the equation

$$\mathbf{s}_j(t) - \mathbf{s}_j(0) = \int_0^t b_j(\mathbf{W}(s))ds, \quad t \in [0, t_{\hat{\gamma}}), \ 1 \leq j \leq 3N, \tag{4.2.6}$$

for

$$b_j(\mathbf{w}) = \begin{cases} -2\pi\Im\Psi_{\mathbf{s}}(z_j, \xi), & 1 \leq j \leq N, \\ -2\pi\Re\Psi_{\mathbf{s}}(z_{j-N}, \xi), & N+1 \leq j \leq 2N, \\ -2\pi\Re\Psi_{\mathbf{s}}(z_{j-2N}^r, \xi), & 2N+1 \leq j \leq 3N. \end{cases} \tag{4.2.7}$$

Here, for $\mathbf{s} = (\mathbf{y}, \mathbf{x}, \mathbf{x}^r) \in \mathcal{S}$, $\Psi_{\mathbf{s}}$ is the BMD-complex Poisson kernel for the domain $D(\mathbf{s}) \in \mathcal{D}$, and $z_j = x_j + iy_j$, $z_j^r = x_j^r + iy_j$ denote the endpoints of the jth slit C_j in $D(\mathbf{s})$, $1 \le j \le N$. It follows that

$$M^j = 0 \quad \text{for } 1 \le j \le 3N, \qquad \langle M^0, M^0 \rangle_t = \int_0^t a_{00}(\mathbf{W}_s) ds \quad \text{for } t \in [0, t_{\dot\gamma}),$$

$b_j(\mathbf{w})$ in (4.2.2) is given by the above expression (4.2.7) for $j \ge 1$ and $a_{ij}(\mathbf{w}) = 0$ for $i + j \ge 1$.

For each $n \ge 1$, define $\widetilde{M}_t^{(n)} := \int_0^{t \wedge \sigma_n} a_{00}(\mathbf{W}_s)^{-1/2} \mathbb{1}_{\{a_{00}(\mathbf{W}_s) > 0\}} dM_s^0$ for $t \ge 0$. Note that $\widetilde{M}^{(n)}$ is a well defined continuous martingale with $\langle \widetilde{M}^{(n)} \rangle_t = t \wedge \sigma_n$. From its definition, $\widetilde{M}_t^{(n+1)} = \widetilde{M}_t^{(n)}$ for every $t \in [0, \sigma_n]$, and so it uniquely defines a continuous process $\widetilde{M}$ on the random time interval $[0, t_{\dot\gamma}) = \cup_{n=1}^\infty [0, \sigma_n]$ so that $\widetilde{M}_t = M_t^{(n)}$ for $t \in [0, \sigma_n]$. In particular, $\{\widetilde{M}_{\sigma_n \wedge n}; n \ge 1\}$ is a martingale with $\langle M \rangle_{\sigma_n \wedge n} = \sigma_n \wedge n$. It follows by the martingale convergence theorem that

$$\widetilde{M}_{t_{\dot\gamma}-} := \lim_{n \to \infty} \widetilde{M}_{\sigma_n \wedge n},$$

exists almost surely on $\{t_{\dot\gamma} \le T\}$ for each $T > 0$ and hence on $\{t_{\dot\gamma} < \infty\}$. Take a standard Brownian motion $\widetilde{B}$ that is independent of the augmented filtration generated by $\mathbf{W}$. For each integer $n \ge 1$, define

$$B_t^{(n)} = \widetilde{M}_t^{(n)} + (\widetilde{B}_t - \widetilde{B}_{t \wedge \sigma_n}).$$

Clearly each $B^{(n)}$ is a continuous martingale with $\langle B^{(n)} \rangle_t = t$ for every $t \ge 0$. Thus Lévy's martingale characterization of Brownian motion (cf. [Ikeda and Watanabe (1981); Revuz and Yor (1999)]), each $B^{(n)}$ is a standard Brownian motion. Define $\widetilde{M}_t = \widetilde{M}_{t_{\dot\gamma}-}$ for any $t \ge t_{\dot\gamma}$ on $\{t_{\dot\gamma} < \infty\}$, and

$$B_t := \widetilde{M}_t + (\widetilde{B}_t - \widetilde{B}_{t \wedge t_{\dot\gamma}}) \quad \text{for } t \ge 0.$$

Since $\lim_{n \to \infty} B_t^{(n)} = B_t$ for every $t \ge 0$, the continuous process B_t is a standard Brownian motion as it has the stationary independent increment of normal distributions. Since for each $n \ge 1$,

$$\int_0^{t \wedge \sigma_n} a_{00}(\mathbf{W}_s)^{-1/2} \mathbb{1}_{\{a_{00}(\mathbf{W}_s) > 0\}} dM_s^0 = \widetilde{M}_t^{(n)} = \widetilde{M}_t = B_t \quad \text{for } t \in [0, \sigma_n],$$

we have

$$M_t^0 = \int_0^t \sqrt{a_{00}}(\mathbf{W}_s) dB_s \quad \text{for } t \in \cup_{n=1}^\infty [0, \sigma_n] = [0, t_{\dot\gamma}).$$

Thus under the condition of **(C.1)** and **(C.2)**, $\mathbf{W}_t = (\xi(t), \mathbf{s}(t))$ satisfies for $t \in [0, t_{\dot\gamma})$,

$$\begin{cases} d\xi(t) = \sqrt{a_{00}(\mathbf{W}_t)} dB_t + b_0(\mathbf{W}_t) dt, \\ d\mathbf{s}_j(t) = b_j(\mathbf{W}_t)) dt, \qquad j = 1, \dots, 3N. \end{cases} \tag{4.2.8}$$

As was introduced in Section 3.1, a real-valued function $u(\mathbf{w}) = u(\xi, \mathbf{s})$ on $\mathbb{R} \times \mathcal{S}$ is called *homogeneous with degree* 0 (resp. -1) if

$$u(c\mathbf{w}) = u(\mathbf{w}) \quad (\text{resp. } u(c\mathbf{w}) = c^{-1} u(\mathbf{w})) \quad \text{for any } c > 0 \text{ and } \mathbf{w} \in \mathbb{R} \times \mathcal{S}.$$

The same definition of the homogeneity is in force for a real-valued function $u(\mathbf{s})$ on $\mathcal{S}$.

Lemma 4.2.1. *Assume that conditions* **(C.1)** *and* **(C.2)** *hold.*

(i) $a_{00}(\mathbf{w})$ *is a homogeneous function of degree* 0, *while* $b_i(\mathbf{w})$ *is a homogeneous function of degree* -1 *for every* $0 \le i \le 3N$.
(ii) *For every* $0 \le j \le 3N$, $\xi \in \mathbb{R}$, $\mathbf{s} \in \mathcal{S}$ *and* $r \in \mathbb{R}$.

$$a_{00}(\xi + r, \mathbf{s} + \widehat{r}) = a_{00}(\xi, \mathbf{s}), \qquad b_j(\xi + r, \mathbf{s} + \widehat{r}) = b_j(\xi, \mathbf{s}). \tag{4.2.9}$$

Proof. (i) By virtue of the scaling property (4.1.18), we have $P_t(\mathbf{w}, E) = P_{c^2t}(c\mathbf{w}, cE)$. Consequently, $P_t f(\mathbf{w}) = P_{c^2t} f^{(c)}(c\mathbf{w})$ and $\mathcal{L}f(\mathbf{w}) = c^2 \mathcal{L} f^{(c)}(c\mathbf{w})$, where $f^{(c)}(\mathbf{w}) = f(\mathbf{w}/c)$. Hence, we get the stated properties of the coefficients a_{ij} and b_i of $\mathcal{L}$.
(ii) By virtue of the homogeneity in the horizontal direction from Theorem 4.1.9, we have $P_t f(\mathbf{w}) = P_t f^r(\mathbf{w} + (r, \widehat{r}))$ so that $\mathcal{L}f(\mathbf{w}) = \mathcal{L}f^r(\mathbf{w} + (r, \widehat{r}))$ where $f^r(\mathbf{w}) = f(\mathbf{w} - (r, \widehat{r}))$. Hence we get (4.2.9). □

Remark 4.2.2. For the functions b_j, $1 \le j \le 3N$, defined by (4.2.7), their homogeneity of degree -1 as well as their homogeneity in the horizontal direction (4.2.9) have been proved by Proposition 3.1.2 using the conformal invariance of BMD Green function directly.

In Lemma 4.2.1, we arrive at them again in a different way from the assumptions that the Jordan arc γ is random with properties **(DMP)**, **(IL)** and that the associated Markov process $(\mathbf{W}_t, \mathbb{P}_{\mathbf{w}})$ satisfies conditions **(C.1)**, **(C.2)**. □

Let

$$\alpha(\mathbf{s}) = \sqrt{a_{00}(0, \mathbf{s})}, \qquad b(\mathbf{s}) = b_0(0, \mathbf{s}), \quad \mathbf{s} \in \mathcal{S}.$$

It follows from Lemma 4.2.1 that $\alpha(\mathbf{s})$ and $b(\mathbf{s})$ are homogeneous functions on $\mathcal{S}$ with degree 0 and -1, respectively. Moreover,

$$\sqrt{a_{00}(\xi, \mathbf{s})} = \alpha(\mathbf{s} - \widehat{\xi}) \quad \text{and} \quad b_0(\xi, \mathbf{s}) = b(\mathbf{s} - \widehat{\xi}).$$

Thus, we have the following from (4.2.8) and Lemma 4.2.1:

Theorem 4.2.3. *Assume that the Jordan arc* γ *is random with properties* **(DMP)**, **(IL)** *and that the associated continuous Markov process* $(\mathbf{W}_t, \mathbb{P}_{\mathbf{w}})$ *on* $\mathbb{R} \times \mathcal{S}$ *satisfies conditions* **(C.1)**, **(C.2)**. *Then the diffusion process* $\mathbf{W} = \{\mathbf{W}_t = (\xi(t), \mathbf{s}(t)); t \in [0, t_\gamma)\}$ *satisfies under* $\mathbb{P}_{(\xi, \mathbf{s})}$ *the following stochastic differential equation:*

$$\xi(t) = \xi + \int_0^t \alpha(\mathbf{s}(s) - \widehat{\xi}(s)) dB_s + \int_0^t b(\mathbf{s}(s) - \widehat{\xi}(s)) ds, \tag{4.2.10}$$

$$\mathbf{s}_j(t) = \mathbf{s}_j + \int_0^t b_j(\xi(s), \mathbf{s}(s)) ds, \quad t \ge 0, \quad 1 \le j \le 3N. \tag{4.2.11}$$

Here, $\{B_t\}$ is the one-dimensional standard Brownian motion. α and b are the homogenous functions on $\mathcal{S}$ of degree 0 and -1, respectively. For each $1 \le j \le 3N$, $b_j(\xi, \mathbf{s})$ is given by (4.2.7), which has the properties that $b_j(\xi, \mathbf{s}) = b_j(0, \mathbf{s} - \widehat{\xi})$ and that $b_j(0, \mathbf{s})$ is a homogeneous function on $\mathcal{S}$ of degree -1.

Note that the solution to SDE (4.2.10) and (4.2.11) is understood in the sense of [Ikeda and Watanabe (1981), Definition 2.1 in Chapter IV] that allows for possible explosions.

Remark 4.2.4 (reduction to $\xi(t) = \sqrt{\kappa}B_t + \xi(0)$ when $D = \mathbb{H}$). We consider the special case that $D = \mathbb{H}$, namely, the case where the slits $K = \bigcup_{j=1}^N C_j \subset \mathbb{H}$ are absent.

Assume that the probability law $\mathbb{P}_{\mathbb{H},z}$, $z \in \partial\mathbb{H}$, on the space $\dot{\Omega}(\mathbb{H})$ of Jordan arcs in $\mathbb{H}$ satisfies the domain Markov property **(DMP)** and the invariance under the linear conformal map **(IL)**. By Theorem 4.1.3, the process $\{\xi(t),\ t \ge 0;\ \mathbb{P}_\xi := \mathbb{P}_{\mathbb{H},\xi},\ \xi \in \partial\mathbb{H}\}$ is a time homogeneous Markov process on $\partial\mathbb{H}$ with continuous sample paths. Its transition function is defined by $P_t f(\xi) = \mathbb{E}_\xi[f(\xi(t))]$, $t \ge 0$, $\xi \in \partial\mathbb{H}$.

Assume that

$$\{P_t\} \text{ satisfies the regularity condition } (\mathbf{C.1}) \text{ with } \mathbb{R} \text{ in place of } \mathbb{R} \times \mathcal{S}. \qquad (4.2.12)$$

Then its generator $\mathcal{L}$ admits the expression

$$\mathcal{L}f(\xi) = \frac{1}{2}a(\xi)f''(\xi) + b(\xi)f'(\xi) + c(\xi)f(\xi), \quad \xi \in \mathbb{R}, \quad f \in C_c^\infty(\mathbb{R}),$$

where a, b, c are functions on $\mathbb{R}$ with $a \ge 0$ and $c \le 0$.

As Lemma 4.2.1 (ii), it follows from Theorem 4.1.9 that the functions a, b are constants. As Lemma 4.2.1 (i), it follows from (4.1.18) that $b(\eta\xi) = \frac{1}{\eta}b(\xi)$ and $c(\eta\xi) = \frac{1}{\eta^2}c(\xi)$ for any $\xi \in \mathbb{R}$ and $\eta > 0$. Hence both b and c vanish. Since the process $\xi(t)$ is non-trivial, this implies that a is strictly positive, which we denote by κ. Then

$$\xi(t) = \sqrt{\kappa}B_t + \xi(0), \qquad (4.2.13)$$

where B_t is the standard Brownian motion on $\mathbb{R}$.

The reduction to (4.2.13) is still possible if we replace the assumption (4.2.12) by the assumption

$$\mathbb{P}_{\mathbb{H},\xi}(t_{\dot{\gamma}} = \infty) = 1, \qquad \xi \in \partial\mathbb{H}, \qquad (4.2.14)$$

which is much simpler than (4.2.12) but stronger than the condition **(C.2)** with $\mathbb{R}$ in place of $\mathbb{R} \times \mathcal{S}$. Indeed, (4.2.14) means the conservativeness of the Markov process $(\xi(t), \mathbb{P}_\xi)$:

$$P_t \mathbb{1}_{\mathbb{H}}(\xi) = \mathbb{P}_\xi(\xi(t) \in \mathbb{H}) = 1 \text{ for any } t > 0 \text{ and } \xi \in \partial\mathbb{H}. \qquad (4.2.15)$$

Further, it follows from Theorem 4.1.9 that P_t is translation invariant:

$$P_t(\xi, A) = P_t(\xi + r, A + r), \ r \in \mathbb{R}, \ A \in \mathcal{B}(\mathbb{R}).$$

Hence $\{\xi(t),\ t \geq 0\}$ is under $\mathbb{P}_0$ a continuous process with stationally independent increment (cf. [Blumenthal and Getoor (1968), p. 17]) so that $\xi(t) = \alpha B_t + bt$ for some constants α, b. Since $\eta^{-1}\xi(\eta^2 t) \sim \xi(t),\ \eta > 0$, under $\mathbb{P}_0$ in view of Theorem 4.1.6, we get $b = 0$.

Thus, we have shown that, given a random Jordan arc γ on the upper half-plane $\mathbb{H}$ with properties **(DMP)** and **(IL)** and satisfying either condition (4.2.12) or (4.2.14), the induced random process $\xi(t)$ on $\partial\mathbb{H}$ is reduced to a family of processes (4.2.13) parametrized by constant $\kappa > 0$.

A remarkable breakthrough was made by Oded Schramm [Schramm (2000)] revealing that the Loewner evolution driven by $\xi(t) = \sqrt{\kappa}B_t,\ \kappa > 0$, called the *stochastic Loewner evolution* (SLE) and denoted by SLE_κ, is the scaling limit of the two-dimensional discrete model of various kinds in statistical physics, under the conformal invariance assumption which has been rigorously established for many discrete models including loop-erased random walk, Ising model, percolation and uniform spanning trees. SLE has now been often called the *Schramm-Loewner evolution* in his honor.

In the next section, we shall consider the KLE driven by the solution $\mathbf{W}(t) = (\xi(t), \mathbf{s}(t))$ of the SDE (4.2.10) and (4.2.11) and call it the stochastic Komatu-Loewner evolution (SKLE) as a generalization of SLE on the simply connected domain $\mathbb{H}$ toward the standard slit domain D.

4.3 Stochastic Komatu-Loewner evolution $\mathrm{SKLE}_{\alpha,b}$

In the preceding section, we have started with a randomized Jordan arc γ and derived a system (4.2.10), (4.2.11) of stochastic differential equations for the random process $(\xi(t), \mathbf{s}(t))$. B_t in the first equation (4.2.10) is the one-dimensional standard Brownian motion and the coefficients in it are given by homogeneous functions α and b on $\mathcal{S}$ of degree 0 and -1, respectively.

In the present section, we shall start with the one-dimensional standard BM B_t and such functions α, b, solve the system (4.2.10), (4.2.11) of SDE in $(\xi(t), \mathbf{s}(t))$ and then consider the associated Komatu-Loewner evolution $\{F_t\}$ following the procedure in Section 3.2. Since the growing hulls $\{F_t\}$ is now a random process, it will be called the stochastic Komatu-Loewner evolution driven by $(\xi(t), \mathbf{s}(t))$.

4.3.1 *SDE with homogeneous coefficients*

Let us consider the following local Lipschitz continuity condition **(L)** for a real-valued function $f = f(\mathbf{s})$ on $\mathcal{S}$:

(L) For any $\mathbf{s}^{(0)} \in \mathcal{S}$ and any finite open interval $J \subset \mathbb{R}$, there exist a neighborhood $U(\mathbf{s}^{(0)})$ of $\mathbf{s}^{(0)}$ in $\mathcal{S}$ and a constant $L > 0$ such that

$$
\begin{aligned}
&|f(\mathbf{s}^{(1)} - \widehat{\xi}) - f(\mathbf{s}^{(2)} - \widehat{\xi})| \leq L\,|\mathbf{s}^{(1)} - \mathbf{s}^{(2)}| \\
&\text{for } \mathbf{s}^{(1)}, \mathbf{s}^{(2)} \in U(\mathbf{s}^{(0)}) \text{ and } \xi \in J,
\end{aligned}
\tag{4.3.1}
$$

where $\widehat{\xi}$ denotes the vector in $\mathbb{R}^{3N}$ whose first N-entries are 0 and the last $2N$ entries are ξ.

Lemma 4.3.1. *If a function f on $\mathcal{S}$ satisfies the condition* **(L)**, *then it holds for any $\mathbf{s}^{(1)},\ \mathbf{s}^{(2)} \in U(\mathbf{s}^{(0)})$ and for any $\xi_1,\ \xi_2 \in J$ that*

$$|f(\mathbf{s}^{(1)} - \widehat{\xi}_1) - f(\mathbf{s}^{(2)} - \widehat{\xi}_2)| \le L\left(|\mathbf{s}^{(1)} - \mathbf{s}^{(2)}| + \sqrt{2N}|\xi_1 - \xi_2|\right). \tag{4.3.2}$$

Proof. Suppose a function f on $\mathcal{S}$ satisfies the condition **(L)**. For any $\mathbf{s}^{(1)},\ \mathbf{s}^{(2)} \in U(\mathbf{s}^{(0)})$ and for any $\xi_1,\ \xi_2 \in J$ with $\xi_1 < \xi_2$, we have

$$\begin{aligned}&|f(\mathbf{s}^{(1)} - \widehat{\xi}_1) - f(\mathbf{s}^{(2)} - \widehat{\xi}_2)| \\ &\quad \le |f(\mathbf{s}^{(1)} - \widehat{\xi}_1) - f(\mathbf{s}^{(2)} - \widehat{\xi}_1)| + |f(\mathbf{s}^{(2)} - \widehat{\xi}_1) - f(\mathbf{s}^{(2)} - \widehat{\xi}_2)|.\end{aligned}$$

Since $\mathbf{s}^{(2)} \in U(\mathbf{s}^{(0)})$, there exists $\delta > 0$ such that $\mathbf{s}^{(2)} - \widehat{\xi} \in U(\mathbf{s}^{(0)})$ for any $\xi \in \mathbb{R}$ with $|\xi| < \delta$. Choose points $r_i,\ 0 \le i \le \ell$, with $r_0 = \xi_1,\ 0 < r_i - r_{i-1} < \delta,\ 1 \le i \le \ell,\ r_\ell = \xi_2$. The first term of the right-hand side of the above inequality is dominated by $L|\mathbf{s}^{(1)} - \mathbf{s}^{(2)}|$. The second term is dominated by $\sum_{i=1}^{\ell} |f(\mathbf{s}^{(2)} - \widehat{r}_i) - f(\mathbf{s}^{(2)} - \widehat{r}_{i-1})| = \sum_{i=1}^{\ell} |f\left((\mathbf{s}^{(2)} - (\widehat{r}_i - \widehat{r}_{i-1})) - \widehat{r}_{i-1}\right) - f((\mathbf{s}^{(2)} - \widehat{r}_{i-1})\,| \le \sum_{i=1}^{\ell} L|\widehat{r}_i - \widehat{r}_{i-1}| = L\sqrt{2N}(\xi_2 - \xi_1)$. □

Throughout the rest of this section, we assume that we are given a non-negative homogeneous function $\alpha(\mathbf{s})$ of $\mathbf{s} \in \mathcal{S}$ with degree 0 and a homogeneous function $b(\mathbf{s})$ of $\mathbf{s} \in \mathcal{S}$ with degree -1 both satisfying the condition **(L)**. A non-negative constant function on $\mathcal{S}$ is a trivial example of such α. A typical example of such a function b will be exhibited in the next chapter.

Theorem 4.3.2. *The SDE* (4.2.10) and (4.2.11) *admits a unique strong solution $\mathbf{W}_t = (\xi(t), \mathbf{s}(t)),\ t \in [0, \zeta)$, where ζ is the time when $\mathbf{W}_t$ approaches the point at infinity of $\mathbb{R} \times \mathcal{S}$.*

Proof. The coefficient $b_j(\xi, \mathbf{s})$ in the equation (4.2.11) is defined by (4.2.7). It admits the expression $\widetilde{b}_j(\mathbf{s} - \widehat{\xi})$ for $\widetilde{b}_j(\mathbf{s}) = b_j(0, \mathbf{s})$ by Proposition 3.1.2 and further $\widetilde{b}_j(\mathbf{s})$ satisfies the condition **(L)** in view of Lemma 3.1.1. Therefore, by virtue of Lemma 4.3.1, every coefficient, say, $f(\xi, \mathbf{s}),\ \xi \in \mathbb{R},\ \mathbf{s} \in \mathcal{S}$, in (4.2.10) and (4.2.11) is locally Lipschitz continuous on $\mathbb{R} \times \mathcal{S}$ $(\subset \mathbb{R}^{3N+1})$ in the following sense: for any $\mathbf{s}^{(0)} \in \mathcal{S}$ and for any finite open interval $J \subset \mathbb{R}$, there exists a ball $U(\mathbf{s}^{(0)}) \subset \mathcal{S}$ centered at $\mathbf{s}^{(0)}$ and a constant L_0 such that

$$|f(\xi_1, \mathbf{s}^{(1)}) - f(\xi_2, \mathbf{s}^{(2)})| \le L_0(|\mathbf{s}^{(1)} - \mathbf{s}^{(2)}| + |\xi_1 - \xi_2|),$$
$$\text{for } \mathbf{s}^{(1)}, \mathbf{s}^{(2)} \in U(\mathbf{s}^{(0)}) \text{ and } \xi_1,\ \xi_2 \in J.$$

Thus, (4.2.10) and (4.2.11) admit a unique local solution. It then suffices to patch together those local solutions just as in [Ikeda and Watanabe (1981), §1 of Chapter V]. □

Proposition 4.3.3. *For* $\mathbf{s} \in \mathcal{S}$, $\xi \in \mathbb{R}$, *let* $\mathbf{W}_t = (\xi(t), \mathbf{s}(t))$ *be the solution of the SDE* (4.2.10), (4.2.11) *with initial value* $(\xi, \mathbf{s})$.

(i) (scaling property) *For any* $c > 0$, *let* $\widetilde{\mathbf{W}}_t = (\widetilde{\xi}(t), \widetilde{\mathbf{s}}(t))$ *be the solution of the SDE* (4.2.10), (4.2.11) *with initial value* $(c\xi, c\mathbf{s})$. *Then*

$$\{c^{-1}\widetilde{\mathbf{W}}_{c^2t},\ t \geq 0\} \text{ has the same distribution as } \{\mathbf{W}_t,\ t \geq 0\}.$$

(ii) (homogeneity in horizontal direction) *For any* $r \in \mathbb{R}$, *let* $(\widetilde{\xi}(t), \widetilde{\mathbf{s}}(t))$ *be the solution of the SDE* (4.2.10), (4.2.11) *with initial value* $(\xi + r, \mathbf{s} + \widehat{r})$. *Then*

$$\{(\widetilde{\xi}(t) - r, \widetilde{\mathbf{s}}(t) - \widehat{r}), t \geq 0\} \text{ has the same distribution as } \{(\xi(t), \mathbf{s}(t))\ t \geq 0\}.$$

Proof. (i) We put $\widetilde{\mathbf{W}}_c(t) = c^{-1}\widetilde{\mathbf{W}}(c^2t) = (\widetilde{\xi}_c(t), \widetilde{\mathbf{s}}_c(t))$ with $\widetilde{\xi}_c(t) = c^{-1}\widetilde{\xi}(c^2t)$, $\widetilde{\mathbf{s}}_c(t) = c^{-1}\widetilde{\mathbf{s}}(c^2t)$. $\widetilde{\mathbf{W}}(t) = (\widetilde{\xi}(t), \widetilde{\mathbf{s}}(t))$ satisfies the equation (4.2.10) with $c\xi$ in place of ξ. Hence, by taking the homogeneity of α, b into account, we get

$$\begin{aligned}
\widetilde{\xi}_c(t) &= \xi + c^{-1}\int_0^{c^2t} \alpha(\mathbf{s}(s) - \widehat{\xi}(s))dB_s + c^{-1}\int_0^{c^2t} b(\mathbf{s}(s) - \widehat{\xi}(s))ds \\
&= \xi + c^{-1}\int_0^{t} \alpha(c(\mathbf{s}_c(s) - \widehat{\xi}_c(s)))dB_{c^2s} + c\int_0^{t} b(c(\mathbf{s}_c(s) - \widehat{\xi}_c(s)))ds \\
&= \xi + \int_0^{t} \alpha(\mathbf{s}_c(s) - \widehat{\xi}_c(s))d\tilde{B}_s + \int_0^{t} b(\mathbf{s}_c(s) - \widehat{\xi}_c(s))ds,
\end{aligned}$$

where $\tilde{B}_s = c^{-1}B_{c^2s}$. Therefore the equation (4.2.10) with a new Brownian motion $\tilde{B}_s$ is satisfied by $\widetilde{\mathbf{W}}_c(t)$. Similarly, (4.2.11) is also satisfied by $\widetilde{\mathbf{W}}_c(t)$.
(ii) This is immediate from the expressions (4.2.10) and (4.2.11) of the SDE and the property of every coefficient, say, $f(\xi, \mathbf{s})$, that $f(\xi + r, \mathbf{s} + \widehat{r}) = f(\xi, \mathbf{s})$ for any $r \in \mathbb{R}$. □

4.3.2 SKLE$_{\alpha,b}$ *and its basic properties*

Given a non-negative homogeneous function $\alpha(\mathbf{s})$ of $\mathbf{s} \in \mathcal{S}$ with degree 0 and a homogeneous function $b(\mathbf{s})$ of $\mathbf{s} \in \mathcal{S}$ with degree -1 both satisfying the condition **(L)**, let $\mathbf{W}_t = (\xi(t), \mathbf{s}(t))$, $t \in [0, \zeta)$, be a solution of the SDE-system (4.2.10), (4.2.11) in Theorem 4.3.2 with initial value $\mathbf{w} = (\xi, \mathbf{s})$. To be precise, let $(\Omega, \mathcal{G}, \mathbb{P})$ be a complete probability space with an increasing family of sub-σ-fields $\mathcal{G}_t$, $t \geq 0$, of $\mathcal{G}$ (a filtration) satisfying the usual conditions (i.e., $\{\mathcal{G}_t\}$ is right-continuous and $\mathbb{P}$-augmented). On $(\Omega, \mathcal{G}, \mathbb{P})$, there exists a one-dimensional $\mathcal{G}_t$-Brownian motion $B_t(\omega)$, $t \geq 0$, with $B_0 = 0$ (cf. [Ikeda and Watanabe (1981), Definition I.7.2]) and $\mathbf{W}_t(\omega) = (\xi(t, \omega), \mathbf{s}(t, \omega))$, $t < \zeta(\omega)$, is an $\mathbb{R} \times \mathcal{S}$-valued $\mathcal{G}_t$-adapted continuous process satisfying the equations (4.2.10) and (4.2.11).

Obviously, $(\xi(t, \omega), \mathbf{s}(t, \omega))$ satisfies the conditions **(I)**, **(II)** in t for each $\omega \in \Omega$ imposed in the beginning of Section 3.2, so that the Komatu-Loewner evolution is well associated with this pair following the procedure in Section 3.2. But we now view the associated family $\{g_t(z), t \in [0, t_z)\}$ of conformal maps and the associated growing $\mathbb{H}$-hulls $\{F_t, t \geq 0\}$ constructed in Theorem 3.2.5 and studied in Theorem 3.2.8 as random processes depending on $\omega \in \Omega$.

Notice that, for $z \in D = D(\mathbf{s})$, $t_z(\omega) \leq \zeta(\omega)$ and $g_t(z)(\omega)$ is defined for $t \in [0, t_z(\omega))$. Theorem 3.2.5 enables us to extend $g_t(z)(\omega)$ in t beyond $t_z(\omega)$ by setting

$$g_t(z)(\omega) = \begin{cases} \lim_{s \uparrow t_z(\omega)} g_s(z)(\omega)(\in \partial\mathbb{H}) & \text{for } \; t \in [t_z(\omega), \zeta(\omega)) \\ \mathbf{0} \in \mathbb{C} & \text{for } \; t \geq \zeta(\omega). \end{cases} \tag{4.3.3}$$

In other words, $g_t(z) = g_t(z)\mathbb{1}_{\{t<t_z\}} + g_{t_z-}(z)\mathbb{1}_{\{t_z \leq t < \zeta\}} + \mathbf{0} \cdot \mathbb{1}_{\{t \geq \zeta\}}$. Furthermore, it follows from Theorem 3.2.5 that

$$D \setminus F_t = \{z \in D : t < t_z\} = \{z \in D : \Im g_t(z) > 0\}, \quad t \geq 0. \tag{4.3.4}$$

Proposition 4.3.4.

(i) *For each $z \in D = D(\mathbf{s})$, $\{g_t(z)(\omega), t \geq 0\}$ is a $\mathcal{G}_t$-adapted process continuous in $t \in [0, \zeta(\omega))$.*

(ii) *For each $t \geq 0$, $\{z \in F_t(\omega)\} = \{t_z(\omega) \leq t\} \in \mathcal{G}_t$ for any $z \in D$.*

To prove Proposition 4.3.4, let us recall basic results concerning filtrations and stopping times from [Karatzas and Shreve (1998), §1.2]. Since our filtration $(\mathcal{G}_t)_{t \in [0,\infty]}$ is right-continuous, we do not distinguish optional and stopping times.

For a $(\mathcal{G}_t)_{t\geq 0}$-stopping time T, a σ-field $\mathcal{G}_T$ is defined by

$$\mathcal{G}_T := \{\, A \in \mathcal{G}_\infty : A \cap \{T \leq t\} \in \mathcal{G}_t \text{ for every } t \geq 0 \,\}$$

[Karatzas and Shreve (1998), Definition 2.12]. If a process $(X_t)_{t\geq 0}$ is progressively measurable with respect to $(\mathcal{G}_t)_{t\geq 0}$, then X_T is $\mathcal{G}_T$-measurable [Karatzas and Shreve (1998), Proposition 2.18]. Since $T + t$ is also a stopping time, the σ-fields $\mathcal{G}'_t := \mathcal{G}_{T+t}$, $t \geq 0$, form a new right-continuous filtration [Karatzas and Shreve (1998), Lemma 2.15 and Problem 2.23]. If there is another stopping time S, then $\mathcal{G}_{S\wedge T} = \mathcal{G}_S \cap \mathcal{G}_T$ holds, and the events $\{S < T\}$, $\{S = T\}$, and $\{S > T\}$ are members of $\mathcal{G}_S \cap \mathcal{G}_T$ [Karatzas and Shreve (1998), Lemma 2.16].

The next lemma follows from the properties collected above. We omit its proof as it is just a routine.

Lemma 4.3.5. *Let T be a $(\mathcal{G}_t)_{t\geq 0}$-stopping time and $\mathcal{G}'_t := \mathcal{G}_{T+t}$ for $t \geq 0$.*

(i) *Let S be another $(\mathcal{G}_t)_{t\geq 0}$-stopping time. Then S and $(S - T) \vee 0$ are $(\mathcal{G}'_t)_{t\geq 0}$-stopping times.*

(ii) *Let T' be a $(\mathcal{G}'_t)_{t\geq 0}$-stopping time. Then $T + T'$ is a $(\mathcal{G}_t)_{t\geq 0}$-stopping time, and $\mathcal{G}'_{T'} = \mathcal{G}_{T+T'}$ holds.*

In the following proof of Proposition 4.3.4, we use Lemma 4.3.5 several times but do not make a repetitive mention of it.

Proof of Proposition 4.3.4. To begin with, we confirm the notation. A point $z \in D$ is fixed throughout this proof. We occasionally indicate the dependence of

$K(t)$ and $\mathcal{X}$ on ω explicitly as $K(t,\omega) = \bigcup_{j=1}^{N} C_j(\mathbf{s}(t,\omega))$ and $\mathcal{X}(\omega)$. Here, the domain $\mathcal{X}$ in $[0,\zeta) \times \mathbb{H}$ is defined as in the second paragraph in Section 3.2. Let $U_\varepsilon^{\mathcal{S}}(\mathbf{s})$ be the ball in $\mathcal{S} \subset \mathbb{R}^{3N}$ with center $\mathbf{s}$ and radius ε. We denote by $B_\varepsilon(z)$ the ε-neighborhood of $z \in \mathbb{H}$ to distinguish it from the symbol $U_\varepsilon^{\mathcal{S}}(\mathbf{s})$.

Our proof is based on a recursive application of the Picard iteration. The first step proceeds as follows: We set $\mathcal{G}_t^1 := \mathcal{G}_t$ and $\rho_1 := 4^{-1} \operatorname{dist}(z, K(0) \cup \partial\mathbb{H}) > 0$. Let τ_1 be the exit time of $\mathbf{s}(t)$ from the ball $U_{\rho_1}^{\mathcal{S}}(\mathbf{s}(0))$. If $\mathbf{s}(t,\omega)$ stays within this ball for all $t < \zeta(\omega)$, then define $\tau_1(\omega) := \zeta(\omega)$. τ_1 is a $(\mathcal{G}_t^1)_{t\geq 0}$-stopping time. We set

$$M_t^1(\omega) := 2\pi \int_0^t \sup_{z' \in B_{\rho_1}(z)} |\Psi_{\mathbf{s}(s)}(z', \xi(s))| \mathbb{1}_{\{s<\tau_1\}}(\omega)\, ds.$$

The process $(M_t^1)_{t\geq 0}$ is non-decreasing, continuous, and $(\mathcal{G}_t^1)_{t\geq 0}$-adapted with $M_0^1 = 0$. We denote by σ_1 the hitting time to ρ_1 by M_t^1 and set $T_1 := \tau_1 \wedge \sigma_1$. By definition, T_1 is a $(\mathcal{G}_t^1)_{t\geq 0}$-stopping time which is positive a.s.

For each $\omega \in \Omega$, we consider the following Picard iteration:

$$w_0^1(t,\omega) = z,$$

$$w_{n+1}^1(t,\omega) = z - 2\pi \int_0^t \Psi_{\mathbf{s}(s,\omega)}(w_n^1(s,\omega), \xi(s,\omega)) \mathbb{1}_{\{s<T_1\}}(\omega)\, ds, \quad n = 0,1,2,\ldots.$$

These processes are well-defined, continuous, $(\mathcal{G}_t^1)_{t\geq 0}$-adapted and satisfy $w_n^1(t,\omega) \in \overline{B_{\rho_1}(z)}$ for all $n \in \mathbb{N}$, $t \geq 0$, and $\omega \in \Omega$. Indeed, the case $n = 0$ is trivial from definition. Assume that w_n^1 enjoys this property. Then

$$|w_{n+1}^1(t) - z| \leq 2\pi \int_0^{T_1} |\Psi_{\mathbf{s}(s)}(w_n^1(s), \xi(s))| \mathbb{1}_{\{s<\tau_1\}}\, ds \leq M_{T_1}^1 \leq \rho_1,$$

which yields $w_{n+1}^1(t) \in \overline{B_{\rho_1}(z)}$. Thus, the induction proves the desired property for all n.

For each $\omega \in \Omega$, $w_n^1(t,\omega)$ converges to $g_t(z)(\omega)$ locally uniformly on the interval $[0, T_1(\omega))$. Indeed, we have defined T_1 so that

$$[0, T_1(\omega) - \varepsilon] \times \overline{B_{\rho_1}(z)} \subset \mathcal{X}(\omega) \quad \text{for any } \varepsilon \in (0, T_1(\omega)),$$

and in view of Proposition 3.2.1 and its proof, $\Psi_{\mathbf{s}(s)}(z, \xi(s))$ satisfies the Lipschitz continuity condition (3.2.4) for (s, z_1), $(s, z_2) \in [0, T_1(\omega) - \varepsilon] \times \overline{B_{\rho_1}(z)}$. Hence the successive approximation $w_n^1(t)$ converges to the solution $g_t(z)$ of the original K-L equation uniformly in $t \in [0, T_1(\omega) - \varepsilon]$ as in the usual theory of ODEs. We set

$$w_\infty^1(t,\omega) := \lim_{n\to\infty} w_n^1(t,\omega) \mathbb{1}_{\{t<T_1\}}(\omega).$$

This process is clearly $(\mathcal{G}_t^1)_{t\geq 0}$-adapted.

We consider the case $t = T_1(\omega)$. If $T_1(\omega) = \zeta(\omega)$, then clearly $t_z(\omega) = \zeta(\omega)$. In this case, we have defined $g_t(z)(\omega) := 0$ for $t \geq \zeta(\omega) = T_1(\omega)$ in (4.3.3). If $T_1(\omega) < \zeta(\omega)$, then we can take ε to be zero in the previous paragraph, and it follows that $g_{T_1(\omega)}(z)(\omega) = \lim_{n\to\infty} w_n^1(T_1(\omega), \omega)$. In summary, the random variable $g_{T_1}(z) = \lim_{n\to\infty} w_n^1(T_1) \mathbb{1}_{\{T_1<\zeta\}}$ is $\mathcal{G}_{T_1}^1 = \mathcal{G}_{T_1}$-measurable.

We move to the next step, in which we consider the successive approximation of the solution $g_t(z)$ after the time T_1. We set $\mathcal{G}_t^2 := \mathcal{G}_{T_1+t}^1$, $t \geq 0$. These σ-fields form a filtration on Ω with the usual conditions. Let

$$\rho_2(\omega) := \frac{1}{4}\operatorname{dist}(g_{T_1}(z)(\omega), K(T_1(\omega),\omega) \cup \partial\mathbb{H}) \cdot \mathbb{1}_{\{T_1<\zeta\}}(\omega).$$

We can observe that ρ_2 is a $\mathcal{G}_0^2 = \mathcal{G}_{T_1}^1$-measurable random variable.

Let τ_2 be the exit time of the shifted process $\mathbf{s}(T_1+t)$ from the (random) ball $U_{\rho_2}^S(\mathbf{s}(T_1))$. As before, if $\mathbf{s}(T_1+t)$ stays within this ball for all $t < \zeta - T_1$, then define $\tau_2 := \zeta - T_1$. In particular, $\tau_2 = 0$ if $T_1 = \zeta$. Noting that $\zeta - T_1(\geq 0)$ is a $(\mathcal{G}_t^2)_{t\geq 0}$-stopping time, we can show that τ_2 is a $(\mathcal{G}_t^2)_{t\geq 0}$-stopping time as follows: For $t \geq 0$, we have

$$\begin{aligned}\{\tau_2 \leq t\} &= \{\zeta - T_1 \leq t\} \cup \bigcap_{n=1}^{\infty} \bigcup_{s\in[0,t]\cap\mathbb{Q}} \{d(\mathbf{s}(T_1+s), \mathbf{s}(T_1)) \geq \rho_2 - n^{-1}\} \\ &\in \mathcal{G}_t^2 \vee \bigvee_{s\in[0,t]\cap\mathbb{Q}} \mathcal{G}_{T_1+s}^1 = \mathcal{G}_t^2 \vee \bigvee_{s\in[0,t]\cap\mathbb{Q}} \mathcal{G}_s^2 = \mathcal{G}_t^2.\end{aligned}$$

Using the random variables ρ_2 and τ_2, we set

$$\begin{aligned}M_t^2 &:= 2\pi \int_0^t \sup_{z'\in B_{\rho_2}(g_{T_1}(z))} |\Psi_{\mathbf{s}(T_1+s)}(z', \xi(T_1+s))| \mathbb{1}_{\{s<\tau_2\}}(\omega)\, ds \\ &= 2\pi \int_0^t \sup_{z'\in\mathbb{Q}^2} \left\{ |\Psi_{\mathbf{s}(T_1+s)}(z', \xi(T_1+s))| \mathbb{1}_{\{z'\in B_{\rho_2}(g_{T_1}(z))\}} \right\} \mathbb{1}_{\{s<\tau_2\}}\, ds.\end{aligned}$$

This process is non-decreasing, continuous, and $(\mathcal{G}_t^2)_{t\geq 0}$-adapted. We denote by σ_2 the hitting time to ρ_2 by M_t^2. Since ρ_2 is $\mathcal{G}_0^2$-measurable, we can prove that σ_2 is a $(\mathcal{G}_t^2)_{t\geq 0}$-stopping time in such a way as we have considered τ_2.

We define a $(\mathcal{G}_t^2)_{t\geq 0}$-stopping time $T_2 := \tau_2 \wedge \sigma_2$, which is positive a.s. on the event $\{T_1 < \zeta\} \in \mathcal{G}_0^2$, and consider the following iteration:

$$\begin{aligned}w_0^2(t,\omega) &= g_{T_1(\omega)}(z)(\omega), \\ w_{n+1}^2(t,\omega) &= g_{T_1(\omega)}(z)(\omega) - 2\pi \int_0^t \Psi_{\mathbf{s}(T_1+s,\omega)}(w_n^2(s,\omega), \xi(T_1+s,\omega)) \mathbb{1}_{\{s<T_2\}}(\omega)\, ds.\end{aligned}$$

As in the first step, the $(\mathcal{G}_t^2)_{t\geq 0}$-adapted process $w_n^2(t,\omega)$ converges to $g_{T_1(\omega)+t}(z)(\omega)$ as $n \to \infty$ locally uniformly on the interval $[0, T_2(\omega))$ for each $\omega \in \Omega$. Moreover, if $(T_1+T_2)(\omega) < \zeta(\omega)$, then $g_{(T_1+T_2)(\omega)}(z)(\omega) = \lim_{n\to\infty} w_n^2(T_1(\omega),\omega)$. Since the events $\{T_1+T_2 = \zeta\}$ and $\{T_1+T_2 < \zeta\}$ belong to the σ-field $\mathcal{G}_{T_1}^2$, the random variable $g_{T_1+T_2}(z) = \lim_{n\to\infty} w_n^2(T_2)\mathbb{1}_{\{T_1+T_2<\zeta\}}$ is $\mathcal{G}_{T_2}^2 = \mathcal{G}_{T_1+T_2}$-measurable.

We now repeat the procedure above to obtain the sequence of filtrations $(\mathcal{G}_t^k)_{t\geq 0}$, $(\mathcal{G}_t^k)_{t\geq 0}$-stopping times T_k, and $\mathcal{G}_t^k$-measurable limit of the successive approximations $w_\infty^k(t,\omega)$ for $k \geq 1$. Let $T_0 := 0$ and $\mathcal{G}_t^0 := \mathcal{G}_t$ for notational convenience. We set $S_l := \sum_{k=0}^{l} T_k$ and $S := \lim_{l\to\infty} S_l$, which are $(\mathcal{G}_t)_{t\geq 0}$-stopping times.

We can show that $S < \zeta$ implies $S = t_z$ by the following pathwise argument: Assume that we have $S(\omega) < \zeta(\omega)$ but $S(\omega) < t_z(\omega)$. Then $g_{S(\omega)}(z)(\omega) \in D_{S(\omega)}$. We set

$$\rho^*(\omega) := \frac{1}{16} \operatorname{dist}(g_{S(\omega)}(z), K(S(\omega)) \cup \partial\mathbb{H}) > 0.$$

There then exists $\delta > 0$ such that $g_t(z) \in B_{\rho^*}(g_S(z))$ and $\mathbf{s}(t) \in U^{\mathcal{S}}_{\rho^*}(\mathbf{s}(S))$ for $t \in (S-\delta, S]$. Since $S = \lim_{l\to\infty} S_l$, the time S_l belongs to this interval from some l on. For such an l, we see from the definition of ρ^*, ρ_{l+1}, and T_{l+1} that $T_{l+1} > S - S_l$ holds. However, this inequality yields $S < S_l + T_{l+1} = S_{l+1}$, a contradiction. Thus, we obtain $S = t_z$.

Since $S = \zeta$ trivially implies $S = t_z$, we get $t_z = S$ on the whole Ω. Hence t_z is a $(\mathcal{G}_t)_{t\geq 0}$-stopping time, which proves Proposition 4.3.4 (ii).

We now prove Proposition 4.3.4 (i), namely, the $(\mathcal{G}_t)_{t\geq 0}$-adaptedness of $g_t(z)$. Let $t \geq 0$ and $B \in \mathcal{B}(\overline{\mathbb{H}})$. It follows from (4.3.3) that

$$\{g_t(z) \in B\} = \{g_t(z) \in B,\ t < S\} \cup \{\lim_{s\uparrow S} g_s(z) \in B,\ S \leq t < \zeta\} \cup \{0 \in B,\ t \geq \zeta\}. \tag{4.3.5}$$

We observe that the three events in the right-hand side belong to $\mathcal{G}_t$ as follows: First, we have

$$\begin{aligned}
&\{g_t(z) \in B,\ t < S\} \\
&= \bigcup_{l=0}^{\infty} (\{g_t(z) \in B\} \cap \{S_l \leq t < S_{l+1}\}) \\
&= \bigcup_{l=0}^{\infty} (\{w^{l+1}_\infty((t - S_l) \vee 0) \in B\} \cap \{(t - S_l) \vee 0 < T_{l+1}\} \cap \{S_l \leq t\}).
\end{aligned}$$

Since $(t - S_l) \vee 0$ is a $(\mathcal{G}^{l+1}_u)_{u\geq 0}$-stopping time, $\{w^{l+1}_\infty((t - S_l) \vee 0) \in B\}$ and $\{(t - S_l) \vee 0 < T_{l+1}\}$ are members of $\mathcal{G}^{l+1}_{(t-S_l)\vee 0} = \mathcal{G}_{S_l+(t-S_l)\vee 0} = \mathcal{G}_{t\vee S_l}$. Hence $\{g_t(z) \in B,\ t < S\}$ belongs to $\mathcal{G}_t$. Next, the random variable $\lim_{s\uparrow S} g_s(z) = \lim_{l\to\infty} w^l_\infty(T_l)$ is $\bigvee_{l=1}^{\infty} \mathcal{G}_{S_l} = \mathcal{G}_S$-measurable. Hence $\{\lim_{s\uparrow S} g_s(z) \in B,\ S \leq t < \zeta\} \in \mathcal{G}_t$. Lastly, $\{0 \in B,\ t \geq \zeta\} \in \mathcal{G}_t$ is obvious. In conclusion, $\{g_t(z) \in B\} \in \mathcal{G}_t$ follows from (4.3.5), which completes the proof of Proposition 4.3.4 (i). □

We call $\{F_t; t \geq 0\}$ or $\{g_t(z),\ t \geq 0\}$ the *stochastic Komatu-Loewner evolution* (SKLE) driven by the solution $\mathbf{W}_t = (\xi(t), s(t))$ of the SDE (4.2.10) and (4.2.11) with coefficients α and b. We designate it as $\mathrm{SKLE}_{\alpha,b}$. Since $\mathrm{SKLE}_{\alpha,b}$ depends also on the initial value $\mathbf{w} = (\xi, \mathbf{s}) \in \mathbb{R} \times \mathcal{S}$ of $\mathbf{W}_t$, we shall denote it occasionally as $\mathrm{SKLE}_{\mathbf{w},\alpha,b}$ or $\mathrm{SKLE}_{\xi,\mathbf{s},\alpha,b}$.

Proposition 4.3.3 combined with Proposition 3.1.2 implies the following properties of the SKLE under scale changes and shifts.

Proposition 4.3.6. *Take any* $\mathbf{s} \in \mathcal{S}$, $\xi \in \mathbb{R}$, $r > 0$ *and* $c \in \mathbb{R}$. *Let* $\{g_t(z),\ t \geq 0\}$, $z \in D = D(\mathbf{s})$, *and* $\{F_t : t \geq 0\}$ *be the* $\mathrm{SKLE}_{\xi,\mathbf{s},\alpha,b}$.

(i) *Let* $\{\widetilde{g}_t(z),\ t \geq 0\}$ *be* $\mathrm{SKLE}_{\xi/r,\mathbf{s}/r,\alpha,b}$. *Then* $\{r\widetilde{g}_{t/r^2}(z/r),\ t \geq 0\}$ *has the same distribution as* $\{g_t(z),\ t \geq 0\}$ *for* $z \in D(\mathbf{s})$.

(ii) *Let* $\{\widetilde{F}_t, t \geq 0\}$ *be* $\mathrm{SKLE}_{\xi/r,\mathbf{s}/r,\alpha,b}$. *Then* $\{r\widetilde{F}_{t/r^2},\ t \geq 0\}$ *has the same distribution as* $\{F_t,\ t \geq 0\}$.

(iii) *Let* $\{\widehat{g}_t(z), t \geq 0\}$ *be* $\mathrm{SKLE}_{\xi+c,\mathbf{s}+c,\alpha,b}$. *Then* $\{\widehat{g}_t(z+c)-c,\ t \geq 0\}$ *has the same distribution as* $\{g_t(z),\ t \geq 0\}$ *for* $z \in D(\mathbf{s})$.

(iv) *Let* $\{\widehat{F}_t, t \geq 0\}$ *be* $\mathrm{SKLE}_{\xi+c,\mathbf{s}+c,\alpha,b}$. *Then* $\{\widehat{F}_t - c,\ t \geq 0\}$ *has the same distribution as* $\{F_t,\ t \geq 0\}$.

Proof. (i) Let $\mathbf{W}(s) = (\xi(s), \mathbf{s}(s))$ be the solution of the SDE (4.2.10) and (4.2.11) with initial value $(\xi, \mathbf{s})$. By Proposition 4.3.3 (i), $\widetilde{\mathbf{W}}(s) := r^{-1}\mathbf{W}(r^2 s)$ has the same distribution as the solution of (4.2.10) and (4.2.11) with initial value $(\xi/r, \mathbf{s}/r)$. Hence $\{\widetilde{g}_t(z), t < t_z\}$ has the same distribution as the solution (denoted by $\widetilde{g}_t(z)$ again) of the Komatu-Loewner equation (3.2.1) driven by $\widetilde{\mathbf{W}}$:

$$\widetilde{g}_t(z/r) - z/r = -2\pi \int_0^t \Psi_{r^{-1}\mathbf{s}(r^2 s)}(\widetilde{g}_s(z/r), r^{-1}\xi(r^2 s))ds,\ z \in D.$$

In view of Theorem 3.2.5 (i), it suffices to show that $h_t(z) := r\widetilde{g}_{t/r^2}(z/r),\ z \in D$, solves the equation (3.2.1).

By the homogeneity (3.1.7) and (3.1.8) of the complex Poisson kernel Ψ,

$$\begin{aligned}
\Psi_{r^{-1}\mathbf{s}(r^2 s)}(\widetilde{g}_s(z), r^{-1}\xi(r^2 s)) &= \Psi_{r^{-1}(\mathbf{s}(r^2 s)-\widehat{\xi}(r^2 s))}(\widetilde{g}_s(z) - r^{-1}\xi(r^2 s), 0) \\
&= r\Psi_{\mathbf{s}(r^2 s)-\widehat{\xi}(r^2 s)}(r\widetilde{g}_s(z) - \xi(r^2 s), 0) \\
&= r\Psi_{\mathbf{s}(r^2 s)}(r\widetilde{g}_s(z), \xi(r^2 s)).
\end{aligned}$$

and so

$$\begin{aligned}
\widetilde{g}_t(z) - z &= -2\pi r \int_0^t \Psi_{\mathbf{s}(r^2 s)}(r\widetilde{g}_s(z), \xi(r^2 s))ds \\
&= -\frac{2\pi}{r} \int_0^{r^2 t} \Psi_{\mathbf{s}(s)}(r\widetilde{g}_{s/r^2}(z), \xi(s))ds.
\end{aligned}$$

Consequently, $h_t(z) - z = -2\pi \int_0^t \Psi_{\mathbf{s}(s)}(h_s(z), \xi(s))ds$.

(ii) $\{\widetilde{F}_t\}$ has the same distribution as the growing $\mathbb{H}$-hulls (denoted by $\widetilde{F}_t$ again) in $D(\mathbf{s}/r)$ associated with $\{\widetilde{g}_t(z/r), z \in D\}$ in the above proof of (i). By (4.3.4), we have $(D/r) \setminus \widetilde{F}_t = \{z/r : z \in D,\ \Im \widetilde{g}_t(z/r) > 0\}$.

On the other hand, $\{F_t, t \geq 0\}$ has the same distribution as the $\mathbb{H}$-hulls associated with $\{h_t(z); t \geq 0\}$ in the above proof of (i) so that

$$\begin{aligned}
D \setminus F_t &\cong \{z \in D : \Im h_t(z) > 0\} = \{z \in D : \Im \widetilde{g}_{(t/r^2)}(z/r) > 0\} \\
&= r\{z/r : z \in D, \Im \widetilde{g}_{(t/r^2)}(z/r) > 0\} = D \setminus (r\widetilde{F}_{t/r^2}).
\end{aligned}$$

(iii) Let $\mathbf{W}(s) = (\xi(s), \mathbf{s}(s))$ be the unique solution of the SDE (4.2.10) and (4.2.11) with initial value $(\xi, \mathbf{s})$. As $b_j(\xi, \mathbf{s}) = b_j(0, \mathbf{s} - \widehat{\xi})$, $\mathbf{W}(t) + c = (\xi(t) + c, \mathbf{s}(t) + c)$ is the unique solution of the SDE (4.2.10) and (4.2.11) with initial value $(\xi + c, \mathbf{s} + c)$.

Hence $\{\widehat{g}_t(z),\ t < t_z\}$ has the same distribution as the solution (denoted by $\widehat{g}_t(z)$ again) of the K-L equation

$$\widehat{g}_t(z) - z = -2\pi \int_0^t \Psi_{\mathbf{s}(t)+\widehat{c}}(\widehat{g}_s(z), \xi(s)+c)ds,\ z \in D(\mathbf{s}+c).$$

In view of second identity in (3.1.8), $k_t(z) := \widehat{g}_t(z+c) - c$ is the unique solution of the Komatu-Loewner equation (3.2.1) for $z \in D(\mathbf{s})$. This implies the conclusion of (iii).

(iv) $\{\widehat{F}_t, t \geq 0\}$ has the same distribution as the $\mathbb{H}$-hulls (denoted by $\{\widehat{F}_t\}$ again) associated with $\{\widehat{g}_t(z), z \in D+c\}$ in (iii). Accordingly $(D+c) \setminus \widehat{F}_t = \{z+c : z \in D,\ \Im\widehat{g}_t(z+c) > 0\}$. On the other hand, $\{F_t, t \geq 0\}$ has the same distribution as the $\mathbb{H}$-hulls associated with $\{k_t(z),\ t \geq 0\}$ so that

$$\begin{aligned} D \setminus F_t &\cong \{z \in D : \Im k_t(z) > 0\} = \{z \in D : \Im\widehat{g}_t(z+c) > 0\} \\ &= \{(z+c) - c : z \in D,\ \Im\widehat{g}_t(z+c) > 0\} = D \setminus (\widehat{F}_t - c). \end{aligned}$$

□

Corresponding statements for SLE_κ, namely, the special case that $D = \mathbb{H}$, $\alpha^2 = \kappa$ for a constant $\kappa > 0$ and $b = 0$ can be found in [Rohde and Schramm (2005), Proposition 2.1].

The following lemma will be utilized in Section 5.4.2.

Lemma 4.3.7.

(i) *For any closed set* $A \subset D = D(\mathbf{s})$ *and* $t \geq 0$, $\{F_t \cap A = \emptyset\} \in \mathcal{G}_t$.

(ii) *For any closed set* $A \subset D$, *define*

$$\tau_A = \sup\{t \in [0,\zeta) : F_t \cap A = \emptyset\}. \tag{4.3.6}$$

Then τ_A *is a* $\{\mathcal{G}_t\}$*-stopping time: for any* $t \geq 0$, $\{t < \tau_A\} \in \mathcal{G}_t$.

(iii) *For any hull* $A \subset D$, *define*

$$\overline{\tau}_A = \sup\{t \in [0,\zeta) : \overline{F}_t \cap \overline{A} = \emptyset\}. \tag{4.3.7}$$

$\overline{\tau}_A$ *is then a* $\{\mathcal{G}_t\}$*-stopping time: for any* $t \geq 0$, $\{t < \overline{\tau}_A\} \in \mathcal{G}_t$.

Proof. (i) Suppose first that A is a compact subset of D. Taking a countable dense subset A_1 of A, we prove that

$$\{F_t(\omega) \cap A = \emptyset\} = \left\{ \inf_{z \in A_1} \Im g_t(z)(\omega) \mathbb{1}_{\{t_z(\omega) > t\}} > 0 \right\}, \tag{4.3.8}$$

whose right-hand side belongs to $\mathcal{G}_t$ by Proposition 4.3.4.

If $F_t(\omega) \cap A = \emptyset$, then $\Im g_t(z)(\omega) > 0$ for any $z \in A$ by (4.3.4). As $g_t(z)(\omega)$ is continuous in z on the compact set $A \subset D \setminus F_t(\omega)$, we have $\min_{z \in A} \Im g_t(z)(\omega) \mathbb{1}_{\{t_z(\omega) > t\}} > 0$, yielding the inclusion $\subset$ in (4.3.8). The converse inclusion $\supset$ is also valid by noting the lower semi-continuity of t_z in z (cf. [Hartman (1964), Theorem V.2.1]) and the uniform continuity of $\Im g_t(z)(\omega)$ in $z \in A$.

For any closed set $A \subset D$, there exist compact sets A_n increasing to A so that $\{F_t \cap A = \emptyset\} = \bigcap_{n=1}^{\infty}\{F_t \cap A_n = \emptyset\} \in \mathcal{G}_t$.

(ii) Take a sequence $t_k > t$ decreasing to t as $k \to \infty$. Then, for any $k_0 \in \mathbb{N}$,

$$\{t < \tau_A\} = \bigcup_{k \geq k_0} \{F_{t_k} \cap A = \emptyset\},$$

which is in $\mathcal{G}_{t_{k_0}}$ by (i). Hence τ_A is an $\{\mathcal{G}_t\}$-stopping time.

(iii) For $\ell \in \mathbb{N}$, denote by A_ℓ the $1/\ell$-neighborhood of a hull A. One can then readily see that

$$\{\overline{F}_t \cap \overline{A} = \emptyset\} = \bigcap_{\ell \in \mathbb{N}} \{F_t \cap \overline{A}_\ell = \emptyset\},$$

which belongs to $\mathcal{G}_t$ by (i). Hence $\overline{\tau}_A$ is an $\{\mathcal{G}_t\}$-stopping time as in the proof of (ii). □

The next proposition will be employed in Section 5.4.2 as well. For $z \in D = D(\mathbf{s})$, we set

$$s_z := \inf\{\, t \in [0, \zeta) : z \notin D_t \,\} = \inf\{\, t \in [0, \zeta) : z \in K(t) \,\} \tag{4.3.9}$$

with the infimums set to be ζ if these sets are empty. Recall that $K(t) = K(t, \omega) = \bigcup_{j=1}^{N} C_j(\mathbf{s}(t, \omega))$ denotes the horizontal slits for the domain D_t. It is not difficult to see that s_z is a $(\mathcal{G}_t)_{t \geq 0}$-stopping time.

Proposition 4.3.8. *The process $g_t^{-1}(z)\mathbb{1}_{\{t<s_z\}}$ is $(\mathcal{G}_t)_{t\geq 0}$-adapted for each $z \in D = D(\mathbf{s})$. Moreover, the function $(z, \omega) \mapsto g_t^{-1}(z)\mathbb{1}_{\{t<s_z\}}(\omega)$ is $\mathcal{B}(D) \times \mathcal{G}_t$-measurable for each t.*

Our proof of Proposition 4.3.8 is based on the following inversion formula:

Lemma 4.3.9. *Let f be a univalent function on a domain $G \subset \mathbb{C}$ and $z \in f(G)$. Suppose that a smooth Jordan curve C in G enjoys $f^{-1}(z) \in \operatorname{ins} C \subset G$. Here, $\operatorname{ins} C$ denotes the bounded region surrounded by C. It holds that*

$$f^{-1}(z) = \frac{1}{2\pi i} \int_C \frac{w f'(w)}{f(w) - z}\, dw.$$

Proof. This lemma is proved just by the change of variables $\tilde{w} = f(w)$ along with Cauchy's integral formula:

$$\frac{1}{2\pi i} \int_C \frac{w f'(w)}{f(w) - z}\, dw = \frac{1}{2\pi i} \int_{f(C)} \frac{f^{-1}(\tilde{w})}{\tilde{w} - z}\, d\tilde{w} = f^{-1}(z).$$

Another proof is given by the residue theorem, as the function $w \mapsto wf'(w)/(f(w) - z)$ has a pole of first order at $w = f^{-1}(z)$ with residue $f^{-1}(z)$. □

Proof of Proposition 4.3.8. To obtain the $\mathcal{G}_t$-measurability of $g_t^{-1}(z)$, we shall express it in terms of $g_t(w)$ in small time intervals using Lemma 4.3.9 and patch them together as in the proof of Proposition 4.3.4.

In the first step, let $\mathcal{G}_t^1 := \mathcal{G}_t$, $\rho_1 := 3^{-1}\,\mathrm{dist}(z, K(0) \cup \partial\mathbb{H})$, $U_1 := \overline{B_{\rho_1}(z)}$, and $V_1 := \overline{B_{2\rho_1}(z)}$. We set

$$T_1 := \inf\{\, t \in [0, \tau_{V_1}) : z \notin g_t(U_1)\}$$

with τ_{V_1} defined by (4.3.6). The infimum is set to be τ_{V_1} if this set is empty. T_1 is a $(\mathcal{G}_t^1)_{t\geq 0}$-stopping time, because for $t > 0$ we have

$$\{T_1 < t\} = \{\tau_{V_1} < t\} \cup \bigcup_{u\in[0,t)\cap\mathbb{Q}} \bigcup_{n=1}^{\infty} \bigcap_{w\in U_1\cap\mathbb{Q}^2} \{|z - g_u(w)| \geq n^{-1}\} \in \mathcal{G}_t = \mathcal{G}_t^1.$$

By Lemma 4.3.9, we have

$$g_t^{-1}(z)\mathbb{1}_{\{t<T_1\}} = \frac{1}{2\pi i}\int_{\partial V_1} \frac{w g_t'(w)}{g_t(w) - z}\mathbb{1}_{\{t<T_1\}}\, dw.$$

As the limit of Riemann sums, the right-hand side is $\mathcal{G}_t = \mathcal{G}_t^1$-measurable. Similarly, putting $S_1 := T_1$, we see that $g_{S_1}^{-1}(z)\mathbb{1}_{\{S_1<s_z\}}$ is $\mathcal{G}_{S_1}$-measurable.

In the next step, let $\mathcal{G}_t^2 := \mathcal{G}_{T_1+t}^1$. We set

$$\rho_2 := 3^{-1}\,\mathrm{dist}(g_{S_1}^{-1}(z)\mathbb{1}_{\{S_1<s_z\}}, K(S_1) \cup F_{S_1} \cup \partial\mathbb{H}),$$
$$U_2 := \overline{B_{\rho_2}(g_{S_1}^{-1}(z)\mathbb{1}_{\{S_1<s_z\}})}, \qquad V_2 := \overline{B_{2\rho_2}(g_{S_1}^{-1}(z)\mathbb{1}_{\{S_1<s_z\}})}.$$

As in the proof of Proposition 4.3.4, we can show that

$$T_2 := \inf\{\, t \in [0, (\tau_{V_2} - S_1) \vee 0) : z \notin g_{S_1+t}(U_2)\mathbb{1}_{\{S_1<s_z\}}\,\}$$

is a $(\mathcal{G}_t^2)_{t\geq 0}$-stopping time. Hence $S_2 := T_1 + T_2 = S_1 + T_2$ is a $(\mathcal{G}_t)_{t\geq 0}$-stopping time. As in the first step, the process $g_{S_1+t}^{-1}(z)\mathbb{1}_{\{t<T_2\}}$ is $(\mathcal{G}_t^2)_{t\geq 0}$-adapted and the random variable $g_{S_2}^{-1}(z)\mathbb{1}_{\{S_2<s_z\}}$ is $\mathcal{G}_{S_2}$-measurable by Lemma 4.3.9 combined with the $\mathcal{G}_{S_1}$-measurability of $g_{S_1}^{-1}(z)\mathbb{1}_{\{S_1<s_z\}}$.

Finally, we repeat the procedure above to obtain filtrations $(\mathcal{G}_t^k)_{t\geq 0}$ and stopping times T_k with respect to them for $k \geq 1$. $S_l := \sum_{k=1}^{l} T_k$ and $S = \lim_{l\to\infty} S_l$ are $(\mathcal{G}_t)_{t\geq 0}$-stopping times. As in the last step of the proof of Proposition 4.3.4, we have $S = s_z$ and then the $\mathcal{G}_t$-measurability of $g_t^{-1}(z)\mathbb{1}_{\{t<s_z\}}$.

To see the joint measurability, we approximate $g_t^{-1}(z)\mathbb{1}_{\{t<s_z\}}(\omega)$ by simple functions, using the collection $\mathcal{Q}_n'$ of all squares of the form $Q = (k2^{-n}, (k+1)2^{-n}] \times (l2^{-n}, (l+1)2^{-n}]$, $k \in \mathbb{Z}$, $l \in \mathbb{Z}_+$. For each $Q \in \mathcal{Q}_n'$, let z_Q be an arbitrary point of Q and $\mathcal{N}(Q)$ be the union of Q itself and eight (five if $l = 0$) squares of $\mathcal{Q}_n'$ surrounding Q. We define an event

$$E'(t, Q) := \{t < \zeta\} \cap \bigcap_{u\in[0,t]\cap\mathbb{Q}} \{\mathcal{N}(Q) \cap K(u) = \emptyset\} \in \mathcal{G}_t.$$

Since the slit motion $\mathbf{s}(t)$ is continuous up to ζ, it is obvious that $Q \times E'(t, Q) \subset \{\, (z, \omega) : t < s_z(\omega)\,\}$. Conversely, if $t < s_z(\omega)$, then for every n large enough, we can find a unique $Q \in \mathcal{Q}'_n$ such that $z \in Q$ and $\mathcal{N}(Q) \cap K(u,\omega) = \emptyset$ for all $u \in [0,t] \cap \mathbb{Q}$. Therefore

$$g_t^{-1}(z)\mathbb{1}_{\{t<s_z\}}(\omega) = \lim_{n\to\infty} \sum_{Q\in\mathcal{Q}'_n} \left(g_t^{-1}(z_Q)\mathbb{1}_{\{t<s_{z_Q}\}}(\omega)\right) \mathbb{1}_Q(z)\mathbb{1}_{E'(t,Q)}(\omega),$$

which yields the $\mathcal{B}(D) \times \mathcal{G}_t$-measurability. □

Chapter 5

KLE and its transformation

5.1 Kernel theorem for subdomains of $\mathbb{H}$

In this section, we introduce the concept of kernel for subdomains of $\mathbb{H}$ and consider the convergence of univalent functions which are not necessarily defined on a common domain. In particular, we establish a version of the classical Carathéodory kernel theorem, which relates the locally uniform convergence of canonical maps to the kernel convergence of their images. Such a geometric consideration will be employed in the next section in the study of Komatu–Loewner evolutions.

First of all, we define the kernel for subdomains of $\mathbb{H}$.

Definition 5.1.1. Let D_n, $n \in \mathbb{N}$, be subdomains of $\mathbb{H}$ such that

$$\bigcap_{n\in\mathbb{N}} D_n \supset \mathbb{H} \setminus \overline{B_\rho}, \tag{5.1.1}$$

for some $\rho > 0$. We define the *kernel* of $(D_n)_{n\in\mathbb{N}}$ as the largest unbounded domain D with the following property: for every compact subset $A \subset D$, there exists $n_A \in \mathbb{N}$ so that $A \subset \bigcap_{n\geq n_A} D_n$.

Assumption (5.1.1) ensures that the kernel D exists uniquely and contains $\mathbb{H}\setminus\overline{B_\rho}$. It is easy to see that the kernel D of $\{D_n\}$ is the unique unbounded connected component of the open set E consisting of all the points $z \in \mathbb{H}$ such that $B(z, \varepsilon_z) \subset \bigcap_{n\geq n_z} D_n$ for some $\varepsilon_z > 0$ and $n_z \in \mathbb{N}$.

Choose any point $w \in \mathbb{H} \setminus \overline{B}_\rho$. The kernel D of $\{D_n\}$ so defined coincides with the kernel of $\{D_n\}$ with respect to the point w in the classical sense (cf. [Conway (1995), Definition 15.4.1]) because D is just the connected component of the open set E containing w. In this book, we always assume (5.1.1) instead of specifying such a base point w.

The following proposition is straightforward from definition:

Proposition 5.1.2. *The kernel of a sequence of subdomains $(D_n)_{n\in\mathbb{N}}$ of $\mathbb{H}$ is contained in the kernel of any subsequence $(D_{n_k})_{k\in\mathbb{N}}$ of $(D_n)_{n\in\mathbb{N}}$.*

Example 5.1.3. Let

$$D_n = \begin{cases} \mathbb{H} & \text{if } n \text{ is even,} \\ \mathbb{H} \setminus \overline{B_1} & \text{if } n \text{ is odd.} \end{cases}$$

The kernel of $(D_n)_{n\in\mathbb{N}}$ is then $\mathbb{H} \setminus \overline{B_1}$, while the kernel of a subsequence $(D_{2n})_{n\in\mathbb{N}}$ is $\mathbb{H}$.

Definition 5.1.4. Let $\{D_n,\ n \in \mathbb{N}\}$ be a sequence of subdomains of $\mathbb{H}$ having property (5.1.1) for some $\rho > 0$ and f_n be an analytic function in D_n for each $n \in \mathbb{N}$. Suppose that a domain D is contained in the kernel of $(D_n)_{n\in\mathbb{N}}$.

(i) The sequence $(f_n)_{n\in\mathbb{N}}$ is said to be *locally bounded* in D if, for any compact subset $A \subset D$ and $n_A \in \mathbb{N}$ with $A \subset \bigcap_{n \geq n_A} D_n$, $(f_n)_{n\geq n_A}$ is uniformly bounded in A. The *locally uniform convergence* of $(f_n)_{n\in\mathbb{N}}$ to f is defined in the same way, and will be abbreviated to

$$f_n \to f \quad \text{(luc)} \quad \text{on} \quad D.$$

(ii) The sequence $(f_n)_{n\in\mathbb{N}}$ is said to be a *normal family* in D if every subsequence of $(f_n)_{n\in\mathbb{N}}$ has a sub-subsequence converging locally uniformly in D.

Note that the convergence of subsequences in D makes sense in view of Proposition 5.1.2. We can also easily check the following criterion for convergence.

Proposition 5.1.5. *$f_n \to f$ (luc) in D if and only if any subsequence of $(f_n)_{n\in\mathbb{N}}$ has a sub-subsequence that converges to f locally uniformly in D.*

The case in which $D_n = D$ for all $n \in \mathbb{N}$ is very familiar to us, and, even if D_n depends on n, well-known theorems listed below hold in the same form. Since their proof is almost the same as in the classical case, we leave the details to the reader.

In the next three theorems, $\{D_n,\ n \in \mathbb{N}\}$ is a sequence of subdomains of $\mathbb{H}$ having property (5.1.1) for some $\rho > 0$ and D is its kernel. For each $n \in \mathbb{N}$, f_n is an analytic function in D_n.

Theorem 5.1.6 (Montel). *$(f_n)_{n\in\mathbb{N}}$ is locally bounded in D if and only if it is a normal family in D (cf. [Conway (1978) Theorem 2.9 in Chapter 7]).*

Theorem 5.1.7 (Vitali). *Let $(f_n)_{n\in\mathbb{N}}$ be locally bounded in D. If there exists a subset $A \subset D$ having an accumulation point in D such that $f_n(z)$ converges as $n \to \infty$ for every $z \in A$, then $(f_n)_{n\in\mathbb{N}}$ converges locally uniformly in D (cf. [Conway (1978), Exercise 4 in §2 of Chapter 7]).*

Theorem 5.1.8 (Hurwitz). *Let f_n be univalent for each $n \in \mathbb{N}$. If $f_n \to f$ (luc) in D and f is non-constant, then f is univalent in D (cf. [Conway (1978), Exercise 10 in §2 of Chapter 7]).*

We now focus on the case where the above f_n's are canonical maps from D_n's. In order to avoid inessential complexities, we extract some properties of canonical maps and define a set $\mathcal{U}(\rho)$ (or $\mathcal{U}_\rho$) for $\rho > 0$ as the totality of conformal mappings $f\colon D \to \widetilde{D}$ with the following properties:

$$D \text{ and } \widetilde{D} \text{ are domains contained in } \mathbb{H}, \quad \text{and} \quad \mathbb{H} \setminus \overline{B_\rho} \subset D; \tag{5.1.2}$$

$$\lim_{z\to\infty}(f(z) - z) = 0; \tag{5.1.3}$$

$$\lim_{z\to\xi} \Im f(z) = 0 \quad \text{for any } \xi \in \partial\mathbb{H} \setminus \overline{B_\rho}. \tag{5.1.4}$$

We also define $\Sigma_0(\rho)$ as the totality of univalent functions g in $\mathbb{C} \setminus \overline{B_\rho}$ with $\lim_{z\to\infty}(g(z) - z) = 0$. The last property of g is equivalent to the Laurent expansion

$$g(z) = z + \sum_{k\geq 1} a_{-k} z^{-k}, \quad z \in \mathbb{C} \setminus \overline{B_\rho}. \tag{5.1.5}$$

The connection between $\mathcal{U}(\rho)$ and $\Sigma_0(\rho)$ is described as follows: Every $f \in \mathcal{U}(\rho)$ extends to a univalent function in $D \cup \Pi D \cup (\mathbb{C} \setminus \overline{B_\rho})$ by the Schwarz reflection principle, and $f|_{\mathbb{C}\setminus\overline{B_\rho}} \in \Sigma_0(\rho)$.

Lemma 5.1.9. *Let $g \in \Sigma_0(\rho)$. Then*

$$g(\mathbb{C} \setminus \overline{B_\rho}) \supset \mathbb{C} \setminus \overline{B_{2\rho}}; \tag{5.1.6}$$

$$|g(z)| \leq 2|z| \quad \text{for } z \in \mathbb{C} \setminus \overline{B_\rho}. \tag{5.1.7}$$

In particular, if $f\colon D \to \widetilde{D}$ belongs to $\mathcal{U}(\rho)$, then $|f(z)| \leq 2(|z| \vee \rho)$ for all $z \in D$.

Proof. Let $\zeta \in \mathbb{C} \setminus g(\mathbb{C} \setminus \overline{B_\rho})$. The function $h_1(z) := \rho(g(\rho z^{-1}) - \zeta)^{-1}$ of $z \in B_1$ is then univalent. Since $g(\rho z^{-1}) = \rho z^{-1} + O(z)$ $(z \to 0)$, it follows that

$$h_1(z) = \frac{z}{1 - \rho^{-1}\zeta z + O(z^2)} = z + \frac{\zeta}{\rho} z^2 + O(z^3) \qquad \text{as } z \to 0.$$

By virtue of this expression, we can apply Bieberbach's theorem (see, e.g., [Conway (1995), Theorem 7.7 in Chapter 14]) to get $|\zeta| \leq 2\rho$, which yields (5.1.6).

Next, let $w \in \mathbb{C}\setminus\overline{B_\rho}$. We consider the function $h_2(z) := w^{-1}g(wz)$ of $z \in \mathbb{C}\setminus B_1$. Since $h_2|_{\mathbb{C}\setminus\overline{B_1}} \in \Sigma_0(1)$, we obtain $h_2(1) \in \overline{B_2}$ from (5.1.6). Hence $|g(w)| \leq 2|w|$, which gives (5.1.7). □

Proposition 5.1.10. *$\Sigma_0(\rho)$ is compact under the topology of locally uniform convergence in $\mathbb{C} \setminus \overline{B_\rho}$.*

Proof. $\Sigma_0(\rho)$ is locally bounded in $\mathbb{C}\setminus\overline{B_\rho}$ by (5.1.7) and hence normal (i.e., relatively compact) by Theorem 5.1.6. To show that it is further closed, we use the following simple observation: for a holomorphic function h in $\mathbb{C} \setminus \overline{B_\rho}$, the coefficients of its Laurent expansion $h(z) = \sum_{k\in\mathbb{Z}} a_k[h] z^k$ are given by

$$a_k[h] = \frac{1}{2\pi i}\int_{\partial B_{2\rho}} \frac{h(z)}{z^{k+1}}\,dz, \quad k \in \mathbb{Z}. \tag{5.1.8}$$

Now, let $(g_n)_{n\in\mathbb{N}}$ be a sequence in $\Sigma_0(\rho)$ which converges to some g locally uniformly in $\mathbb{C}\setminus\overline{B_\rho}$. It follows from (5.1.8) that $\lim_{n\to\infty} a_k[g_n] = a_k[g]$ for every $k \in \mathbb{Z}$. Hence $g(z)$ has the Laurent expansion of the form (5.1.5). Moreover, g is univalent in $\mathbb{C}\setminus\overline{B_\rho}$ by Hurwitz's theorem because it is not constant. Thus, $g \in \Sigma_0(\rho)$, which implies that $\Sigma_0(\rho)$ is closed. □

Proposition 5.1.11. *Any sequence of conformal mappings $f_n\colon D_n \to \widetilde{D}_n$, $n \in \mathbb{N}$, in $\mathcal{U}(\rho)$ has a subsequence which converges locally uniformly in the kernel D of $(D_n)_{n\in\mathbb{N}}$, and the limit $f\colon D \to f(D)$ belongs to $\mathcal{U}(\rho)$.*

Proof. The sequence $(f_n)_{n\in\mathbb{N}}$ is locally bounded in D by Lemma 5.1.9 and hence normal in D by Theorem 5.1.6. Therefore, it remains to show that the limit of any convergent subsequence of $(f_n)_{n\in\mathbb{N}}$ belongs to $\mathcal{U}(\rho)$. To this end, it suffices to prove that, if $(f_n)_{n\in\mathbb{N}}$ itself is convergent, then its limit f belongs to $\mathcal{U}(\rho)$. Note that, in this case, $(f_n|_{\mathbb{C}\setminus\overline{B_\rho}})_{n\in\mathbb{N}}$ is a sequence in $\Sigma_0(\rho)$ and has a subsequence which converges to some $g \in \Sigma_0(\rho)$ locally uniformly in $\mathbb{C}\setminus\overline{B_\rho}$ by Proposition 5.1.10. It is then clear that $f = g$ in $\mathbb{H}\setminus\overline{B_\rho}$. From this we can easily deduce that $f\colon D \to f(D)$ enjoys (5.1.2), (5.1.3) and (5.1.4). Hence $f \in \mathcal{U}(\rho)$. □

Lemma 5.1.12. *Let $f\colon D \to \widetilde{D}$ be a conformal mapping in $\mathcal{U}(\rho)$ for some $\rho > 0$. Then*

(i) $\mathbb{H}\setminus\overline{B_{2\rho}} \subset f(\mathbb{H}\setminus\overline{B_\rho}) \subset \widetilde{D}$;
(ii) $f^{-1}\colon \widetilde{D} \to D$ *is in* $\mathcal{U}(2\rho)$.

Proof. (i) follows immediately from (5.1.6). We prove (ii) here. Among the properties (5.1.2)–(5.1.4) defining $\mathcal{U}(2\rho)$, the first one has already been obtained in (i). As for the second one, note that $\lim_{\zeta\to\infty} f^{-1}(\zeta) = \infty$. We have

$$\lim_{\zeta\to\infty}(f^{-1}(\zeta) - \zeta) = \lim_{z\to\infty}(z - f(z)) = 0.$$

In order to get the last property (5.1.4) with ρ replaced by 2ρ, we note that, for any $\rho' > \rho$, the interval $[\rho', +\infty)$, which is regarded as a subset of $\partial\mathbb{H}$, is mapped onto $[f(\rho'), +\infty)$ by f. This is because $[\rho', +\infty) \ni r \mapsto f(r) \in \partial\mathbb{H}$ is a simple curve and because $\lim_{r\to+\infty} f(r) = +\infty$ in $\mathbb{R}$ follows from the Laurent expansion of f. Moreover, $f(\rho') \leq 2\rho'$ by (5.1.7). Since $\rho' > \rho$ is arbitrary, we have $f^{-1}((2\rho, +\infty)) \subset (\rho, +\infty) \subset \partial\mathbb{H}$. In the same way, we see that $f^{-1}((-\infty, -2\rho)) \subset (-\infty, -\rho) \subset \partial\mathbb{H}$. Thus, we obtain (5.1.4) with ρ replaced by 2ρ. □

We now state part of the kernel theorem for space $\mathcal{U}(\rho)$.

Proposition 5.1.13. *Let $f_n\colon D_n \to \widetilde{D}_n$, $n \in \mathbb{N}$, be conformal mappings which belong to $\mathcal{U}(\rho)$ for some $\rho > 0$ and D be the kernel of $(D_n)_{n\in\mathbb{N}}$. Suppose that $(f_n)_{n\in\mathbb{N}}$ converges to some f locally uniformly in D. Then $f(D)$ is the kernel of $(\widetilde{D}_n)_{n\in\mathbb{N}}$, and $f_n^{-1} \to f^{-1}$ as $n \to \infty$ locally uniformly in $f(D)$.*

Note that by Proposition 5.1.11, the limiting function f is a conformal map in $\mathcal{U}(\rho)$. The main step of the proof of Proposition 5.1.13 is the following lemma.

Lemma 5.1.14. *Under the assumption of Proposition* 5.1.13, *let* $\widetilde{D}$ *be the kernel of* $(\widetilde{D}_n)_{n\in\mathbb{N}}$. *Then*

(i) $f(D) \subset \widetilde{D}$;
(ii) $(f_n^{-1})_{n\in\mathbb{N}}$ *converges to some univalent function* g *locally uniformly in* $\widetilde{D}$, *and* $g|_{f(D)} = f^{-1}$.

Proof. (i) Let A be a compact subset of $f(D)$. We take a relatively compact subdomain V of $f(D)$ with smooth boundary so that $A \subset V$. Then $\delta := 2^{-1}\operatorname{dist}(A, \partial V) > 0$. By definition,

$$|f(z) - \zeta| > \delta \quad \text{for } z \in \partial f^{-1}(V) \text{ and } \zeta \in A. \tag{5.1.9}$$

Since $f_n \to f$ uniformly in the compact set $\overline{f^{-1}(V)} \subset D$, there exists $n_A \in \mathbb{N}$ such that

$$|f_n(z) - f(z)| < \delta \quad \text{for } n \geq n_A \text{ and } z \in \overline{f^{-1}(V)}. \tag{5.1.10}$$

By virtue of (5.1.9), (5.1.10), and the identity

$$f_n(z) - \zeta = (f(z) - \zeta) + (f_n(z) - f(z)),$$

Rouché's theorem implies that, for every $\zeta \in A$ and $n \geq n_A$, the function $z \mapsto f_n(z) - \zeta$ has exactly one zero in $f^{-1}(V)$. In particular, $A \subset f_n(f^{-1}(V)) \subset \widetilde{D}_n$ for any $n \geq n_A$. Since $A \subset f(D)$ was arbitrary, we obtain $f(D) \subset \widetilde{D}$.

(ii) $(f_n^{-1})_{n\in\mathbb{N}}$ is a sequence in $\mathcal{U}(2\rho)$ by Lemma 5.1.12 and hence a normal family in $\widetilde{D}$ by Proposition 5.1.11. Let $(f_{n_k}^{-1})_{k\in\mathbb{N}}$ be a convergent subsequence. By Proposition 5.1.11, its limit g is a univalent function on $\widetilde{D}$ in $\mathcal{U}(2\rho)$. We claim that $g|_{f(D)} = f^{-1}$. Let $z \in D$. Since $f_{n_k}(z) \to f(z)$ as $k \to \infty$, there exists $k_0 \in \mathbb{N}$ such that $A := \{f_{n_k}(z) : k \geq k_0\} \cup \{f(z)\}$ is a compact subset of $f(D)$. Then

$$\begin{aligned} |g(f(z)) - z| &\leq \lim_{k\to\infty} |g(f(z)) - g(f_{n_k}(z))| + \limsup_{k\to\infty} |g(f_{n_k}(z)) - f_{n_k}^{-1}(f_{n_k}(z))| \\ &\leq \lim_{k\to\infty} \sup_{\zeta\in A} |g(\zeta) - f_{n_k}^{-1}(\zeta)| = 0. \end{aligned}$$

Thus, $g(f(z)) = z$ for every $z \in D$; in other words, $g|_{f(D)} = f^{-1}$. Next, let h be the limit of another convergent subsequence of $(f_n^{-1})_{n\in\mathbb{N}}$. Then by the same reasoning, we have $g|_{f(D)} = f^{-1} = h|_{f(D)}$. Since $f(D)$ is open and $\widetilde{D}$ is connected, it follows that $g = h$ in $\widetilde{D}$. This completes the proof of (ii). □

Proof of Proposition 5.1.13. We have shown in Lemma 5.1.14 (ii) that $f_n^{-1} \to g$ as $n \to \infty$ locally uniformly in $\widetilde{D}$. Then by Lemma 5.1.14 (i) with the roles of f_n and f_n^{-1} interchanged, we obtain $g(\widetilde{D}) \subset D$. Since $g|_{f(D)} = f^{-1}$, we have $D = g(f(D)) \subset g(\widetilde{D}) \subset D$ and, in particular, $\widetilde{D} = f(D)$. The remainder of the proposition then follows again from Lemma 5.1.14. □

In order to state the other part of the kernel theorem, we define the convergence of domains D_n, rather than conformal maps f_n on them.

Definition 5.1.15. Let D_n, $n \in \mathbb{N}$, be subdomains of $\mathbb{H}$ satisfying (5.1.1) for some $\rho > 0$. We say that $(D_n)_{n\in\mathbb{N}}$ *converges to its kernel* D if every subsequence of $(D_n)_{n\in\mathbb{N}}$ has the same kernel D. We use $D_n \to D$ as $n \to \infty$ to denote this convergence.

Definition 5.1.15 is called the kernel convergence or the convergence in Carathéodory's sense. We remark that, even if $(D_n)_{n\in\mathbb{N}}$ has D as its kernel, it does not necessarily converge to D. Clearly, the sequence $(D_n)_{n\in\mathbb{N}}$ in Example 5.1.3 does not converge in the kernel sense.

The following proposition treats the case that the sequence $(D_n)_{n\in\mathbb{N}}$ is monotone:

Proposition 5.1.16. *Let $(D_n)_{n\in\mathbb{N}}$ be a sequence of subdomains of $\mathbb{H}$ which enjoy (5.1.1) for some $\rho > 0$.*

(i) *If $(D_n)_{n\in\mathbb{N}}$ is increasing, then $D_n \to \bigcup_{m\in\mathbb{N}} D_m$ as $n \to \infty$.*
(ii) *If $(D_n)_{n\in\mathbb{N}}$ is decreasing, then its kernel is the largest domain D satisfying*

$$\mathbb{H} \setminus \overline{B_\rho} \subset D \subset \bigcap_{n\in\mathbb{N}} D_n.$$

Moreover, $D_n \to D$ as $n \to \infty$.

Proof. (i) Let $D := \bigcup_{n\in\mathbb{N}} D_n$, which is trivially a subdomain of $\mathbb{H}$ which enjoys (5.1.1). We take any subsequence $(D_{n_k})_{k\in\mathbb{N}}$ of $(D_n)_{n\in\mathbb{N}}$. Since $(D_n)_{n\in\mathbb{N}}$ is increasing, $\bigcup_{k\in\mathbb{N}} D_{n_k} = \bigcup_{n\in\mathbb{N}} D_n = D$. By definition, the kernel of $(D_{n_k})_{k\in\mathbb{N}}$ is a subset of D. On the other hand, if A is a compact subset of D, then there exists $k' \in \mathbb{N}$ such that $A \subset D_{n_{k'}} \subset D$, as $(D_{n_k})_{k\in\mathbb{N}}$ is an open covering of A. Thus, D is exactly the kernel of $(D_{n_k})_{k\in\mathbb{N}}$. This shows that $D_n \to D$ as $n \to \infty$.

(ii) Let D' be the largest domain in $E := \bigcap_{n\in\mathbb{N}} D_n$ that contains $\mathbb{H} \setminus \overline{B_\rho}$. (Notice that E itself may not be connected or open.) We take any subsequence $(D_{n_k})_{k\in\mathbb{N}}$ of $(D_n)_{n\in\mathbb{N}}$. Since $(D_n)_{n\in\mathbb{N}}$ is decreasing, $\bigcap_{k\in\mathbb{N}} D_{n_k} = \bigcap_{n\in\mathbb{N}} D_n = E$. Every compact subset of D' is trivially a subset of E, and thus D' is contained in the kernel of $(D_{n_k})_{k\in\mathbb{N}}$ by definition.

We now assume that D' does not coincide with the kernel. Since the kernel is a domain containing $\mathbb{H} \setminus \overline{B_\rho}$, it must have a point $z \notin E$ by the maximality of D'. Then for every $k' \in \mathbb{N}$, the compact set $\{z\}$ does not lie in $\bigcap_{k\geq k'} D_{n_k}$, because the last intersection is again E. However, this contradicts Definition 5.1.1. Hence D' coincides with the kernel D of $(D_{n_k})_{k\in\mathbb{N}}$. This means that $D = D'$ and $D_n \to D$. □

We now establish the kernel theorem for $\mathcal{U}(\rho)$ in the form employed in the next section.

Theorem 5.1.17. *Let $f_n\colon D_n \to \widetilde{D}_n$, $n \in \mathbb{N}$, be conformal mappings belonging to $\mathcal{U}(\rho)$ for some $\rho > 0$. Suppose that $D_n \to D$ as $n \to \infty$.*

(i) *If*

$$(f_n)_{n\in\mathbb{N}} \text{ converges to some } f \text{ locally uniformly in } D, \tag{5.1.11}$$

then

$$(\widetilde{D}_n)_{n\in\mathbb{N}} \text{ converges to its kernel } \widetilde{D}. \tag{5.1.12}$$

In this case, $\widetilde{D} = f(D)$, and $f_n^{-1} \to f^{-1}$ locally uniformly in $f(D)$.

(ii) *If D is finitely connected, then* (5.1.12) *implies* (5.1.11).

Proof. (i) Assume (5.1.11). For any subsequence $(n_k)_{k\in\mathbb{N}} \subset \mathbb{N}$, the sequence $(D_{n_k})_{k\in\mathbb{N}}$ has D as its kernel by assumption. Hence $(\widetilde{D}_{n_k})_{k\in\mathbb{N}}$ has $\widetilde{D} = f(D)$ as its kernel by Proposition 5.1.13. Thus, (5.1.12) holds. The other statement also follows from Proposition 5.1.13.

(ii) Suppose that D is finitely connected and that (5.1.12) holds. Since D enjoys (5.1.1), it must be of the form $(\mathbb{H} \setminus \bigcup_{j=1}^N A_j) \setminus F$, where A_j, $j = 1, 2, \ldots, N$, are disjoint compact continua, F is an $\mathbb{H}$-hull with $F \subset \mathbb{H} \setminus \bigcup_{j=1}^N A_j$. Hence there exists a unique canonical map g_F from D onto a standard slit domain by Proposition 2.1.2.

Now, assume that (5.1.11) is false. By Proposition 5.1.11, there exist at least two subsequences which converge to distinct limits $f^{(1)}, f^{(2)} \in \mathcal{U}(\rho)$ locally uniformly in D. By virtue of (i) applied to these two subsequences, both $f^{(1)}$ and $f^{(2)}$ are conformal mappings from D onto $\widetilde{D}$. Then, as in the proof of Lemma 5.1.12, we can see that the composite $g_F \circ (f^{(1)})^{-1} \circ f^{(2)}$ is a canonical map from D onto a standard slit domain, which must coincide with g_F itself by the uniqueness. This implies $f^{(1)} = f^{(2)}$, a contradiction. Hence, we obtain (5.1.11). □

We close this section with a remark on continuous parameters. For a family of domains $D_t \subset \mathbb{H}$, $t \in \mathbb{R}$, which enjoy (5.1.1) for some $\rho > 0$, we say that D_t *converges* to some D as $t \to \alpha$ if $D_{t_n} \to D$ as $n \to \infty$ for any sequence $(t_n)_{n\in\mathbb{N}}$ with $t_n \to \alpha$. Then, with an obvious modification, Theorem 5.1.17 holds also for the continuous parameter t in place of discrete parameter n.

5.2 Continuously growing hulls and KLE

Recall the notion of the Komatu-Loewner evolution (KLE) $\{F_t\}$ on a standard slit domain D driven by a continuous real-valued function $\xi(t)$ introduced at the end of Section 3.2. The aim of this section is to give a geometric characterization of a family $\{F_t\}$ of growing hulls to be a KLE by using the kernel convergence.

Here is our setting. Let $\{F_t\}_{t\in[0,t_0)}$ be a family of growing hulls in a standard slit domain $D = \mathbb{H} \setminus K$, where $K := \bigcup_{j=1}^N C_j$. Let $g_t = g_{F_t}\colon D \setminus F_t \to D_t$ be the corresponding canonical map with half-plane capacity $a_t = \mathrm{hcap}^D(F_t)$. Denote the slits of D_t by $\{C_j(t),\ 1 \leq j \leq N\}$. The labels of these slits are put so that $g_t(C_j^p) = C_j(t)^p$ holds in the sense of boundary correspondence described in the last two paragraphs of Section 2.1.1. The endpoints of $C_j(t)$, $1 \leq j \leq N$, are specified by $\mathbf{s}(t) \in \mathcal{S}$ with the space $\mathcal{S}$ defined by (3.1.1). We also set $K(t) := \bigcup_{j=1}^N C_j(t)$. Moreover, we define

$$g_{t,s} = g_s \circ g_t^{-1} \colon D_t \to D_s \setminus g_s(F_t \setminus F_s) \quad \text{for } 0 \leq s < t < t_0,$$

whose inverse function $g_{t,s}^{-1}$ coincides with the canonical map $g^{D_s}_{g_s(F_t \setminus F_s)}$ of the hull $g_s(F_t \setminus F_s)$ in D_s; see Figure 5.1. Here, the hull property of $g_s(F_t \setminus F_s)$ is observed as follows: The domain $(D \setminus F_s) \setminus F_t = D \setminus F_t$ is $(N+1)$-connected. Since the conformal map $g_s \colon D \setminus F_s \to D_s$ preserves the degree of connectivity, $D_s \setminus g_s(F_t \setminus F_s) = g_s((D \setminus F_s) \setminus F_t)$ is also $(N+1)$-connected. On account of the boundary correspondence by g_s, the N inner boundary components of $D_s \setminus g_s(F_t \setminus F_s)$ are the slits of D_s. Filling the slits results in a simply connected domain $\mathbb{H} \setminus g_s(F_t \setminus F_s)$. Hence $g_s(F_t \setminus F_s)$ is an $\mathbb{H}$-hull.

We can apply the results in Section 5.1 to this setting. In fact, let $\rho_t := \sup\{|z| : z \in F_t \cup K\}$, which is non-decreasing in t by definition. The function g_t then belongs

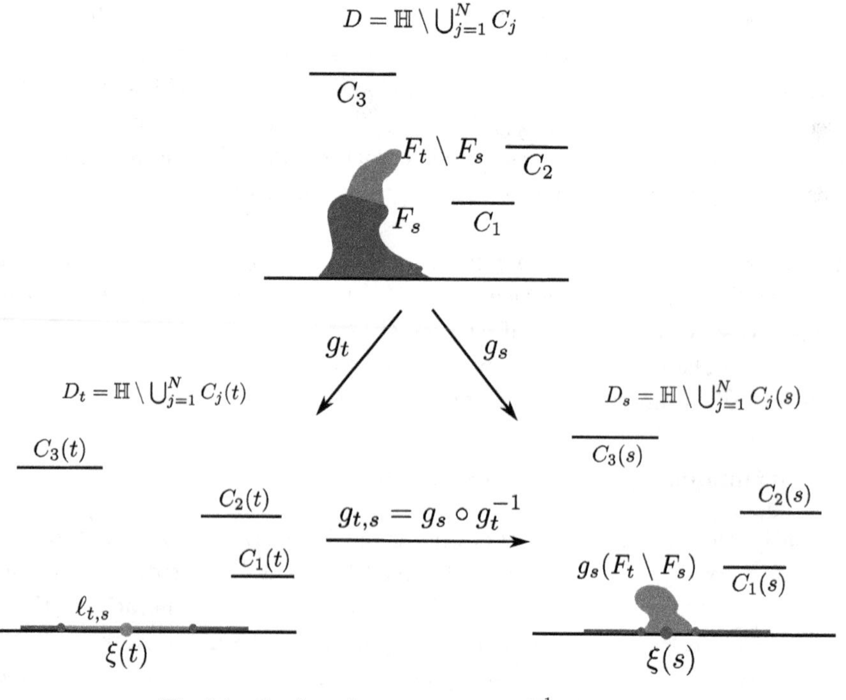

Fig. 5.1 Conformal map $g_{t,s} = g_s \circ g_t^{-1} : D_t \to D_s$.

to $\mathcal{U}(\rho_t)$, and $\mathbb{H} \setminus \overline{B_{2\rho_t}} \subset D_t$ holds by (5.1.6). Further, for $0 \le s < t < t_0$, (5.1.6) with $g = g_s$ and $\rho = \rho_t$ yields that

$$\mathbb{H} \setminus \overline{B_{2\rho_t}} \subset g_s(D \setminus \overline{B_{\rho_t}}) \subset g_s(D \setminus F_t) = D_s \setminus g_s(F_t \setminus F_s).$$

Thus, the functions g_s, g_t, $g_{t,s}$ and their inverse functions all belong to $\mathcal{U}(2\rho_t)$.

We now introduce the following concept of continuity of growing hulls.

Definition 5.2.1. We say that $\{F_t\}_{t\in[0,t_0)}$ is *left* (resp. *right*) *continuous in* D *at* $s \in (0, t_0)$ (resp. $s \in [0, t_0)$) if $D \setminus F_t \to D \setminus F_s$ as $t \uparrow s$ (resp. $t \downarrow s$) in the sense of kernel convergence introduced in Section 5.1.

Recall that there is another right continuity (3.2.27) with limit point, which appeared as a property of KLEs at the end of Section 3.2. Namely, the family $\{F_t\}_{t\in[0,t_0)}$ of growing hulls is said to be right continuous at $t \in [0, t_0)$ with limit $\xi(t) \in \partial\mathbb{H}$ if

$$\bigcap_{\varepsilon>0} \overline{g_t(F_{t+\varepsilon} \setminus F_t)} = \{\xi(t)\}. \tag{5.2.1}$$

If (5.2.1) holds for every $t \in [0, t_0)$, by identifying a point $x + i0 \in \partial\mathbb{H}$ with the real number $x \in \mathbb{R}$ we may view $\xi(t)$ to be a real-valued function of t and say that $\{F_t\}_{t\in[0,t_0)}$ *has a driving function* $\xi(t)$. (Note that we do not assume the continuity of $\xi(t)$ in the definition of driving function.) The next lemma shows that the property (5.2.1) is stronger than the right continuity of $\{F_t\}$ in the sense of Definition 5.2.1.

Lemma 5.2.2. *Suppose that* $\{F_t\}_{t\in[0,t_0)}$ *has a driving function* $\xi(t)$, *i.e., enjoys* (5.2.1) *for all* $t \in [0, t_0)$. *Then the following are true:*

(i) *The half-plane capacity* a_t *is strictly increasing and right continuous in* t.
(ii) $\{F_t\}$ *is right continuous in* D *at every* t *in the sense of Definition 5.2.1.*

Proof. (i) Let $\varepsilon > 0$. As mentioned above, we have

$$g_{t+\varepsilon}(z) - g_t(z) = g^{D_t}_{g_t(F_{t+\varepsilon}\setminus F_t)}(g_t(z)) - g_t(z), \quad z \in D, \tag{5.2.2}$$

which implies

$$a_{t+\varepsilon} - a_t = \mathrm{hcap}^{D_t}(g_t(F_{t+\varepsilon} \setminus F_t)), \tag{5.2.3}$$

where the right-hand side denotes the half-plane capacity of $g_t(F_{t+\varepsilon} \setminus F_t)$ relative to D_t. It follows from this and Theorem 2.2.4 (iii) that a_t is strictly increasing.

By (5.2.3) and Theorem 2.2.4 (ii), we have the expression

$$a_{t+\varepsilon} - a_t = \frac{2\rho}{\pi} \int_0^\pi \mathbb{E}^{t,*}_{\rho e^{i\theta}} \left[\Im Z^{t,*}_{\sigma_{g_t(F_{t+\varepsilon}\setminus F_t)}} ; \sigma_{g_t(F_{t+\varepsilon}\setminus F_t)} < \infty \right] \sin\theta d\theta,$$

in terms of the BMD $\mathbb{Z}^{t,*} = (Z^{t,*}_s, \mathbb{P}^{t,*}_z)$ on D^*_t, where $\rho > 0$ is positive with $B(0,\rho) \supset g_t(F_{t+\varepsilon} \setminus F_t) \cup K(t)$. Since $\Im Z^{t,*}_{\sigma_{g_t(F_{t+\varepsilon}\setminus F_t)}} \mathbb{1}_{\{\sigma_{g_t(F_{t+\varepsilon}\setminus F_t)}<\infty\}}$ is uniformly bounded and tends to 0 as $\varepsilon \downarrow 0$ on account of (5.2.1), we get the right continuity of a_t; $\lim_{\varepsilon\downarrow 0} a_{t+\varepsilon} = a_t$.

(ii) By the preceding observation, we have, for a fixed $t_1 \in (t, t_0)$,

$$\{g_{t+\varepsilon,t}^{-1}\}_{\varepsilon\in(0,t_1-t]} \subset \mathcal{U}(2\rho_{t_1}). \tag{5.2.4}$$

Accordingly, by virtue of (i) and the uniform bound (2.2.16), we get

$$g_{t+\varepsilon,t}^{-1}(z) = g_{g_t(F_{t+\varepsilon}\setminus F_t)}^{D_t}(z) \to z \quad \text{as } \varepsilon \downarrow 0 \text{ for } z \in \mathbb{H} \setminus \overline{B_{6\rho_{t_1}}}. \tag{5.2.5}$$

On the other hand, since $D_t \setminus g_t(F_{t+\varepsilon} \setminus F_t) \to D_t$ as $\varepsilon \downarrow 0$ by Proposition 5.1.16 (i), (5.2.4) combined with Lemma 5.1.9 implies that the family $\{g_{t+\varepsilon,t}^{-1}\}_{\varepsilon\in(0,t_1-t]}$ is locally bounded in the kernel D_t. Therefore, it follows from (5.2.5) and Vitali's theorem that $g_{t+\varepsilon,t}^{-1}(z) \to z$ (luc) in the whole D_t.

Theorem 5.1.17 (i) applied to $g_{t+\varepsilon,t}^{-1}$ now implies that $D_{t+\varepsilon} \to D_t$ and $g_{t+\varepsilon,t}(z) \to z$ (luc) in D_t as $\varepsilon \downarrow 0$. In particular, the latter gives

$$g_{t+\varepsilon}^{-1} = g_t^{-1} \circ g_{t+\varepsilon,t} \to g_t^{-1} \text{ (luc) in } D_t.$$

We then apply Theorem 5.1.17 (i) to $g_{t+\varepsilon}^{-1}$ to obtain $D \setminus F_{t+\varepsilon} \to D \setminus F_t$ as $\varepsilon \downarrow 0$, which proves the right continuity of $\{F_t\}$ in D. □

The next two lemmas are direct consequences of Definition 5.2.1, which are fundamental in our approach based on kernel convergence.

Lemma 5.2.3.

(i) *Assume that* $\{F_s\}_{s\in[0,t_0)}$ *is left continuous in* D *at* $t \in (0, t_0)$*. Then, as* $s \uparrow t$*,* $g_s \to g_t$ (luc) *on* $D \setminus F_t$*,* $\mathbf{s}(s) \to \mathbf{s}(t)$ *in* $\mathcal{S}$*,* $D_s \to D_t$*,* $g_s^{-1} \to g_t^{-1}$ (luc) *on* D_t*, and* $a_s \to a_t$*.*

(ii) *Assume that* $\{F_t\}_{t\in[0,t_0)}$ *is right continuous in* D *at* $s \in [0, t_0)$*. Then, as* $t \downarrow s$*,* $g_t \to g_s$ (luc) *on* $D \setminus F_s$*,* $\mathbf{s}(t) \to \mathbf{s}(s)$ *in* $\mathcal{S}$*,* $D_t \to D_s$*,* $g_t^{-1} \to g_s^{-1}$ (luc) *on* D_s*, and* $a_t \to a_s$*.*

Before proving Lemma 5.2.3, let us make a technical remark. In the second paragraph of this section, we have defined $\mathbf{s}(t)$ via the boundary correspondence $C_j^{\mathfrak{p}}(\mathbf{s}(t)) = g_t(C_j^{\mathfrak{p}})$. This definition does not say any continuity of $\mathbf{s}(t)$ in $t \in [0, t_0)$. In such a general situation, we should note that $D_s \to D_t$ $(s \to t)$ does not imply $\mathbf{s}(s) \to \mathbf{s}(t)$ in $\mathcal{S}$. (The converse implication is clearly true.)

To understand what a problem is, we consider the example in which $\widetilde{\mathbf{s}} = \lim_{s\to t} \mathbf{s}(s)$ exists and $D_s \to D_t$ holds. Then $D(\widetilde{\mathbf{s}}) = D_t = D(\mathbf{s}(t))$ follows, but this identity implies only the coincidence of the "unlabelled" slits $\bigcup_{j=1}^N C_j(\widetilde{\mathbf{s}}) = \bigcup_{j=1}^N C_j(\mathbf{s}(t))$. Since the permutation of the labels is possible, one cannot conclude the identity of the labelled slits $\widetilde{\mathbf{s}} = \mathbf{s}(t)$.

In the proof of Lemma 5.2.3, we make use of the left (or right) continuity of g_t, which itself is a conclusion of the lemma, to show that the permutation of the labels does not occur.

Proof of Lemma 5.2.3. We shall only present a proof of (i). The proof of (ii) is quite similar.

We first show that $g_s \to g_t$ (luc) on $D \setminus F_t$ as $s \uparrow t$. From the third paragraph of this section, it follows that $\bigcup_{s\in[0,t]} K(s) \subset \overline{B_{2\rho_t}}$. In particular, $\{\mathbf{s}(s), s \in [0,t]\}$ is bounded in $\mathbb{R}^{3N}$. Now let us consider an arbitrary sequence in $[0,t]$ increasing to t. We extract its subsequence $(s_n)_{n\in\mathbb{N}}$ convergent in $\overline{\mathcal{S}}$ (the closure is taken in $\mathbb{R}^{3N}$). The limit $\widetilde{\mathbf{s}} := \lim_{n\to\infty} \mathbf{s}(s_n)$ does not necessarily belong to $\mathcal{S}$. Nevertheless, $\widetilde{\mathbf{s}}$ corresponds to a "slit" domain $D(\widetilde{\mathbf{s}})$ some of whose slits may degenerate, i.e., be absorbed in another slit or reduced to singletons. As is easily seen, the sequence $(D(\mathbf{s}(s_n)))_n$ has $D(\widetilde{\mathbf{s}})$ as its kernel, and $D(\mathbf{s}(s_n)) \to D(\widetilde{\mathbf{s}})$. Also $D \setminus F_s \to D \setminus F_t$ as $s \uparrow t$ by assumption. Thus by Theorem 5.1.17 (ii), $g_{s_n} \in \mathcal{U}(\rho_t)$ converges to a conformal mapping $\widetilde{g}$ (luc) on $D \setminus F_t$ as $n \to \infty$, and $\widetilde{g}(D \setminus F_t) = D(\widetilde{\mathbf{s}})$. Since conformal mappings preserve the non-degeneracy of multiply connected domains, $D(\widetilde{\mathbf{s}})$ is a non-degenerate slit domain. Also $\widetilde{g} \in \mathcal{U}(\rho_t)$ by Proposition 5.1.11. Thus, $\widetilde{g}$ is a canonical map from $D\setminus F_t$. The uniqueness of canonical maps (Proposition 2.1.2) yields $\widetilde{g} = g_t$. In summary, we have shown that any sequence in $[0,t]$ increasing to t has a subsequence $(s_n)_{n\in\mathbb{N}}$ such that $g_{s_n} \to g_t$ (luc) on $D \setminus F_t$. Hence $g_s \to g_t$ (luc) on $D \setminus F_t$ as $s \uparrow t$.

We next show that $\mathbf{s}(s) \to \mathbf{s}(t)$ in $\mathcal{S}$. To this end, it suffices to prove $\widetilde{\mathbf{s}} = \mathbf{s}(t)$ for the subsequential limit $\widetilde{\mathbf{s}}$ in the preceding paragraph. As $D(\widetilde{\mathbf{s}}) = g_t(D \setminus F_t) = D(\mathbf{s}(t))$, the unlabelled slits coincide with each other. We shall confirm that the labels are not permuted.

Fix $j \in \{1,2,\dots,n\}$. Let η_1 and η_2 be simple closed curves in $D \setminus F_t$ with the properties $C_j \subset \operatorname{ins}\eta_1$, $\overline{\operatorname{ins}\eta_1} \subset \operatorname{ins}\eta_2$, and $\bigcup_{k\neq j} C_k \cap \overline{\operatorname{ins}\eta_2} = \emptyset$. We know that $g_{s_n} \to g_t$ as $n \to \infty$ uniformly on $\overline{\operatorname{ins}\eta_2} \setminus \operatorname{ins}\eta_1$. Hence $\overline{\operatorname{ins} g_{s_n}(\eta_1)} \subset \operatorname{ins} g_t(\eta_2)$ for sufficiently large n. Since we have $C_j(\mathbf{s}(s_n)) \subset \operatorname{ins} g_{s_n}(\eta_1)$, $C_j(\mathbf{s}(t)) \subset \operatorname{ins} g_t(\eta_2)$, and $\bigcup_{k\neq j} C_k(\mathbf{s}(t)) \cap \overline{\operatorname{ins} g_t(\eta_2)} = \emptyset$, it follows that $C_j(\widetilde{\mathbf{s}}) = C_j(\mathbf{s}(t))$. Thus $\widetilde{\mathbf{s}} = \mathbf{s}(t)$.

It is now clear that $D_s \to D_t$ as $s \uparrow t$. One can also see this by applying Theorem 5.1.17 (i) to g_s. By the same theorem, $g_s^{-1} \to g_t^{-1}$ (luc) on D_t as $s \uparrow t$.

Finally, we prove $\lim_{s\uparrow t} a_s = a_t$. We regard g_s as an element of $\Sigma_0(\rho_t)$ by the Schwarz reflection. $\Sigma_0(\rho_t)$ is locally bounded by (5.1.7), and we have already shown that $\lim_{s\uparrow t} g_s(z) = g_t(z)$ for $z \in D \setminus F_t$. Hence g_s converges to g_t locally uniformly in $\mathbb{C} \setminus \overline{B_{\rho_t}}$ by Vitali's theorem. In particular, this convergence is uniform on the circle $\partial B_{2\rho_t}$. Since $a_s = a_{-1}[g_s]$ with $a_k[g_s]$ being the k-th coefficient of the Laurent expansion of $g_s(z)$ around infinity, we have $\lim_{s\uparrow t} a_s = a_t$ by (5.1.8) with $h = g_s$ and $\rho = \rho_t$. □

Lemma 5.2.4. *Suppose that $\{F_t\}_{t\in[0,t_0)}$ be continuous in D over $[0,t_0)$ and $t_1 \in (0,t_0)$. Then the function $(t,z) \mapsto g_t(z)$ is jointly continuous in $[0,t_1] \times (D \setminus F_{t_1}) \cup K^p$.*

Proof. By Lemma 5.2.3, the family $\{g_t\}_{t\in[0,t_1]}$ is continuous with respect to the locally uniform convergence topology in $D \setminus F_{t_1}$. Hence the joint continuity of $g_t(z)$ in $[0,t_1] \times (D \setminus F_{t_1})$ is obvious.

Let $R_+ := \{z \in \mathbb{C} : x_j < \Re z < x_j^r,\ y_j < \Im z < y_j + \delta\}$ with $\delta > 0$ sufficiently small and R_- be the reflection of R_+ across C_j^0. $\{g_t\}_{t\in[0,t_1]} \subset \mathcal{U}(\rho_{t_1})$ is uniformly bounded in R_+ (not "locally") by Lemma 5.1.9. Hence it extends to the uniformly bounded family of analytic functions g_t^+, $t \in [0, t_1]$, in the rectangle $R := R_+ \cup C_j^+ \cup R_-$. Therefore, the continuity of $\{g_t\}_{t\in[0,t_1]}$ with respect to the locally uniform convergence in R_+ implies the continuity of $\{g_t^+\}_{t\in[0,t_1]}$ with respect to the locally uniform convergence in R by virtue of Vitali's theorem. This yields the joint continuity of $g_t(z)$ in $[0, t_1] \times (R_+ \cup C_j^+)$ and similarly in $[0, t_1] \times (R_- \cup C_j^-)$.

Finally, by using the map $\psi(z) = (z - z_j)^{1/2}$, $z \in B(z_j, \varepsilon) \setminus C_j$, appearing in the proof of Theorem 2.3.3, one can show in an analogous manner to the above the joint continuity of $g_t(z)$ in $[0, t_1] \times B(z_j, \varepsilon) \cap (D \cup C_j^p)$. The same property of $g_t(z)$ also holds for the right endpoint z_j^r of C_j in place of z_j. □

For a continuously growing hulls with continuous driving function, we can derive the corresponding Komatu-Loewner equation. To this end, we shall implement almost the same proof as that of Proposition 3.3.2 in Section 3.3. While, in that proof, we employed the joint continuity of the canonical map $g_t(z)$ as a solution to the Komatu-Loewner ODE (see Lemma 3.3.5), we here use the continuity resulting from the continuity of hulls $\{F_t\}$.

Proposition 5.2.5. *Suppose that $\{F_t\}_{t\in[0,t_0)}$ is a family of continuously growing hulls (in the sense of Definition 5.2.1) with continuous driving function $\xi(t)$ in the sense of (5.2.1) and with half-plane capacity $a_t = 2t$. Then the KL equation*

$$\frac{d}{dt} g_t(z) = -2\pi \Psi_{\mathbf{s}(t)}(g_t(z), \xi(t)), \quad g_0(z) = z, \tag{5.2.6}$$

holds for $t \in [0, t_0)$ and $z \in (D \setminus F_t) \cup K^p$.

Proof. Let $0 \le s < t < t_0$ and $\ell_{t,s}$ be the set of all limit points of $g_{t,s}^{-1}(z)$ as z approaches $\overline{g_s(F_t \setminus F_s)}$. Then $\ell_{t,s}$ is a compact subset of $\partial\mathbb{H}$, and $\lim_{z\to x} \Im g_{t,s}(z) = 0$ for $x + i0 \in \partial\mathbb{H} \setminus \ell_{t,s}$. In exactly the same way as the proof of Lemma 3.3.4, we then have

$$g_s(z) - g_t(z) = \int_{\ell_{t,s}} \Psi_{\mathbf{s}(t)}(g_t(z), x) \Im g_{t,s}(x + i0)\, dx, \quad z \in D \setminus F_t, \tag{5.2.7}$$

$$2(s - t) = a_s - a_t = -\frac{1}{\pi} \int_{\ell_{t,s}} \Im g_{t,s}(x + i0)\, dx. \tag{5.2.8}$$

The identity (5.2.7) holds also for $z \in K^p$ by virtue of Lemma 5.2.4 and Theorem 1.9.1 (ii).

For a fixed s, we shall take limit as $t \downarrow s$ in (5.2.7) and (5.2.8) to obtain the KL right-differential equation. To this end, let us examine the behavior of $g_{t,s}(z)$ near $z = \xi(s)$ closely.

For any $\varepsilon_0 > 0$ with $B(\xi(s), 2\varepsilon_0) \cap \mathbb{H} \subset D_s$, by (5.2.1) there exists $\delta_1 > 0$ with $s + \delta_1 < t_0$ such that

$$\overline{g_s(F_t \setminus F_s) \cup \Pi g_s(F_t \setminus F_s)} \subset B(\xi(s), \varepsilon_0) \quad \text{for any } t \in (s, s+\delta_1).$$

In particular, $\ell_{t,s}$ is in the interior of the region bounded by the Jordan curve $g_{t,s}^{-1}(\partial B(\xi(s), 2\varepsilon_0))$. Then the family $\{g_{t,s}^{-1}, t \in [s, s+\delta_1)\}$, whose elements are extended analytically to $(D_s \cup \Pi D_s \cup \partial\mathbb{H}) \setminus \overline{B(\xi(s), \varepsilon_0)}$, is locally bounded on this domain by virtue of Lemma 5.1.9. Moreover, we have $\lim_{t\downarrow s} g_{t,s}^{-1}(z) = z$ for $z \in D_s$ as was already seen in the proof of Lemma 5.2.2 (ii). Vitali's theorem thus implies that $\lim_{t\downarrow s} g_{t,s}^{-1}(z) = z$ locally uniformly in $(D_s \cup \Pi D_s \cup \partial\mathbb{H}) \setminus \overline{B(\xi(s), \varepsilon_0)}$. In particular, this convergence occurs uniformly in $z \in \partial B(\xi(s), 2\varepsilon_0)$: there exists $\delta_2 \in (0, \delta_1)$ with

$$|g_{t,s}^{-1}(z) - z| < \varepsilon_0 \quad \text{for any } z \in \partial B(\xi(s), 2\varepsilon_0) \text{ and for any } t \in (s, s+\delta_2). \tag{5.2.9}$$

By (5.2.9), the diameter of $g_{t,s}^{-1}(\partial B(\xi(s), 2\varepsilon_0))$ is less than $6\varepsilon_0$. We therefore get, for any $x \in \ell_{t,s}$,

$$|\xi(s) - x| \le |\xi(s) - z| + |z - g_{t,s}^{-1}(z)| + |g_{t,s}^{-1}(z) - x| < 9\varepsilon_0, \quad t \in (s, s+\delta_2). \tag{5.2.10}$$

with z chosen arbitrarily from $\partial B(\xi(s), 2\varepsilon_0)$.

On the other hand, using the continuity of $\mathbf{s}(t) \in \mathcal{S}$ proven in Lemma 5.2.3, we can show that

$$\Psi_{\mathbf{s}(t)}(z, x) \quad \text{is jointly continuous in} \quad (t, z, x) \tag{5.2.11}$$

as in the proof of Theorem 2.5.1. Fix $z \in (D \setminus F_s) \cup K^p$. Since $g_t(z)$ is continuous in t, $\Psi_{\mathbf{s}(t)}(g_t(z), x)$ is continuous in t and $x \in \partial\mathbb{H}$. Thus, for any $\varepsilon > 0$, there exist $\delta_3 > 0$ and $\varepsilon_0 > 0$ such that

$$|\Psi_{\mathbf{s}(t)}(g_t(z), x) - \Psi_{\mathbf{s}(s)}(g_s(z), \xi(s))| < \varepsilon \tag{5.2.12}$$

for any $t \in (s, s+\delta_3)$ and for any $x \in \partial\mathbb{H}$ with $|x - \xi(s)| < 9\varepsilon_0$. For this ε_0, let $\delta_2 > 0$ be such that (5.2.9) holds and set $\delta_0 = \delta_2 \wedge \delta_3$. It then follows from (5.2.7), (5.2.8), (5.2.10), and (5.2.12) that

$$\left| \frac{g_s(z) - g_t(z)}{2(s-t)} + \pi \Psi_{\mathbf{s}(s)}(g_s(z), \xi(s)) \right| < \varepsilon. \quad \text{for any} \quad t \in (s, s+\delta_0).$$

This gives

$$\frac{d^+}{ds} g_s(z) = -2\pi \Psi_{\mathbf{s}(s)}(g_s(z), \xi(s)), \tag{5.2.13}$$

where the left-hand side denotes the right derivative of $g_s(z)$ in s.

Finally, we consider s as a variable. The function $g_s(z)$ is continuous in s by Lemma 5.2.4, and $\xi(s)$ is assumed to be continuous. Thus, the right-hand side of (5.2.13) is continuous in s as in (5.2.12). We can then use the fact stated in [Lawler (2005), Lemma 4.3] that, if a continuous function $u\colon [0, t_0) \to \mathbb{C}$ has continuous right derivative $\partial_t^+ u(t)$ on $[0, t_0)$, then u is in fact of class C^1 on $(0, t_0)$. Owing to this fact, the left-hand side of (5.2.13) can be replaced by the genuine derivative, which completes the proof of (5.2.6). □

Proposition 5.2.6. *Let $\{F_t\}_{t\in[0,t_0)}$ be growing hulls in the upper half-plane $\mathbb{H}$ with half-plane capacity $a^{\mathbb{H}}(F_t) = 2t$ and $g_t^{\mathbb{H}}$ be the canonical map from $\mathbb{H} \setminus F_t$ for $t \in [0, t_0)$. Assume that $\{F_t\}$ has a continuous driving function $\xi(t)$, i.e., (5.2.1) holds for $g_t^{\mathbb{H}}$ in place of g_t, and that $\{F_t\}$ is continuous in $\mathbb{H}$, i.e., Definition 5.2.1 holds for $\mathbb{H}$ in place of D. Then the Loewner equation*

$$\frac{dg_t^{\mathbb{H}}(z)}{dt} = \frac{2}{g_t^{\mathbb{H}}(z) - \xi(t)}, \quad g_0(z) = z, \tag{5.2.14}$$

holds for $t \in [0, t_0)$ and $z \in \mathbb{H} \setminus F_t$.

This is a special case of Proposition 5.2.5 that $K = \emptyset$ and $D = \mathbb{H}$. In fact, $D_t = \mathbb{H}$ and $\Psi_{D_t}(z, \zeta) = \Psi_{\mathbb{H}}(z, \zeta) = -\frac{1}{\pi}\frac{1}{z-\zeta}$ in this case. Instead of the equalities (5.2.7) and (5.2.8), the Loewner equation (5.2.14) can be also derived from the estimate (2.2.22) of Theorem 2.2.5 (iv) just as in [Lawler (2005), Proposition 4.4] and [Chen, Fukushima and Suzuki (2017), Proposition 2.3].

Using Proposition 5.2.5, we can now give a geometric characterization of KLEs that does not explicitly involve the KL equation.

Theorem 5.2.7. *Suppose that $\xi(t)$ is continuous. Then a family $\{F_t\}$ of growing hulls on a standard slit domain D is KLE driven by ξ if and only if it is continuously growing with driving function ξ in the sense of* (5.2.1) *and with half-plane capacity* $a_t = 2t$.

Proof. (i) (The "only if" part.) Assume that $\{F_t\}$ is a KLE driven by ξ. Thus, along with a continuous ξ, let $\mathbf{s}(t) \in \mathcal{S}$ be the solution of the slit motion equation

$$\mathbf{s}_j(t) - \mathbf{s}_j(0) = \int_0^t b_j(\xi(s), \mathbf{s}(s))ds, \qquad 1 \leq j \leq 3N, \tag{5.2.15}$$

for $b_j(\xi, \mathbf{s})$, $1 \leq j \leq 3N$, $\xi \in \partial\mathbb{H}$, $\mathbf{s} \in \mathcal{S}$, defined by (3.1.3), and let $g_t(z)$ be the unique solution of the KL equation (5.2.6) for $z \in D$.

F_t is defined by (3.2.22) in terms of the maximal time interval of the existence of the solution to (5.2.6). It then holds by Theorem 3.2.5 (ii) and Theorem 3.2.8 (i) that

$$g_t \text{ is a canonical map from } D \setminus F_t \text{ and } D(\mathbf{s}(t)) = g_t(D \setminus F_t). \tag{5.2.16}$$

By Theorem 3.2.8 (iii), $\xi(t)$ is the continuous driving function of $\{F_t\}$ in the sense of (5.2.1) so that $\{F_t\}$ is right continuous in the sense of Definition 5.2.1 by Lemma 5.2.2 (ii). Furthermore the half-plane capacity a_t of F_t equals $2t$ in view of Theorem 3.3.1. Hence it suffices to deduce the left continuity of $\{F_t\}$.

We fix $t \in (0, t_0)$ and an increasing sequence $(s_n)_{n\in\mathbb{N}} \subset [0, t]$ with $\lim_{n\to\infty} s_n = t$. Since $\mathbf{s}(s)$ is continuous in s as the solution of the ODE (5.2.15), $D_{s_n} \to D_t$ as

$n \to \infty$ in view of (3.1.2). By Proposition 5.1.11, $(s_n)_{n\in\mathbb{N}}$ has a subsequence $(s'_n)_{n\in\mathbb{N}}$ such that $(g^{-1}_{s'_n})_{n\in\mathbb{N}} \subset \mathcal{U}(2\rho_t)$ converges to a conformal mapping $\widetilde{f}\colon D_t \to \widetilde{f}(D_t)$ locally uniformly in D_t and $\widetilde{f} \in \mathcal{U}(2\rho_t)$. By Theorem 5.1.17 (i), $D \setminus F_{s'_n} \to \widetilde{f}(D_t)$, and $g_{s'_n} \to \widetilde{f}^{-1}$ locally uniformly in $\widetilde{f}(D_t)$. Note that $D \setminus F_t \subset \widetilde{f}(D_t)$ because $D \setminus F_t$ is a domain contained in $\bigcap_n (D \setminus F_{s'_n})$ so that Proposition 5.1.16 (ii) applies. Since $\lim_{n\to\infty} g_{s'_n}(z) = g_t(z)$ for $z \in D \setminus F_t$ by the s-continuity of $g_s(z)$ as the solution of (5.2.6), it follows that $\widetilde{f}^{-1} = g_t$ in $D \setminus F_t$. Now, if we assume $\widetilde{f}(D_t) \setminus (D \setminus F_t) \neq \emptyset$, then

$$\begin{aligned} D_t = \widetilde{f}^{-1}(\widetilde{f}(D_t)) &= g_t(D \setminus F_t) \cup \widetilde{f}^{-1}(\widetilde{f}(D_t) \setminus (D \setminus F_t)) \\ &= D_t \cup \widetilde{f}^{-1}(\widetilde{f}(D_t) \setminus (D \setminus F_t)), \end{aligned}$$

a contradiction. Hence $\widetilde{f}(D_t) = D \setminus F_t$ must be true, which yields $\widetilde{f} = g_t^{-1}$ as $\lim_{z\to\infty}(\widetilde{f}(z) - z) = 0$. Since the original sequence $(s_n)_{n\in\mathbb{N}}$ is arbitrary, we finally conclude that $g_s^{-1} \to g_t^{-1}$ locally uniformly in D_t as $s \uparrow t$. By Theorem 5.1.17 (i), we have $D \setminus F_s \to D \setminus F_t$, which proves the left continuity of $\{F_t\}$.

(ii) (The "if" part.) Conversely, assume that $\{F_t\}$ is a continuously growing hulls in D with driving function ξ and half-plane capacity $a_t = 2t$. We define the corresponding symbols as we have done at the beginning of this section. By Proposition 5.2.5, the canonical map $g_t(z)$ enjoys the ODE (5.2.6) for any $z \in (D \setminus F_t) \cup K^p$. Therefore, the only thing remaining is to show that $\mathbf{s}(t)$ satisfies (5.2.15). This is done in almost the same way as in Theorem 2.6.4.

We recall that, in Section 2.6, the proof of Theorem 2.6.4 relies on three Lemmas 2.6.1–2.6.3 preceding it. In proving these lemmas, the key was Theorem 2.3.6 and Theorem 2.5.1, which ensure the joint continuity of $g_t(z)$ and $\Psi_{D_t}(z, \zeta)$, respectively. In the present case, Lemma 5.2.4 and (5.2.11) can be utilized in place of Theorem 2.3.6 and Theorem 2.5.1, respectively, and therefore we can obtain the assertions corresponding to Lemmas 2.6.1–2.6.3 in the same way except for (iii), (vi), and (vii) of Lemma 2.6.1. Thus, it suffices to prove these remaining assertions in the present setting. As their proofs are similar, we prove (iii) only.

Let us recall from (iii) of Lemma 2.6.1 what we should prove:

$$\begin{aligned} &(g_t^{\pm})'(z) \text{ are differentiable in } t \in [0, t_0), \\ &\text{and } \partial_t (g_t^{\pm})'(z) \text{ are continuous in } (t, z) \in [0, t_0) \times R. \end{aligned} \tag{5.2.17}$$

Here, $g_t^{\pm}$ and R are defined as in the proof of Lemma 5.2.4. Below, we treat only g_t^{+} for notational simplicity.

Let $0 \leq s < t < t_0$ and $\eta_t(z, x) := \Psi_{\mathbf{s}(t)}(g_t(z), x)$ analogously to Lemma 2.6.1 (ii). We have mentioned that Lemma 2.6.1 (i), (ii) are still applicable to the present case, so that $g_t(z)$, $\eta_t(z, x)$ on R_+ extend to analytic functions $g_t^{+}(z)$, $\eta_t^{+}(z, x)$ on R by Schwarz reflection. Thus, in the same way as the proof of Lemma 2.6.1 (iii), we can get from (5.2.7)

$$(g_s^{+})'(z) - (g_t^{+})'(z) = \int_{\ell_{t,s}} (\eta_t^{+}(z, x))' \Im g_{t,s}(x + i0)\, dx, \quad z \in R. \tag{5.2.18}$$

$(\eta_t^+(z,x))'$ is continuous in $(t,z,x) \in [s,t_0) \times R \times \partial\mathbb{H}$ by Lemma 2.6.1 (ii) applied to the present case. Therefore, for any $\varepsilon > 0$, there exist $\delta > 0$ and $\varepsilon_0 > 0$ such that

$$\left|(\eta_t^+(z,x))' - (\eta_s^+(z,\xi(s)))'\right| < \varepsilon$$

for any $t \in (s, s+\delta)$ and for any $x \in \partial\mathbb{H}$ with $|x - \xi(s)| < 9\varepsilon_0$. We now make the same reasoning as in the proof of Proposition 5.2.5, using (5.2.8) and (5.2.18), to find $\delta_0 \in (0,\delta)$ such that, for $t \in (s, s+\delta_0)$,

$$\left|\frac{(g_s^+)'(z) - (g_t^+)'(z)}{2(s-t)} + \pi(\eta_s^+(z,\xi(s)))'\right| < \varepsilon.$$

This yields

$$\partial_s^+(g_s^+)'(z) = -2\pi(\eta_s^+(z,\xi(s)))'.$$

Here, $(g_s^+)'(z)$ and the right-hand side of this equation are both continuous in $s \in [0,t_0)$. The former continuity follows from the locally uniform convergence of $(g_s^+)_s$ on R, which was observed in the proof of Lemma 5.2.4, and the Cauchy integral formula for $(g_s^+)'$. Thus, we can apply [Lawler (2005), Lemma 4.3] to replace the left-hand side of the last equation with the genuine derivative. Now the joint continuity of $(\eta_s^+(z,\xi(s)))'$ proves (5.2.17).

Let $\{F_t'\}$ be the KLE driven by ξ. F_t' is defined by (3.2.22) in terms of the solution $g_t(z)$ to (5.2.6). Then $D \setminus F_t' = g_t^{-1}(D(\mathbf{s}(t))$ by Theorem 3.2.5. Since $D \setminus F_t = g_t^{-1}(D(\mathbf{s}(t))$ is clear from definition, we have $F_t = F_t'$, namely, $\{F_t\}$ is the KLE driven by ξ. □

The following result can be regarded as a special case of Theorem 5.2.7 with $K = \emptyset$ and $D = \mathbb{H}$. The same proof works except that, in the proof of the "if" part, one shall use Proposition 5.2.6 instead of Proposition 5.2.5.

Theorem 5.2.8. *Suppose that $\xi(t)$ is continuous. A family $\{F_t\}$ of growing hulls on the upper half-plane $\mathbb{H}$ is a Loewner evolution driven by ξ, namely, F_t is defined by* (3.2.22) *in terms of the unique solution of the Loewner equation*

$$\frac{d}{dt}g_t(z) = \frac{2}{g_t(z) - \xi(t)} \qquad g_0(z) = z \in \mathbb{H}, \tag{5.2.19}$$

if and only if $\{F_t\}$ is a continuously growing hulls in $\mathbb{H}$ with driving function ξ and half-plane capacity $a_t^{\mathbb{H}} = 2t$.

5.3 Invariance of KLE under univalent map

We continue to work under the setting of the preceding section. A purpose of this section is to show that the Komatu-Loewner evolution (KLE) is invariant under a certain class of univalent transformations of the standard slit domain D up to appropriate reparametrizations.

As preliminaries, we summarize some topological aspects of $\mathbb{H}$-hulls.

Lemma 5.3.1. *Let F be a bounded and relatively closed set in $\mathbb{H}$, and denote by $\mathbb{H}^-$ the lower half-plane $\{z \in \mathbb{C} : \Im z < 0\}$. The set F is an $\mathbb{H}$-hull if and only if $\mathbb{H} \setminus F$ is connected (i.e., it is a domain) and either of the following holds:*

(i) $\mathbb{H} \setminus F$ *is simply-connected.*
(ii) $F \cup \overline{\mathbb{H}^-} \cup \{\infty\}$ *is connected (in the Riemann sphere $\overline{\mathbb{C}}$).*
(iii) $F \cup \partial\mathbb{H} \cup \{\infty\}$ *is connected (in $\overline{\mathbb{C}}$).*
(iv) $F \cup \partial\mathbb{H}$ *is connected.*

Proof. Since (i) is the definition of $\mathbb{H}$-hull, what we have to prove is the equivalence of (i)–(iv). Among these, (i) $\Leftrightarrow$ (ii) is a classical fact. See, e.g., [Conway (1995), Proposition 13.1.1]. In addition, (iv) $\Rightarrow$ (iii) $\Rightarrow$ (ii) holds, as the union of two connected sets with non-empty intersection is again connected. Hence it suffices to prove (ii) $\Rightarrow$ (iv). Below we prove its contraposition.

Assume that (iv) does not hold, i.e., there exist two open sets which separate $F \cup \partial\mathbb{H}$. Since $\partial\mathbb{H}$ is connected, it is contained in exactly one of these two open sets, say O_1. We may then assume that the other open set O_2 is a bounded set in $\mathbb{H}$ since F is so. By this boundedness, we can take an open neighborhood O_3 of ∞ in $\overline{\mathbb{C}}$ which is disjoint from O_2. Now, the two open sets $O_1 \cup \mathbb{H}^- \cup O_3$ and O_2 separate $F \cup \overline{\mathbb{H}^-} \cup \{\infty\}$, which implies that (ii) does not hold. □

In what follows, we say that a set $A \subset \overline{\mathbb{C}}$ *separates* two points if these points are contained in different components of $\overline{\mathbb{C}} \setminus A$. The next lemma is adapted from [Newman (1961), Theorem 9.2 in Chapter V].

Lemma 5.3.2. *Let A_1 and A_2 be closed sets in $\overline{\mathbb{C}}$ with $A_1 \cap A_2$ being connected (possibly empty). If two points are not separated by A_1 or by A_2, then they are not separated by $A_1 \cup A_2$.*

Using this lemma, we discuss the union of two $\mathbb{H}$-hulls.

Proposition 5.3.3. *Let F_1 and F_2 be $\mathbb{H}$-hulls with $(F_1 \cap F_2) \cup \partial\mathbb{H}$ connected. Then $F_1 \cup F_2$ is an $\mathbb{H}$-hull.*

Proof. By assumption, the closed sets $F_1 \cup \overline{\mathbb{H}^-} \cup \{\infty\}$ and $F_2 \cup \overline{\mathbb{H}^-} \cup \{\infty\}$ have a connected intersection. Hence it follows from Lemma 5.3.2 that $\mathbb{H} \setminus (F_1 \cup F_2)$ is a domain. Since it is clear that $(F_1 \cup F_2) \cup \partial\mathbb{H} \cup \{\infty\}$ is connected, $F_1 \cup F_2$ is an $\mathbb{H}$-hull by Lemma 5.3.1. □

We now establish a fundamental result of this section.

Theorem 5.3.4. *Suppose that D and $\widetilde{D}$ are either a standard slit domain or the upper half-plane $\mathbb{H}$, respectively, and the degree of connectivity of $\widetilde{D}$ is equal to or smaller than that of D. Suppose that V is a subdomain of D and h is a univalent function from V into $\widetilde{D}$ satisfying one of the following conditions:*

(a) *$V = D \setminus A$ for a hull $A \subset D$, and h is the canonical map g_A from V onto $\widetilde{D}$.*
(b) *$V = D$, $\widetilde{D}$ is a domain obtained from D by filling some of its slits, and h is the identity map from D into $\widetilde{D}$.*

Let $\{F_t, t \in [0, t_0)\}$, $t_0 > 0$, be a KLE on D driven by a continuous function $\xi(t)$. Assume in case **(a)** *that*

$$\overline{F}_t \cap \overline{A} = \emptyset, \quad \text{for any} \quad t \in [0, t_0). \tag{5.3.1}$$

Then the following holds (see Figure 5.2).

(i) *$\{h(F_t),\ t \in [0, t_0)\}$ is a family of continuously growing hulls in $\widetilde{D}$.*
(ii) *Let g_t be the canonical map from $D \setminus F_t$ onto a standard slit domain D_t, and $\widetilde{g}_t$ be the canonical map from $\widetilde{D} \setminus h(F_t)$. Set $h_t = \widetilde{g}_t \circ h \circ g_t^{-1}$ with the domain of definition being $D_t \setminus g_t(A)$ in case* **(a)** *and D_t in case* **(b)**. *Then h_t can be extended by Schwarz reflection to be univalent on the domain*

$$G_t := \begin{cases} (D_t \cup \Pi D_t \cup \partial \mathbb{H}) \setminus (\overline{g_t(A)} \cup \Pi g_t(A)) & \text{in case } \mathbf{(a)} \\ D_t \cup \Pi D_t \cup \partial \mathbb{H} & \text{in case } \mathbf{(b)}, \end{cases} \tag{5.3.2}$$

and $\xi(t) \in G_t$. Here $\Pi z := \overline{z}$ for $z \in \mathbb{C}$. The restriction to $G_t \cap \partial \mathbb{H}$ of the extended function h_t is an infinitely differentiable real-valued function.
(iii) *$h_t(\xi(t))$ is a driving function of $\{h(F_t)\}$ in the sense that the relation (5.2.1) holds for $\widetilde{g}_t$, $h(F_t)$ and $h_t(\xi(t))$ in place of g_t, F_t and $\xi(t)$.*

Proof. (i) In the case **(a)**, $F_t \cup A$ is an $\mathbb{H}$-hull in D by Proposition 5.3.3. Hence $D \setminus (F_t \cup A)$ is an $(N+1)$-connected domain with inner boundaries given by N slits.

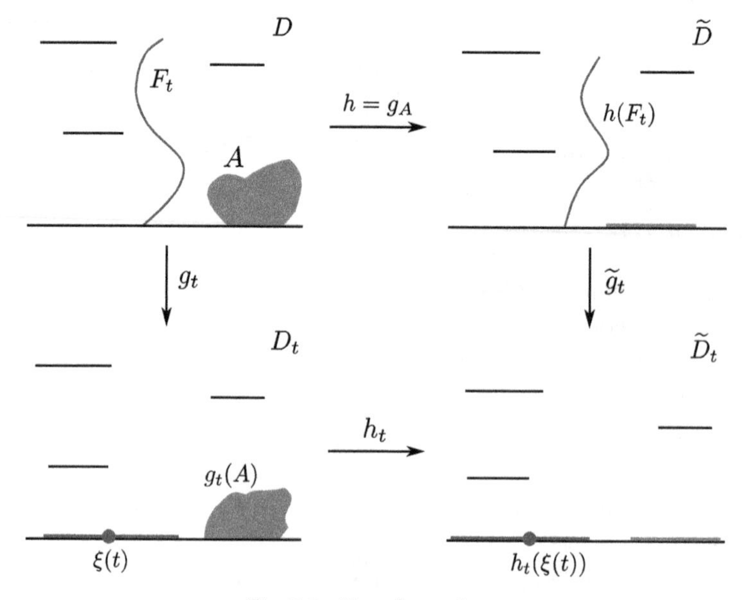

Fig. 5.2 Transformations.

Taking the boundary correspondence induced by $h = g_A$ into account, $\widetilde{D} \setminus h(F_t) = h(D \setminus (F_t \cup A))$ is also an $(N+1)$-connected domain with inner boundaries given by N slits. Filling these slits, we see that $\mathbb{H} \setminus h(F_t)$ is a simply-connected domain, i.e., $h(F_t)$ is an $\mathbb{H}$-hull.

We show the continuity of $\{h(F_t)\}$. By Theorem 5.2.7, $\{F_t\}$ is a family of continuously growing hulls in D. Let $\{t_n\}_{n\in\mathbb{N}} \subset [0, t_0)$ be a sequence with $t = \lim_{n\to\infty} t_n \in [0, t_0)$. Since $D \setminus F_{t_n} \to D \setminus F_t$ as $n \to \infty$, we have $D \setminus (F_{t_n} \cup A) \to D \setminus (F_t \cup A)$ by the definition of kernel. Applying Theorem 5.1.17 (i) to the conformal mappings $h|_{D\setminus(F_{t_n}\cup A)}$, we see that

$$\widetilde{D} \setminus h(F_{t_n}) = h(D \setminus (F_{t_n} \cup A)) \to h(D \setminus (F_t \cup A)) = \widetilde{D} \setminus h(F_t) \quad \text{as } n \to \infty.$$

Hence $\{h(F_t)\}_{t\in[0,t_0)}$ is continuous.

In the case **(b)**, h is just an identity map, and hence $h(F_t) = F_t$ is a hull. Actually, the definition of the continuity of $\{F_t\}_{t\in[0,t_0)}$ depends on in what domain we consider F_t, but one can easily observe that $\{F_t\}$ is also continuous in $\widetilde{D}$.

(ii) In the case **(a)**, $g_t(A)$ can be seen to be an $\mathbb{H}$-hull in D_t in the same way as in the first paragraph in the proof of (i). Let f_t be the canonical map from $D_t \setminus g_t(A)$. Then $h_t = f_t$ because both $f_t \circ g_t$ and $\widetilde{g}_t \circ g_A$ are the unique canonical map from $D \setminus (F_t \cup A)$ and so they coincide. Hence h_t extends to G_t by Schwarz reflection to be a univalent map in view of Theorem 2.2.3. Under the assumption on $\xi(t)$, (5.2.1) is fulfilled by Theorem 3.2.8. Hence the assumption (5.3.1) implies $\xi(t) \in G_t$.

We next consider the case **(b)**. Below we fix $t \in (0, t_0)$.

We take a Jordan arc Γ with endpoints $z_1 = -\xi_0 + i0$, $z_2 = \xi_0 + i0 \in \partial\mathbb{H}$ for $\xi_0 > 0$ so that $\Gamma \setminus \{z_1, z_2\} \subset D$. Let $\ell \subset \partial\mathbb{H}$ be the line segment with endpoints z_1, z_2 and U be the domain enclosed by $\Gamma \cup \ell$. For every sufficiently large ξ_0, we can choose such a Γ so that

$$F_t \subset U \subset D,\ U \setminus F_t \text{ is connected, } \overline{F}_t \cap \partial\mathbb{H} \subset \ell, \text{ and } \overline{U} \cap K = \emptyset.$$

Here, the connectedness of $U \setminus F_t$ can be concluded from Lemma 5.3.2, because any two points in $U \setminus F_t$ are separated neither by $F_t \cup \overline{\mathbb{H}^-} \cup \{\infty\}$ nor by $\Gamma \cup \ell$.

The canonical map g_t from $D \setminus F_t$ onto $D_t = g_t(D \setminus F_t)$ can be extended analytically across $\partial\mathbb{H} \setminus \overline{F}_t$ by Schwarz reflection in view of Theorem 2.2.3 so that $g_t(z) \in \partial\mathbb{H}$ is well defined for $z \in \partial\mathbb{H} \setminus \overline{F}_t$. Denote by U_t the domain enclosed by $g_t(\Gamma)$ and the line segment $\ell_t \subset \partial\mathbb{H}$ with endpoints $g_t(z_1)$, $g_t(z_2)$.

For $\xi + i0 \in \partial\mathbb{H} \setminus \overline{F}_t$, we denote $\Re g_t(\xi + i0) \in \mathbb{R}$ simply by $g_t(\xi)$. Fix $\rho > 0$ with $B_\rho \cap \mathbb{H} \supset F_t \cup K$. By virtue of the uniform bound (2.2.16) in Theorem 2.2.4,

$$|g_t(\xi) - \xi| \le \frac{L}{\xi^2} \quad \text{for any} \quad \xi \in \mathbb{R} \quad \text{with} \quad |\xi| > 3\rho,$$

where $L = 3(1+\pi)c_{1/2}a_{F_t}\rho$ for the half-plane capacity a_{F_t} of F_t relative to D.

This bound of $g_t(\xi)$ readily yields the implication

$$\xi > \sqrt[3]{10L} \vee (6\rho) \quad \Longrightarrow \quad g_t(\xi/2) \in (0, g_t(\xi)) \subset (g_t(-\xi), g_t(\xi)).$$

Consequently, if the endpoints $-\xi_0 + i0$, $\xi_0 + i0$ of $\ell \subset \partial\mathbb{H}$ is chosen so that

$$\xi_0 \ > \ \sqrt[3]{10L} \vee (6\rho), \tag{5.3.3}$$

then there exist points $z_0 \in U \setminus F_t$ in a neighborhood of $\xi_0/2$ and $w_0 \in U_t$ in a neighborhood of $g_t(\xi_0/2)$ with $g_t(z_0) = w_0$.

We now choose ξ_0 satisfying (5.3.3). Then

$$g_t(U \setminus F_t) = U_t. \tag{5.3.4}$$

Indeed, for any $z \in U \setminus F_t$, there is a continuous arc $\gamma_{z_0,z} \subset U \setminus F_t$ connecting z_0 and z as $U \setminus F_t$ is connected. $g_t(\gamma_{z_0,z})$ is then a continuous arc connecting $w_0 = g_t(z_0)$ and $w = g_t(z)$ that does not cross $g_t(\Gamma)$ so that $w = g_t(z) \in U_t$. Conversely, for any $w \in U_t$, there is a continuous arc $\gamma_{w_0,w} \subset U_t$ connecting $w_0 = g_t(z_0)$ and w. $g_t^{-1}(\gamma_{w_0,w})$ is then a continuous arc connecting $z_0 = g_t^{-1}(w_0)$ and $z = g_t^{-1}(w)$ that does not cross Γ so that $z \in U \setminus F_t$ and $w = g_t(z) \in g_t(U \setminus F_t)$, yielding (5.3.4).

Next, let $\widetilde{g}_t$ be the canonical map from $\widetilde{D} \setminus h(F_t) = \widetilde{D} \setminus \widetilde{F}_t$ and $\widetilde{a}_{\widetilde{F}_t}$ be the associated half-plane capacity. Let $\widetilde{U}_t$ be the domain enclosed by $\widetilde{g}_t(\Gamma) = \widetilde{g}_t(h(\Gamma))$ and the line segment in $\partial\mathbb{H}$ with endpoints $\widetilde{g}_t(-\xi_0 + i0)$, $\widetilde{g}_t(\xi_0 + i0)$. In the same way as above, we have $\widetilde{g}_t(U \setminus F_t) = \widetilde{U}_t$ provided that ξ_0 satisfies (5.3.3) for $\widetilde{L} = 3(1+\pi)c_{1/2}\widetilde{a}_{F_t}\rho$ in place of L.

Combining this with (5.3.4), we conclude that, if $\xi_0 > L_0 := \sqrt[3]{10L} \wedge \sqrt[3]{10\widetilde{L}} \wedge (6\rho)$, then

$$h_t(U_t) = \widetilde{U}_t, \tag{5.3.5}$$

namely, h_t is a conformal map from the Jordan domain U_t onto a Jordan domain $\widetilde{U}_t$. Consequently h_t extends continuously to a homeomorphism from ℓ_t onto $\widetilde{\ell}_t$ (cf. [Conway (1995), Corollary 14.5.7]). In particular

$$\lim_{z \in \mathbb{H}, z \to \zeta} \Im h_t(z) = 0, \quad \text{for any} \quad \zeta \in \ell_t. \tag{5.3.6}$$

Since ξ_0 can be taken arbitrarily larger than L_0, h_t can be extended to the region (5.3.2) to be univalent there by Schwarz reflection.

(iii) From (5.2.1), we obtain

$$\bigcap_{\varepsilon > 0} \overline{\widetilde{g}_t(h(F_{t+\varepsilon} \setminus F_t))} = \bigcap_{\varepsilon > 0} \overline{h_t(g_t(F_{t+\varepsilon} \setminus F_t))} = h_t(\xi(t)) \tag{5.3.7}$$

so that $h_t(\xi(t))$ is a driving function of $\{h(F_t)\}$. The proof of the theorem is now complete. □

We shall keep the setting of Theorem 5.3.4 in the rest of this section. We saw in Theorem 5.3.4 (iii) that $\xi(t)$ sits in the domain G_t defined by (5.3.2). The following much stronger assertion will be of crucial use in the sequel.

Proposition 5.3.5. *Fix any* $s \in [0, t_0)$.

(i) *There exist* $0 < r_1 < r_2$ *and* $\delta > 0$ *such that*

$$\{\xi(u); u \in (s-\delta, s+\delta) \cap [0, t_0)\} \subset B(\xi(s), r_1), \tag{5.3.8}$$

$$\overline{B(\xi(s), r_2)} \subset \bigcap_{u \in (s-\delta, s+\delta) \cap [0, t_0)} G_u. \tag{5.3.9}$$

(ii) $h_t(z)$, $h'_t(z)$, $h''_t(z)$ *are jointly continuous in* $(t, z) \in (s-\delta, s+\delta) \cap [0, t_0) \times B(\xi(s), r_2)$.

Proof. We give a proof only in case **(a)** in Theorem 5.3.4.

(i) For each $t \in [0, t_0)$, $g_t(z)$, $z \in D \setminus F_t$, is extended to $H_t = (D \setminus F_t) \cup \Pi(D \setminus F_t) \cup (\partial\mathbb{H} \setminus \overline{F}_t)$ to be analytic by Schwarz reflection on account of Theorem 2.2.3. For $s \in [0, t_0)$, take $t_1 \in (s, t_0)$ and let $\rho_{t_1} = \sup_{z \in F_{t_1} \cup K} |z|$. Then $g_u\big|_{D \setminus F_u} \in \mathcal{U}_{\rho_{t_1}}$ for any $u \in [0, t_1]$ so that $g_u(z)$ is locally bounded in $z \in H_{t_1}$ uniformly in $u \in [0, t_1]$ due to Lemma 5.1.9.

On the other hand, we see from Lemma 5.2.3 that $\lim_{t \to s} g_t(z) = g_s(z)$, $z \in D \setminus F_{t_1}$. Therefore $\lim_{t \to s} g_t(z) = g_t(z)$ locally uniformly in $z \in H_{t_1}$ by Vitali's theorem. In particular, on account of the assumption (5.3.1),

$$\lim_{t \to s} g_t(z) = g_s(z) \quad \text{uniformly in} \quad z \in \overline{A} \cup \Pi A. \tag{5.3.10}$$

Furthermore, owing to Lemma 5.2.4, $\lim_{t \to s} g_t(z) = g_s(z)$ uniformly in $z \in K^p$.

As $\xi(s) \in G_s$, one can therefore find $r_2 > 0$ and $\delta > 0$ for which (5.3.9) is satisfied. Take any $r_1 \in (0, r_2)$. Since $\xi(t)$ is continuous, one can make (5.3.8) to hold by choosing $\delta > 0$ smaller than the above taken one if necessary.

(ii) As was observed in the proof of Theorem 5.3.4 (ii), h_u is the canonical map from $D_u \setminus g_u(A)$. We set

$$\rho = \sup\{|g_u(z)| : u \in (s-\delta, s+\delta),\ z \in K^p \cup A\}.$$

ρ is finite by Lemma 5.2.4 and (5.3.1). $h_u \in \mathcal{U}_\rho$ for any $u \in (s-\delta, s+\delta)$ so that $h_u(z)$ is locally bounded in $z \in B(\xi(s), r_2)$ uniformly in $u \in (s-\delta, s+\delta)$ by Lemma 5.1.9.

On the other hand, Lemma 5.2.3 and Theorem 5.3.4 (i) imply that, as $t \to u$, $g_t^{-1} \to g_u^{-1}$ (luc) on D_u and $\widetilde{g}_t \to \widetilde{g}_u$ (luc) on $\widetilde{D} \setminus h(F_u)$ so that $h_t = \widetilde{g}_t \circ h \circ g_t^{-1} \to h_u$ (luc) on D_u. In particular $\lim_{t \to u} h_t(z) = h_u(z)$ for each $u \in (s-\delta, s+\delta)$ and $z \in B(\xi(s), r_2) \cap \mathbb{H}$. Therefore, by Vitali's theorem again, $h_t \to h_u$ as $t \to u \in (s-\delta, s+\delta)$ locally uniformly in $B(\xi(s), r_2)$ and, in particular, $h_t(z)$ is jointly continuous in $(t, z) \in (s-\delta, s+\delta) \times B(\xi(s), r_2)$. As $h_t(z)$ is analytic in z, such joint continuity is inherited by $h'_t(z)$ and $h''_t(z)$. □

Proposition 5.3.6. *The half-plane capacity* $\widetilde{a}_t$ *of* $h(F_t)$ *relative to* $\widetilde{D}$ *is differentiable in* t *and*

$$\frac{d\widetilde{a}_t}{dt} = 2\, h'_t(\xi(t))^2, \quad t \in [0, t_0). \tag{5.3.11}$$

Proof. It suffices to prove that the half-plane capacity $\widetilde{a}_t$ of $h(F_t)$ relative to $\widetilde{D}$ is right-differentiable in t and

$$\frac{d^+\widetilde{a}_t}{dt} = 2\, h_t'(\xi(t))^2, \quad t \in [0, t_0), \tag{5.3.12}$$

the left-hand side denoting the right derivative. In fact, as $h(F_t)$ is continuously growing by Theorem 5.3.4 (i), $\widetilde{a}_t$ is continuous in t according to Lemma 5.2.3. Further, the right-hand side of (5.3.12) is continuous in t by virtue of Proposition 5.3.5. Hence (5.3.12) implies (5.3.11) by [Lawler (2005), Lemma 4.3].

For a standard slit domain $D = \mathbb{H} \setminus \bigcup_{j=1}^N C_j$ and a hull $F \subset D$, the canonical map from $D \setminus F$ and the half-plane capacity of F relative to D will be designated by g_F^D and $a^D(F)$, respectively, to indicate their dependence on D. The half-plane capacity of F relative to the upper half-plane $\mathbb{H}$ will be denoted by $a^{\mathbb{H}}(F)$.

For $\xi \in \partial\mathbb{H}$, we denote by $D - \xi$ the standard slit domain obtained from D by the horizontal shift by $-\xi$ of each slit C_j. Obviously we have for each $\xi \in \partial\mathbb{H}$

$$\begin{aligned} g_{F-\xi}^{D-\xi}(z) &= g_F^D(z+\xi) - \xi, \quad z \in (D-\xi) \setminus (F-\xi), \\ \text{and} \quad a^{D-\xi}(F-\xi) &= a^D(F). \end{aligned} \tag{5.3.13}$$

Fix $t \in [0, t_0)$. For $\varepsilon > 0$, we put $F_{t,\varepsilon} = g_t(F_{t+\varepsilon} \setminus F_t)$ and $\widetilde{F}_{t,\varepsilon} = \widetilde{g}_t(h(F_{t+\varepsilon}) \setminus h(F_t))$. It then holds that

$$a^{D_t}(F_{t,\varepsilon}) - a^{\mathbb{H}}(F_{t,\varepsilon}) = o(\varepsilon), \quad a^{\widetilde{D}_t}(\widetilde{F}_{t,\varepsilon}) - a^{\mathbb{H}}(\widetilde{F}_{t,\varepsilon}) = o(\varepsilon). \tag{5.3.14}$$

Indeed, it follows from (5.2.1) and (5.3.7) that $\mathrm{rad}(F_{t,\varepsilon} - \xi(t)) = o(1)$ and $\mathrm{rad}(\widetilde{F}_{t,\varepsilon} - h_t(\xi(t))) = o(1)$ as $\varepsilon \downarrow 0$. Here, we have put $\mathrm{rad}(B) = \sup_{z\in B}|z|$ for a set $B \subset \mathbb{H}$. Hence Theorem A.8.1 in the Appendix applies in getting

$$a^{D_t}(F_{t,\varepsilon}) = a^{D_t - \xi(t)}(F_{t,\varepsilon} - \xi(t)) = a^{\mathbb{H}}(F_{t,\varepsilon} - \xi(t)) + o(\varepsilon) = a^{\mathbb{H}}(F_{t,\varepsilon}) + o(\varepsilon).$$

The second property of (5.3.14) is obtained similarly.

We now notice that $a^D(F_t) = 2t$ by Theorem 5.2.7 and $\widetilde{F}_{t,\varepsilon} = h_t(F_{t,\varepsilon})$. By (5.3.14) and (5.2.3), $a^{\mathbb{H}}(F_{t,\varepsilon}) = a_{t+\varepsilon}^D - a_t^D + o(\varepsilon) = 2\varepsilon + o(\varepsilon)$, so that we can apply [Lawler, Schramm and Werner (2001), Lemma 2.8] to $F_{t,\varepsilon}$ and the univalent function h_t in obtaining from [Lawler, Schramm and Werner (2001), (2.7)]

$$a^{\mathbb{H}}(\widetilde{F}_{t,\varepsilon}) = |h_t'(\xi(t))|^2(2\varepsilon + o(\varepsilon)). \tag{5.3.15}$$

It follows from (5.2.3), (5.3.14) and (5.3.15) that

$$\widetilde{a}_{t+\varepsilon} - \widetilde{a}_t = a^{\widetilde{D}_t}(\widetilde{F}_{t,\varepsilon}) = a^{\mathbb{H}}(\widetilde{F}_{t,\varepsilon}) + o(\varepsilon) = 2\varepsilon|h_t'(\xi(t))|^2 + o(\varepsilon) \quad \text{as } \varepsilon \to 0,$$

yielding (5.3.12). □

Theorem 5.3.4 and Proposition 5.3.6 along with Theorem 5.2.7 and Theorem 5.2.8 enable us to transform by h a KLE on a standard slit domain D to a KLE on $\widetilde{D}$ (a Loewner evolution on $\mathbb{H}$ in case **(b)** with $\widetilde{D} = \mathbb{H}$) as will be explained below precisely.

Under the setting of Theorem 5.3.4, $\{h(F_t),\ t \in [0, t_0)\}$ is a family of continuously growing hulls in $\widetilde{D}$ with the continuous driving function $h_t(\xi(t))$. The half-plane capacity $\widetilde{a}_t$ of $h(F_t)$ satisfies $\frac{d\widetilde{a}_t}{dt} = 2h_t'(\xi(t))^2$ by (5.3.11). In particular, $\widetilde{a}_t$ is continuous. It is strictly increasing by Lemma 5.2.2 (i).

We can now make the half-plane capacity reparametrization (time change) of $h(F_t)$ by

$$\check{F}_t = h\left(F_{\widetilde{a}^{-1}(2t)}\right), \quad t \in [0, \check{t}_0) \text{ for } \check{t}_0 = \frac{1}{2}\widetilde{a}(t_0-), \tag{5.3.16}$$

so that the half-plane capacity of the hull $\check{F}_t$ equals $2t$. In accordance with this, the canonical map $\widetilde{g}_t$ from $\widetilde{D} \setminus h(F_t)$ and the driving function $h_t(\xi(t))$ of $\{h(F_t)\}$ are time changed into

$$\check{g}_t = \widetilde{g}_{\widetilde{a}^{-1}(2t)} \quad \text{and} \quad \check{\xi}(t) = h_{\widetilde{a}^{-1}(2t)}(\xi(\widetilde{a}^{-1}(2t))), \tag{5.3.17}$$

respectively. Moreover, in the case that $\widetilde{D} \subsetneq \mathbb{H}$, the slit motion $\widetilde{\mathbf{s}}(t)$ induced by $\widetilde{g}_t$ is time changed into

$$\check{\mathbf{s}}(t) = \widetilde{\mathbf{s}}_{\widetilde{a}^{-1}(2t)}. \tag{5.3.18}$$

Proposition 5.3.7. *Consider the case that $\widetilde{D} \subsetneq \mathbb{H}$ in* Theorem 5.3.4.

(i) *The reparametrization $\{\check{F}_t; t \in [0, \check{t}_0)\}$ of $\{h(F_t); t \in [0, t_0)\}$ defined by* (5.3.16) *is the KLE on $\widetilde{D}$ driven by the time change $\check{\xi}(t)$ of $h_t(\xi(t))$ defined by* (5.3.17). *In particular, $\check{\mathbf{s}}(t)$ and $\check{g}_t$ satisfy the ODEs* (5.2.15) *and* (5.2.6) *with $\xi(t)$ being replaced by $\check{\xi}(t)$, respectively.*

(ii) *$\widetilde{g}_t(z)$ satisfies the ODE*

$$\frac{d}{dt}\widetilde{g}_t(z) = -2\pi h_t'(\xi(t))^2 \Psi_{\widetilde{\mathbf{s}}(t)}(\widetilde{g}_t(z), h_t(\xi(t))), \quad \widetilde{g}_0(z) = z \in \widetilde{D}. \tag{5.3.19}$$

Proof. (i) By Proposition 5.3.5 (ii), $h_t(\xi(t))$ is continuous in t and so is its time change $\check{\xi}(t)$. $\{\check{F}_t; t \in [0, \check{t}_0)\}$ is a family of continuously growing hulls in $\widetilde{D}$ with continuous driving function $\check{\xi}$ and the half-plane capacity of $\check{F}_t$ equals $2t$. Hence (i) follows from Theorem 5.2.7.

(ii) We have by (i) that

$$\widetilde{g}_{\widetilde{a}^{-1}(2s)}(z) - z = -2\pi \int_0^s \Psi_{\widetilde{\mathbf{s}}^{-1}(2v)}(\widetilde{g}_{\widetilde{a}^{-1}(2v)}(z), h_{\widetilde{a}^{-1}(2v)}(\xi(\widetilde{a}^{-1}(2v))))dv,$$

which is equal to

$$-\pi \int_0^{\widetilde{a}^{-1}(2s)} \frac{d\widetilde{a}_u}{du} \Psi_{\widetilde{\mathbf{s}}(u)}(\widetilde{g}_u(z), h_u(\xi(u)))du,$$

by the change of variable $2v = \widetilde{a}(u)$. Replacement of $\widetilde{a}^{-1}(2s)$ by t along with (5.3.11) yields (5.3.19). □

The next lemma can be shown in the same way as above but using Theorem 5.2.8 instead of Theorem 5.2.7.

Lemma 5.3.8. *Consider the case that $\widetilde{D} = \mathbb{H}$ in* Theorem 5.3.4.

(i) $\{\check{F}_t; t \in [0, \check{t}_0)\}$ *defined by* (5.3.16) *is the Loewner evolution on* $\mathbb{H}$ *driven by* $\check{\xi}(t)$. *In particular,* $\check{g}_t$ *and* $\check{\xi}(t)$ *satisfy the ODE* (5.2.19) *with* $\check{\xi}(t)$ *in place of* $\xi(t)$.
(ii) $\widetilde{g}_t(z)$ *satisfies the ODE*

$$\frac{d}{dt}\widetilde{g}_t(z) = \frac{2h_t'(\xi(t))^2}{\widetilde{g}_t(z) - h_t(\xi(t))} \quad \text{with } \widetilde{g}_0(z) = z \in \mathbb{H}. \tag{5.3.20}$$

5.4 Transformation of SKLE$_{\alpha,b}$

Proposition 5.3.7 and Lemma 5.3.8 illustrate how a Komatu-Loewner evolution (KLE) on a standard slit domain D is transformed by a univalent map h of Theorem 5.3.4 to a KLE (or a Loewner evolution) on a domain $\widetilde{D}$ of a possibly lower degree of connectivity. In this and next sections, we investigate how a stochastic Komatu-Loewner evolution SKLE$_{\alpha,b}$ on D introduced in Section 4.3.2 is transformed under such a map. In this case, it is driven by a random process adapted to the filtration of the one-dimensional Brownian motion B_t. This section is devoted to a study of how this random process is transformed by a univalent map h.

5.4.1 *BMD domain constant* b_{BMD}

Recall the collection $\mathcal{D}$ of all labelled standard slit domains introduced in the second half of Section 2.3.2. The space $\mathcal{D}$ is a metric space equipped with distance (2.3.15), which can be identified with the space $\mathcal{S}$ defined by (3.1.1) as a topological space. We write $\mathbf{s}(D)$ (resp. $D(\mathbf{s})$) the element in $\mathcal{S}$ (resp. $\mathcal{D}$) corresponding to $D \in \mathcal{D}$ (resp. $\mathbf{s} \in \mathcal{S}$). In particular, for $\mathbf{s} = (\mathbf{y}, \mathbf{x}, \mathbf{x}^r) \in \mathcal{S}$, $z_j = x_j + iy_j$ and $z_j^r = x_j^r + iy_j$ are the left endpoint and the right endpoint of the j-th slit C_j of $D(\mathbf{s}) \in \mathcal{D}$.

For each standard slit domain $D \in \mathcal{D}$, let $\Psi(z, \xi) = \Psi_D(z, \xi)$, $z \in D$, $\xi \in \partial\mathbb{H}$, be the BMD-complex Poisson kernel on D. Define

$$b_{\mathrm{BMD}}(\xi, D) = 2\pi \lim_{z \to \xi} \left(\Psi_D(z, \xi) + \frac{1}{\pi} \frac{1}{z - \xi} \right), \quad \xi \in \mathbb{R}. \tag{5.4.1}$$

In view of the proof of Lemma 1.9.2, $b_{\mathrm{BMD}}(\xi, D)$ is well defined as a real number. It indicates a discrepancy of the slit domain D from $\mathbb{H}$ relative to BMD. We occasionally write $b_{\mathrm{BMD}}(\xi, D)$ as $b_{\mathrm{BMD}}(\xi, \mathbf{s})$ in terms of the slit $\mathbf{s} = \mathbf{s}(D) \in \mathcal{S}$ of D. We set $b_{\mathrm{BMD}}(\mathbf{s}) = b_{\mathrm{BMD}}(0, \mathbf{s})$ and call it the *BMD domain constant* of $D = D(\mathbf{s})$.

Lemma 5.4.1.

(i) $b_{\mathrm{BMD}}(\mathbf{s})$, $\mathbf{s} \in \mathcal{S}$, *is a homogeneous function of degree* -1 *on* $\mathcal{S}$.
(ii) $b_{\mathrm{BMD}}(\xi, \mathbf{s}) = b_{\mathrm{BMD}}(\mathbf{s} - \widehat{\xi})$ *for* $\mathbf{s} \in \mathcal{S}$ *and* $\xi \in \mathbb{R}$. *Here* $\widehat{\xi}$ *denotes the vector in* $\mathbb{R}^{3N}$ *whose first* N *entries are* 0 *and the last* $2N$ *entries are* ξ.
(iii) $b_{\mathrm{BMD}}(\mathbf{s})$ *satisfies the local Lipschitz continuity condition* **(L)** *introduced in the beginning of Section 4.3.1.*

Proof. (i) The scaling property (3.1.7) of $\Psi_{\mathbf{s}}(z,\xi)$ implies that, for any $\mathbf{s}\in\mathcal{S}$, $\xi\in\mathbb{R}$ and $c>0$,

$$b_{\rm BMD}(c\mathbf{s}) = 2\pi\lim_{z\to 0}\left(\Psi_{c\mathbf{s}}(z,0)+\frac{1}{\pi z}\right) = 2\pi\lim_{z\to 0}\left(\frac{1}{c}\Psi_{\mathbf{s}}(c^{-1}z,0)+\frac{1}{\pi z}\right)$$
$$= \frac{2\pi}{c}\lim_{z\to 0}\left(\Psi_{\mathbf{s}}(c^{-1}z,0)+\frac{1}{\pi c^{-1}z}\right) = c^{-1}b_{\rm BMD}(\mathbf{s}).$$

(ii) The translation invariance property (3.1.8) of $\Psi_{\mathbf{s}}(z,\xi)$ implies that, for any $\eta\in\mathbb{R}$,

$$2\pi\left(\Psi_{\mathbf{s}}(z,\xi)+\frac{1}{\pi}\frac{1}{z-\xi}\right) = 2\pi\left(\Psi_{\mathbf{s}+\widehat{\eta}}(z+\eta,\xi+\eta)+\frac{1}{\pi}\frac{1}{(z+\eta)-(\xi+\eta)}\right).$$

By letting $z\to\xi$, we get $b_{\rm BMD}(\xi,\mathbf{s}) = b_{\rm BMD}(\xi+\eta,\mathbf{s}+\widehat{\eta})$, which equals $b_{\rm BMD}(\mathbf{s}-\widehat{\xi})$ by setting $\eta=-\xi$.

(iii) For any $\mathbf{s}_0\in\mathcal{S}$, choose $\widehat{\mathcal{D}}_0$, $\widehat{\mathcal{D}}_{00}\subset\mathcal{D}$ as in the proof of Lemma 3.1.1 so that any D, $\widetilde{D}\in\widehat{\mathcal{D}}_{00}$ satisfies $d(D,\widetilde{D})<\varepsilon_0$ for $\varepsilon_0>0$ appearing in the statement of Theorem 2.4.1. Consider any finite interval $J=(c,d)\subset\partial\mathbb{H}$ $(c<d)$, and choose $\delta>0$ with $R=\{x+iy: c<x<d,\ 0\le y\le\delta\}\subset\mathbb{H}\setminus\bigcup_{j=1}^N\overline{V}_j$. By taking a compact set Q with $R\subset Q\subset\overline{\mathbb{H}}$, we get from Theorem 2.4.1 that, for any D, $\widetilde{D}\in\widehat{\mathcal{D}}_{00}$,

$$2\pi\left|\left(\Psi(z,\xi)+\frac{1}{\pi}\frac{1}{z-\xi}\right)-\left(\widetilde{\Psi}(z,\xi)+\frac{1}{\pi}\frac{1}{z-\xi}\right)\right|$$
$$=2\pi|\Psi(z,\xi)-\widetilde{\Psi}(z,\xi)|\le 2\pi L_{Q,J}d(D,\widetilde{D}),$$

holding for any $z\in R\setminus J$ and $\xi\in J$.

By letting $z\to\xi\in J$ and taking (ii) into account, we arrive at (4.3.1) for $f(\mathbf{s})=b_{\rm BMD}(\mathbf{s})$, $\mathbf{s}^{(1)}=\mathbf{s}(D)$ and $\mathbf{s}^{(2)}=\mathbf{s}(\widetilde{D})$. □

When $D=\mathbb{H}$, namely, when $K=\emptyset$, we set

$$b_{\rm BMD}(\xi,\mathbb{H})=0,\qquad \xi\in\mathbb{R}. \tag{5.4.2}$$

5.4.2 *Transformation of driving process of* SKLE$_{\alpha,b}$

In the rest of this chapter, we work under the setting of Theorem 5.3.4 on domains $D,\widetilde{D},V$ and a univalent map $h:V\mapsto\widetilde{D}$, where the pair (V,h) satisfies either the condition **(a)** or **(b)** there. Moreover, as the family of growing hulls $\{F_t\}$ we shall take an SKLE$_{\alpha,b}$, so that each F_t is a random set.

To be precise, given a non-negative homogeneous function $\alpha(\mathbf{s})$ of $\mathbf{s}\in\mathcal{S}$ with degree 0 and a homogeneous function $b(\mathbf{s})$ of $\mathbf{s}\in\mathcal{S}$ with degree -1 both satisfying the condition **(L)** of (4.3.1), denote by $(\xi(t),\mathbf{s}(t))\in\mathbb{R}\times\mathcal{S}$, $t\in[0,\zeta)$, the solution of the SDE system (4.2.10) having coefficients α,b and (4.2.11) with initial value $(\xi,\mathbf{s})\in\mathbb{R}\times\mathcal{S}$ for $\mathbf{s}$ with $D(\mathbf{s})=D$. More specifically, we assume that, on a probability space $(\Omega,\mathcal{G},\mathbb{P})$ with a right continuous increasing family $\{\mathcal{G}_t\}$ of sub-σ-fields of $\mathcal{G}$ (a filtration), there exists a one-dimensional $\{\mathcal{G}_t\}$-Brownian motion $B_t(\omega)$, $t \geq 0$, $\omega \in \Omega$, with $B_0 = 0$, and $(\xi(t,\omega), \mathbf{s}(t,\omega))$, $t < \zeta(\omega)$, is an $\mathbb{R} \times \mathcal{S}$-valued $\{\mathcal{G}_t\}$-adapted continuous process satisfying (4.2.10)-(4.2.11). Let $\{F_t, t \in [0,\zeta)\}$ be the SKLE$_{\alpha,b}$ driven by $(\xi(t), \mathbf{s}(t))$ as is introduced in Section 4.3.2. SKLE$_{\alpha,b}$ can be regarded as a specific KLE on the standard slit domain D driven by $\xi(t)$.

In relation to the pair (V, h), we define a random time $t_0 = t_0(\omega)$ by

$$t_0 := \begin{cases} \overline{\tau}_A & \text{in case } \textbf{(a)} \\ \zeta & \text{in case } \textbf{(b)}, \end{cases} \tag{5.4.3}$$

where $\overline{\tau}_A = \sup\{t \in [0,\zeta) : \overline{F}_t \cap \overline{A} = \emptyset\}$. By Lemma 4.3.7, $t_0 = t_0(\omega)$ is a $\{\mathcal{G}_t\}$-stopping time.

Condition (5.3.1) in Theorem 5.3.4 is then fulfilled for $t \in [0, t_0(\omega))$ so that $\{h(F_t), t \in [0,t_0)\}$ is a family of continuously growing hulls in $\widetilde{D}$ with a continuous driving function (process) $\{\widetilde{\xi}(t) := h_t(\xi(t)),\ t \in [0,t_0)\}$ having half-plane capacity $\widetilde{a}_t$ with $\dfrac{d\widetilde{a}_t}{dt} = 2h_t'(\xi(t))^2$ by virtue of Proposition 5.3.6. Here $h_t = \widetilde{g}_t \circ h \circ g_t^{-1}$ for the canonical map g_t (resp. $\widetilde{g}_t$) from $D \setminus F_t$ (resp. $\widetilde{D} \setminus h(F_t)$). Furthermore, in the case that $\widetilde{D} \neq \mathbb{H}$ (resp. $\widetilde{D} = \mathbb{H}$), $\widetilde{g}_t(z)$ satisfies the ODE (5.3.19) (resp. (5.3.20)) according to Proposition 5.3.7 (resp. Lemma 5.3.8).

Proposition 5.4.2.

(i) $\widetilde{g}_t(h(z))\mathbb{1}_{\{t<t_0,\, z\notin F_t\}}$ *is* $\mathcal{G}_t$*-measurable for every* $z \in V$.
(ii) $\mathbb{1}_{\{t<t_0\}}h_t(\xi(t))$, $\mathbb{1}_{\{t<t_0\}}h_t'(\xi(t))$ *and* $\mathbb{1}_{\{t<t_0\}}h_t''(\xi(t))$ *are* $\mathcal{G}_t$*-measurable.*

This proposition legitimates the stochastic integral with respect to the Brownian motion B_t appearing in the next theorem.

Theorem 5.4.3. *The driving process* $\left\{\widetilde{\xi}(t) := h_t(\xi(t)),\ t \in [0,t_0)\right\}$ *of the transformed hulls* $\{h(F_t),\ t \in [0,t_0)\}$ *on* $\widetilde{D}$ *of the* SKLE$_{\alpha,b}$*-hulls* $\{F_t,\ t \in [0,t_0)\}$ *on* D *satisfies*

$$\begin{aligned} d\widetilde{\xi}(t) &= h_t'(\xi(t))\left(b(\mathbf{s}(t) - \widehat{\xi}(t)) + b_{\mathrm{BMD}}(\xi(t), \mathbf{s}(t))\right)dt \\ &\quad + \frac{1}{2}h_t''(\xi(t))\left(\alpha(\mathbf{s}(t) - \widehat{\xi}(t))^2 - 6\right)dt \\ &\quad - h_t'(\xi(t))^2 b_{\mathrm{BMD}}(\widetilde{\xi}(t), h_t(\mathbf{s}(t)))dt \\ &\quad + h_t'(\xi(t))\alpha(\mathbf{s}(t) - \widehat{\xi}(t))dB_t. \end{aligned} \tag{5.4.4}$$

Here $\widehat{\xi}(t)$ *denotes the vector in* $\mathbb{R}^{3N}$ *whose first* N *entries are* 0 *and the last* $2N$ *entries are* $\xi(t)$.

In the rest of this section, we shall prove Proposition 5.4.2 and Theorem 5.4.3 only for the case **(a)** that $V = D \setminus A$ for a hull $A \subset D$. Recall that h is the canonical map g_A from $D \setminus A$ onto $\widetilde{D} = \mathbb{H} \setminus \bigcup_{j=1}^N \widetilde{C}_j$, $t_0 = \overline{\tau}_A$, g_t is the canonical map from

$D \setminus F_t$ and $\widetilde{g}_t$ is the canonical map from $\widetilde{D} \setminus g_A(F_t)$. A similar and actually simpler proof works in the case **(b)** that $V = D$, h is the identity map from D into $\widetilde{D}$, and $t_0 = \zeta$.

Proof of Proposition 5.4.2. (i) It suffices to prove that

$$\Im \widetilde{g}_t(g_A(z)) \mathbb{1}_{\{t < \overline{\tau}_A,\ z \notin F_t\}} \text{ is } \mathcal{G}_t\text{-measurable for each } z \in D \setminus A, \tag{5.4.5}$$

because the assertion (i) with $V = D \setminus A$, $h = g_A$ and $t_0 = \overline{\tau}_A$ follows from (5.4.5) in an analogous manner to the proof of Lemma 4.1.2 (ii).

We show (5.4.5) by making use of the representation (2.2.1) of $\Im \widetilde{g}_t(\widetilde{z})$, $\widetilde{z} \in \widetilde{D} \setminus g_A(F_t)$, in terms of the BMD $\widetilde{\mathbf{Z}}^* = (\widetilde{Z}^*_t, \widetilde{\mathbb{P}}^*_{\widetilde{z}})$ on $\widetilde{D}^* = \widetilde{D} \cup \{\widetilde{c}^*_1, \cdots, \widetilde{c}^*_N\}$ obtained from the ABM on $\mathbb{H}$ by rendering each slit $\widetilde{C}_j$ into a singleton $\widetilde{c}^*_j$. For $r > 0$, let

$$\mathbb{H}_r = \{z \in \mathbb{H} : \Im z < r\} \quad \text{and} \quad \Gamma_r = \{z \in \mathbb{H} : \Im z = r\}.$$

Take $r_0 > 0$ with $\mathbb{H}_{r_0} \supset \bigcup_{j=1}^N \widetilde{C}_j$ and denote by σ_r the hitting time of Γ_r by $\widetilde{Z}^*$.

According to (2.2.1),

$$\begin{aligned} &\Im \widetilde{g}_t(g_A(z))(\omega) \mathbb{1}_{\{t < \overline{\tau}_A,\ z \notin F_t\}} \\ &= \lim_{r \to \infty} r \widetilde{\mathbb{P}}^*_{\widetilde{z}} \left(\sigma_r < \sigma_{g_A(F_t(\omega))}\right) \mathbb{1}_{\{t < \overline{\tau}_A(\omega),\ z \notin F_t,\ g_A(F_t(\omega)) \subset \mathbb{H}_r\}}, \end{aligned}$$

where $\widetilde{z} := g_A(z) \in \widetilde{D} \setminus g_A(F_t)$ for $z \in D \setminus A$. As

$$\widetilde{\mathbb{P}}^*_{\widetilde{z}} \left(\sigma_r < \sigma_{g_A(F_t)}\right) = \widetilde{\mathbb{P}}^*_{\widetilde{z}} \left(g_A(F_t) \cap (\widetilde{Z}^*_{[0,\sigma_r]} \cap \widetilde{D}) = \emptyset,\ \sigma_r < \infty\right),$$

it is enough to show that, for each $r > r_0$,

$$\mathbb{1}_{\{t < \overline{\tau}_A(\omega),\ z \notin F_t,\ g_A(F_t(\omega)) \subset \mathbb{H}_r\}} \widetilde{\mathbb{P}}^*_{\widetilde{z}} \left(g_A(F_t(\omega)) \cap (\widetilde{Z}^*_{[0,\sigma_r]} \cap \widetilde{D}) = \emptyset,\ \sigma_r < \infty\right) \tag{5.4.6}$$

is $\mathcal{G}_t$-measurable in $\omega \in \Omega$.

To this end, let the BMD $\widetilde{\mathbf{Z}}^* = (\widetilde{Z}^*_t, \widetilde{\mathbb{P}}^*_{\widetilde{z}})$ be defined on a base probability space $(\widetilde{\Omega}, \widetilde{\mathcal{G}})$. Consider the direct product

$$\left(\Omega \times \widetilde{\Omega},\ \mathcal{G}_t \times \widetilde{\mathcal{G}},\ \mathbb{P} \times \widetilde{\mathbb{P}}^*_{\widetilde{z}}\right)$$

and set

$$\Lambda = \left\{(\omega, \widetilde{\omega}) \in \Gamma_{t,A,r,z} \times \widetilde{\Omega} : g_A(F_t(\omega)) \cap \widetilde{Z}^*_{[0,\sigma_r(\widetilde{\omega})]} \cap \widetilde{D} = \emptyset \text{ and } \sigma_r(\widetilde{\omega}) < \infty\right\},$$

where $\Gamma_{t,A,r,z} = \{\omega \in \Omega : t < \overline{\tau}_A(\omega),\ z \notin F_t(\omega),\ g_A(F_t(\omega)) \subset \mathbb{H}_r\} \in \mathcal{G}_t$. If $\Lambda \in \mathcal{G}_t \times \widetilde{\mathcal{G}}$, then, by the Fubini theorem, $\widetilde{\mathbb{P}}^*_{\widetilde{z}}(\Lambda_\omega)$ is $\mathcal{G}_t$-measurable for the ω-section $\Lambda_\omega = \{\widetilde{\omega} \in \widetilde{\Omega} : (\omega, \widetilde{\omega}) \in \Lambda\}$, yielding (5.4.6).

We have

$$\begin{aligned} \Lambda &= \Big\{(\omega, \widetilde{\omega}) \in \Gamma_{t,A,r,z} \times \widetilde{\Omega} : \widetilde{Z}^*_u(\widetilde{\omega}) \in \widetilde{D} \setminus g_A(F_t)(\omega) \\ &\qquad \text{for any } u \in [0, \sigma_r(\widetilde{\omega})] \text{ with } \widetilde{Z}^*_u(\widetilde{\omega}) \in \widetilde{D}\Big\} \\ &= \Big\{(\omega, \widetilde{\omega}) \in \Gamma_{t,A,r,z} \times \widetilde{\Omega} : g_A^{-1}(\widetilde{Z}^*_u(\widetilde{\omega})) \in D \setminus F_t(\omega) \\ &\qquad \text{for any } u \in [0, \sigma_r(\widetilde{\omega})] \text{ with } g_A^{-1}(\widetilde{Z}^*_u(\widetilde{\omega})) \in D\Big\}. \end{aligned}$$

Notice that, for each $\zeta \in D$, $\{\omega \in \Omega : \zeta \in F_t(\omega)\} \in \mathcal{G}_t$ by Proposition 4.3.4 (ii). Taking the continuity of $g_A^{-1}(\widetilde{Z}_u^*(\widetilde{\omega}))$ in u into account, we have

$$\Lambda = \bigcup_n \bigcup_{s \in \mathbb{Q}_+} \Bigg[\{(\omega, \widetilde{\omega}) \in \Gamma_{t,A,r,z} \times \widetilde{\Omega} : \sigma_r(\widetilde{\omega}) \le s\} \\ \cap \bigcap_{u \in [0,s] \cap \mathbb{Q}_+} \Big(\{(\omega, \widetilde{\omega}) : \widetilde{Z}_u^*(\widetilde{\omega}) \in \{\widetilde{c}_1^*, \dots, \widetilde{c}_N^*\}\} \\ \cup \bigcap_{\zeta \in \mathbb{Q}^2} \{(\omega, \widetilde{\omega}) : \zeta \in F_t(\omega),\ g_A^{-1}(\widetilde{Z}_u^*(\widetilde{\omega})) \in D,\ |\zeta - g_A^{-1}(\widetilde{Z}_u^*(\widetilde{\omega}))| > 1/n\} \Big) \Bigg]$$

where $\mathbb{Q}^2 = \mathbb{Q} \times \mathbb{Q}$ and $\mathbb{Q}_+$ is the set of the positive rational numbers. Consequently $\Lambda \in \mathcal{G}_t \times \widetilde{\mathcal{G}}$ as desired.

(ii) Fix $t > 0$. We only present a proof for the $\mathcal{G}_t$-measurability of $h_t(\xi(t))\mathbb{1}_{\{t < \overline{\tau}_A\}}$. The proof for the other assertions is similar.

We begin with recalling Proposition 4.3.8: The function $g_t^{-1}(z)\mathbb{1}_{\{t < s_z\}}(\omega)$ is $\mathcal{B}(D) \times \mathcal{G}_t$-measurable, where s_z is defined by (4.3.9). Hence

$$\{ (z, \omega) : t < s_z(\omega),\ z \notin g_t(A) \} \\ = \{ (z, \omega) : t < s_z(\omega),\ g_t^{-1}(z) \notin A \} \in \mathcal{B}(D) \times \mathcal{G}_t.$$

Through the proof of the $\mathcal{G}_t$-adaptedness of $\widetilde{g}_t(g_A(z))\mathbb{1}_{\{t < t_z\}}$ in (i), the $\mathcal{B}(D \setminus A) \times \mathcal{G}_t$-measurability of the function $(z, \omega) \mapsto \widetilde{g}_t(g_A(z))\mathbb{1}_{\{t < t_z\}}(\omega)$ can also be obtained as in the proof of Lemma 4.1.2 (ii). Hence, the composite

$$\widetilde{g}_t(g_A(g_t^{-1}(z)))\mathbb{1}_{\{t < t_{g_t^{-1}(z)},\ t < s_z,\ g_t^{-1}(z) \notin A\}}(\omega) = h_t(z)\mathbb{1}_{\{t < s_z,\ z \notin g_t(A)\}}(\omega)$$

is $\mathcal{B}(D) \times \mathcal{G}_t$-measurable. Here, we have dropped the condition $t < t_{g_t^{-1}(z)}$ because $g_t^{-1}(z) \in D \setminus F_t$ holds trivially on the event $\{t < s_z\}$.

We next set $y_0(t) := \min\{ y_j(t)\mathbb{1}_{\{t < \zeta\}} : 1 \le j \le N \}$. As each $y_j(t)$ is continuous and decreasing up to ζ, it follows on the event $\{t < \zeta\}$ that $z \in \mathbb{H}_{y_0(t)}$ implies $z \in \bigcap_{u \in [0,t]} D_u$, which further means $t < s_z$. We also recall from Theorem 5.3.4 (iii) that, on the event $\{t < \overline{\tau}_A\}$, there exists a domain G_t with the properties $\xi(t) \in G_t$ and $G_t \cap \mathbb{H} \subset D_t \setminus g_t(A)$. Therefore, on the event $\{t < \overline{\tau}_A\}$, $\mathcal{G}_t$-measurable random variables $z_n := \xi(t) + in^{-1}$, $n \ge 1$, satisfy $t < s_{z_n}$ and $z_n \notin g_t(A)$ from some n_0 on. Hence

$$h_t(\xi(t))\mathbb{1}_{\{t < \overline{\tau}_A\}} = \lim_{n \to \infty} h_t(z_n)\mathbb{1}_{\{t < s_{z_n},\ z_n \notin g_t(A)\}}\mathbb{1}_{\{t < \overline{\tau}_A\}}.$$

This equality shows that $h_t(\xi(t))\mathbb{1}_{\{t < \overline{\tau}_A\}}$ is $\mathcal{G}_t$-measurable. □

We now turn to the proof of Theorem 5.4.3. By making use of Proposition 5.3.5 shown in the preceding section, we shall prepare two propositions.

Denote by ∂_t the partial derivative in t. It follows from (5.2.6) that the inverse map g_t^{-1} of g_t satisfies

$$\partial_t g_t^{-1}(z) = 2\pi (g_t^{-1})'(z)\Psi_{\mathbf{s}(t)}(z, \xi(t)), \quad g_0^{-1}(z) = z \in D_t.$$

This together with (5.3.19) yields that, for $z \in D_t \setminus g_t(A)$,

$$\begin{aligned}\partial_t h_t(z) &= (\partial_t \widetilde{g}_t)(h \circ g_t^{-1}(z)) + (\widetilde{g}_t \circ h)'(g_t^{-1}(z))\partial_t(g_t^{-1}(z)) \\ &= -2\pi h_t'(\xi(t))^2 \widetilde{\Psi}_t(\widetilde{g}_t \circ h \circ g_t^{-1}(z), h_t(\xi(t))) \\ &\quad + 2\pi(\widetilde{g}_t \circ h)'(g_t'(z))(g_t^{-1})'(z)\Psi_{\mathbf{s}(t)}(z, \xi(t)),\end{aligned}$$

namely,

$$\begin{aligned}\partial_t h_t(z) &= -2\pi h_t'(\xi(t))^2 \widetilde{\Psi}_t(h_t(z), h_t(\xi(t))) \\ &\quad + 2\pi h_t'(z)\Psi_{\mathbf{s}(t)}(z, \xi(t)), \qquad z \in D_t \setminus g_t(A),\end{aligned} \tag{5.4.7}$$

where $\widetilde{\Psi}_t$ is the complex Poisson kernel of the standard slit domain $D(\widetilde{\mathbf{s}}(t)) = \widetilde{g}_t(\widetilde{D} \setminus h(F_t)) =: \widetilde{D}_t$. So we may write $\widetilde{\Psi}_t$ as $\Psi_{h_t(\mathbf{s}(t))}$. See Figure 5.2.

Proposition 5.4.4. *Fix any $s \in [0, \overline{\tau}_A)$. Take $0 < r_1 < r_2$ and $\delta > 0$ so that (5.3.8) and (5.3.9) hold. Then $h_t(z)$ is differentiable in $t \in (s - \delta, s + \delta)$ for each $z \in B(\xi(s), r_2)$, and $\partial_t h_t(z)$ is jointly continuous in $(t, z) \in (s-\delta, s+\delta) \times B(\xi(s), r_2)$. In particular, $\partial_t h_t(z)$ is uniformly bounded in $(t, z) \in (s-\delta/2, s+\delta/2) \times B(\xi(s), r_1)$. Furthermore, it holds for $t \in (s - \delta, s + \delta)$ that*

$$\begin{aligned}(\partial_t h_t)(\xi(t)) &= h_t'(\xi(t)) b_{\mathrm{BMD}}(\xi(t), \mathbf{s}(t)) \\ &\quad - h_t'(\xi(t))^2 b_{\mathrm{BMD}}(h_t(\xi(t)), h_t(\mathbf{s}(t)) - 3h_t''(\xi(t)).\end{aligned} \tag{5.4.8}$$

Proof. Take any $r \in (r_1, r_2)$ and denote the circle $\partial B(\xi(s), r)$ by C. Denote the right-hand side of (5.4.7) by $k(t, z) = k(t, z, \omega)$. For $t \in (s - \delta, s + \delta)$, $k(t, z)$ is extended in z from $C \cap \mathbb{H}$ to $C \cap \overline{\mathbb{H}}$ and

$$k(t, z) \text{ is jointly continuous in } (t, z) \in (s - \delta, s + \delta) \times (C \cap \overline{\mathbb{H}}). \tag{5.4.9}$$

In fact, as $\mathbf{s}(t)$ is continuous, the same proof of Theorem 2.5.1 works to verify that the complex Poisson kernel $\Psi_{\mathbf{s}(t)}(z, \zeta)$ of $\mathbf{s}(t)$ is jointly continuous in the sense formulated in Theorem 2.5.1. In view of (5.3.8), this particularly implies that $\Psi_{\mathbf{s}(t)}(z, \xi(t))$ satisfies (5.4.9). Hence so does the second term of $k(t, z)$ by Proposition 5.3.5 (ii). Since $\{h(F_t)\}$ is a family of continuously growing hulls by Theorem 5.3.4 (i), $\widetilde{\mathbf{s}}(t)$ is continuous by Lemma 5.2.3. Consequently, in an analogous manner to the above, we can see that the first term of $k(t, z)$ also satisfies (5.4.9).

By means of the Schwarz reflection of $\widetilde{\Psi}_t$ and $\Psi_{\mathbf{s}(t)}$, $k(t, z)$ can then be extended to $(t, z) \in (t - \delta, t + \delta) \times (C \cap \mathbb{H}^-)$ to be jointly continuous there, and the identity (5.4.7) extends to $(t, z) \in (s - \delta, s + \delta) \times (C \setminus \partial\mathbb{H})$.

Expressing $(h_u(z) - h_t(z))/(u - t)$, $z \in B(\xi(s), r)$, u, $t \in (s - \delta, s + \delta)$, by the Cauchy integral formula around the circle C and letting $u \to t$, we see from (5.4.9) that $h_t(z)$ is differentiable in t for each $z \in B(\xi(s), r)$ and $\partial_t h_t(z)$ is analytic in $z \in B(\xi(s), r)$ and jointly continuous in $(t, z) \in (s - \delta, s + \delta) \times B(\xi(s), r)$. In particular, $(\partial_t h_t)(\xi(t))$ can be evaluated explicitly by $\lim_{z \to \xi(t),\, z \in \mathbb{H}} \partial_t h_t(z)$. Indeed, by the definition (5.4.1) of $b_{\mathrm{BMD}}(\xi, \mathbf{s})$, we obtain from (5.4.7)

$$\begin{aligned}(\partial_t h_t)(\xi(t)) &= h_t'(\xi(t)) b_{\mathrm{BMD}}(\xi(t), \mathbf{s}(t)) \\ &\quad - h_t'(\xi(t))^2 b_{\mathrm{BMD}}(h_t(\xi(t)), h_t(\mathbf{s}(t))) + \lim_{z \to \xi(t),\, z \in \mathbb{H}} J(t, z),\end{aligned}$$

where

$$J(t,z) = \frac{2h'_t(\xi(t))^2}{h_t(z) - h_t(\xi(t))} - \frac{2h'_t(z)}{z - \xi(t)}.$$

By using the Taylor expansion of $h_t(z)$ around $z = \xi(t)$, we readily get

$$\lim_{z \to \xi(t),\, z \in \mathbb{H}} J(t,z) = -3h''_t(\xi(t)),$$

which yields (5.4.8). □

For each $s \in [0, \overline{\tau}_A)$, the constants $r_1 > 0$ and $\delta > 0$ appearing in Proposition 5.3.5 and Proposition 5.4.4 depend on s. We put

$$I_s = (s - \delta/2, s + \delta/2) \cap [0, \overline{\tau}_A), \qquad B_s = B(\xi(s), r_1).$$

We have then, for each $s \in [0, \overline{\tau}_A)$,

1°. $h_t(z)$ is differentiable in $t \in I_s$ for each $z \in B_s$,
2°. $h_t(z)$, $\partial_t h_t(z)$, $h'_t(z)$, $h''_t(z)$ are jointly continuous and bounded in $(t,z) \in I_s \times B_s$,
3°. $\{\xi(u) : u \in \overline{I}_s\} \subset B_s$,
4°. $(\partial_t h_t)(\xi(t))$ satisfies (5.4.8) for $t \in I_s$.

The process $\xi(t)$ is a continuous semi-martingale adapted to the Brownian filtration $\{\mathcal{G}_t\}$ as a solution of the SDE (4.2.10). Although $h_t(z)$ is a random function, we have the following version of generalized Itô's formula for the composite process $\widetilde{\xi}(t)\mathbb{1}_{\{t<t_0\}} = h_t(\xi(t))\mathbb{1}_{\{t<t_0\}}$, which is also $\{\mathcal{G}_t\}$-adapted by Proposition 5.4.2.

When the hull A is non-empty, for $\ell \in \mathbb{N}$, let $\tau_\ell = \sup\{t \in [0,\zeta); F_t \cap \overline{A}_\ell = \emptyset\}$, where A_ℓ denotes $1/\ell$-neighborhood of the hull A. Note that $\tau_\ell < \overline{\tau}_A$, $\tau_\ell \uparrow \overline{\tau}_A$ as $\ell \to \infty$ and each τ_ℓ is a $\{\mathcal{G}_t\}$-stopping time in view of Lemma 4.3.7. When $A = \emptyset$, we define $\tau_\ell = \zeta$ for every $\ell \in \mathbb{N}$. We also choose relatively compact open sets G_k, $k \in \mathbb{N}$, of $\mathbb{R} \times \mathcal{S}$ so that $\overline{G}_k \subset G_{k+1}$ for all k and $G_k \uparrow \mathbb{R} \times \mathcal{S}$ as $k \to \infty$. Let σ_k be the exit time of $(\xi(t), \mathbf{s}(t))$ from G_k:

$$\sigma_k = \inf\{\, t > 0 : (\xi(t), \mathbf{s}(t)) \in (\mathbb{R} \times \mathcal{S}) \setminus G_k \,\}, \quad k \in \mathbb{N}.$$

Note that σ_k increases to the lifetime ζ as $k \to \infty$.

Proposition 5.4.5. *It holds for each $T > 0$ and k, $\ell \in \mathbb{N}$ that*

$$\widetilde{\xi}(T \wedge \tau_\ell \wedge \sigma_k) - \widetilde{\xi}(0) = \int_0^{T\wedge\tau_\ell\wedge\sigma_k} (\partial_u h_u)(\xi(u))du + \int_0^{T\wedge\tau_\ell\wedge\sigma_k} h'_u(\xi(u))d\xi(u)$$
$$+ \frac{1}{2}\int_0^{T\wedge\tau_\ell\wedge\sigma_k} h''_u(\xi(u))d\langle\xi\rangle(u). \tag{5.4.10}$$

Here, the second term of the right-hand side of the identity is the stochastic integral and $\langle\xi\rangle(u)$ in the third term is the quadratic variation process of $\xi(u)$.

Proof. We fix $T > 0$ and set $\sigma = \sigma_k \wedge \tau_\ell$ for fixed $k,\ \ell \in \mathbb{N}$. Consider the partition Δ of the compact time interval $[0, T \wedge \sigma]$:

$$\Delta := \{0 = t_0 < \cdots < t_i < t_{i+1} < \cdots < t_n = T \wedge \sigma\}.$$

We have

$$\begin{aligned} \widetilde{\xi}(T \wedge \sigma) - \widetilde{\xi}(0) &= \sum_{t_i \in \Delta} (\widetilde{\xi}(t_{i+1}) - \widetilde{\xi}(t_i)) = \sum_{t_i \in \Delta} (h_{t_{i+1}}(\xi(t_{i+1})) - h_{t_i}(\xi(t_i))) \\ &=: \mathrm{I}_\Delta + \mathrm{II}_\Delta, \end{aligned} \tag{5.4.11}$$

where

$$\mathrm{I}_\Delta = \sum_{t_i \in \Delta} (h_{t_{i+1}}(\xi(t_{i+1}) - h_{t_i}(\xi(t_{i+1})) \quad \text{and} \quad \mathrm{II}_\Delta = \sum_{t_i \in \Delta} (h_{t_i}(\xi(t_{i+1}) - h_{t_i}(\xi(t_i)).$$

Since $[0, T \wedge \sigma] \subset \bigcup_{j=1}^m I_{s_j}$ for some $s_1, \cdots, s_m \in [0, T \wedge \sigma]$, we can use properties $\mathbf{1}^\circ, \mathbf{2}^\circ, \mathbf{3}^\circ$ on (I_{s_j}, B_{s_j}) for each $1 \le j \le m$ to conclude that

$$\mathrm{I}_\Delta = \sum_{t_i \in \Delta} \int_{t_i}^{t_{i+1}} (\partial_u h_u)(\xi(t_{i+1}))du = \int_0^{T \wedge \sigma} (\partial_u h_u)(\xi^\Delta(u))du,$$

where $\xi^\Delta(u) = \xi(t_{i+1})$ for $u \in [t_i, t_{i+1})$, and that

$$\lim_{|\Delta| \to 0} \mathrm{I}_\Delta = \int_0^{T \wedge \sigma} (\partial_u h_u)(\xi(u))du, \quad \text{a.s.} \tag{5.4.12}$$

Note that, according to (4.2.10), $\xi(t) = M^0(t) + \eta(t)$ with

$$M^0(t) = \int_0^t \alpha(\mathbf{s}(s) - \widehat{\xi}(s))dB_s \quad \text{and} \quad \eta(t) = \xi + \int_0^t b(\mathbf{s}(s) - \widehat{\xi}(s))ds.$$

By assumption, $\alpha(\xi, \mathbf{s}) = \alpha(\mathbf{s} - \widehat{\xi})$ and $b(\xi, \mathbf{s}) = b(\mathbf{s} - \widehat{\xi})$ satisfy the local Lipschitz condition (4.3.1) on $\mathbb{R} \times \mathcal{S}$ and so

$$\sup_{(\xi, \mathbf{s}) \in G_k} |\alpha(\xi, \mathbf{s})| =: c_1 < \infty, \qquad \sup_{(\xi, \mathbf{s}) \in G_k} |b(\xi, \mathbf{s})| =: c_2 < \infty.$$

We now use Itô's formula for the continuous semi-martingale $\xi(t)$ (cf. [Kunita (1990), Theorem 2.3.11]) to obtain

$$\begin{aligned} \mathrm{II}_\Delta &= \sum_{t_i \in \Delta} \left(\int_{t_i}^{t_{i+1}} h'_{t_i}(\xi(u))d\xi(u) + \frac{1}{2} \int_{t_i}^{t_{i+1}} h''_{t_i}(\xi(u))d\langle \xi \rangle(u) \right) \\ &= \int_0^{T \wedge \sigma} h'_{[u]_\Delta}(\xi(u))dM^0(u) \\ &\quad + \int_0^{T \wedge \sigma} h'_{[u]_\Delta}(\xi(u))d\eta(u) + \frac{1}{2} \int_0^{T \wedge \sigma} h''_{[u]_\Delta}(\xi(u))d\langle M^0 \rangle(u) \\ &=: \mathrm{III}_\Delta + \mathrm{IV}_\Delta + \mathrm{V}_\Delta, \end{aligned}$$

where $[u]_\Delta = t_i$ for $u \in [t_i, t_{i+1})$.

By taking the above bounds of α, b into account and making a similar consideration to the proof of (5.4.12), we obtain almost surely that

$$\left|\mathrm{IV}_\Delta - \int_0^{T\wedge\sigma} h'_u(\xi(u))d\eta(u)\right| \le c_2 \int_0^{T\wedge\tau_\ell} |h'_{[u]_\Delta}(\xi(u)) - h'_u(\xi(u))|du \\ \to 0 \quad \text{as } |\Delta| \to 0, \tag{5.4.13}$$

and

$$\left|\mathrm{V}_\Delta - \frac{1}{2}\int_0^{T\wedge\sigma} h''_u(\xi(u))d\langle M^0\rangle(u)\right| \\ \le \frac{c_1^2}{2}\int_0^{T\wedge\tau_\ell} (h''_{[u]_\Delta}(\xi(u)) - h''_u(\xi(u)))^2 du \to 0 \quad \text{as } |\Delta| \to 0. \tag{5.4.14}$$

Concerning the stochastic integral III_Δ, we have almost surely that

$$\left\langle \int_0^{\cdot} \mathbb{1}_{\{u\le\sigma\}} \left[h'_{[u]_\Delta}(\xi(u)) - h'_u(\xi(u))\right] dM^0(u)\right\rangle_T \\ \le c_1^2 \int_0^{T\wedge\tau_\ell} \left(h'_{[u]_\Delta}(\xi(u)) - h'_u(\xi(u))\right)^2 du \to 0,$$

as $|\Delta| \to 0$, and so III_Δ converges to $\int_0^{T\wedge\sigma} h'_u(\xi(u))dM^0(u)$ in probability as $|\Delta| \to 0$ owing to [Kunita (1990), Theorem 2.2.15]. This combined with (5.4.11), (5.4.12), (5.4.13) and (5.4.14) gives the desired formula (5.4.10). □

Proof of Theorem 5.4.3. It follows from Proposition 5.4.5 that

$$d\widetilde{\xi}(t) = (\partial_t h_t)(\xi(t))dt + \left(h'_t(\xi(t))b(\mathbf{s}(t) - \widehat{\xi}(t)) + \frac{1}{2}h''_t(\xi(t))\alpha(\mathbf{s}(t) - \widehat{\xi}(t))^2\right)dt \\ + h'_t(\xi(t))\alpha(\mathbf{s}(t) - \widehat{\xi}(t))dB_t, \quad t \in (0, t_0).$$

We then substitute (5.4.8) into this to get (5.4.4). □

5.5 Invariance of $\mathrm{SKLE}_{\sqrt{6},-b_{\mathrm{BMD}}}$ and equivalence of $\mathrm{SKLE}_{\sqrt{\kappa},b}$ to SLE_κ

We continue to work under the setting of Section 5.4.2. That is, we consider the domains D, $\widetilde{D}$, V and a univalent map $h : V \mapsto \widetilde{D}$ satisfying either condition **(a)** or condition **(b)** as in Theorem 5.3.4, along with $\mathrm{SKLE}_{\alpha,b}$ $\{F_t\}$ on D driven by the solution $(\xi(t), \mathbf{s}(t))$ of the SDE system (4.2.10) and (4.2.11).

The random time $t_0 = t_0(\omega)$ is defined by (5.4.3). The family of image hulls $\{h(F_t),\ t \in [0, t_0)\}$ by h has the driving function (process) $\widetilde{\xi}(t) = h_t(\xi(t))$ whose differential has been shown to satisfy the equation (5.4.4) in Theorem 5.4.3.

We make a half-plane capacity reparametrization

$$\check{F}_t = h(F_{\widetilde{a}^{-1}(2t)}) \quad \text{for } t \in [0, \check{t}_0) \text{ where } \check{t}_0 := \widetilde{a}(t_0)/2, \tag{5.5.1}$$

of $h(F_t)$ so the half-plane capacity of $\check{F}_t$ is $2t$. Here, in view of Proposition 5.3.6,

$$\widetilde{a}(t) = 2\int_0^t h_u'(\xi(u))^2 du \quad \text{for } t < t_0, \tag{5.5.2}$$

which is continuous and strictly increasing. The family of hulls $\{\check{F}_t, t \in [0, \check{t}_0)\}$ has the driving function (process)

$$\check{\xi}(t) = \widetilde{\xi}(\widetilde{a}^{-1}(2t)), \quad t \in [0, \check{t}_0). \tag{5.5.3}$$

Lemma 5.5.1. *There exists a one-dimensional Brownian motion $\check{B}_t$ so that for $t \in [0, \check{t}_0)$,*

$$\begin{aligned} d\check{\xi}(t) &= \check{h}_t'(\xi^0(t))^{-1}\left(b(\mathbf{s}^0(t) - \widehat{\xi^0}(t)) + b_{\mathrm{BMD}}(\xi^0(t), \mathbf{s}^0(t))\right) dt \\ &+ \frac{1}{2}\check{h}_s''(\xi^0(t))\check{h}_t'(\xi^0(t))^{-2}\left(\alpha(\mathbf{s}^0(t) - \widehat{\xi^0}(t))^2 - 6\right) dt \\ &- b_{\mathrm{BMD}}(\check{\xi}(t), \check{\mathbf{s}}(t))dt + \alpha(\mathbf{s}^0(t) - \widehat{\xi^0}(t))d\check{B}_t. \end{aligned} \tag{5.5.4}$$

Here

$$\check{\mathbf{s}}(t) = h_{\widetilde{a}^{-1}(2t)}(\mathbf{s}(\widetilde{a}^{-1}(2t))), \tag{5.5.5}$$

which is the same as (5.3.18). *Further*

$$\check{h}_t'(z) = h_{\widetilde{a}^{-1}(2t)}'(z), \quad \check{h}_t''(z) = h_{\widetilde{a}^{-1}(2t)}''(z), \tag{5.5.6}$$

and

$$\xi^0(t) = \xi(\widetilde{a}^{-1}(2t)), \quad \mathbf{s}^0(t) = \mathbf{s}(\widetilde{a}^{-1}(2t)). \tag{5.5.7}$$

Notice that h_t is univalent on G_t by Theorem 5.3.4 and consequently h_t' never vanishes there.

Proof. We give a proof only in case **(a)** that $V = D \setminus A$ for a hull $A \subset D$ and h is the canonical map from V onto $\widetilde{D}$. The proof for the case **(b)** is essentially the same and simpler.

Let τ_ℓ, σ_k, $\ell, k \in \mathbb{N}$, be stopping times considered in Proposition 5.4.5. Since $\tau_\ell \wedge \sigma_k \uparrow t_0$ as ℓ, $k \to \infty$, it suffices to prove (5.5.4) on $[0, \sigma)$ for $\sigma = \tau_\ell \wedge \sigma_k$ with fixed ℓ, $k \in \mathbb{N}$.

By making a finite covering of the interval $[0, T\wedge\sigma]$ as in the proof of Proposition 5.4.5, we have

$$\int_0^T \mathbb{1}_{\{u \le \sigma\}} h_u'(\xi(u))^2 du = \int_0^{T\wedge\sigma} h_u'(\xi(u))^2 du < \infty \quad \text{a.s. for any } T > 0. \tag{5.5.8}$$

Hence, by the same procedure as in [Ikeda and Watanabe (1981), p. 52], the stochastic integral

$$M_t = \int_0^t \mathbb{1}_{\{u \le \sigma\}} h_u'(\xi(u)) dB_u, \quad t > 0,$$

is well defined as a continuous locally square-integrable martingale with the quadratic variation process

$$\langle M\rangle_t = \int_0^{t\wedge\sigma} h_u'(\xi(u))^2 du = \frac{\widetilde{a}(t\wedge\sigma)}{2}, \quad t > 0. \tag{5.5.9}$$

According to [Ikeda and Watanabe (1981), Theorem 7.3' in Chapter II], there exists then a one-dimensional Brownian motion $\check{B}_t$ such that

$$M_{\widetilde{a}^{-1}(2t)} = \check{B}_t \quad \text{for } 0 \le t < \check{\sigma} := \widetilde{a}(\sigma)/2. \tag{5.5.10}$$

It now follows from (5.4.4) that for $t \in [0, \check{\sigma})$,

$$\begin{aligned}
\check{\xi}(t) - \check{\xi}(0) &= \int_0^{\widetilde{a}^{-1}(2t)} h_u'(\xi(u)) \Big(b(\mathbf{s}(u) - \widehat{\xi}(u)) + b_{\mathrm{BMD}}(\xi(u), \mathbf{s}(u))\Big)\, du \\
&\quad + \frac{1}{2}\int_0^{\widetilde{a}^{-1}(2t)} h_u''(\xi(u)) \Big(\alpha(\mathbf{s}(u) - \widehat{\xi}(u))^2 - 6\Big)\, du \\
&\quad - \int_0^{\widetilde{a}^{-1}(2t)} b_{\mathrm{BMD}}(\widetilde{\xi}(u), h_u(\mathbf{s}(u))) \frac{d\widetilde{a}(u)}{2} \\
&\quad + \int_0^{\widetilde{a}^{-1}(2t)} \alpha(\mathbf{s}(u) - \widehat{\xi}(u)) dM_u.
\end{aligned}$$

Set $u = \widetilde{a}^{-1}(2v)$ so that $du = h_u'(\xi(u))^{-2}\, dv$. Then for $t \in [0, \check{\tau}_\ell)$,

$$\begin{aligned}
\check{\xi}(t) - \check{\xi}(0) &= \int_0^t \check{h}_v'(\xi(v))^{-1} \Big(b(\mathbf{s}^0(v) - \widehat{\xi}^0(v)) + b_{\mathrm{BMD}}(\xi^0(v), \mathbf{s}^0(v))\Big)\, dv \\
&\quad + \frac{1}{2}\int_0^t h_v''(\xi^0(v)) h_v'(\xi^0(v))^{-2} \Big(\alpha(\mathbf{s}^0(v) - \widehat{\xi}^0(v))^2 - 6\Big)\, dv \\
&\quad - \int_0^t b_{\mathrm{BMD}}(\check{\xi}(v), \check{\mathbf{s}}(v)) dv + \int_0^t \alpha(\mathbf{s}^0(v) - \widehat{\xi}^0(v)) d\check{B}_v,
\end{aligned}$$

yielding (5.5.4). □

Theorem 5.5.2 (Conformal invariance of SKLE$_{\sqrt{6},-b_{\mathrm{BMD}}}$). *Suppose $\{F_t\}$ is an SKLE$_{\sqrt{6},-b_{\mathrm{BMD}}}$ on D. Define $t_0 = t_0(\omega)$ by (5.4.3). Then the half-plane capacity reparametrization $\{\check{F}_t,\ t \in [0, \check{t}_0)\}$ defined by (5.5.1) of the family of image hulls $\{h(F_t),\ t \in [0, t_0)\}$ by h is SKLE$_{\sqrt{6},-b_{\mathrm{BMD}}}$ on $\widetilde{D}$ when $\widetilde{D} \subsetneq \mathbb{H}$ and SLE$_6$ when $\widetilde{D} = \mathbb{H}$.*

Proof. Assume that $\{F_t\}$ is an SKLE$_{\sqrt{6},-b_{\mathrm{BMD}}}$ on D. Then the driving process $\{\check{\xi}(t),\ t \in [0, \check{t}_0)\}$ of $\{\check{F}_t,\ t \in [0, \check{t}_0)\}$ satisfies

$$d\check{\xi}(t) = -b_{\mathrm{BMD}}(\check{\xi}(t), \check{\mathbf{s}}(t))dt + \sqrt{6}d\check{B}_t, \quad \text{on } [0, \check{t}_0), \tag{5.5.11}$$

by virtue of Lemma 5.5.1.

When $\widetilde{D} \subsetneq \mathbb{H}$, Proposition 5.3.7 asserts that the canonical map $\check{g}_t$ from $\widetilde{D} \setminus \check{F}_t$ and the induced slit motion $\check{\mathbf{s}}(t)$ with $D(\check{\mathbf{s}}(t)) = \check{g}_t(\widetilde{D})$ satisfy the ODE

$$\check{\mathbf{s}}_j(t) - \check{\mathbf{s}}_j(0) = \int_0^t b_j(\check{\xi}(u), \check{\mathbf{s}}(u))du, \quad 1 \le j \le 3N, \quad t < \check{t}_0, \tag{5.5.12}$$

$$\frac{d}{dt}\check{g}_t(z) = -2\pi\Psi_{\check{\mathbf{s}}(t)}(\check{g}_t(z), \check{\xi}(t)), \quad \check{g}_0(z) = z \in \widetilde{D}, \quad t < \check{t}_0. \tag{5.5.13}$$

(5.5.11), (5.5.12) and (5.5.13) mean that $\{\check{F}_t,\ t \in [0, \check{t}_0)\}$ is $\mathrm{SKLE}_{\sqrt{6}, -b_{\mathrm{BMD}}}$ on $\widetilde{D}$.

When $\widetilde{D} = \mathbb{H}$, we see from Lemma 5.3.8 (i) and (5.5.11) with $b_{\mathrm{BMD}} = 0$ that $\{\check{F}_t,\ t \in [0, \check{t}_0)\}$ is SLE_6. □

In case **(a)** with $D = \mathbb{H}$, this theorem was discovered by [Lawler, Schramm and Werner (2001)] under the term *locality of* SLE_6.

Taking $\widetilde{D} = \mathbb{H}$ and h the identity map from D into $\mathbb{H}$, the last part of the above Theorem in particular yields the following result.

Theorem 5.5.3. $\mathrm{SKLE}_{\sqrt{6}, -b_{\mathrm{BMD}}}$ *reparametrized by* (5.5.1) *has the same distribution as* SLE_6 *over the random time interval* $[0, \check{t}_0)$.

Theorem 5.5.4. *For any positive constant* α *and a homogeneous function* $b(\mathbf{s})$ *of* $\mathbf{s} \in \mathcal{S}$ *with degree* -1 *satisfying the condition* **(L)** *of* (4.3.1), *let* $\check{t}_0$ *be defined by* (5.5.1) *with* $t_0 = \zeta$. *Then on the random time interval* $[0, \check{t}_0)$, *the law of* $\mathrm{SKLE}_{\alpha,b}$ *reparametrized by* (5.5.1) *is absolutely continuous with respect to that of an* SLE_{α^2}. *More precisely, there exists a sequence of stopping times* $\{\sigma_n; n \ge 1\}$ *increasing to* $\check{t}_0$ *such that* $\mathrm{SKLE}_{\alpha,b}$ *reparametrized by* (5.5.1) *has the same distribution as* SLE_{α^2} *over random time interval* $[0, \sigma_n]$ *under a suitable Girsanov transform for every* $n \ge 1$.

Proof. By (5.5.4) of Lemma 5.5.1,

$$d\check{\xi}(t) = \alpha\left(d\check{B}_t + \beta(t)dt\right)$$

on $[0, \check{t}_0)$ with $t_0 = \zeta$, where $\check{B}$ is a one-dimensional Brownian motion and

$$\beta(t) := \alpha^{-1}\check{h}_t'(\xi^0(s))^{-1}\left(b(\mathbf{s}^0(t) - \widehat{\xi}^0(t)) + b_{\mathrm{BMD}}(\xi^0(t), \mathbf{s}^0(t))\right)$$
$$+ \frac{\alpha - 6}{2\alpha}\check{h}_s''(\xi^0(t))\check{h}_t'(\xi^0(t))^{-2} - \alpha^{-1} b_{\mathrm{BMD}}(\check{\xi}(t), \check{\mathbf{s}}(t)).$$

For every integer $n \ge 1$, let

$$\sigma_n := \inf\left\{t \in [0, \check{t}_0) : \int_0^t \beta(s)^2 ds \ge n\right\} \wedge \check{t}_0,$$

which is a stopping time with respect to the filtration $\{\widetilde{\mathcal{G}}_t := \mathcal{G}_{\widetilde{a}^{-1}(2t)}; t \ge 0)\}$ and increases to $\check{t}_0$ as $n \to \infty$. On each $\widetilde{\mathcal{G}}_{\sigma_n}$, we define a probability measure $\mathbf{Q}_n$ by

$$\frac{d\mathbf{Q}_n}{d\mathbb{P}} = \exp\left(-\int_0^{\sigma_n} \beta(s)d\check{B}_s - \frac{1}{2}\int_0^{\sigma_n} \beta(s)^2 ds\right).$$

By Girsanov's theorem (see, e.g., [Revuz and Yor (1999), Theorem VIII.1.7]), $\{\check{\xi}_t; t \in [0, \sigma_n]\}$ is an α-multiple of a standard Brownian motion under the probability measure $\mathbf{Q}_n$. Consequently, $\{\check{F}_t; t \in [0, \sigma_n]\}$ is SLE_{α^2} on the random time interval $[0, \sigma_n]$. As $\mathbf{Q}_{n+1}\big|_{\mathcal{G}_{\sigma_n}} = \mathbf{Q}_n$ for every $n \geq 1$, by Kolmogorov's extension theorem there is a probability measure $\mathbf{Q}$ on $\sigma(\cup_{n\geq 1}\mathcal{G}_{\sigma_n})$ so that $\mathbf{Q}\big|_{\mathcal{G}_{\sigma_n}} = \mathbf{Q}_n$ for every $n \geq 1$. This establishes the desired assertions. □

When α is a positive constant, it follows from Theorem 5.5.4 and [Rohde and Schramm (2005)] that $\mathrm{SKLE}_{\alpha,b}$ is generated by a continuous curve γ and that the curve γ is simple when $\alpha \leq 2$, self-intersecting when $2 < \alpha \leq 2\sqrt{2}$ and space-filling when $\alpha > 2\sqrt{2}$.

Appendix

A.1 Dirichlet problem and α-order exit distribution

Let $\mathbf{Z} = \{Z_t, t \geq 0;\ \mathbb{P}_z, z \in \mathbb{C}\}$ be the planar Brownian motion and U be an open subset of $\mathbb{C}$ such that $\mathbb{C} \setminus U$ is non-polar. We fix $\alpha \geq 0$ and set

$$\mathcal{H}_U^\alpha f(z) = \mathbb{E}_z\left[e^{-\alpha\tau_U} f(Z_{\tau_U})\right], \quad z \in U,$$

where τ_U is the exit time of $\mathbf{Z}$ from U and f is a bounded measurable function on ∂U.

The function $\mathcal{H}_U^\alpha f(z)$ is the weighted average of f by the α-order exit distribution $\mathbb{E}_z\left[e^{-\alpha\tau_U}; Z_{\tau_U} \in dy\right]$ of $\mathbf{Z}$ starting from z. When $\alpha = 0$, $\mathcal{H}_U^0 f(z) = \mathbb{E}_z\left[f(Z_{\tau_U})\right]$ is a harmonic function of $z \in U$.

Theorem A.1.1. $\lim_{z \in U,\, z \to w} \mathcal{H}_U^\alpha f(z) = f(w)$ *whenever* $w \in \partial U \cap (\mathbb{C} \setminus U)^r$ *and* f *is continuous at* w.

Proof. Let $\tau_{B(z,\varepsilon)} := \inf\{t > 0 : |Z_t - z| > \varepsilon\}$. By the strong Markov property of $\mathbf{Z}$, we have for any $t > 0$ and $z \in \mathbb{C}$,

$$\begin{aligned}
\mathbb{P}_z(\tau_{B(z,\varepsilon)} \leq t) &\leq \mathbb{P}_z(|Z_t - z| > \varepsilon/2) + \mathbb{P}_z(\tau_{B(z,\varepsilon)} < t \text{ and } |Z_t - z| \leq \varepsilon/2) \\
&\leq \mathbb{P}_z(|Z_t - z| > \varepsilon/2) + \mathbb{P}_z(\tau_{B(z,\varepsilon)} \leq t \text{ and } |Z_t - Z_{\tau_{B(z,\varepsilon)}}| \geq \varepsilon/2) \\
&\leq \mathbb{P}_z(|Z_t - z| > \varepsilon/2) \\
&\quad + \mathbb{E}_z\left[\mathbb{1}_{\{\tau_{B(z,\varepsilon)} \leq t\}} \mathbb{P}_{Z_{\tau_{B(z,\varepsilon)}}}\left(|Z_{t-\tau_{B(z,\varepsilon)}} - Z_0| \geq \varepsilon/2\right)\right] \\
&\leq 2 \sup_{x \in \mathbb{C}, s \in (0,t]} \mathbb{P}_x(|Z_s - x| \geq \varepsilon/2) \\
&= 2 \sup_{s \in (0,t]} \mathbb{P}_0(|Z_s| \geq \varepsilon/2) = 2\mathbb{P}_0(|Z_t| \geq \varepsilon/2).
\end{aligned}$$

It follows that

$$\begin{aligned}
\limsup_{t \downarrow 0} \sup_{z \in \mathbb{C}} \mathbb{P}_z\left(\sup_{s \in [0,t]} |Z_s - z| > \varepsilon\right) &= \limsup_{t \downarrow 0} \sup_{z \in \mathbb{C}} \mathbb{P}_z(\tau_{B(z,\varepsilon)} < t) \\
&\leq \lim_{t \downarrow 0} 2\mathbb{P}_0(|Z_t| \geq \varepsilon/2) = 0.
\end{aligned} \tag{A.1.1}$$

On the other hand, $\mathbf{Z}$ has the strong Feller property in the sense that $\mathbb{E}_z[g(Z_t)]$ is continuous in z for any bounded measurable function g. Consequently

$$\mathbb{P}_z(\tau_U > t) \text{ is upper semi-continuous in } z \in \mathbb{C}, \tag{A.1.2}$$

as it is the decreasing limit of $\mathbb{E}_z\left[\mathbb{P}_{Z_{t_n}}(\tau_U > t - t_n)\right]$ as $t_n \downarrow 0$.

Assume that $w \in \partial U$ is regular for $\mathbb{C} \setminus U$, namely, $\mathbb{P}_w(\tau_U = 0) = 1$. Take any bounded measurable function f on ∂U that is continuous at w. Then for any $\varepsilon > 0$ there is $\delta > 0$ so that $|f(y) - f(w)| < \varepsilon$ for any $y \in B_\delta(w) \cap \partial U$. Denoting τ_U by τ, we have for $z \in U$,

$$\begin{aligned}
&|\mathcal{H}_U^\alpha f(z) - f(w)| \qquad\qquad (A.1.3)\\
&= \Big|\mathbb{E}_z\left[e^{-\alpha\tau}\mathbb{1}_{B(w,\delta)}(Z_\tau)(f(Z_\tau) - f(w))\right]\\
&\quad + \mathbb{E}_z\left[e^{-\alpha\tau}\mathbb{1}_{\mathbb{C}\setminus B(w,\delta)}(Z_\tau)(f(Z_\tau) - f(w))\right] - \mathbb{E}_z[1 - e^{-\alpha\tau}]f(w)\Big|\\
&\le \varepsilon + 2\|f\|_\infty (1 - \mathbb{P}_z(Z_\tau \in B_\delta(w)) + |f(w)| \left(1 - \mathbb{E}_z[e^{-\alpha\tau}]\right). \qquad (A.1.4)
\end{aligned}$$

By (A.1.1), there is $t > 0$ so that

$$1 - e^{-\alpha t} < \varepsilon \quad \text{and} \quad \mathbb{P}_z\left(\sup_{s \le t}|Z_s - z| \ge \delta/2\right) < \varepsilon \quad \text{for all } z \in U. \tag{A.1.5}$$

As $\mathbb{P}_w(\tau > t) = 0$, (A.1.2) implies that there is some $\delta_0 \in (0, \delta/2)$ so that

$$\mathbb{P}_z(\tau > t) < \varepsilon \quad \text{for all } z \in B_{\delta_0}(w). \tag{A.1.6}$$

This together with (A.1.5) yields that

$$\mathbb{P}_z\left(\tau < t \text{ and } \sup_{s \le t}|Z_s - z| < \delta/2\right) > 1 - 2\varepsilon \quad \text{for every } z \in U \cap B_{\delta_0}(w).$$

Since the left-hand side is dominated by $\mathbb{P}_z(|z - Z_\tau| < \delta/2)$, we obtain

$$\mathbb{P}_z(|w - Z_\tau| < \delta) > 1 - 2\varepsilon \quad \text{for all } z \in U \cap B_{\delta_0}(w). \tag{A.1.7}$$

We further get from (A.1.5) and (A.1.6) that for $z \in B_{\delta_0}(w)$,

$$\begin{aligned}
1 - \mathbb{E}_z[e^{-\alpha\tau}] &\le 1 - \mathbb{E}_z[e^{-\alpha\tau}; \tau \le t] \le 1 - e^{-\alpha t}\mathbb{P}_z(\tau \le t)\\
&\le (1 - e^{\alpha t}) + \mathbb{P}_z(\tau > t) < 2\varepsilon,
\end{aligned}$$

which together with (A.1.4) and (A.1.7) yields that $|\mathcal{H}_U^\alpha f(z) - f(w)| \le \varepsilon + 6\varepsilon\|f\|_\infty$ for all $z \in U \cap B_\delta(w)$. This shows that $\mathcal{H}_U^\alpha f(z)$ is continuous at $w \in \partial U$ with value $f(w)$. □

A.2 Conformal invariance of ABM

Let D and $\widetilde{D}$ be domains in $\mathbb{C}$ and ψ be a conformal map from D onto $\widetilde{D}$. Denote by $\mathbf{Z}^0 = (Z_t^0, \zeta^0, \mathbb{P}_z^0)$ and $\widetilde{\mathbf{Z}}^0 = (\widetilde{Z}_t^0, \widetilde{\zeta}^0, \widetilde{\mathbb{P}}_z^0))$ the absorbed Brownian motions (ABM) on D and $\widetilde{D}$, respectively. Set

$$A_t = \int_0^{t\wedge\zeta^0} |\psi'(Z_s^0)|^2 ds, \qquad t \geq 0,$$

which is continuous in $t \geq 0$ and strictly increasing in $t \in [0, \zeta^0]$ because $\psi'(z) \neq 0$ on D. Let $\{\tau_t,\ t \in [0, A_{\zeta^0}]\}$ be the inverse of $\{A_t,\ t \in [0, \zeta^0]\}$ (that is, $\tau_t = \inf\{s > 0 : A_s > t\}$), and $\check{\mathbf{Z}}^0 = (\check{Z}_t^0, \check{\zeta}^0, \mathbb{P}_z^0)$ be the time change of $\mathbf{Z}^0$ by A_t, namely,

$$\check{Z}_t^0 = Z_{\tau_t}^0 \quad \text{and} \quad \check{\zeta}^0 = A_{\zeta^0}.$$

Theorem A.2.1. *For each $z \in D$,*

$$\{\psi(\check{Z}_t^0),\ t \in [0, \check{\zeta}^0)\} \ \textit{under}\ \mathbb{P}_z^0 \ \textit{is identical in law}$$
$$\textit{with}\ \left\{\widetilde{Z}_t^0,\ t \in [0, \widetilde{\zeta}^0)\right\} \ \textit{under}\ \widetilde{\mathbb{P}}_{\psi(z)}^0.$$

Proof. Let $B_t = (B_t^1, B_t^2)$, $t \geq 0$, be the standard two dimensional Brownian motion defined on a filtered probability space $(\Omega, \mathcal{B}, \mathbb{P})$ with $B_0 = \mathbf{0}$. Fix $z \in D$ and put $Z_t = z + B_t^1 + iB_t^2$, $t \geq 0$, which is the planar Brownian motion starting from z.

Consider a strictly increasing sequence of relatively compact open sets $\{D_n\}$ such that $z \in D_1$, $\partial D_n \subset D_{n+1}$ and $\bigcup_{n=1}^\infty D_n = D$. Put $\{\widetilde{D}_n = \psi(D_n),\ n \geq 1\}$, which is increasing and satisfies $\psi(z) \in \widetilde{D}_1$, $\partial\widetilde{D}_n \subset \widetilde{D}_{n+1}$ and $\bigcup_{n=1}^\infty \widetilde{D}_n = \widetilde{D}$. Define $\sigma_n = \inf\{t > 0 : Z_t \in \partial D_n\}$. As $n \to \infty$, σ_n increases to ζ^0, the exit time of Z_t from D. Let $u = \Re\psi$ and $v = \Im\psi$. Define for $t \in [0, \zeta^0)$,

$$N_t^1 = \int_0^t u_x(Z_s) dB_s^1 + \int_0^t u_y(Z_s) dB_s^2,$$
$$N_t^2 = \int_0^t v_x(Z_s) dB_s^1 + \int_0^t v_y(Z_s) dB_s^2.$$

Note that N_t^1 and N_t^1 are well defined and are continuous local martingales on $[0, \zeta^0)$ with

$$\langle N^1\rangle_t = \langle N^2\rangle_t = A_t \quad \text{and} \quad \langle N^1, N^2\rangle_t = 0 \quad \text{for every } t \in [0, \zeta^0).$$

This is because for each $n \geq 1$, $t \mapsto N_{t\wedge\sigma_n}^1$ and $t \mapsto N_{t\wedge\sigma_n}^2$ are continuous square-integrable martingales on $[0, \infty)$ with

$$\langle N_{\cdot\wedge\sigma_n}^1\rangle_t = \langle N_{\cdot\wedge\sigma_n}^2\rangle_t = A_{t\wedge\sigma_n} \quad \text{and} \quad \langle N_{\cdot\wedge\sigma_n}^1, N_{\cdot\wedge\sigma_n}^2\rangle_t = 0 \quad \text{for every } t \geq 0.$$

By a martingale representation theorem [Ikeda and Watanabe (1981), Theorem II.7.3'], there exists a planar Brownian motion $\widetilde{B}_t = \widetilde{B}_t^1 + i\widetilde{B}_t^2, t \geq 0$, on an enlarged filtered probability space $(\widetilde{\Omega}, \widetilde{\mathcal{B}}, \widetilde{\mathbb{P}})$ such that

$$\widetilde{B}_t = N_{\tau_t}^1 + iN_{\tau_t}^2 \quad \text{for any } t \in [0, A_{\zeta^0}). \tag{A.2.1}$$

On the other hand, for each $n \geq 1$, by Itô's formula ([Ikeda and Watanabe (1981), Theorem II.5.1]) and the Cauchy-Riemann equation, that

$$\psi(Z_{t\wedge\sigma_n}) - \psi(z) = N^1_{t\wedge\sigma_n} + iN^2_{t\wedge\sigma_n} \tag{A.2.2}$$

Set $\widetilde{Z}_t = \psi(z) + \widetilde{B}_t$, $t \geq 0$, which is a planar Brownian motion starting at $\psi(z)$. It follows from (A.2.1) and (A.2.2) that

$$\widetilde{Z}_t = \psi(Z_{\tau_t}) \quad \text{for any } t \in [0, A_{\zeta^0}). \tag{A.2.3}$$

Since $\widetilde{Z}_t = \psi(Z_{\tau_t}) \in \psi(D_n) = \widetilde{D}_n$ for $t \in [0, A_{\sigma_n})$ and $\widetilde{Z}_{A_{\sigma_n}} = \psi(Z_{\sigma_n}) \in \psi(\partial D_n) = \partial\widetilde{D}_n$ by (A.2.3), we have $A_{\sigma_n} = \inf\{t > 0 : \widetilde{Z}_t \notin \widetilde{D}_n\}$. It follows that

$$A_{\zeta^0} = \lim_{n\to\infty} A_{\sigma_n} = \inf\{t > 0 : \widetilde{Z}_t \notin \widetilde{D}\},$$

is the exit time of the planar Brownian motion $\widetilde{Z}$ from $\widetilde{D}$. This together with (A.2.3) gives the desired result. □

A.3 Construction of BMD starting at every point of D^*

In this section, we present two different proofs of Theorem 1.3.4 that refines the construction of a BMD from starting at quasi-every point of D^* in Proposition 1.3.3 to starting at every point of D^*.

First proof. We follow a method in [Fukushima and Tanaka (2005), §4.5] which treated the case where $N = 1$ and a process starting at a_1^* was constructed by piecing together excursions around a_1^* by a Poisson point process. We start with an ABM $(\Omega^0, X_t^0(\omega^0), t \geq 0; \mathbb{P}_z^0, z \in D)$ on D with lifetime ζ^0 defined on a certain sample path space Ω^0 with transition function p_t^0, along with the BMD's $(\Omega^j, X_t^j(\omega^j), \zeta^j, \mathbb{P}^j_{a_j^*})$, $1 \leq j \leq N$, on D^* defined on certain sample path spaces Ω^j with initial distributions $\delta_{a_j^*}$, $1 \leq j \leq N$, and transition function p_t^*, which have been well constructed by Proposition 1.3.3. As $\mathbb{C} \setminus D$ is non-polar for the planar Brownian motion, we can assume that $\zeta^0(\omega) < \infty$ for any $\omega \in \Omega^0$. See Section 1.1.

For convenience, we assume that Ω^0 contains extra points $\eta_1, \ldots, \eta_N$ with $\mathbb{P}_z^0(\{\eta_j\}) = 0$ for every $z \in D$ and $\mathbb{P}^0_{a_j^*}(\{\eta_j\}) = 1$ for $1 \leq j \leq N$.

Define

$$\widetilde{\Omega} = \Omega^0 \times \Omega^1 \times \cdots \times \Omega^N, \quad \widetilde{\mathbb{P}}_z = \mathbb{P}_z^0 \times \mathbb{P}^1_{a_1^*} \times \cdots \times \mathbb{P}^N_{a_N^*}, \quad z \in D^*,$$

and, for $\widetilde{\omega} = (\omega^0, \omega^1, \ldots, \omega^N)$,

(1) when $\omega^0 \in \Omega^0 \setminus \{\eta_1, \ldots, \eta_N\}$,

$$\widetilde{X}_t(\widetilde{\omega}) := \begin{cases} X_t^0(\omega^0) & \text{for } 0 \leq t < \zeta^0(\omega^0), \\ X^j_{t-\zeta^0(\omega^0)}(\omega^j) & \text{for } \zeta^0(\omega^0) \leq t < \zeta^0(\omega^0) + \zeta^j(\omega^j) \\ & \text{if } X^0_{\zeta^0(\omega^0)-} \in A_j,\ 1 \leq j \leq N, \end{cases}$$

(2) when $\omega^0 = \eta_j$, $\widetilde{X}_t(\widetilde{\omega}) := X_t^j(\omega^j)$ for $0 \le t < \zeta^j(\omega^j)$, $1 \le j \le N$.

The lifetime $\widetilde{\zeta}(\widetilde{\omega})$ of $\widetilde{X}_t(\widetilde{\omega})$ is defined by

$$\widetilde{\zeta}(\widetilde{\omega}) = \begin{cases} \zeta^0(\omega^0) & \text{if } \omega^0 \in \Omega^0 \setminus \{\eta_1, \cdots, \eta_N\} \\ & \text{and } X^0_{\zeta^0(\omega^0)-} \in \partial E, \\ \zeta^0(\omega^0) + \zeta^j(\omega^j) & \text{if } \omega^0 \in \Omega^0 \setminus \{\eta_1, \cdots, \eta_N\} \\ & \text{and } X^0_{\zeta^0(\omega^0)-} \in \partial A_j,\ 1 \le j \le N, \\ \zeta^j(\omega^j) & \text{if } \ \omega^0 = \eta_j, \ \ 1 \le j \le N. \end{cases}$$

Let ∂ be the point at infinity of D^*. We set $\widetilde{X}_t(\widetilde{\omega}) = \partial$ for $t \ge \widetilde{\zeta}(\widetilde{\omega})$ and any numerical function f on D^* is extended to $D^* \cup \{\partial\}$ by setting $f(\partial) = 0$.

For the process $\widetilde{Z} = (\widetilde{\Omega}, \widetilde{X}_t(\widetilde{\omega}), \widetilde{\zeta}(\widetilde{\omega}), \widetilde{\mathbb{P}}_z)$ on D^* so defined, let

$$\widetilde{p}_t(z, B) = \widetilde{\mathbb{P}}_z(\widetilde{X}_t \in B), \quad t \ge 0,\ z \in D^*,\ B \in \mathcal{B}(D^*). \tag{A.3.1}$$

By the above definition of $\widetilde{Z}$, we have for $B \in \mathcal{B}(D^*)$,

$$\begin{cases} \widetilde{p}_t(z, B) = p_t^0(z, B) \\ \qquad + \sum_{j=1}^N \int_0^t \mathbb{P}_z^0 \left(X^0_{\zeta^0(\omega^0)-} \in A_j,\ \zeta^0(\omega^0) \in ds \right) p^*_{t-s}(a_j^*, B), \\ \qquad\qquad z \in D, \\ \widetilde{p}_t(a_j^*, B) = p_t^*(a_j^*, B), \quad 1 \le j \le N. \end{cases} \tag{A.3.2}$$

Consequently, if we define, for $\alpha > 0$ and $f \in b\mathcal{B}(D^*)$,

$$\widetilde{G}_\alpha f(z) = \int_0^\infty e^{-\alpha t} \widetilde{p}_t f(z) dt, \quad z \in D^*, \tag{A.3.3}$$

then,

$$\begin{cases} \widetilde{G}_\alpha f(z) = G_\alpha^0 f(z) + \sum_{j=1}^N u_\alpha^{(j)}(z) G_\alpha^* f(a_j^*), & z \in D, \\ \widetilde{G}_\alpha f(a_j^*) = G_\alpha^* f(a_j^*), \quad 1 \le j \le N, \end{cases} \tag{A.3.4}$$

where

$$G_\alpha^0 f(z) = \int_0^\infty e^{-\alpha t} p_t^0 f(z) dt, \qquad z \in D,$$

$$G_\alpha^* f(a_j^*) = \int_0^\infty e^{-\alpha t} p_t^* f(z) dt, \qquad 1 \le j \le N.$$

Lemma A.3.1.

(i) *It holds that, for* $f \in b\mathcal{B}(D^*)$,

$$\widetilde{p}_t f(z) = p_t^* f(z) \quad \textit{for any } t \ge 0 \textit{ and } z \in D^* \setminus \mathcal{N}, \tag{A.3.5}$$

where p_t^* *is the transition function of* $Z^*\big|_{D^* \setminus \mathcal{N}}$ *defined by* (1.3.18).

(ii) $\{\widetilde{G}_\alpha,\ \alpha > 0\}$ *defined by* (A.3.3) *enjoys the following properties:*

$$\widetilde{G}_\alpha f(z) - \widetilde{G}_\beta f(z) + (\alpha - \beta) \widetilde{G}_\alpha \widetilde{G}_\beta f(z) = 0 \tag{A.3.6}$$

for $\alpha, \beta > 0$, $z \in D^*$, $f \in b\mathcal{B}(D^*)$, *and*

$$\widetilde{G}_\alpha(C_\infty(D^*)) \subset C_\infty(D^*). \tag{A.3.7}$$

Proof. (i) By the strong Markov property of $Z^*\big|_{D^*\setminus\mathcal{N}}$, we have for $z \in D^* \setminus \mathcal{N}$ and $B \in \mathcal{B}(D^* \setminus \mathcal{N})$,

$$p_t^*(z,B) = p_t^{*,0}(z,B) + \sum_{j=1}^N \int_0^t \mathbb{P}_z^{*,0}\left(Z^0_{\zeta^{*,0}-} = a_j^*,\ \zeta^{*,0} \in ds\right) p_{t-s}^*(a_j^*, B),$$

which along with (A.3.2) and Proposition 1.3.3 (i) yields (A.3.5).

(ii) The resolvent $G_\alpha^*(z,B) = \int_0^\infty e^{-\alpha t} p_t^*(z,B)dt$ of $Z^*\big|_{D^*\setminus\mathcal{N}}$ satisfies the resolvent equation

$$G_\alpha^* f(z) - G_\beta^* f(z) + (\alpha-\beta) G_\alpha^* G_\beta^* f(z) = 0,\ \alpha,\beta > 0,\ z \in D^* \setminus \mathcal{N},\ f \in b\mathcal{B}(D^*),$$

so that (A.3.6) holds for any $z \in D^* \setminus \mathcal{N}$ by (A.3.5). Since $G_\alpha^0 f$ is a difference of α-excessive function relative to the ABM on D and $u_\alpha^{(j)}$ is bounded continuous on D by Lemma 1.3.1, we get from the identity (A.3.4) the equation (A.3.6) holding for every $z \in D^*$.

In view of the identity (A.3.4) and the properties of $u_\alpha^{(j)}$ shown by Lemma 1.3.1, it suffices to prove that

$$G_\alpha^0 f \in C_\infty(D) \quad \text{for any} \quad f \in C_\infty(D^*) \tag{A.3.8}$$

for the proof of (A.3.7).

Let G_α be the resolvent of the planar BM Z. G_α satisfies $G_\alpha(C_\infty(\mathbb{C})) \subset C_\infty(\mathbb{C})$. For $f \in C_\infty(D^*)$, there is a function $g \in C_\infty(\mathbb{C})$ such that $f = g$ on D. Then $G_\alpha^0 f(z) = G_\alpha^0 g(z)$, $z \in D$, and

$$G_\alpha^0 f(z) = G_\alpha g(z) - \mathbf{H}_D^\alpha(G_\alpha g)(z), \quad z \in D. \tag{A.3.9}$$

Choose $g_n \in C_c(\mathbb{C})$ which converges uniformly to g. As $G_\alpha g_n \in H^1(\mathbb{C})$, $\mathbf{H}_D^\alpha(G_\alpha g_n) \in H^1(\mathbb{C})$ is α-harmonic on D in the distribution sense. Since $G_\alpha g_n$ is a difference of α-excessive functions relative to Z, so is $\mathbf{H}_D^\alpha(G_\alpha g_n)$. Hence $\mathbf{H}_D^\alpha(G_\alpha g_n)(z)$ can be verified to be continuous in $z \in D$ just as we did for $u_\alpha^{(i)}(z)$ in the proof of Lemma 1.3.1. $\mathbf{H}_D^\alpha(G_\alpha g)(z)$ is also continuous in $z \in D$ as the uniform convergent limit of $\mathbf{H}_D^\alpha(G_\alpha g_n)(z)$.

Thus, we see from (A.3.9) that $G_\alpha^0 f(z)$ is continuous in $z \in D$ and $\lim_{z\to\infty,\ z\in D} |G_\alpha^0 f(z)| \le \lim_{z\to\infty} G_\alpha|g|(z) = 0$. By virtue of Theorem A.1.1 and the property (**Z.1**) of the planar BM, we further have $\lim_{z\in D, z\to w\in\partial D} G_\alpha^0 f(z) = G_\alpha g(w) - G_\alpha g(w) = 0$, yielding (A.3.8). □

Theorem A.3.2. *$\widetilde{Z} = (\widetilde{\Omega}, \widetilde{X}_t, \widetilde{\zeta}, \widetilde{\mathbb{P}}_z)_{z\in D^*}$ is a BMD on D^*.*

Proof. Observe that $t \in [0, \widetilde{\zeta}(\widetilde{\omega})) \mapsto \widetilde{X}_t(\widetilde{\omega}) \in D^*$ is continuous $\widetilde{\mathbb{P}}_z$-a.s. for every $z \in D^*$. Hence, $\widetilde{p}_t f(z) = \widetilde{\mathbb{E}}_z[f(\widetilde{X}_t)]$ is right continuous in t for any $f \in C_\infty(D^*)$.

For $f \in C_\infty(D^*)$, it holds that

$$\int_0^\infty e^{-\alpha t}\left(\int_0^\infty e^{-\beta s}\widetilde{p}_{t+s} f(z) ds\right) dt = \int_0^\infty e^{-\alpha t}\left(\widetilde{p}_t(\widetilde{G}_\beta f)(z)\right) dt,$$

because the left-hand side equals $(\alpha-\beta)^{-1}(\widetilde{G}_\beta f(z)-\widetilde{G}_\alpha f(z)\}$, which in turn equals $\widetilde{G}_\alpha\widetilde{G}_\beta f(z)$ by the equation (A.3.6). The functions inside brackets $(\dots)$ of the both sides are right continuous in $t\geq 0$ on account of (A.3.7) so that

$$\int_0^\infty e^{-\beta s}\widetilde{p}_{t+s}f(z)ds = \widetilde{p}_t(\widetilde{G}_\beta f)(z) = \int_0^\infty e^{-\beta s}\widetilde{p}_t(\widetilde{p}_s f)(z)ds.$$

Since both $\widetilde{p}_{t+s}f(z)$, $\widetilde{p}_t(\widetilde{p}_s f)(z)$ are right continuous in $s\geq 0$, we obtain the semi-group property for $\widetilde{p}_t$:

$$\widetilde{p}_{t+s}f(z) = \widetilde{p}_t(\widetilde{p}_s f)(z), \quad t\geq 0, \quad s\geq 0, \quad z\in D^*. \tag{A.3.10}$$

We now prove that $\widetilde{Z}$ is a Markov process on D^* with transition function $\widetilde{p}_t$, or equivalently, that, for any $0<s_1<s_2<\cdots,s_n$, $f_1,f_2,\dots,f_n\in b\mathcal{B}(D^*)$ and $z\in D^*$,

$$\widetilde{\mathbb{E}}_z\left[f_1(\widetilde{X}_{s_1})f_2(\widetilde{X}_{s_2})\cdots f_n(\widetilde{X}_{s_n})\right] = \widetilde{p}_{s_1}f_1\widetilde{p}_{s_2-s_1}f_2\cdots\widetilde{p}_{s_n-s_{n-1}}f_n(z). \tag{A.3.11}$$

Here, on the right-hand side, we employ an abbreviated notation for the repeated operations:

$$\widetilde{p}_{s_1}\left(f_1\widetilde{p}_{s_2-s_1}\left(f_2\cdots\widetilde{p}_{s_{n-1}-s_{n-2}}(\widetilde{p}_{s_n-s_{n-1}}f_n)\right)\right)(z).$$

Identity (A.3.11) is valid for $z=a_j^*$, $1\leq j\leq N$. Indeed, by using the Markov property of $Z^*\big|_{D^*\setminus\mathcal{N}}$ and by noting that $\mathcal{N}$ is properly exceptional for $Z^*\big|_{D^*\setminus\mathcal{N}}$,

$$\begin{aligned}
&\widetilde{\mathbb{E}}_{a_j^*}\left[f_1(\widetilde{X}_{s_1})f_2(\widetilde{X}_{s_2})\cdots f_n(\widetilde{X}_{s_n})\right] = \mathbb{E}^j_{a_j^*}\left[f_1(X^j_{s_1})f_2(X^j_{s_2})\cdots f_n(X^j_{s_n})\right]\\
&= \mathbb{E}^j_{a_j^*}\left[f_1\cdot\mathbb{1}_{D^*\setminus\mathcal{N}}(X^j_{s_1})f_2\cdot\mathbb{1}_{D^*\setminus\mathcal{N}}(X^j_{s_2})\cdots f_n\cdot\mathbb{1}_{D^*\setminus\mathcal{N}}(X^j_{s_n})\right]\\
&= p^*_{s_1}f_1\cdot\mathbb{1}_{D^*\setminus\mathcal{N}}p^*_{s_2-s_1}f_2\cdot\mathbb{1}_{D^*\setminus\mathcal{N}}\cdots p^*_{s_n-s_{n-1}}f_n\cdot\mathbb{1}_{D^*\setminus\mathcal{N}}(a_j^*),
\end{aligned}$$

which is equal to the right-hand side of (A.3.11) with $z=a_j^*$ on account of (A.3.5).

The Markov property (A.3.11) for $z\in D$ follows from the Markov property of the ABM Z^0 on D and the validity of (A.3.11) for $z=a_j^*$, $1\leq j\leq N$. To see this, we put, for any $0<s_1<s_2<\cdots<s_n$ and $f_1,f_2,\dots,f_n\in b\mathcal{B}(D^*)$,

$$I_k = \widetilde{\mathbb{E}}_z\left[f_1(\widetilde{X}_{s_1})\cdots f_{k-1}(\widetilde{X}_{s_{k-1}})f_k(\widetilde{X}_{s_k})\cdots f_n(\widetilde{X}_{s_n});\ s_{k-1}<\sigma_{K^*}\leq s_k\right],$$

for $1\leq k<n$ with $s_0=0$ and $J=\widetilde{\mathbb{E}}_z\left[f_1(\widetilde{X}_{s_1})\cdots f_n(\widetilde{X}_{s_n});\ s_n<\sigma_{K^*}\right]$.

Using the definition of $\widetilde{Z}$, the validity of (A.3.11) for $z = a_j^*$, $1 \leq j \leq N$, and the identity (A.3.2) successively, we are led to

$$
\begin{aligned}
I_k &= \mathbb{E}_z^0\Big[f_1(X_{s_1}^0)\cdots f_{k-1}(X_{s_{k-1}}^0) \\
&\qquad \times \sum_{j=1}^N \int_0^{s_k-s_{k-1}} \mathbb{P}^0_{X^0_{s_{k-1}}}(X^0_{\zeta^0-} \in A_j,\ \zeta^0 \in ds) \\
&\qquad \times \widetilde{\mathbb{E}}_{a_j^*}[f_k(\widetilde{X}_{s_k-s_{k-1}-s})\cdots f_n(\widetilde{X}_{s_n-s_{n-1}-s})]\Big] \\
&= \mathbb{E}_z^0\Big[f_1(X_{s_1}^0)\cdots f_{k-1}(X_{s_{k-1}}^0) \\
&\qquad \times \sum_{j=1}^N \int_0^{s_k-s_{k-1}} \mathbb{P}^0_{X^0_{s_{k-1}}}(X^0_{\zeta^0-} \in A_j,\ \zeta^0 \in ds) \\
&\qquad \times \widetilde{p}_{s_k-s_{k-1}-s} f_k \widetilde{p}_{s_{k+1}-s_k} f_{k+1}\cdots \widetilde{p}_{s_n-s_{n-1}} f_n(a_j^*)\Big] \\
&= \mathbb{E}_z^0\Big[f_1(X_{s_1}^0)\cdots f_{k-1}(X_{s_{k-1}}^0) \\
&\qquad \times (\widetilde{p}_{s_k-s_{k-1}} - p^0_{s_k-s_{k-1}}) f_k \widetilde{p}_{s_{k+1}-s_k} f_{k+1}\cdots \widetilde{p}_{s_n-s_{n-1}} f_n(X^0_{s_{k-1}})\Big].
\end{aligned}
$$

By the Markov property of the ABM Z^0 on D, we thus get

$$
\begin{aligned}
I_k &= p^0_{s_1} f_1 \cdots p^0_{s_{k-1}-s_{k-2}} f_{k-1} \widetilde{p}_{s_k-s_{k-1}} f_k \widetilde{p}_{s_{k+1}-s_k} f_{k+1} \cdots \widetilde{p}_{s_n-s_{n-1}} f_n(z) \\
&\quad - p^0_{s_1} f_1 \cdots p^0_{s_{k-1}-s_{k-2}} f_{k-1} p^0_{s_k-s_{k-1}} f_k \widetilde{p}_{s_{k+1}-s_k} f_{k+1} \cdots \widetilde{p}_{s_n-s_{n-1}} f_n(z).
\end{aligned}
$$

Clearly, we also have

$$
J = \mathbb{E}_z^0\left[f_1(X_{s_1}^0)\cdots f_n(X_{s_n}^0)\right] = p^0_{s_1} f_1 \cdots p^0_{s_n-s_{n-1}} f_n(z).
$$

Hence we arrive at desired (A.3.11) for $z \in D$ because its left-hand side equals $\sum_{k=1}^n I_k + J$.

Since $\widetilde{G}_\alpha f(\widetilde{X}_t)$ is right continuous in $t \in [0, \widetilde{\zeta})$ $\widetilde{\mathbb{P}}_z$-a.s. for every $z \in D^*$ and for any $f \in C_\infty(D^*)$ by (A.3.7), $\widetilde{Z}$ is a strong Markov process on D^* in view of [Blumenthal and Getoor (1968), Theorem I.8.11].

As Z^* is associated with the Dirichlet form $(\mathcal{E}^*, \mathcal{F}^*)$, Z^* is m-symmetric and so is $\widetilde{Z}$ on account of (A.3.5). According to the construction of $\widetilde{Z}$ in the above, its part process on D is the ABM on D starting at every point of D. Therefore, $\widetilde{Z}$ is a BMD on D^* starting at every point of D^*. □

Second proof. Let $\mathbf{Z}^* = (Z_t^*, \zeta^*, \mathbb{P}_z^*, z \in D^* \setminus \mathcal{N})$ be an m-symmetric Hunt process associated with the regular strongly local Dirichlet form $(\mathcal{E}^*, \mathcal{F}^*)$ on $L^2(D^*; m)$ given by (1.3.7)-(1.3.8), where $\mathcal{N}$ is a properly exceptional set for $\mathbf{Z}^*$. As each $a_i^* \in K^*$ is of positive $\mathcal{E}^*$-capacity, $\mathcal{N} \subset D$. Since $(\mathcal{E}^*, \mathcal{F}^*)$ is strongly local, $\mathbf{Z}^*$ is a diffusion process that admits no killings inside D^*. We show that Z^* can be refined to start from every point in D; in other words, the properly exceptional set $\mathcal{N}$ can be taken as the empty set. The idea is that for every $x \in D$, we run an ABM Z^E in E starting from x till it hits $K = \cup_{j-1}^N A_j$; if at that time the process is at A_j, we identify A_j with a_j^* and run a copy of $\mathbf{Z}^*$ starting from a_j^*. This patching-together procedure gives a refinement of BMD that starts from every point in D^*. We rigorously carry out this procedure in the following.

Let $D_\partial^* = D^* \cup \{\partial\}$ be a one-point compactification of D^*. Denote by Ω be the space of continuous functions on $[0, \infty)$ that take values on D_∂^* so that $\omega(t) = \partial$ for all $t \geq \zeta(\omega) := \inf\{s \geq 0 : \omega(s) = \partial\}$. On Ω, there is a family of shifting operators $\{\theta_t; t \geq 0\}$ so that $(\theta_t\omega)(s) = \theta(t+s)$. For each $x \in D^* \setminus \mathcal{N}$, let $\mathbb{P}_x$ be the probability law induced on Ω by the process $\mathbf{Z}^*$. In this way, we can identify the Hunt process $\mathbf{Z}^*$ with the canonical process defined on Ω by $Z_t^*(\omega) = \omega(t)$. Let $\mathcal{F}_t^0 := \sigma\{\omega(s); s \leq t\}$, the natural filtration generated by the canonical process, and $\mathcal{F}_t := \cap_{s>t}\mathcal{F}_s^0$. Let $\sigma^* := \sigma_{K^*} = \inf\{t \geq 0 : Z_t^* \in K^*\}$, and $\mathcal{F}_{\sigma^*} = \{A \in \mathcal{F}_\infty : A \cap \{\sigma^* \leq t\} \in \mathcal{F}_t\}$. Denote by $\mathbb{P}_x^D$ the law of the stopped ABM $\{Z_{t\wedge\tau_D^E}^E; t \geq 0]\}$ induced on Ω by identifying A_j with a_j^*, where $\tau_D^E = \{t \geq 0 : Z_t^E \in K\}$. Consider the following collection of subsets of Ω:

$$\mathcal{C} = \{\cup_{i=1}^n (A_{i,1} \cap (A_{i,2} \circ \theta_{\sigma^*})) : \ n \geq 1, A_{i,1} \in \mathcal{F}_{\sigma^*} \text{ and } A_{i,2} \in \mathcal{F}_\infty \text{ for } 1 \leq i \leq n\}.$$

Here $A_{i,2} \circ \theta_{\sigma^*} := \{\omega \in \Omega : \theta_{\sigma^*(\omega)}\omega \in A_{i,2}\}$. Clearly $\mathcal{C}$ is an algebra that generates $\mathcal{F}_\infty$. Any $A \in \mathcal{C}$ can be expressed as a disjoint union of $A_{i,1} \cap (A_{i,2} \circ \theta_{\sigma^*})$ for some $A_{i,1} \in \mathcal{F}_{\sigma^*}$ and $A_{i,2} \in \mathcal{F}_\infty$ with $1 \leq i \leq n$. For such A, define for $z \in D$,

$$\widetilde{\mathbb{P}}_z(A) = \sum_{i=1}^n \sum_{j=1}^N \mathbb{P}_z^D \left(A_{i,1} \cap \{\omega \in \Omega : \omega(\sigma^*) = a_j^*\}\right) \mathbb{P}_{a_j^*}(A_{i,2}). \tag{A.3.12}$$

It is easy to see that $\widetilde{\mathbb{P}}_z$ is a probability measure on the algebra $\mathcal{C}$. By the Carathéodory's extension theorem, $\widetilde{\mathbb{P}}_z$ has a unique extension to $\mathcal{F}_\infty$, which will still be denoted by $\widetilde{\mathbb{P}}_z$. We see from (A.3.12) that $\widetilde{\mathbb{P}}_z = \mathbb{P}_z^D$ on $\mathcal{F}_\sigma^*$ and

$$\widetilde{\mathbb{P}}_z \left[A \circ \theta_{\sigma^*} | \mathcal{F}_{\sigma^*}\right] = \sum_{j=1}^N \mathbb{P}_{a_j^*}(A) 1_{\{\omega_{\sigma^*} = a_j^*\}} \quad \text{for } A \in \mathcal{F}_\infty.$$

Thus under $\widetilde{\mathbb{P}}_z$, the canonical process behaves as an ABM in D starting from z before it hits K^* and then proceeds as BMD starting from K^*. By the strong Markov property of Z^*, we have $\mathbb{P}_z = \widetilde{\mathbb{P}}_z$ for every $x \in D \setminus \mathcal{N}$. For $x \in \mathcal{N}$, we re-define $\mathbb{P}_z$ by $\widetilde{\mathbb{P}}_z$. We next show that for each $z \in \mathcal{N}$, the canonical process Z^* has the strong Markov property under $\mathbb{P}_z$. Let T be a finite stopping time with respect to the filtration $\{\mathcal{F}_t, t \geq 0\}$. By a result of Courrége-Priouret (see [Ikeda, Nagasawa and Watanabe (1966), Lemma 2.3] or [Courrége and Priouret (1964)]), there exists a random variable $\eta(\omega, \omega')$ defined on $\Omega \times \Omega$ that is $\mathcal{F}_{\sigma^*} \times \mathcal{F}_\infty$-measurable so that

(i) for each fixed $\omega \in \{T > \sigma^*\}$, $\eta(\omega, \cdot)$ is an $\{\mathcal{F}_t\}_{t\geq 0}$-stopping time;
(ii) $T(\omega) = \sigma^*(\omega) + \eta(\omega, \theta_{\sigma^*(\omega)}\omega)$ for $\omega \in \{T > \sigma^*\}$.

As $\eta(\omega,\omega')$ is $\mathcal{F}_{\sigma^*}\times\mathcal{F}_\infty$-measurable, we can refine η to have the property (see [Galmarino (1963)]) that $\eta(\omega_1,\omega')=\eta(\omega_2,\omega')$ for $\omega_1,\omega_2\in\Omega$ with $\omega_1(t)=\omega_2(t)$ for all $t\in[0,\sigma^*(\omega)]$; that is, $\eta(\omega,\omega')$ depends on ω only through $\omega([0,\sigma^*(\omega)])$. From this and using a measure theoretical argument, one can show (cf. [Courrége and Priouret (1964)]) that $\mathcal{F}_T\cap\{\sigma^*<T\}$ is the σ-field on $\{\sigma^*<T\}$ generated by the sets of the form

$$A(\omega)=A_1(\omega)\cap A_2(\omega,\theta_{\sigma^*}(\omega)), \tag{A.3.13}$$

where $A_1\in\mathcal{F}_{\sigma*}$ and $A_2(\omega,\omega')\in\mathcal{F}_{\sigma*}\times\mathcal{F}_\infty$ so that $A_2(\omega,\cdot)\in\mathcal{F}_{\eta(\omega,\cdot)}$ for each $\omega\in\mathcal{F}_{\sigma^*}$.

Let $x\in\mathcal{N}$. Clearly, for any bounded $\xi\in\mathcal{F}_\infty$ and $A\in\mathcal{F}_T$,

$$\mathbb{E}_z\left[\xi\circ\theta_T;A\cap\{T\le\sigma^*\}\right]=\mathbb{E}_z\left[\mathbb{E}_{Z_T^*}[\xi];A\cap\{T\le\sigma^*\}\right]. \tag{A.3.14}$$

For $A\in\mathcal{F}_T\cap\{\sigma^*<T\}$ of the form (A.3.13), we have by the definition of $\mathbb{P}_z=\widetilde{\mathbb{P}}_z$ for $z\in\mathcal{N}$,

$$\begin{aligned}
&\mathbb{E}_z\left[\xi\circ\theta_T;A\cap\{T>\sigma^*\}\right]\\
&=\mathbb{E}_z\left[\xi\circ\theta_{\sigma^*(\omega)+\eta(\omega,\theta_{\sigma^*}\omega)}\omega;\ A_1(\omega)\cap A_2(\omega,\theta_{\sigma^*}(\omega))\cap\{T(\omega)>\sigma^*(\omega)\}\right]\\
&=\mathbb{E}_z\left[\mathbb{E}_{Z^*_{\sigma^*(\omega)}}\left[\xi\circ\theta_{\eta(\omega_0,\cdot)};A_2(\omega_0,\cdot)\right]\big|_{\omega_0=\omega};\ A_1(\omega)\cap\{T(\omega)>\sigma^*(\omega)\}\right]\\
&=\mathbb{E}_z\left[\mathbb{E}_{Z_T^*}[\xi];A\cap\{T>\sigma^*\}\right],
\end{aligned}$$

where in the last equality we used the fact that $Z^*_{\sigma^*}\in K^*$ and the strong Markov property of Z^* under each $\mathbb{P}_{a_j^*}$. Consequently, we have for any $\xi\in\mathcal{F}_\infty$ and $A\in\mathcal{F}_T$,

$$\mathbb{E}_z\left[\xi\circ\theta_T;A\cap\{T>\sigma^*\}\right]=\mathbb{E}_z\left[\mathbb{E}_{Z_T^*}[\xi];A\cap\{T>\sigma^*\}\right].$$

This combined with (A.3.14) establishes the strong Markov property of Z^* under $\mathbb{P}_z$. Therefore, $\{Z^*,\mathbb{P}_z,z\in D^*\}$ is a BMD on D^* that it can start from every point in D^*. □

A.4 Probabilistic representation of $\Im g_F$

In this section, we give the proof of Theorem 2.2.1 through a series of lemmas. Take $r>0$ large enough so that the stripe $\mathbb{H}_r:=\{z=x+iy:0<y<r\}$ contains $F\cup K$. Let $D_r:=\mathbb{H}_r\backslash K$. Then $D_r\backslash F=\mathbb{H}_r\backslash(F\cup K)$. Recall that $\Gamma_r:=\{z=x+iy:\ y=r\}$.

Lemma A.4.1. *Define*

$$v_{D_r\backslash F}(z):=\mathbb{P}_z^{\mathbb{H}}\left(\sigma_{\Gamma_r}<\sigma_{F\cup K}\right),\quad z\in D_r\setminus F. \tag{A.4.1}$$

Then the limit

$$v(z)=\lim_{r\to\infty}r\,v_{D_r\backslash F}(z),\quad z\in D, \tag{A.4.2}$$

exists and admits the expression (2.2.3).

Proof. Obviously

$$v_{\mathbb{H}_r}(z) := \mathbb{P}_z^{\mathbb{H}}(Z_{\sigma_{\Gamma_r}} < \infty) = y/r \quad \text{for } z = x + iy \in \mathbb{H}_r.$$

By the strong Markov property of the ABM $\mathbf{Z}^{\mathbb{H}}$ on $\mathbb{H}$,

$$v_{\mathbb{H}_r}(z) = v_{D_r\setminus F}(z) + \mathbb{E}_z^{\mathbb{H}}\left[v_{\mathbb{H}_r}(Z^{\mathbb{H}}_{\sigma_{F\cup K}}); \sigma_{F\cup K} < \sigma_{\Gamma_r}\right].$$

Consequently, the limit in (A.4.2) exists and satisfies (2.2.3). □

Consider the BMD $\mathbf{Z}^{\mathbb{H}\setminus F,*} = (Z_t^{\mathbb{H}\setminus F,*}, \mathbb{P}_z^{\mathbb{H}\setminus F,*})$ on $(\mathbb{H} \setminus F \cup K) \cup K^*$ obtained from the ABM on $\mathbb{H}\setminus F$ by rendering each compact continuum $A_j \subset \mathbb{H}\setminus F$ into one point a_j^*, $1 \le j \le N$. By virtue of Theorem 1.4.1, $\mathbf{Z}^{\mathbb{H}\setminus F,*}$ is nothing but the part process of the BMD $\mathbf{Z}^*$ in Theorem 2.2.1 on $(\mathbb{H} \setminus F \cup K) \cup K^*$, namely, $\mathbf{Z}^{\mathbb{H}\setminus F,*}$ is obtained from $\mathbf{Z}^*$ by killing upon σ_F.

In particular, ν_i and q_{ij}^* defined by (2.2.5) and (2.2.7) for $1 \le i, j \le N$ can be rewritten in terms of $Z^{\mathbb{H}\setminus F,*}$ as

$$\nu_i(dz) = \mathbb{P}_{a_i^*}^{\mathbb{H}\setminus F,*}\left(Z_{\sigma_{\eta_i}}^{\mathbb{H}\setminus F,*} \in dz; \sigma_{\eta_i} < \infty\right), \tag{A.4.3}$$

$$q_{ij}^* = \begin{cases} \mathbb{P}_{a_i^*}^{\mathbb{H}\setminus F,*}(\sigma_{K^*} < \infty,\ Z_{\sigma_{K^*}}^{\mathbb{H}\setminus F,*} = a_j^*)/(1 - R_i^*) & \text{if } i \ne j, \\ 0 & \text{if } i = j, \end{cases} \tag{A.4.4}$$

where $\{R_i^*,\ 1 \le i \le N\}$ are defined in the same way as (2.2.6). Note that for each $1 \le i \le N$, $0 < R_i^* < 1$ and

$$\sum_{j:j\ne i} q_{ij}^* = \frac{\int_{\eta_i} \mathbb{P}_z^{\mathbb{H}\setminus F,*}(\sigma_{K^*} < \infty)\nu_i(dz) - R_i^*}{1 - R_i^*} < 1. \tag{A.4.5}$$

This is because $\varphi(z) := \mathbb{P}_z^{\mathbb{H}\setminus F,*}(\sigma_{K^*} < \infty) = \mathbb{P}_z^{\mathbb{H}\setminus F}(\sigma_K < \infty)$ is harmonic in $z \in \mathbb{H} \setminus (F \cup K)$ and $\varphi(z) \le \mathbb{P}_z^{\mathbb{H}}(\sigma_K < \infty) \to 0$ as $\Im z \to 0$, and thus $\varphi(z) < 1$ for any $z \in \mathbb{H}\setminus(F\cup K)$ by Lemma 1.1.3. In a similar way, we see that $R_i^* \in (0, 1)$, $1 \le i \le N$. Hence, $I - Q^*$ admits the inverse M with positive entries

$$M_{ij} = \sum_{n=0}^{\infty} q_{ij}^{*n} \quad \text{where} \quad q_{ij}^{*n} = (Q^*)_{ij}^n,\ 1 \le i, j \le N. \tag{A.4.6}$$

Lemma A.4.2. *The limit $v^*(z)$ in (2.2.1) exists for $z \in D^*\setminus F$ and satisfies (2.2.2) together with (2.2.3) and (2.2.4).*

Proof. Put

$$v_r^*(z) := \mathbb{P}_z^{\mathbb{H}\setminus F,*}(\sigma_{\Gamma_r} < \infty), \quad z \in (\mathbb{H}_r \setminus F \cup K) \cup K^*. \tag{A.4.7}$$

Observe that $v_r^*(z) = \mathbb{P}_z^*(\sigma_{\Gamma_r} < \sigma_F)$ for $z \in (\mathbb{H}_r \setminus F \cup K) \cup K^*$. We claim that

$$v_r^*(z) = v_{D_r\setminus F}(z) + \sum_{j=1}^{N} \mathbb{P}_z^{\mathbb{H}\setminus F,*}\left(\sigma_{K^*} < \sigma_{\Gamma_r},\ Z_{\sigma_{K^*}}^{\mathbb{H}\setminus F,*} = a_j^*\right) v_r^*(a_j^*)$$
$$\text{for every } z \in \mathbb{H}_r \setminus F \cup K. \tag{A.4.8}$$

Indeed, if we write $v_r^*(z)$ as $\mathrm{I}+\mathrm{II}$, where

$$\mathrm{I} = \mathbb{P}_z^{\mathbb{H}\setminus F,*}\left(\sigma_{\Gamma_r} < \sigma_{K^*}\right) \quad \text{and} \quad \mathrm{II} = \mathbb{P}_z^{\mathbb{H}\setminus F,*}\left(\sigma_{K^*} < \sigma_{\Gamma_r} < \infty\right),$$

then

$$\mathrm{I} = \mathbb{P}_z^{\mathbb{H}\setminus F}(\sigma_{\Gamma_r} < \sigma_K) = \mathbb{P}_z^{\mathbb{H}}(\sigma_{\Gamma_r} < \sigma_{F\cup K}) = v_{D_r\setminus F}(z),$$

and, by the strong Markov property of $\mathbf{Z}^{\mathbb{H}\setminus F,*}$,

$$\begin{aligned}
\mathrm{II} &= \mathbb{P}_z^{\mathbb{H}\setminus F,*}\left(\sigma_{K^*} < \sigma_{\Gamma_r},\ \sigma_{\Gamma_r}(\theta_{\sigma_{K^*}}\omega) < \infty\right) \\
&= \mathbb{E}_z^{\mathbb{H}\setminus F,*}\left[\mathbb{P}^{\mathbb{H}\setminus F,*}_{Z^{\mathbb{H}\setminus F,*}_{\sigma_{K^*}}}\left(\sigma_{\Gamma_r} < \infty\right); \sigma_{K^*} < \sigma_{\Gamma_r}\right] \\
&= \mathbb{E}_z^{\mathbb{H}\setminus F,*}\left[v_r^*\left(Z^{\mathbb{H}\setminus F,*}_{\sigma_{K^*}}\right);\ \sigma_{K^*} < \sigma_{\Gamma_r}\right],
\end{aligned}$$

yielding (A.4.8).

Since $v_r^*(a_i^*) = \int_{\eta_i} v_r^*(z)\,\nu_i(dz)$, and

$$\int_{\eta_i} \mathbb{P}_z^{\mathbb{H}\setminus F,*}\left(\sigma_{K^*} < \sigma_{\Gamma_r},\ Z^{\mathbb{H}\setminus F,*}_{\sigma_{K^*}} = a_j^*\right)\nu_i(dz) = \Pi_{ij}^{r*}, \quad j \neq i,$$

where $\Pi_{ij}^{r*} = \mathbb{P}_{a_i^*}^{\mathbb{H}\setminus F,*}\left(\sigma_{K^*} < \sigma_{\Gamma_r},\ Z^{\mathbb{H}\setminus F,*}_{\sigma_{K^*}} = a_j^*\right)$, we get by integrating the both sides of (A.4.8) with respect to $\nu_i(dz)$ that

$$v_r^*(a_i^*) = \int_{\eta_i} v_{D_r\setminus F}(z)\,\nu_i(dz) + \sum_{j\neq i} \Pi_{ij}^{r*} v_r^*(a_j^*) + R_i^{r*} v_r^*(a_i^*), \tag{A.4.9}$$

where R_i^{r*} is defined by (2.2.6) with $F \cup \Gamma_r$ in place of F.

Define for $i \neq j$, $q_{ij}^{r*} := \Pi_{ij}^{r*}/(1-R_i^{r*})$, and

$$f_r^*(a_i^*) := \frac{1}{1-R_i^{r*}} \int_{\eta_i} v_{D_r\setminus F}(z)\,\nu_i(dz), \quad 1 \le i \le N.$$

In the same way as we saw for q_{ij}^* in the above, $q_{ij}^{r*} \in (0,1)$ for $i \neq j$ and

$$\sum_{j:j\neq i} q_{ij}^{r*} = \frac{\int_{\eta_i} \mathbb{P}_z^{\mathbb{H}\setminus F,*}(\sigma_{K^*} < \sigma_{\Gamma_r})\nu_i(dz) - R_i^{r*}}{1-R_i^{r,*}} < 1, \quad 1 \le i \le N. \tag{A.4.10}$$

The equation (A.4.9) can be rewritten as

$$v_r^*(a_i^*) - \sum_{j:j\neq i} q_{ij}^{r*} v_r^*(a_j^*) = f_r^*(a_i^*) \quad \text{for } 1 \le i \le N. \tag{A.4.11}$$

In view of (A.4.10), the equation (A.4.11) admits a unique solution

$$v_r^*(a_i^*) = \sum_{j=1}^N M_{ij}^r\, f_r^*(a_j^*) \quad \text{for } 1 \le i \le N. \tag{A.4.12}$$

Here M_{ij}^r, $1 \le i, j \le N$, are the entries of the inverse of $I - Q^{r*}$ for the matrix Q^{r*} with off-diagonal elements q_{ij}^{r*}, $i \neq j$ and zero diagonal elements.

Observe that R_i^{r*} and Π_{ij}^{r*} increase to R_i^* and $\mathbb{P}_{a_i^*}^{\mathbb{H}\setminus F,*}(\sigma_{K^*} < \infty, Z_{\sigma_{K^*}}^{\mathbb{H}\setminus F,*} = a_j^*)$, respectively, as $r \uparrow \infty$ for every $1 \le i, j \le N$. Accordingly, q_{ij}^{r*} and M_{ij}^{r*} increase to q_{ij}^* and M_{ij}^*, respectively, as $r \uparrow \infty$. Furthermore, by Lemma A.4.1,

$$\lim_{r\to\infty} r f_r^*(a_i^*) = \frac{1}{1-R_i^*}\int_{\eta_i} v(z)\,\nu_i(dz), \quad 1 \le i \le N.$$

Therefore, we deduce from (A.4.12) that the limit $v^*(a_i^*) := \lim_{r\to\infty} r v_r^*(a_i^*)$ exists and satisfies (2.2.4). Moreover, it follows from Lemma A.4.1 and (A.4.8) that the limit $v^*(z) = \lim_{r\to\infty} r v_r^*(z)$ exists for every $z \in D$ and satisfies (2.2.2). □

The next lemma concerns the limiting behaviors of $v^*(z)$ as $z \to \infty$.

Lemma A.4.3.

$$\lim_{z\to\infty} |v^*(z) - \Im z| = 0. \tag{A.4.13}$$

$$\lim_{y\to\infty} v_x^*(x+iy) = 0 \quad \text{uniformly in } x \in \mathbb{R}. \tag{A.4.14}$$

$$\lim_{x\to\pm\infty} |v_y^*(x+iy) - 1| = 0 \quad \text{uniformly in } y > 0. \tag{A.4.15}$$

Proof. We get from (2.2.2) and (2.2.3)

$$|v^*(z) - \Im z| \le M\mathbb{P}_z^{\mathbb{H}}(\sigma_{K\cup F} < \infty), \quad \text{where } M = (\max_{z\in K\cup F} \Im z) \vee (\max_{1\le j\le N} v^*(a_j^*)).$$

Choose $R > 0$ with $B(0,R) \supset K \cup F$. Then $\mathbb{P}_z^*(\sigma_{K\cup F} < \infty)$ is dominated on $\mathbb{H} \setminus \overline{B(0,R)}$ by the left-hand side of (1.1.16) with $\theta_1 = 0$ and $\theta_2 = \pi$. Consequently, we get (A.4.13) from Proposition 1.1.2.

Take $\ell_1 > 0$ such that $\mathbb{H}_{\ell_1} \supset F \cup K$. $h(z) = v^*(z) - \Im z$ is a harmonic function on $\mathbb{H}_{\ell_1}^+ = \{z \in \mathbb{H} : \Im z > \ell_1\}$ satisfying (A.4.13). So h satisfies the identity (1.1.52). We then get (A.4.14) from

$$|p_x(x-\xi, y-\ell_1)| \le \frac{2}{(y-\ell_1)} p(x-\xi, y-\ell_1) \quad \text{for} \quad y > \ell_1.$$

Property (A.4.15) follows from another Poisson integral representation (1.1.53) of $h(z) = v^*(z) - \Im z$. □

The function v^* on D^* constructed in Lemma A.4.2 is obviously $\mathbf{Z}^*$-harmonic. In view of Theorem 1.6.5, $-v^*\big|_D$ admits a harmonic conjugate u^* on D uniquely up to an additive constant so that $f(z) = u^*(z) + iv^*(z)$, $z \in D$, is an analytic function. Moreover, by virtue of Lemma 2.2.2, f can be extended by Schwarz reflection to be analytic on $\mathbb{C} \setminus [(K \cup \overline{F}) \cup \Pi(K \cup \overline{F})]$, where $\Pi z := \bar{z}$.

The function u^* admits the expression (1.6.6) in terms of v^* up to a real additive constant independent of the choice of a rectifiable curve C connecting z_0 with z. In particular, for a large $\ell > 0$ with $R_\ell = \{x + iy : |x| < \ell,\ 0 < y < \ell\} \supset K \cup F$,

$$u^*(x+iy) - u^*(i\ell) = \int_0^x v_y^*(\xi + i\ell)d\xi - \int_\ell^y v_x^*(x+i\eta)d\eta \quad \text{for } x \in \mathbb{R},\ y > \ell. \tag{A.4.16}$$

$$u^*(x+iy) - u^*(\ell+i0) = -\int_0^y v_x^*(\ell+i\eta)d\eta + \int_\ell^x v_y^*(\xi+iy)d\xi \quad \text{for } x > \ell,\ y > 0. \tag{A.4.17}$$

The next lemma concerns the limiting behaviors of $f(z)$ as $z \to \infty$.

Lemma A.4.4. *It holds that*

$$\lim_{y\to\infty} \frac{f(iy)}{iy} = 1. \tag{A.4.18}$$

Furthermore,

$$\limsup_{z\to\infty, z\in D} \frac{|f(z)|}{|z|} \leq 2. \tag{A.4.19}$$

Proof. The real part of $f(iy)/(iy)$ equals $v^*(iy)/y$, which tends to 1 as $y \to \infty$ by (A.4.13); while its imaginary part equals $-u^*(iy)/y = -u^*(i\ell)/y + \frac{1}{y}\int_\ell^y v_x^*(i\eta)d\eta$, which tends to 0 as $y \to \infty$ by (A.4.14). This establishes (A.4.18).

We next show that

$$\limsup_{z\to\infty} |u^*(z)/z| \leq 1, \tag{A.4.20}$$

which combined with (A.4.13) implies (A.4.19).

Properties (A.4.14) and (A.4.15) imply that, for any $\varepsilon > 0$, there exist $\ell > 0$ such that

$$|v_x^*(x+i\eta)| < \varepsilon \quad \text{for any } x \in \mathbb{R} \text{ and for any } \eta \geq \ell,$$

$$|v_y^*(\xi+iy)| < 1+\varepsilon \quad \text{for any } y > 0 \text{ and } |\xi| \geq \ell.$$

Let $M_1 = \sup_{|\xi|\leq\ell} |v_y^*(\xi+i\ell)|$. We get from (A.4.16)

$$\frac{|u^*(z)|}{|z|} \leq \frac{|u^*(i\ell)|}{|z|} + \frac{M_1\ell + |x|(1+\varepsilon)}{|z|} + \frac{|y-\ell|}{|z|}\varepsilon,$$

and so $\limsup_{y\to\infty} \sup_{x\in\mathbb{R}} \frac{|u^*(z)|}{|z|} \leq 1+2\varepsilon$. By letting $M_2 = \sup_{\eta\in[0,\ell]} |v_x^*(\ell+i\eta)|$, we obtain from (A.4.17)

$$\frac{|u^*(z)|}{|z|} \leq \frac{|u^*(\ell+i0)|}{|z|} + \frac{M_2\ell + |y|\varepsilon}{|z|} + \frac{|x-\ell|}{|z|}(1+\varepsilon),$$

and so $\limsup_{x\to\infty} \sup_{y>0} \frac{|u^*(z)|}{|z|} \leq 1$. Similarly, we obtain $\limsup_{x\to-\infty} \sup_{y>0} \frac{|u^*(z)|}{|z|} \leq 1$, arriving at (A.4.20). □

Lemma A.4.5. *u^* can be chosen uniquely in such a way that $f = u^* + iv^*$ satisfies* (2.2.8).

Proof. We have seen that f extends to be an analytic function on $\mathbb{C} \setminus ((K \cup \overline{F}) \cup \Pi(K \cup \overline{F}))$, which will still be denoted as f.

Let $g(z) := zf(1/z)$. Then g is analytic and bounded in $B(0, \varepsilon) \setminus \{\mathbf{0}\}$ for some $\varepsilon > 0$ by (A.4.19). So $\mathbf{0}$ is a removable singularity of $g(z)$ and $\lim_{z \to \mathbf{0}} g(z) = a_0$ exists. By (A.4.18), $\lim_{y \downarrow 0} g(-iy) = 1$ so that $a_0 = 1$ and g can be expanded near zero as $g(z) = 1 + a_1 z + a_2 z^2 + \cdots$. Therefore,

$$f(z) = z + a_1 + \frac{a_2}{z} + \cdots \qquad \text{as } z \to \infty. \tag{A.4.21}$$

Since $\Im f = 0$ on $\partial(\mathbb{H} \setminus \overline{F})$, the coefficients $a_1, a_2, \ldots$ are real numbers. Thus f satisfies (2.2.8) if and only if $\lim_{z \to \infty}(f(z) - z) = 0$. The functions f and u^* are uniquely determined by v^* under this condition. □

The proof of Theorem 2.2.1 is now complete.

A.5 Proper maps and their degree

Recall that a continuous map $f : X \to Y$ between topological spaces is called *proper* if the pre-images of compact sets are compact. Intuitively, if X and Y are subsets of larger spaces, this means that the boundary of X maps into the boundary of Y (though it is not required that f extends continuously to the boundary). In rather general situations (for instance smooth orientation preserving maps between manifolds with boundary, see, e.g., [Milnor (1965)]), such maps have the property that every $y \in Y$ has the same number of pre-images $x \in X$ (counted according to multiplicity). In this section, we formulate and prove a simple version of this principle, in the setting of analytic functions that is suitable for our purpose. We allow ∞ to be in the domain and in the range of f and adopt the usual definition that a function f (defined in a neighborhood of ∞ with $f(\infty) = \infty$) is *analytic* if $1/f(1/z)$ is analytic in a neighborhood of 0.

Lemma A.5.1. *Suppose D_1 and D_2 are two connected open subsets of the Riemann sphere $\overline{\mathbb{C}}$ and f is non-constant and analytic in D_1. If f is a proper map between D_1 and D_2 (that is, if $f(D_1) \subset D_2$ and if $f^{-1}(K)$ is a compact subset of D_1 whenever K is a compact subset of D_2), then $f(D_1) = D_2$ and there is a finite number d such that every $w \in D_2$ has precisely d pre-images in D_1, counting multiplicity.*

Proof. Fix $w_0 \in f(D_1)$. Let w_1 be an arbitrary point in D_2. We first assume that neither w_0 nor w_1 are critical values (that is, we assume $f' \neq 0$ for every pre-image, if there is one, of w_0 and w_1). Since f is non-constant, it has only isolated critical points so there is a simple curve $\gamma \subset D_2$ with $\gamma(0) = w_0$ and $\gamma(1) = w_1$ that is disjoint from the set of critical values. For every pre-image z_0 of w_0, the branch g_0 of the inverse function f^{-1} with $g_0(w_0) = z_0$, defined in a neighborhood of w_0, can be analytically continued along γ and yields a branch g_1 of f^{-1} near w_1.

This is where both the assumption on the critical values, and the assumption on properness are used: During the process of analytic continuation, the curve

$f^{-1}(\gamma(t))$ cannot escape from D_1 by properness, and one can always analytically continue further because one does not meet critical values. Formally, one considers the subset $S = \{s \in [0,1] : g_0 \text{ can be analytically continued along } \gamma[0,s]\}$ of the interval $[0,1]$ and shows that it is both open and closed. This in particular shows that $w_1 \in f(D_1)$, proving that $f(D_1) = D_2$ since for the case when w_1 is a critical value we automatically have $w_1 \in f(D_1)$.

Conversely, continuing g_1 along the reversed curve γ^{-1} leads us back to g_0. Thus we have a bijection between the sets $f^{-1}(w_0)$ and $f^{-1}(w_1)$ when neither w_0 nor w_1 are critical values. Note that every level set of a non-trivial analytic function has no accumulation points. As f is a proper map, $f^{-1}(\{w_0\})$ is a compact subset of D_1 so it can only have finite many points. If a pre-image z_0 (or z_1) of w_0 (or w_1) is a critical point (so that the local degree of f at z_0 is more than 1), simply replace w_0 (or w_1) by nearby points w_0' or w_1' and use the fact that the number of pre-images of w_0' near z_0 equals the local degree of f at z_0. □

The number d in the above lemma is called the *degree* of the map f. Now we will show that the assumption $f(D_1) \subset D_2$ can be removed if the degree is one and if the complement of D_2 has empty interior. For a function f and an open set $D \subset \overline{\mathbb{C}}$, we define by

$$f(\partial D) = \bigcap_{K \Subset D} \overline{f(D \setminus K)}, \tag{A.5.1}$$

the set of limit points of f as z approaches ∂D (the intersection is over all compact subsets of D). It is easy to see that any proper map f from D_1 onto D_2 satisfies $f(\partial D_1) = \partial D_2$. The next theorem goes in the opposite direction. Notice that we do not assume a priori that $f(D_1) \subset D_2$.

Theorem A.5.2. *Suppose that D_1 and D_2 are two connected open subsets of the Riemann sphere $\overline{\mathbb{C}}$, and f is non-constant and analytic in D_1. Assume that the complement of D_2 has empty interior, that*

$$f(\partial D_1) = \partial D_2, \tag{A.5.2}$$

and that there is one point $w_0 \in D_2$ that has precisely one pre-image z_0 under f (counting multiplicity). Then f is a conformal map from D_1 onto D_2.

Proof. Let $D \subset D_1$ be any non-empty connected component of $f^{-1}(D_2)$. Then $f : D \to D_2$ is proper.

To see this, suppose K is a compact subset of D_2 and $\{z_n; n \geq 1\}$ is a sequence in $f^{-1}(K) \cap D$. We need to show that every sub-sequential limit of $\{z_n; n \geq 1\}$ is in D. Assume to the contrary that a subsequence $\{z_{n_k}; k \geq 1\}$ of $\{z_n; n \geq 1\}$ converges to a point $z \in \partial D$ as $k \to \infty$. Since $f(x_n) \in K$ and K is compact, by taking a further subsequence if needed, we can and do assume that $f(x_{n_k})$ is convergent to some point $w \in K$ as $k \to \infty$. If $z \in D_1$, then $f(z) \in K \subset D_2$ and thus a neighborhood $U(z)$ of z is contained in some connected component $\widetilde{D}$

of $f^{-1}(D_2)$. As $z_n \in D \cap U(z)$ for large enough n, $\widetilde{D} = D$, a contradiction to $z \in \partial D$. If $z \in \partial D_1$, then $f(z) := \lim_{k\to\infty} f(z_{n_k}) = w \in f(\partial D_1) = \partial D_2$, which is a contradiction as $w \in K \subset D_2$. This proves that $f : D \to D_2$ is proper.

Now Lemma A.5.1 implies that $f(D) = D_2$ and every $w \in D_2$ has the same finite number of pre-images in D. This number is trivially at least one, and it is at most one by assumption on w_0. Thus the degree of f on D equals one, and there is only one such component D (in other words, $D = f^{-1}(D_2)$). In particular, f is a conformal bijection between D and D_2.

It remains to show that $D = D_1$. Suppose not, then there would be a point $z_1 \in D_1 \setminus D$. By the assumption of the connectivity of D_1, we may assume that $z_1 \in D_1 \cap \partial D$, because otherwise take a continuous curve $\gamma \subset D_1$ joining z_1 to a point $z_2 \in D$ and we may replace z_1 by the first point in γ that belongs to $\overline{D}$.

As $z_1 \in D_1$, $w_1 = f(z_1)$ is well defined. Since f is a conformal map from D onto D_2 and $z_1 \in \partial D$, it follows that $w_1 \in \partial D_2$. Due to the assumption $f(\partial D_1) = \partial D_2$, there is a sequence $\zeta_n \in D_1$ and a point $z^* \in \partial D_1$ such that $\zeta_n \to z^*$ and $f(\zeta_n) \to w_1$ as $n \to \infty$. Notice that $z^* \neq z_1$ because $z_1 \in D_1$.

Now define a new sequence ζ_n' as follows: If $f(\zeta_n) \in D_2$, set $\zeta_n' = \zeta_n$. If $f(\zeta_n) \notin D_2$, choose any point ζ_n' in D_1 close to ζ_n (say $|\zeta_n - \zeta_n'| < 1/n$) in such a way that $|f(\zeta_n) - f(\zeta_n')| < 1/n$ and $f(\zeta_n') \in D_2$. This can be done because f is an open mapping and the complement of D_2 has empty interior by our assumption.

Since $f(\zeta_n') \in D_2$ and f is bijection between $D = f^{-1}(D_2)$ and D_2, it holds that $\zeta_n' \in D$. Since $f(\zeta_n') \to w_1$ and f maps any open neighborhood of z_1 to an open neighborhood of w_1, the equation $f(z) = f(\zeta_n')$ for a large n has two solutions in D, one being $z = \zeta_n'$ and another one near z_1. We thus obtain a contradiction to the injectivity of f on D. □

Notice that the last claim $D = D_1$ is not true if the degree is more than one, as for instance the map $f(z) = z^2$ with $D_1 = D_2 = \mathbb{C} \setminus [0, 1]$ has $D = \mathbb{C} \setminus [-1, 1]$.

A.6 Green function of ABM under perturbation of standard slit domain

In this section, we present a proof for Proposition 2.4.3. Its assertions (i) and (ii) are established by the next lemma. Recall that $\mathcal{A}_\varepsilon$ is the second order elliptic differential operator of divergence form defined by (2.4.9).

Lemma A.6.1. *The function $g(z, w, \varepsilon)$ defined by (2.4.8) is a fundamental solution of $\mathcal{A}_\varepsilon$ in the sense of (2.4.10). The coefficients $A_{k\ell}^{(\varepsilon)}$ of $\mathcal{A}_\varepsilon$ admit expressions*

$$A_{k\ell}^{(\varepsilon)} = \frac{1}{2}\delta_{k\ell} + \varepsilon b_{k\ell}^{(\varepsilon)} \quad \textit{for } 1 \le k, \ell \le 2, \tag{A.6.1}$$

where $b_{k\ell}^{(\varepsilon)}$, $1 \le k, \ell \le 2$, are smooth functions on $\mathbb{H}$ with $b_{k\ell}^{(\varepsilon)} = b_{\ell k}^{(\varepsilon)}$ vanishing on $(\mathbb{H} \setminus \cup_{i=1}^N \overline{V}_i) \cup (\cup_{i=1}^N U_i)$ which together with their derivatives are uniformly bounded in $\varepsilon \in (0, \varepsilon_0)$, $D \in \mathcal{D}_0$ and $\widetilde{D} = \widetilde{f}_\varepsilon(D) \in \mathcal{D}$.

Proof. Recall the map $\widetilde{f}_\varepsilon$ defined by (2.4.7) and write $\widetilde{f}_\varepsilon(x_1+ix_2)$ as $\widetilde{x}_1+i\widetilde{x}_2$. Denote the Jacobian determinant $\frac{\partial(\widetilde{x}_1,\widetilde{x}_2)}{\partial(x_1,x_2)}$ of the map $\widetilde{f}_\varepsilon$ by $J=J(x_1,x_2)$. Let $(\widetilde{\mathcal{E}},\widetilde{\mathcal{F}})$ and $\widetilde{\mathcal{A}}$ be the Dirichlet form and the L^2-generator on $L^2(\widetilde{D}):=L^2(\widetilde{D};dx)$ of the ABM on $\widetilde{D}$, respectively. Then $\widetilde{u}\in\mathcal{D}(\widetilde{\mathcal{A}})$ and $\widetilde{\mathcal{A}}\widetilde{u}=\widetilde{f}\in L^2(\widetilde{D})$ if and only if $\widetilde{u}\in H^1_0(\widetilde{D})$ and $\widetilde{\mathcal{E}}(\widetilde{u},\widetilde{v})=-\int_{\widetilde{D}}\widetilde{f}\widetilde{v}d\widetilde{x}_1d\widetilde{x}_2$. It follows from

$$
\begin{aligned}
&\widetilde{\mathcal{E}}(\widetilde{u},\widetilde{v})\\
&=\frac{1}{2}\int_{\widetilde{D}}\sum_{j=1}^{2}\frac{\partial\widetilde{v}}{\partial\widetilde{x}_j}\frac{\partial\widetilde{u}}{\partial\widetilde{x}_j}d\widetilde{x}_1d\widetilde{x}_2=\frac{1}{2}\int_{D}\sum_{j=1}^{2}\sum_{k,\ell=1}^{2}\frac{\partial\widetilde{v}}{\partial x_k}\frac{\partial x_k}{\partial\widetilde{x}_j}\frac{\partial\widetilde{u}}{\partial x_\ell}\frac{\partial x_\ell}{\partial\widetilde{x}_j}J(x_1,x_2)dx_1dx_2\\
&=\sum_{k,\ell=1}^{2}\int_D A^{(\varepsilon)}_{k\ell}\frac{\partial\widetilde{v}}{\partial x_k}\frac{\partial\widetilde{u}}{\partial x_\ell}dx_1dx_2=-\int_D\widetilde{v}(x_1,x_2)\mathcal{A}_\varepsilon\widetilde{u}(x_1,x_2)dx_1dx_2, \qquad (A.6.2)
\end{aligned}
$$

and $\int_{\widetilde{D}}\widetilde{f}\widetilde{v}d\widetilde{x}_1d\widetilde{x}_2=\int_D\widetilde{f}(x_1,x_2)\widetilde{v}(x_1,x_2)J(x_1,x_2)dx_1dx_2$ for every $\widetilde{v}\in C^1_c(\widetilde{D})$ that

$$
(\widetilde{\mathcal{A}}\widetilde{u})(x_1,x_2)=J^{-1}\mathcal{A}_\varepsilon\widetilde{u}(x_1,x_2). \qquad (A.6.3)
$$

On the other hand, if we define $\widetilde{G}\widetilde{f}(\widetilde{z})=\int_{\widetilde{D}}\widetilde{G}(\widetilde{z},\widetilde{w})\widetilde{f}(\widetilde{w})d\widetilde{w}$, then

$$
\widetilde{G}\widetilde{f}(\widetilde{z})=\int_D g(z,w,\varepsilon)(\widetilde{f}\circ\widetilde{f}_\varepsilon\cdot J)(w)dw_1dw_2=:g_\varepsilon(\widetilde{f}\circ\widetilde{f}_\varepsilon\cdot J)(z).
$$

Since $\widetilde{\mathcal{A}}(\widetilde{G}\widetilde{f})(\widetilde{z})=-\widetilde{f}(\widetilde{z})$, we have by (A.6.3) that $J^{-1}\mathcal{A}_\varepsilon\cdot g_\varepsilon(\widetilde{f}\circ\widetilde{f}_\varepsilon\cdot J)(z)=-\widetilde{f}\circ\widetilde{f}_\varepsilon(z)$. This establishes (2.4.10) by taking $f=\widetilde{f}\circ\widetilde{f}_\varepsilon\cdot J$.

The stated expression and properties of coefficients of $\mathcal{A}_\varepsilon$ follow from (2.4.7), (2.4.9) and the uniform boundedness of the coefficients of the linear map (2.4.6). In particular, $b^{(\varepsilon)}_{k\ell}$, $1\le k,\ell\le N$, vanish on $\mathbb{H}\setminus\bigcup_{i=1}^N\overline{V}_i$ because $\widetilde{f}_\varepsilon$ is an identity map there, and on each U_i as well because $q(x_1,x_2)$ is constant there. □

To derive the perturbation formulae (2.4.12) and (2.4.13), we first construct an appropriate parametrix for the elliptic differential operator $\mathcal{A}_\varepsilon$ by following the method of interior variations presented in Section 15.1 of Garabedian's book [Garabedian (1964)].

Denote by $a^{(\varepsilon)}=(a^{(\varepsilon)}_{k\ell})_{1\le k,\ell\le 2}$ the inverse matrix of $A^{(\varepsilon)}=(A^{(\varepsilon)}_{k\ell})_{1\le k,\ell\le 2}$. Since $\det A^{(\varepsilon)}=1/4$, we have

$$
a^{(\varepsilon)}_{11}=2+4\varepsilon b^{(\varepsilon)}_{22},\quad a^{(\varepsilon)}_{22}=2+4\varepsilon b^{(\varepsilon)}_{11}\quad\text{and}\quad a^{(\varepsilon)}_{12}=a^{(\varepsilon)}_{21}=-4\varepsilon b^{(\varepsilon)}_{12}. \qquad (A.6.4)
$$

Define

$$
\Gamma(z,\zeta)=\frac{1}{2}\sum_{i,j=1}^{2}a^{(\varepsilon)}_{ij}(\zeta)(x_i-\zeta_i)(x_j-\zeta_j),\quad\text{where } z=x_1+ix_2 \text{ and } \zeta=\zeta_1+i\zeta_2. \qquad (A.6.5)
$$

The function $-\frac{1}{2\pi}\log\Gamma(z,\zeta)$ has the same logarithmic singularity as the fundamental solution $g(z,\zeta,\varepsilon)$ of the elliptic differential operator $\mathcal{A}_\varepsilon$ (cf. [Garabedian (1964), (5.80)]).

Recall the constant b_0 defined in (2.4.16). We fix an arbitrary $\ell_0 \in (0, b_0]$ and consider a smooth non-positive real function $\alpha(t)$, $t \in \mathbb{R}$, with

$$\alpha(0) = -1/(2\pi), \quad \alpha(t) = 0 \ \text{ if } \ t \notin (-\ell_0^2, \ell_0^2). \tag{A.6.6}$$

Let

$$\begin{cases} P_\varepsilon(z, \zeta) = \alpha(|z - \zeta|^2) \log \Gamma(z, \zeta), \\ P_0(z, \zeta) = \alpha(|z - \zeta|^2) \log |z - \zeta|^2, \quad z, \zeta \in \overline{\mathbb{H}}, \ z \neq \zeta. \end{cases} \tag{A.6.7}$$

For a function $u(z, \zeta, \varepsilon)$, $z, \zeta \in D \cup \partial\mathbb{H}$, $z \neq \zeta$, $\varepsilon \in (0, \varepsilon_0)$, we write $u(z, \zeta, \varepsilon) = O(\varepsilon/r)$ with $r = |z - \zeta|$ if

$$|u(z, \zeta, \varepsilon)| \leq \varepsilon \frac{M_1}{|z - \zeta|} + \varepsilon M_2, \quad z, \ \zeta \in D \cup \partial\mathbb{H}, \tag{A.6.8}$$

for positive constants M_1, M_2 independent of $\varepsilon \in (0, \varepsilon_0)$, $D \in \mathcal{D}_0$ and $\widetilde{D} = \widetilde{f}_\varepsilon(D) \in \mathcal{D}$.

In what follows, the set $\bigcup_{i=1}^N (\overline{V}_i \backslash U_i)$ will be denoted as Λ. When there are more than one variable present, we use notation $\mathcal{A}_{\varepsilon,z}$ to emphasize that the differential operator $\mathcal{A}$ is applied to the z-variable.

Lemma A.6.2. *$G(z, \zeta) - P_0(z, \zeta) + P_\varepsilon(z, \zeta)$ is a parametrix of the operator $\mathcal{A}_\varepsilon = \mathcal{A}_{\varepsilon,z}$ in a specific sense that*

$$\mathcal{A}_{\varepsilon,z}(G(z, \zeta) - P_0(z, \zeta) + P_\varepsilon(z, \zeta)) = O(\varepsilon/r), \quad \textit{where } r := |z - \zeta|. \tag{A.6.9}$$

Proof. It suffices to show that, with $r := |z - \zeta|$,

$$\mathcal{A}_{\varepsilon,z} P_\varepsilon(z, \zeta) - \frac{1}{2}\Delta P_0 = O(\varepsilon/r), \tag{A.6.10}$$

and

$$(\mathcal{A}_{\varepsilon,z} - \frac{1}{2}\Delta)(G - P_0) = O(\varepsilon/r). \tag{A.6.11}$$

Note that they imply (A.6.9) because $\Delta_z G(z, \zeta) = 0$ for $z \neq \zeta$.

Designating $A_{ij}^{(\varepsilon)}$ by A_{ij}, it holds that

$$\mathcal{A}_{\varepsilon,z} P_\varepsilon(z, \zeta) - \frac{1}{2}\Delta_z P_0(z, \zeta) = \mathrm{I}_\varepsilon + \mathrm{II}_\varepsilon + \mathrm{III}_\varepsilon + \mathrm{IV}_\varepsilon + \mathrm{V}_\varepsilon, \tag{A.6.12}$$

with

$$\begin{cases} \mathrm{I}_\varepsilon = \alpha(|z - \zeta|^2) \sum_{ij=1}^2 A_{ij}(z)(\log \Gamma(z, \zeta))_{x_i x_j}, \\ \mathrm{II}_\varepsilon = \sum_{i,j=1}^2 A_{ij,x_i}(z)[\alpha(|z - \zeta|^2) \log \Gamma(z, \zeta)]_{x_j}, \\ \mathrm{III}_\varepsilon = \sum_{i=1}^2 \alpha(|z - \zeta|^2)_{x_i x_i} [A_{ii}(z) \log \Gamma(z, \zeta) - \frac{1}{2} \log |z - \zeta|^2], \\ \mathrm{IV}_\varepsilon = 2\alpha'(|z - \zeta|^2)[\sum_{i,j=1}^2 A_{ij}(z)(x_i - \zeta_i)(\log \Gamma(z, \zeta))_{x_j} - 2], \\ \mathrm{V}_\varepsilon = \sum_{i \neq j} A_{ij}(z) \alpha(|z - \zeta|^2)_{x_i x_j} \log \Gamma(z, \zeta). \end{cases}$$

In fact, it follows from

$$\mathcal{A}_{\varepsilon,z}P_\varepsilon(z,\zeta) = \sum_{i,j=1}^{2} A_{ij}(z)[\alpha(|z-\zeta|^2)\log\Gamma(z,\zeta)]_{x_ix_j} + \sum_{i,j=1}^{2} A_{ij,x_i}(z)[\alpha(|z-\zeta|^2)\log\Gamma(z,\zeta)]_{x_j},$$

that

$$\mathcal{A}_{\varepsilon,z}P_\varepsilon(z,\zeta) - \mathrm{I}_\varepsilon - \mathrm{II}_\varepsilon - \mathrm{V}_\varepsilon = \sum_{i=1}^{2}\alpha(|z-\zeta|^2)_{x_ix_i}A_{ii}(z)\log\Gamma(z,\zeta) + 2\alpha'(|z-\zeta|^2)\sum_{i,j=1}^{2}A_{ij}(z)(x_i-\zeta_i)(\log\Gamma(z,\zeta))_{x_j}.$$

On the other hand,

$$\frac{1}{2}\Delta_z P_0(z,\zeta) = \frac{1}{2}\sum_{i=1}^{2}\alpha(|z-\zeta|^2)_{x_ix_i}\log|z-\zeta|^2 + 4\alpha'(|z-\zeta|^2),$$

yielding (A.6.12).

By using the decomposition (A.6.12), we shall show (A.6.10). We designate by $\eta_k(z,\zeta),\ z,\zeta\in D,\ k=1,2,\ldots$, quantities bounded uniformly in $\varepsilon, D\in\mathcal{D}_0, \widetilde{D}\in\mathcal{D}$. As for I_ε, we have

$$\sum_{ij=1}^{2} A_{ij}(z)(\log\Gamma(z,\zeta))_{x_ix_j} = \frac{\sum_{i,j=1}^{2}A_{ij}(z)a_{ij}(\zeta)\Gamma(z,\zeta) - \sum_{i,j=1}^{2}A_{ij}(z)\Gamma_{x_i}(z,\zeta)\Gamma_{x_j}(z,\zeta)}{\Gamma(z,\zeta)^2}. \tag{A.6.13}$$

Since

$$A_{ij}(z) = A_{ij}(\zeta) + \varepsilon\sum_{k=1}^{2}C_{i,j,k}(z,\zeta)(x_k-\zeta_k), \tag{A.6.14}$$

with $C_{i,j,k}$ involving only derivatives of b_{ij}, we get from (A.6.4) that

$$\sum_{i,j=1}^{2}A_{ij}(z)a_{ij}(\zeta) = 2 + \varepsilon\eta_1(z,\zeta)|z-\zeta|. \tag{A.6.15}$$

We also have from (A.6.4) that

$$\Gamma(z,\zeta) = |z-\zeta|^2(1+\varepsilon\eta_2(z,\zeta)). \tag{A.6.16}$$

Furthermore, it follows from (A.6.4) and (A.6.14) that

$$\sum_{i,j=1}^{2}A_{ij}(z)\Gamma_{x_i}(z,\zeta)\Gamma_{x_j}(z,\zeta) = \sum_{k,\ell=1}^{2}\sum_{i,j=1}^{2}A_{ij}(z)a_{ik}(\zeta)a_{j\ell}(\zeta)(x_k-\zeta_k)(x_\ell-\zeta_\ell) = 2\Gamma(z,\zeta) + \varepsilon\eta_3(z,\zeta)|z-\zeta|^3. \tag{A.6.17}$$

By (A.6.15)–(A.6.17), the right hand side of (A.6.13) equals

$$\varepsilon\frac{\eta_1(z,\zeta)|z-\zeta|\Gamma(z,\zeta)-\eta_3(z,\zeta)|z-\zeta|^3}{\Gamma(z,\zeta)^2}=\varepsilon\eta_4(z,\zeta)\frac{1}{|z-\zeta|},$$

yielding that $\mathrm{I}_\varepsilon = O(\varepsilon/r)$.

Since $A_{ij,x_i}(z)$, $1 \leq i,j \leq 2$, and $A_{ij}(z)$, $i \neq j$, are of the type $\varepsilon\eta(z)$ for $\eta(z)$ uniformly bounded in $\varepsilon > 0$, $D \in \mathcal{D}_0$, $\widetilde{D} \in \mathcal{D}$ in view of Lemma A.6.1, we conclude from (A.6.16) that $\mathrm{II}_\varepsilon = O(\varepsilon/r)$, $\mathrm{V}_\varepsilon = O(\varepsilon/r)$. As it follows from (A.6.16) that $\log\Gamma(z,\zeta) - \log|z-\zeta|^2 = \varepsilon\eta_5(z,\zeta)$, we also obtain $\mathrm{III}_\varepsilon = O(\varepsilon/r)$ using (A.6.1). Finally (A.6.14) and (A.6.1) imply that

$$\sum_{i,j=1}^{2} A_{ij}(z)(x_i-\zeta_i)(\log\Gamma(z,\zeta))_{x_j} = \frac{2|z-\zeta|^2+\varepsilon\eta_6(z,\zeta)|z-\zeta|^3}{\Gamma(z,\zeta)}.$$

Hence $IV_\varepsilon = O(\varepsilon/r)$ in view of (A.6.16).

Finally, we verify (A.6.11). By Lemma A.6.1, $\mathcal{A}_\varepsilon - \frac{1}{2}\Delta$ is equal to $\varepsilon\mathcal{B}^{(\varepsilon)}$, where $\mathcal{B}^{(\varepsilon)}$ is the differential operator defined by (2.4.11). By (1.1.12), the left hand side of (A.6.11) can be written as $\varepsilon\mathcal{B}_z^{(\varepsilon)}u(z,\zeta)+\varepsilon\mathcal{B}_z^{(\varepsilon)}S(z,\zeta)$, where

$$u(z,\zeta) = -\left(\alpha(|z-\zeta|^2)+\frac{1}{2\pi}\right)\log|z-\zeta|^2 \quad\text{and}\quad S(z,\zeta)=\frac{1}{2\pi}\mathbb{E}_z\left[\log|Z_{\tau_D}-\zeta|^2\right].$$

Here $(Z_t,\mathbb{P}_z)_{\{z\in\mathbb{C}\}}$ is the planar Brownian motion. By taking (A.6.6) into account, we can readily verify that $\varepsilon\mathcal{B}_z^{(\varepsilon)}u(z,\zeta) = O(\varepsilon/r)$. Note that the function $S(z,\zeta)$ depends on D. Since the coefficients of $\mathcal{B}^{(\varepsilon)}$ are supported by $\Lambda = \bigcup_{j=1}^N(\overline{V}_j\setminus U_j)$ and $S(z,\zeta)=S(\zeta,z)$ in view of (1.1.15),

$$|\mathcal{B}_z^{(\varepsilon)}S(\zeta,z)| \leq M\mathbb{P}_\zeta(\sigma_K<\infty) \leq M\mathbb{P}_\zeta(\sigma_{\cup_{j=1}^N\overline{V}_j}<\infty) \quad \text{for } \zeta\in D,$$

where $M := \sup\left\{\frac{1}{2\pi}|\mathcal{B}_z^{(\varepsilon)}\log|w-z|^2| : z\in\Lambda,\ w\in K,\ \varepsilon\in(0,\varepsilon_0),\ D\in\mathcal{D}_0,\ \widetilde{D}\in\mathcal{D}\right\}$, which is finite by (2.4.2) and Lemma A.6.1. Hence $\varepsilon\mathcal{B}_z^{(\varepsilon)}S(\zeta,z)=O(\varepsilon/r)$. □

Denote by W_{ℓ_0} the ℓ_0-neighborhood of Λ. We choose $\ell_0 > 0$ so that $W_{\ell_0}\subset\mathbb{H}$ and $\overline{W}_{\ell_0}\cap(\cup_{j=1}^N C_j)=\emptyset$.

Lemma A.6.3. *For any $w,\zeta\in D$, $w\neq\zeta$, it holds that with $z = x_1+ix_2$,*

$$g(\zeta,w,\varepsilon) = G(w,\zeta)-P_0(w,\zeta)+P_\varepsilon(w,\zeta) + \int_{W_{\ell_0}} g(z,w,\varepsilon)\mathcal{A}_{\varepsilon,z}\left(G-P_0+P_\varepsilon\right)(z,\zeta)dx_1dx_2. \quad \text{(A.6.18)}$$

Proof. It follows from (2.4.17) that the self-adjoint elliptic differential operator $\mathcal{A}_\varepsilon$ admits Green's second formula

$$\int_E(v\mathcal{A}_\varepsilon u - u\mathcal{A}_\varepsilon v)dx_idx_2 = \int_{\partial E}\Xi_\varepsilon[u,v]ds, \quad \text{(A.6.19)}$$

where E is a bounded domain in $\mathbb{H}$ with smooth boundary ∂E and

$$\Xi_\varepsilon[u,v] = \sum_{k,\ell=1}^{2} A_{k\ell}^{(\varepsilon)}\left(v\frac{\partial u}{\partial x_k}\frac{\partial x_\ell}{\partial\mathbf{n}} - u\frac{\partial v}{\partial x_k}\frac{\partial x_\ell}{\partial\mathbf{n}}\right). \quad \text{(A.6.20)}$$

Here $\mathbf{n}$ is the unit outward normal vector field at ∂E for E.

We fix $w, \zeta \in D$ with $w \neq \zeta$. We take a large $\ell > 0$ such that the rectangle $R_\ell = \{x_1 + ix_2 \in \mathbb{H} : |x_1| < \ell,\ 0 < x_2 < \ell\}$ contains the points w, ζ and the set $\cup_{i=1}^N \overline{V}_i$. For each $1 \le i \le N$, we choose a smooth Jordan curve γ_i surrounding C_i in such a way that $\gamma_i \subset U_i \setminus \overline{W}_{\ell_0}$ and that $w, \zeta \notin \mathrm{ins}\gamma_i$. We apply the identity (A.6.19) to

$$\begin{cases} E = R_\ell \setminus (\bigcup_{i=1}^N \mathrm{ins}\gamma_i) \setminus \overline{B_\delta(w)} \setminus \overline{B_\delta(\zeta)}, \\ u(z) = G(z,\zeta) - P_0(z,\zeta) + P_\varepsilon(z,\zeta), \\ v(z) = g(z,w,\varepsilon). \end{cases}$$

for a sufficiently small $\delta > 0$.

By Lemma A.6.1, $\mathcal{A}_\varepsilon v = 0$ on E. We have the implication

$$z \in E \setminus W_{\ell_0} \implies u(z) = G(z,\zeta) \text{ and } \mathcal{A}_\varepsilon u(z) = 0. \tag{A.6.21}$$

Indeed, if $\zeta \in \Lambda$ and $z \in \mathbb{H} \setminus W_{\ell_0}$, then $|z - \zeta| > \ell_0$ so that $\alpha(|z-\zeta|^2) = 0$ and $u(z) = G(z,\zeta)$. If $\zeta \in \mathbb{H} \setminus \Lambda$, then by Lemma A.6.1 and (A.6.4) $a_{ij}^{(\varepsilon)}(\zeta) = 2\delta_{ij}$ and so $P_\varepsilon(z,\zeta) = P_0(z,\zeta)$ and $u(z) = G(z,\zeta)$. Since $A_{k\ell}^{(\varepsilon)} = \frac{1}{2}\delta_{k\ell}$ on $\mathbb{H} \setminus \Lambda$ in view of Lemma A.6.1, we obtain from (A.6.19) and (A.6.20) that

$$\begin{aligned} \int_{W_{\ell_0}} v(z)\mathcal{A}_\varepsilon u(z)dx_1dx_2 &= \frac{1}{2}\int_{\Sigma_\ell}\left(v\frac{\partial u}{\partial \mathbf{n}} - \frac{\partial v}{\partial \mathbf{n}}u\right)ds + \frac{1}{2}\sum_{i=1}^N \int_{\gamma_i}\left(v\frac{\partial u}{\partial \mathbf{n}} - \frac{\partial v}{\partial \mathbf{n}}u\right)ds \\ &\quad + \int_{\partial B_\delta(w)} \Xi_\varepsilon[u,v]ds + \int_{\partial B_\delta(\zeta)} \Xi_\varepsilon[u,v]ds \\ &=: \mathrm{I}_\ell + \frac{1}{2}\sum_{i=1}^N \mathrm{II}(\gamma_i) + \mathrm{III}_\delta + \mathrm{IV}_\delta, \end{aligned} \tag{A.6.22}$$

where $\Sigma_\ell = \partial R_\ell \setminus \partial\mathbb{H}$.

On $\mathbb{H} \setminus \cup_{i=1}^N \overline{V}_i$, $u(z) = G(z,\zeta)$, $v(z) = g(z,w,\varepsilon)$, and both are harmonic and converge to 0 when $|z| \to \infty$ as they are dominated by $G^{\mathbb{H}}$ of (1.1.12). Hence we can use Lemma 1.1.7 to conclude that $\lim_{\ell\to\infty} \mathrm{I}_\ell = 0$.

On $U_i \setminus C_i \setminus \{\zeta\}$, $i \le i \le N$, we again have $u(z) = G(z,\zeta)$ which is harmonic there and extends continuously to $U_i \setminus \{\zeta\}$ by setting its value on C_i to be 0 (see the proof of Proposition 1.1.5 (i)). $v(z) = g(z,w,\varepsilon) = \widetilde{G}(\widetilde{f}_\varepsilon(z), \widetilde{f}_\varepsilon(w))$ has the same property on $U_i \setminus \{\widetilde{f}_\varepsilon(w)\}$. By (2.4.3), we may assume $\widetilde{f}_\varepsilon(w) \notin C_i$.

Following the proof of Proposition 1.1.5 (iii), one can take as γ_i the component γ_i^n of ∂D_{a_n} surrounding C_i and contained in U_i for the set $D_{a_n} = \{z \in D : G(z,\zeta) > a_n\}$, where a_n is a sequence decreasing to 0 chosen in a way that ∂D_{a_n} avoids the critical points of $G(\cdot,\zeta)$. We show that $\lim_{n\to\infty} \mathrm{II}(\gamma_i^n) = 0$, where

$$\mathrm{II}(\gamma_i^n) = \mathrm{II}_1(\gamma_i^n) - \mathrm{II}_2(\gamma_i^n), \quad \mathrm{II}_1(\gamma_i^n) = \int_{\gamma_i^n} v\frac{\partial u}{\partial \mathbf{n}}ds \quad \text{and} \quad \mathrm{II}_2(\gamma_i^n) = \int_{\gamma_i^n} \frac{\partial v}{\partial \mathbf{n}}uds.$$

Since $\lim_{n\to\infty} \mathrm{dist}(\gamma_i^n, C_i) = 0$, we find for any $\eta > 0$ a positive integer n_0 so that $v(z) < \eta$ for all $z \in \gamma_i^n$ with $n \ge n_0$. As $\frac{\partial u}{\partial \mathbf{n}}$ never changes sign on each γ_i^n, we get by Proposition 1.1.5 (iii) that $|\mathrm{II}_1(\gamma_i^n)| \le 2\eta$ for every $n \ge n_0$. On the other hand, (2.4.19) and Proposition 1.1.5 (iii) lead us to

$$\mathrm{II}_2(\gamma_i^n) = a_n \int_{\gamma_i^n} \frac{\partial v}{\partial \mathbf{n}} ds = a_n \int_{\widetilde{\gamma}_i^n} \frac{\partial \widetilde{G}(z, \widetilde{f}_\varepsilon(w))}{\partial \mathbf{n}_z} ds = 2a_n \widetilde{\varphi}^{(i)}(\widetilde{f}_\varepsilon(w)),$$

which tends to zero as $n \to \infty$.

By the remark made right below (A.6.5), one can replace $v(z) = g(z, w, \varepsilon)$ by $\widehat{g}(z, w) = -\frac{1}{2\pi} \log \Gamma(z, w)$ in computing the integral III_δ readily to conclude that it tends to $-u(w) = -G(w, \zeta) + P_0(w, \zeta) - P_\varepsilon(w, \zeta)$ as $\delta \downarrow 0$. As for the integral IV_δ, replacing $u(z) = G(z, \zeta) - P_0(z, \zeta) + P_\varepsilon(z, \zeta)$ by $\widehat{g}(z, \zeta)$ of the same singularity at ζ, we see that it tends to $v(\zeta) = g(\zeta, w, \varepsilon)$ as $\delta \downarrow 0$. □

If we write

$$\begin{cases} \widehat{K}_\varepsilon(\zeta, z) = \mathcal{A}_{\varepsilon,z}\left(G - P_0 + P_\varepsilon\right)(z, \zeta) \\ G_\varepsilon(\zeta, w) = G(w, \zeta) - P_0(w, \zeta) + P_\varepsilon(w, \zeta), \end{cases}$$

then the identity (A.6.18) is converted into a Fredholm type integral equation:

$$g(\zeta, w, \varepsilon) = G_\varepsilon(\zeta, w) + \int_{W_{\ell_0}} \widehat{K}_\varepsilon(\zeta, z) g(z, w, \varepsilon) dx_1 dx_2. \tag{A.6.23}$$

In view of (A.6.4) and (A.6.7), we have

$$|P_\varepsilon(w, \zeta) - P_0(w, \zeta)| \le -2\varepsilon\alpha(|w - \zeta|^2)\left(|b_{11}^{(\varepsilon)}(\zeta)| + |b_{22}^{(\varepsilon)}(\zeta)| + |b_{12}^{(\varepsilon)}(\zeta)|\right),$$

so that

$$P_\varepsilon(w, \zeta) - P_0(w, \zeta) = \varepsilon\eta_1^{(\varepsilon)}(w, \zeta) \quad \text{for } w,\ \zeta \in \overline{\mathbb{H}}, \tag{A.6.24}$$

where $\eta_1^{(\varepsilon)}(w, \zeta)$ a continuous function on $\overline{\mathbb{H}} \times \overline{\mathbb{H}}$ bounded uniformly in $\varepsilon \in (0, \varepsilon_0)$, $D \in \mathcal{D}_0$ and $\widetilde{f}_\varepsilon(D)$.

For a function $u(\zeta, w)$, $\zeta, w \in D \cup \partial\mathbb{H}$, we define $\|u\|_\infty = \sup_{\zeta, w \in D \cup \partial\mathbb{H}} |u(\zeta, w)|$. If we write $(\widehat{K}_\varepsilon G_\varepsilon)(\zeta, w) = \int_{W_{\ell_0}} \widehat{K}_\varepsilon(\zeta, z) G_\varepsilon(z, w) dx_1 dx_2$, then

$$|(\widehat{K}_\varepsilon G_\varepsilon)(\zeta, w)| \le \int_{W_{\ell_0}} |\widehat{K}_\varepsilon(\zeta, z)|(G^{\mathbb{H}}(z, w) + |P_\varepsilon(z, w) - P_0(z, w)|) dx_1 dx_2,$$

and we can verify using Lemma A.6.2, (1.1.12) and (A.6.24) that

$$\|\widehat{K}_\varepsilon G_\varepsilon\|_\infty \le \varepsilon C_1 \quad \text{for } C_1 \text{ independent of } \varepsilon \in (0, \varepsilon_0),\ D \in \mathcal{D}_0 \text{ and } \widetilde{f}_\varepsilon(D).$$

By Lemma A.6.2, we also have $\int_{W_{\ell_0}} |\widehat{K}_\varepsilon(\zeta, z)| dx_1 dx_2 \le \varepsilon C_2$ for any $\zeta \in D \cup \partial\mathbb{H}$, where $C_2 > 0$ is a constant independent of $\varepsilon \in (0, \varepsilon_0)$, $D \in \mathcal{D}_0$ and $\widetilde{f}_\varepsilon(D)$. Hence $\|\widehat{K}_\varepsilon^{(2)} G_\varepsilon\|_\infty \le \varepsilon^2 C_1 C_2$ for $\widehat{K}_\varepsilon^{(2)} G_\varepsilon(\zeta, w) = \widehat{K}_\varepsilon(\widehat{K}_\varepsilon G_\varepsilon)(\zeta, w)$. Similarly, we have $\|\widehat{K}_\varepsilon^{(n)} G_\varepsilon\|_\infty \le \varepsilon^n C_1 C_2^{n-1}$ for every $n \ge 1$.

Denote $\varepsilon_0 \wedge (1/(2C_2))$ by $\widehat{\varepsilon}_0$. For $0 < \varepsilon < \widehat{\varepsilon}_0$,

$$\sum_{n=1}^{\infty} \|\widehat{K}_\varepsilon^{(n)} G_\varepsilon\|_\infty \le \sum_{n=1}^{\infty} \varepsilon^n C_1 C_2^{n-1} < 2\varepsilon C_1,$$

and so the convergence

$$\sum_{n=1}^{\infty} \widehat{K}_\varepsilon^{(n)} G_\varepsilon(\zeta, w) = \varepsilon\eta_2^{(\varepsilon)}(\zeta, w), \tag{A.6.25}$$

is uniform on $D \cup \partial\mathbb{H} \times D \cup \partial\mathbb{H}$, where $\varepsilon\eta_2^{(\varepsilon)}(\zeta, w)$ is a continuous function there that is uniformly bounded in $\varepsilon \in (0, \widehat{\varepsilon}_0)$, $D \in \mathcal{D}_0$ and $\widetilde{f}_\varepsilon(D)$.

Finally we verify that

$$\sup_{\zeta, w\in D} \int_{W_{\ell_0}} |\widehat{K}_\varepsilon(\zeta, z)| g(z, w.\varepsilon) dx_1 dx_2 = \varepsilon C_3, \quad \text{for all } \varepsilon \in (0, \widetilde{\varepsilon}_0), \tag{A.6.26}$$

for some $\widetilde{\varepsilon}_0 \in (0, \widehat{\varepsilon}_0)$ and for $C_3 > 0$ independent of $D \in \mathcal{D}_0$ and $\widetilde{D} = \widetilde{f}_\varepsilon(D) \in \mathcal{D}$. On account of the domination $g(z, w, \varepsilon) \leq G^{\mathbb{H}}(\widetilde{f}_\varepsilon(z), \widetilde{f}_\varepsilon(w))$, it is enough to prove that, for a bounded open set $U \subset \mathbb{H}$ containing $\overline{W}_{\ell_0}$, there are constants $\widetilde{\varepsilon}_0 \in (0, \widehat{\varepsilon}_0)$ and $C_4 > 0$ independent of $D \in \mathcal{D}_0$ and $\widetilde{D} = \widetilde{f}_\varepsilon(D) \in \mathcal{D}$ so that

$$\sup_{\zeta\in D, w\in U} \int_U |\widehat{K}_\varepsilon(\zeta, z)| \left|\log|\widetilde{f}_\varepsilon(z) - \widetilde{f}_\varepsilon(w)|\right| dx_1 dx_2 = \varepsilon C_4 \quad \text{for every } \varepsilon \in (0, \widetilde{\varepsilon}_0), \tag{A.6.27}$$

It follows from (2.4.7) that $|\widetilde{f}_\varepsilon(z) - \widetilde{f}_\varepsilon(w)|^2/|z-w|^2 \leq 1 + \varepsilon M$ for every $\varepsilon \in (0, \varepsilon_0)$, where M is a constant that is independent of $\varepsilon \in (0, \varepsilon_0)$, $D \in \mathcal{D}_0$ and $\widetilde{D} \in \mathcal{D}$. Let $\widetilde{\varepsilon}_0 = \widehat{\varepsilon}_0 \wedge (1/2M)$. Then, for $\varepsilon \in (0, \widetilde{\varepsilon}_0)$,

$$\left|\log|\widetilde{f}_\varepsilon(z) - \widetilde{f}_\varepsilon(w)|\right| \leq |\log|z-w|| + (\log 2)/2 \quad \text{for every } z, w \in D,$$

which combined with Lemma A.6.2 implies (A.6.27).

By (A.6.26), for each $\varepsilon \in (0, \widetilde{\varepsilon}_0)$, $|\widehat{K}_\varepsilon^{(n+1)} g(\zeta, w, \varepsilon)| \leq \varepsilon^{(n+1)} C_3 C_2^n \to 0$ as $n \to \infty$,. Thus one can solve the integral equation (A.6.23) in $g(\zeta, w, \varepsilon)$ and, by setting $\eta^{(\varepsilon)}(\zeta, w) = \eta_1^{(\varepsilon)}(w, \zeta) + \eta_2^{(\varepsilon)}(\zeta, w)$, one gets the following from (A.6.24) and (A.6.25).

Lemma A.6.4. *For any ζ, $w \in D$ with $\zeta \neq w$,*

$$g(\zeta, w, \varepsilon) = G(\zeta, w) + \varepsilon\eta^{(\varepsilon)}(\zeta, w), \quad \varepsilon \in (0, \widetilde{\varepsilon}_0), \tag{A.6.28}$$

where $\eta^{(\varepsilon)}(\zeta, w)$ is a continuous function on $(D \cup \partial\mathbb{H}) \times (D \cup \partial\mathbb{H})$ that is uniformly bounded in $\varepsilon \in (0, \widetilde{\varepsilon}_0)$, $D \in \mathcal{D}_0$ and $\widetilde{D} = \widetilde{f}_\varepsilon(D) \in \mathcal{D}$.

Take $\zeta \in \mathbb{H} \setminus \Lambda$ in (A.6.18). By noting that $P_\varepsilon(z, \zeta) = P_0(z, \zeta)$ and letting $\ell_0 \downarrow 0$, we arrive at (2.4.12) as $\mathcal{A}_{\varepsilon,z} G(z, \zeta) = \varepsilon \mathcal{B}_z^{(\varepsilon)} G(z, \zeta)$ for $z \in \Lambda$. By substituting (A.6.28) into (2.4.12), we obtain (2.4.13). This completes the proof for the assertions (iii) and (iv) of Proposition 2.4.3.

We finally prove the assertions (v)–(viii) of Proposition 2.4.3 by making use of the probabilistic potential theory for Dirichlet forms about the relationship between hitting probabilities, equilibrium potentials and capacities. Such relations have already been used in the proof of Lemmas 1.7.2 and 2.4.8.

(v): The Green function G can be expressed by the Green function $G^{\mathbb{H}}$ of the ABM $\mathbf{Z}^{\mathbb{H}}$ as

$$G(z,\zeta)=G^{\mathbb{H}}(z,\zeta)-S_0(z,\zeta),\quad \text{where } S_0(z,\zeta)=\mathbb{E}_z^{\mathbb{H}}\left[G^{\mathbb{H}}(Z^{\mathbb{H}}_{\sigma_K},\zeta);\sigma_K<\infty\right]. \tag{A.6.29}$$

By the expression (1.1.12) of $G^{\mathbb{H}}$ and Proposition 2.4.3 (ii), $B_z^{(\varepsilon)}G^{\mathbb{H}}(z,\zeta)$ is uniformly bounded on $\Lambda\times J$ in $\varepsilon\in(0,\varepsilon_0)$, $D\in\mathcal{D}_0$ and $\widetilde{f}_\varepsilon(D)$. As $S_0(z,\zeta)=S_0(\zeta,z)$ by the symmetry (1.1.15) of G, we have, for any $\varepsilon\in(0,\varepsilon_0)$, $z\in\Lambda$ and $\zeta\in\mathbb{H}\setminus\bigcup_{j=1}^N V_j$,

$$|\mathcal{B}_z^{(\varepsilon)}S_0(z,\zeta)|\le M\,\mathbb{P}_\zeta^{\mathbb{H}}(\sigma_{\overline{U}}<\infty)\quad \text{for } M=\sup_{\substack{\varepsilon\in(0,\varepsilon_0),w\in K,\xi\in\Lambda\\ D\in\mathcal{D}_0,\widetilde{D}\in\mathcal{D}}}|\mathcal{B}_\xi^{(\varepsilon)}G^{\mathbb{H}}(w,\xi)|<\infty,$$

where $U=\bigcup_{j=1}^N U_j$. Just as (1.7.7), we have $\mathbb{P}_\zeta^{\mathbb{H}}(\sigma_{\overline{U}}<\infty)=\int_{\overline{U}}G^{\mathbb{H}}(\zeta,w)\mu(dw)$ for any $\zeta\in\mathbb{H}$, where μ is a finite measure concentrated on $\overline{U}$ with $\mu(\overline{U})=\mathrm{Cap}_0^{\mathbb{H}}(\overline{U})$. Since both $\mathcal{B}_z^{(\varepsilon)}S_0(z,\zeta)$ and $\mathbb{P}_\zeta^{\mathbb{H}}(\sigma_{\overline{U}}<\infty)$ vanish when $\zeta\in\partial\mathbb{H}$, we deduce from above that

$$\left|\mathcal{B}_z^{(\varepsilon)}\frac{\partial}{\partial\zeta_2}S_0(z,\zeta)\right|\le M\sup_{w\in\overline{U},\,\zeta\in J}\frac{\partial}{\partial\zeta_2}G^{\mathbb{H}}(\zeta,w)\cdot\mathrm{Cap}_0^{\mathbb{H}}(\overline{U}),\quad \zeta\in J.$$

(vi): Let $G^{D^i}(z,\zeta)$ be the Green function of the ABM on $D^i=D\cup C_i=\mathbb{H}\setminus(K\setminus C_i)$. Then

$$G^{D^i}(z,\zeta)=G^{\mathbb{H}}(z,\zeta)-\mathbb{E}_\zeta^{\mathbb{H}}\left[G^{\mathbb{H}}(Z^{\mathbb{H}}_{\sigma_{K\setminus C_i}},z);\ \sigma_{K\setminus C_i}<\infty\right],\quad z,\zeta\in D^i.$$

In view of Lemma 1.7.2, $\varphi^{(i)}$ admits an expression

$$\varphi^{(i)}(z)=G^{D^i}\nu_i(z)=\int_{C_i}G^{D^i}(z,\zeta)\nu_i(d\zeta),\quad z\in D^i,$$

for some finite positive measure ν_i concentrated on C_i. Hence

$$\varphi^{(i)}(z)=G^{\mathbb{H}}\nu_i(z)-\mathbb{E}_{\nu_i}^{\mathbb{H}}\left[G^{\mathbb{H}}(Z^{\mathbb{H}}_{\sigma_{K\setminus C_i}},z);\ \sigma_{K\setminus C_i}<\infty\right],\quad z\in D^i. \tag{A.6.30}$$

Consequently, we have for the same constant M as in the proof of **(v)** that

$$|\mathcal{B}_z^{(\varepsilon)}\varphi^{(i)}(z)|\le 2M\,\nu_i(C_i),\quad \varepsilon\in(0,\varepsilon_0),\ z\in\Lambda. \tag{A.6.31}$$

For an open set $G\subset\mathbb{H}$, denote by $\mathrm{Cap}_0(B;G)$ the 0-order capacity of $B\subset G$ relative to $(H^1_{0,e}(G),\frac{1}{2}\mathbf{D})$. It increases as B increases or G decreases. Moreover, we have $\mathrm{Cap}_0(C_i,D^i)=\nu_i(C_i)$. Hence (A.6.31) leads us to a uniform bound

$$|\mathcal{B}_z^{(\varepsilon)}\varphi^{(i)}(z)|\le 2M\,\mathrm{Cap}_0(\overline{U}_i;\mathbb{H}\setminus(\cup_{k\neq i}\overline{V}_k)),\quad \varepsilon\in(0,\varepsilon_0),\ z\in\Lambda. \tag{A.6.32}$$

(vii): We again use the identity (A.6.29) and the symmetry $S_0(z,\zeta)=S_0(\zeta,z)$. Let $b_0>0$ be defined by (2.4.16) and choose $\alpha>0,\beta\ge b_0$ such that $\Lambda\subset(-\alpha,\alpha)\times(b_0,\beta)$. By (1.1.12), we see that the first integrals in (2.4.14) for $k=1,2$, evaluated for $G^{\mathbb{H}}(\zeta,z)$ in place of $G(\zeta,z)$ are bounded by $\dfrac{2}{\pi}\dfrac{|x_1-\zeta_1|}{(x_1-\zeta_1)^2+b_0^2}\le\dfrac{1}{\pi b_0}$ and $\dfrac{2}{\pi b_0}$, respectively. The second integrals evaluated for $G^{\mathbb{H}}(\zeta, z)$ and $k=1,2$, are bounded by $\frac{4}{\pi b_0}$ and $\frac{2}{b_0}$, respectively.

On the other hand, we have for $\zeta \in \mathbb{H}$ and $(x_1, x_2) \in \mathbb{R} \times (0, \infty)$,

$$\int_0^\infty \left| \frac{\partial}{\partial x_1} G^{\mathbb{H}}(\zeta, x_1 + ix_2) \right| dx_2 \le 1 \quad \text{and} \quad \int_{-\infty}^\infty \left| \frac{\partial}{\partial x_2} G^{\mathbb{H}}(\zeta, x_1 + ix_2) \right| dx_1 \le 2. \tag{A.6.33}$$

Since $|z - \zeta| > b \ge b_0$ for $(z, \zeta) \in \Lambda \times K$, we have for any $\zeta \in K$,

$$\begin{aligned} &\int_{-\infty}^\infty \mathbb{1}_\Lambda(z) \left| \frac{\partial}{\partial x_1} G^{\mathbb{H}}(\zeta, z) \right| dx_1 \le \frac{4}{\pi b_0^2}\alpha^2 \quad \text{and} \\ &\int_0^\infty \mathbb{1}_\Lambda(z) \left| \frac{\partial}{\partial x_2} G^{\mathbb{H}}(\zeta, z) \right| dx_2 \le \frac{2}{\pi b_0^2}\beta^2. \end{aligned} \tag{A.6.34}$$

Denote by c the maximum of the bounds in (A.6.34) and 2. Then each of the four integrals in (2.4.14) by using (A.6.29) and the symmetry of S_0 admits a bound $\frac{4}{\pi b_0} + c \frac{\partial}{\partial \zeta_2} \mathbb{P}_\zeta^{\mathbb{H}}(\sigma_{\overline{U}} < \infty)$, which is in turn dominated by

$$\frac{4}{\pi b_0} + c \sup_{w \in \overline{U},\, \zeta \in J} \frac{\partial}{\partial \zeta_2} G^{\mathbb{H}}(\zeta, w)\, \mathrm{Cap}_0^{\mathbb{H}}(\overline{U}).$$

as in the proof of **(v)** above.

(viii): By virtue of (A.6.30), (A.6.33) and (A.6.34), we see that each of the four integrals in (2.4.15) is dominated by $2c\,\nu_j(C_j) \le 2c\,\mathrm{Cap}_0\left(\overline{U}_j; \mathbb{H} \setminus (\cup_{k \ne j} \overline{V}_k)\right)$.

The proof for Proposition 2.4.3 is now complete. □

A.7 BMD Poisson kernel under small perturbation of $\mathbb{H}$

The main result of this section is Theorem A.7.7. We prepare several lemmas for its proof. Theorem A.7.7 will be applied in the next section in getting a comparison of half-plane capacities.

For $\rho > 0$, define $B_\rho = \{z \in \mathbb{C} : |z| < \rho\}$, $\mathbb{H}_\rho = \mathbb{H} \setminus \overline{B}_\rho$.

Lemma A.7.1. *Let U be an open subset of $\mathbb{H}_\rho$ with $\overline{U} \supset \partial\mathbb{H}_\rho$ and f be a harmonic function on U with $\lim_{z \to \zeta, z \in U} f(z) = 0$ for any $\zeta \in \partial\mathbb{H}_\rho$. Then f extends to a harmonic function on a neighborhood of $\partial\mathbb{H}_\rho$ in $\mathbb{C}$ across $\partial\mathbb{H}_\rho$.*

Proof. By Schwarz reflection, f can be extended to a harmonic function (denoted by f again) on $V = U \cup \Pi U \cup (\partial\mathbb{H} \setminus [-\rho, \rho]) \subset \mathbb{C} \setminus \overline{B}_\rho$ and $\lim_{z \to \zeta, z \in V} f(z) = 0$ for any $\zeta \in \partial B_\rho$.

Now $\gamma = \partial B_\rho \setminus \{-i\rho\}$ is an analytic Jordan arc (see the paragraph below the proof of Lemma 1.1.4). Indeed, by the Möbius transformation $\psi(z) = i\frac{z - i\rho}{z + i\rho}$, γ is sent onto $\partial\mathbb{H}$, V is sent into $\mathbb{H}$ and B_ρ is sent to the lower half plane $\mathbb{H}_-$. Therefore,

f admits a harmonic extension from V to a neighborhood of γ in $\mathbb{C}$ across γ in view of [Garnett and Marshall (2005), Lemma II.2.4]. $\square$

For $\varepsilon > 0$, let $\mathbf{Z}^{\mathbb{H}_\varepsilon} = (Z_t^{\mathbb{H}_\varepsilon}, \zeta^{\mathbb{H}_\varepsilon}, \mathbb{P}_z^{\mathbb{H}_\varepsilon})$ be the ABM on $\mathbb{H}_\varepsilon$. The Poisson kernel for $\mathbb{H}$ is given by $K_{\mathbb{H}}(z,\zeta) = \dfrac{1}{\pi}\dfrac{\Im z}{|z-\zeta|^2}$ with $z \in \mathbb{H}$ and $\zeta \in \partial\mathbb{H}$.

Lemma A.7.2. *For $0 \le \theta_1 < \theta_2 \le \pi$ and $\varepsilon > 0$, let $\Gamma^\varepsilon_{\theta_1,\theta_2} = \{\varepsilon e^{i\theta} : \theta \in (\theta_1,\theta_2)\}$. Then*

$$\mathbb{P}_z^{\mathbb{H}_\varepsilon}\left(Z^{\mathbb{H}_\varepsilon}_{\zeta^{\mathbb{H}_\varepsilon}-} \in \Gamma^\varepsilon_{\theta_1,\theta_2}\right) = \varepsilon\int_{\theta_1}^{\theta_2} K_{\mathbb{H}_\varepsilon}(z,\varepsilon e^{i\theta})d\theta, \quad z \in \mathbb{H}_\varepsilon, \tag{A.7.1}$$

for

$$K_{\mathbb{H}_\varepsilon}(z,\varepsilon e^{i\theta}) = 2K_{\mathbb{H}}(z,0)\left(1 + c_1(\theta,z,\varepsilon)\frac{\varepsilon}{|z|}\right)\sin\theta, \quad z \in \mathbb{H}_\varepsilon, \tag{A.7.2}$$

where $c_1(\theta,z,\varepsilon)$ is a continuous function of $(\theta,z,\varepsilon) \in (0,\pi) \times \bigcup_{\delta>0}(\mathbb{H}_\delta \times \{\delta\})$ with

$$M_{\varepsilon_0,\varepsilon_1} := \sup_{\theta\in(0,\pi),\ |z|>\varepsilon_1,\ \varepsilon\in(0,\varepsilon_0)} |c_1(\theta,z,\varepsilon)| < \infty \quad \textit{for any } 0 < \varepsilon_0 < \varepsilon_1. \tag{A.7.3}$$

Proof. Put $\rho = \varepsilon$ in (1.1.16). Then the left-hand side of (A.7.1) equals

$$\varepsilon\frac{2}{\pi}\frac{\Im z}{|z|^2}\int_{\theta_1}^{\theta_2}\left(1 + c(\theta,\vartheta,\eta)\frac{\varepsilon}{|z|}\right)\sin\theta d\theta,$$

so that (A.7.1) and (A.7.2) follow by setting $c_1(\theta,z,\varepsilon) = c(\theta,\vartheta,\eta)$. The continuity and the bound (A.7.3) for c_1 follow from the continuity and the bound (1.1.17) for c. $\square$

In what follows, we fix a standard slit domain $D = \mathbb{H} \setminus K$ with $K = \bigcup_{j=1}^N C_j$. Let $\mathbf{Z}^D = (Z_t^D, \zeta^D, \mathbb{P}_z^D)$ be the ABM on D, $G^D(z,\zeta)$ its Green function and

$$K_D(z,\zeta) := -\frac{1}{2}\frac{\partial}{\partial \mathbf{n}_\zeta}G^D(z,\zeta), \quad z \in D,\ \zeta \in \partial\mathbb{H}, \tag{A.7.4}$$

its Poisson kernel. According to (1.1.28), for any $z \in D$ and $f \in C_b(\partial\mathbb{H})$,

$$\mathbb{E}_z^D\left[f(Z^D_{\zeta^D-}); Z^D_{\zeta^D-} \in \partial\mathbb{H}\right] = \int_{\partial\mathbb{H}} K_D(z,\zeta)f(\zeta)s(d\zeta), \tag{A.7.5}$$

where $s(d\zeta)$ is the Lebesgue measure on $\partial\mathbb{H}$.

For $\varepsilon > 0$ with $\mathbb{H} \cap \overline{B}_\varepsilon \subset D$, let $\mathbf{Z}^{D_\varepsilon} = (Z_t^{D_\varepsilon}, \zeta^{D_\varepsilon}, \mathbb{P}_z^{D_\varepsilon})$ be the ABM on $D_\varepsilon := D \setminus \overline{B}_\varepsilon$ and $G^{D_\varepsilon}(z,\zeta)$ its Green function.

Lemma A.7.3. *For a fixed $z \in D_\varepsilon$, the function $G^{D_\varepsilon}(z,\zeta)$ of ζ extends across $\partial\mathbb{H}_\varepsilon$ to be a C^∞-function on a neighborhood of $\partial\mathbb{H}_\varepsilon$ on $\mathbb{C}$. Define*

$$K_{D_\varepsilon}(z,\zeta) = -\frac{1}{2}\frac{\partial}{\partial \mathbf{n}_\zeta}G^{D_\varepsilon}(z,\zeta), \quad z \in D_\varepsilon,\ \ \zeta \in \partial\mathbb{H}_\varepsilon. \tag{A.7.6}$$

Then

$$\mathbb{E}_z^{D_\varepsilon}\left[f(Z^{D_\varepsilon}_{\zeta^{D_\varepsilon}-}); Z^{D_\varepsilon}_{\zeta^{D_\varepsilon}-} \in \partial\mathbb{H}_\varepsilon\right] = \int_{\partial\mathbb{H}_\varepsilon} K_{D_\varepsilon}(z,\zeta)f(\zeta)s(d\zeta) \tag{A.7.7}$$
$$\textit{for } z \in D_\varepsilon \textit{ and } f \in C_b(\partial\mathbb{H}_\varepsilon).$$

Proof. In the same way as the proof of Proposition 1.1.5 (i), we see that $v(\zeta) = G_{D_\varepsilon}(z,\zeta)$ is harmonic in $D_\varepsilon \setminus \{z\}$ and continuously extendable to $\partial\mathbb{H}_\varepsilon$ by setting $v(\zeta) = 0$, $\zeta \in \partial\mathbb{H}_\varepsilon$. By Lemma A.7.1, v extends to a neighborhood of $\partial\mathbb{H}_\varepsilon$ in $\mathbb{C}$ to be harmonic and hence infinitely differentiable there.

Identity (A.7.7) can be proved in the same way as the proof of Proposition 1.1.5 (iv) for D_ε and $\partial\mathbb{H}_\varepsilon$ in place of D and $\partial\mathbb{H}$ there. □

Lemma A.7.4. *Take any* $0 < \varepsilon_0 < \varepsilon_1$ *with* $B_{\varepsilon_1} \cap \mathbb{H} \subset D$. *It holds that, for any* $z \in D$ *with* $|z| > \varepsilon_1$ *and any* $\varepsilon \in (0, \varepsilon_0)$,

$$K_{D_\varepsilon}(z, \varepsilon e^{i\theta}) = \left[2K_D(z,0) + c_2(\theta, z, \varepsilon)\frac{\varepsilon}{|z|}\right]\sin\theta, \quad \theta \in (0,\pi), \tag{A.7.8}$$

with

$$\sup_{\theta\in(0,\pi),\ |z|>\varepsilon_1,\ \varepsilon\in(0,\varepsilon_0)} |c_2(\theta,z,\varepsilon)| < \infty. \tag{A.7.9}$$

Proof. By (A.7.1), (A.7.7) and the strong Markov property,

$$K_{D_\varepsilon}(z,\varepsilon e^{i\theta}) = K_{\mathbb{H}_\varepsilon}(z,\varepsilon e^{i\theta}) - \mathbb{E}_z^{\mathbb{H}_\varepsilon}\left[K_{\mathbb{H}_\varepsilon}(Z_{\sigma_K},\varepsilon e^{i\theta}); \sigma_K < \infty\right] = \mathrm{I} + \mathrm{II},$$

where

$$\mathrm{I} = K_{\mathbb{H}_\varepsilon}(z,\varepsilon e^{i\theta}) - \mathbb{E}_z^{\mathbb{H}}\left[K_{\mathbb{H}_\varepsilon}(Z_{\sigma_K},\varepsilon e^{i\theta}); \sigma_K < \infty\right],$$

$$\mathrm{II} = \mathbb{E}_z^{\mathbb{H}}\left[K_{\mathbb{H}_\varepsilon}(Z_{\sigma_K},\varepsilon e^{i\theta}); \sigma_K < \infty\right] - \mathbb{E}_z^{\mathbb{H}_\varepsilon}\left[K_{\mathbb{H}_\varepsilon}(Z_{\sigma_K},\varepsilon e^{i\theta}); \sigma_K < \infty\right].$$

It follows from (A.7.2) that $\mathrm{I} = \mathrm{I}_1 + \mathrm{I}_2 + \mathrm{I}_3$ with

$$\begin{cases} \mathrm{I}_1 = 2\left[K_{\mathbb{H}}(z,0) - \mathbb{E}_z^{\mathbb{H}}\left[K_{\mathbb{H}}(Z_{\sigma_K},0); \sigma_K < \infty\right]\right]\sin\theta, \\ \mathrm{I}_2 = 2K_{\mathbb{H}}(z,0)c_1(\theta,z,\varepsilon)\frac{\varepsilon}{|z|}\sin\theta, \\ \mathrm{I}_3 = -2\mathbb{E}_z^{\mathbb{H}}\left[K_{\mathbb{H}}(Z_{\sigma_K},0)c_1(\theta,Z_{\sigma_K},\varepsilon); \sigma_K < \infty\right]\frac{\varepsilon}{|z|}\sin\theta. \end{cases}$$

We have $\mathrm{I}_1 = 2K_D(z,0)\sin\theta$ in view of (1.1.27) and

$$\sup_{\theta\in[0,\pi],\ |z|>\varepsilon_1,\ \varepsilon<\varepsilon_0} |\mathrm{I}_2|\frac{|z|}{\varepsilon}\frac{1}{\sin\theta} \le \frac{2}{\pi\varepsilon_1}M_{\varepsilon_0,\varepsilon_1} < \infty,$$

by (A.7.3). Since $C = \sup_{z\in K} K_{\mathbb{H}}(z,0)$ is finite and $|Z_{\sigma_K}| > \varepsilon_1$, we get $|\mathrm{I}_3| \le 2CM_{\varepsilon_0,\varepsilon_1} \cdot \frac{\varepsilon}{|z|} \cdot \sin\theta$ from (A.7.3).

As $\mathrm{II} = \mathbb{E}_z^{\mathbb{H}}\left[K_{\mathbb{H}_\varepsilon}(Z_{\sigma_K},\varepsilon e^{i\theta}); \sigma_{\overline{B}_\varepsilon} < \sigma_K\right]$, we get $|\mathrm{II}| \le 2C(1 + M_{\varepsilon_0,\varepsilon_1})\sin\theta$ $\mathbb{P}_z^{\mathbb{H}}(\sigma_{\overline{B}_\varepsilon} < \infty)$ from (A.7.2). On the other hand, it follows from (1.1.16) and (1.1.17) that $\mathbb{P}_z^{\mathbb{H}}(\sigma_{\overline{B}_\varepsilon} < \infty) \le 2\frac{\varepsilon}{|z|}(1 + M_{\varepsilon_0,\varepsilon_1})$, and so

$$\sup_{\theta\in(0,\pi),\ |z|>\varepsilon_1,\varepsilon\in(0,\varepsilon_0)} |\mathrm{II}|\frac{|z|}{\varepsilon}\frac{1}{\sin\theta} < \infty.$$

□

Let $\{\varphi^{(j)},\ 1 \le j \le N\}$ be the harmonic basis for D; that is,

$$\varphi^{(j)}(z) = \mathbb{P}_z^{\mathbb{H}}\left(\sigma_K < \infty,\ Z_{\sigma_K}^{\mathbb{H}} \in C_j\right), \quad z \in D,\ 1 \le j \le N.$$

We put $D^j = D \cup C_j (= \mathbb{H} \setminus \bigcup_{k \ne j} C_k)$ and let $\mathbf{Z}^{D^j} = (Z_t^{D^j}, \zeta^{D^j}, \mathbb{P}_z^{D^j})$ be the ABM on D^j. Then $\varphi^{(j)}(z) = \mathbb{P}_z^{D^j}(\sigma_{C_j} < \infty)$ for $z \in D^j$. Denote by $G_\alpha^{D^j}(z, z')$ and $G^{D^j}(z, z')$ the α-order resolvent density and the 0-order resolvent density (the Green function) of $\mathbf{Z}^{D^j}$, respectively, that are defined by (1.1.3) with D^j in place of D. We have then

Lemma A.7.5. *For each $1 \le j \le N$, there exists a unique finite positive measure ν_j concentrated on C_j such that*

$$\varphi^{(j)}(z) = \int_{C_j} G^{D^j}(z, w)\nu_j(dw) \quad \text{for every } z \in D^j. \tag{A.7.10}$$

This is just the restatement of Lemma 1.7.2 with C_j in place of A_j for $1 \le j \le N$.

Define the harmonic basis $\varphi_\varepsilon^{(i)}(z),\ z \in D_\varepsilon,\ 1 \le i \le N$, for the perturbed slit domain $D_\varepsilon = \mathbb{H}_\varepsilon \setminus K$ with $K = \bigcup_{i=1}^N C_i$ by

$$\varphi_\varepsilon^{(i)}(z) = \mathbb{P}_z^{\mathbb{H}_\varepsilon}\left(\sigma_K < \infty,\ Z_{\sigma_K}^{\mathbb{H}_\varepsilon} \in C_i\right), \quad z \in D_\varepsilon.$$

In the same way as the proof of Lemma 1.1.4, $\varphi_\varepsilon^{(i)}$ is harmonic on D_ε and continuously extendable to $\partial\mathbb{H}_\varepsilon$ by setting its value there to be zero. By Lemma A.7.1, $\varphi_\varepsilon^{(i)}$ extends to a neighborhood of $\partial\mathbb{H}_\varepsilon$ in $\mathbb{C}$ to be C^∞ there.

Let $\mathbf{Z}^{*,\varepsilon} = (Z_t^{*,\varepsilon}, \zeta^{*,\varepsilon}, \mathbb{P}_z^{*,\varepsilon})$ be the BMD on $D_\varepsilon^* = D_\varepsilon \cup \{c_1^*, \ldots, c_N^*\}$ obtained from the ABM $\mathbf{Z}^{\mathbb{H}_\varepsilon}$ on $\mathbb{H}_\varepsilon$ by rendering each slit C_i into a single point c_i^*. The Poisson kernel of $\mathbf{Z}^{*,\varepsilon}$ will be denoted by $K_{D_\varepsilon}^*(z, \zeta)$. The next lemma is a counterpart of (iv) and (v) of Proposition 1.7.4 with $\mathbf{Z}^{*,\varepsilon}$ in place of $\mathbf{Z}^*$.

Lemma A.7.6. *It holds that*

$$\mathbb{P}_z^{*,\varepsilon}\left(\zeta^{*,\varepsilon} < \infty,\ Z_{\zeta^{*,\varepsilon}-}^{*,\varepsilon} \in \partial\mathbb{H}_\varepsilon\right) = 1 \quad \text{for any } z \in D_\varepsilon^*, \tag{A.7.11}$$

and

$$\mathbb{E}_z^{*,\varepsilon}\left[f(Z_{\zeta^{*,\varepsilon}-}^{*,\varepsilon})\right] = \int_{\partial\mathbb{H}_\varepsilon} K_{D_\varepsilon}^*(z, \zeta) f(\zeta) s(d\zeta) \quad \textit{for } z \in D_\varepsilon \textit{ and } f \in C_b(\partial\mathbb{H}_\varepsilon), \tag{A.7.12}$$

where

$$K_{D_\varepsilon}^*(z, \zeta) = K_{D_\varepsilon}(z, \zeta) - \sum_{i,j=1}^N \varphi_\varepsilon^{(i)}(z)\, b_{ij}^\varepsilon\, \frac{\partial}{\partial \mathbf{n}_\zeta}\varphi_\varepsilon^{(j)}(\zeta), \quad z \in D_\varepsilon,\ \zeta \in \partial\mathbb{H}_\varepsilon. \tag{A.7.13}$$

Here (b_{ij}^ε) is the inverse matrix of (a_{ij}^ε) whose entry is the period of $\varphi_\varepsilon^{(i)}$ around C_j defined by (1.1.21).

Proof. We state an outline in several steps.
1. In exactly the same way as the proof of Proposition 1.1.5 (iii), one can show by using the approximation of D_ε by sets of the form $\{\zeta \in D_\varepsilon : G^{D_\varepsilon}(z,\zeta) > a\}$ that

$$-2\varphi_\varepsilon^{(i)}(z) = \text{period of } G^{D_\varepsilon}(z,\cdot) \text{ around } C_i. \tag{A.7.14}$$

2. Let $G^*_{D_\varepsilon}$ be the 0-order resolvent of BMD $\mathbf{Z}^{*,\varepsilon}$: $G^*_{D_\varepsilon}f(z) = \mathbb{E}^{*,\varepsilon}_z\left[\int_0^\infty f(Z^{*,\varepsilon}_s)ds\right]$, $z \in D^*_\varepsilon$. Then $G^*_{D_\varepsilon}f(z) = \int_{D_\varepsilon} G^*_{D_\varepsilon}(z,\zeta)f(\zeta)m(d\zeta)$, $z \in D_\varepsilon$, with

$$G^*_{D_\varepsilon}(z,\zeta) = G^{D_\varepsilon}(z,\zeta) + 2\sum_{i,j=1}^N \varphi_\varepsilon^{(i)}(z)b^\varepsilon_{ij}\varphi_\varepsilon^{(j)}(\zeta), \quad z,\zeta \in D_\varepsilon. \tag{A.7.15}$$

This can be proved in the same way as the proof of Theorem 1.7.1 using (A.7.14).
3. Since $\varphi_\varepsilon^{(j)}(z) \le \varphi^{(j)}(z)$,

$$\begin{aligned}
-\int_{\partial\mathbb{H}\setminus\overline{B}_\varepsilon} \frac{\partial\varphi_\varepsilon^{(j)}(\zeta)}{\partial\mathbf{n}_\zeta}s(d\zeta) &= \int_{\mathbb{R}\setminus[-\varepsilon,\varepsilon]} \frac{\partial\varphi_\varepsilon^{(j)}(x+iy)}{\partial y}\Big|_{y=0}dx \\
&\le \int_{\mathbb{R}\setminus[-\varepsilon,\varepsilon]} \frac{\partial\varphi^{(j)}(x+iy)}{\partial y}\Big|_{y=0}dx,
\end{aligned}$$

which is dominated by $2\mathrm{Cap}_0^{\mathbb{H}}(C_j)$ in view of Lemma 1.7.3. Hence, we have

$$-\int_{\partial\mathbb{H}_\varepsilon} \frac{\partial\varphi_\varepsilon^{(j)}(\zeta)}{\partial\mathbf{n}_\zeta}s(d\zeta) < \infty. \tag{A.7.16}$$

4. (A.7.11) can be proved in the same way as the proof of Proposition 1.7.4 (iv) using Theorem 1.4.2.
5. (A.7.12) can be proved in the same way as the proof of Proposition 1.7.4 (v) using (A.7.7), (A.7.15) and (A.7.16). □

According to Proposition 1.7.4,

$$\mathbb{E}^*_z\left[f(Z^*_{\zeta^*-})\right] = \int_{\partial\mathbb{H}} K^*_D(z,\zeta)f(\zeta)s(d\zeta), \quad z \in D, \quad f \in C_b(\partial\mathbb{H}), \tag{A.7.17}$$

where $K^*_D(z,\zeta)$ is the BMD Poisson kernel for D having the representation

$$K^*_D(z,\zeta) = K_D(z,\zeta) - \sum_{i,j=1}^N \varphi^{(i)}(z)\, b_{ij}\, \frac{\partial}{\partial\mathbf{n}_\zeta}\varphi^{(j)}(\zeta). \quad z \in D, \quad \zeta \in \partial\mathbb{H}. \tag{A.7.18}$$

Theorem A.7.7. *Take any $0 < \varepsilon_0 < \varepsilon_1$ with $B_{\varepsilon_1} \cap \mathbb{H} \subset D$. It holds that, for any $z \in D$ with $|z| > \varepsilon_1$, any $\varepsilon \in (0,\varepsilon_0)$ and any $f \in C_b(\partial\mathbb{H}_\varepsilon)$ vanishing on $\partial\mathbb{H}\setminus[-\varepsilon,\varepsilon]$,*

$$\mathbb{E}^*_z\left[f(Z^*_{\sigma_{\partial B_\varepsilon\cap\mathbb{H}}});\ \sigma_{\partial B_\varepsilon\cap\mathbb{H}} < \infty\right] = \varepsilon\int_0^\pi K^*_{D_\varepsilon}(z,\varepsilon e^{i\theta})f(\varepsilon e^{i\theta})d\theta. \tag{A.7.19}$$

Furthermore

$$K^*_{D_\varepsilon}(z,\varepsilon e^{i\theta}) = \left[2K^*_D(z,0) + c_3(\theta,z,\varepsilon)\frac{\varepsilon}{|z|}\right]\sin\theta, \quad \theta \in (0,\pi), \tag{A.7.20}$$

with

$$\sup_{\theta\in(0,\pi),\ |z|>\varepsilon_1,\ \varepsilon\in(0,\varepsilon_0)} |c_3(\theta,z,\varepsilon)| < \infty. \tag{A.7.21}$$

Proof. By the localization property of BMD, $\mathbf{Z}^{*,\varepsilon}$ on $D_\varepsilon^* = D_\varepsilon \cup \{a_1^*, \dots, a_N^*\}$ can be obtained from $\mathbf{Z}^*$ on $D^* = D \cup \{a_1^*, \dots, a_N^*\}$ by killing upon hitting the set $\partial B_\varepsilon \cap \mathbb{H}$. Hence (A.7.19) follows from (A.7.12). The proof of (A.7.20) and (A.7.21) will be carried out in five steps.

1. It holds that, for any $z \in D$ with $|z| > \varepsilon_1$ and any $\varepsilon \in (0, \varepsilon_0)$,

$$\begin{cases} \varphi_\varepsilon^{(i)}(z) = \varphi^{(i)}(z) + c_4(z,\varepsilon)\frac{\varepsilon}{|z|} \quad \text{where} \\ c_4(z,\varepsilon) \text{ is a continuous function of } (z,\varepsilon) \text{ with } \sup_{|z|>\varepsilon_1,\, \varepsilon\in(0,\varepsilon_0)} |c_4(z,\varepsilon)| < \infty. \end{cases} \tag{A.7.22}$$

To see this, put $v^{(i)}(z,\varepsilon) = \varphi_\varepsilon^{(i)}(z) - \varphi^{(i)}(z)$. Then, for z and ε as above,

$$v^{(i)}(z,\varepsilon) = -\mathbb{E}_z^{\mathbb{H}}\left[\varphi^{(i)}(Z_{\sigma_{\partial B_\varepsilon \cap \mathbb{H}}})\ ;\ \sigma_{\partial B_\varepsilon \cap \mathbb{H}} < \infty\right].$$

We get from (A.7.1)

$$v^{(i)}(z,\varepsilon) = -\varepsilon \int_0^\pi \varphi^{(i)}(\varepsilon e^{i\theta}) K_{\mathbb{H}_\varepsilon}(z, \varepsilon e^{i\theta}) d\theta,$$

which is a continuous function of (z,ε) by (A.7.2). Furthermore, (A.7.2) and (A.7.3) imply the bound

$$|v^{(i)}(z,\varepsilon)| \le \mathbb{P}_z^{\mathbb{H}}(\sigma_{\partial B_\varepsilon \cap \mathbb{H}} < \infty) \le 4\varepsilon K_{\mathbb{H}}(z,0)(1 + M_{\varepsilon_0,\varepsilon_1}),$$

yielding (A.7.22).

2. It holds that, for each $1 \le i, j \le N$ and any $\varepsilon \in (0,\varepsilon_0)$,

$$\begin{cases} a_{ij}^\varepsilon = a_{ij} + c_{ij}(\varepsilon)\cdot\varepsilon \quad \text{where} \\ c_{ij}(\varepsilon) \text{ is a continuous function of } \varepsilon \text{ with } \sup_{\varepsilon\in(0,\varepsilon_0)} |c_{ij}(\varepsilon)| < \infty. \end{cases} \tag{A.7.23}$$

Let γ be a smooth Jordan curve surrounding C_j so that ins $\gamma \supset C_j$ and $\overline{\text{ins}\,\gamma} \cap C_k = \emptyset$ for $k \ne j$. Take analytic Jordan curves γ_1, γ_2 with

$$\text{ins}\,\gamma_1 \supset \text{ins}\,\gamma \supset \text{ins}\,\gamma_2 \supset C_j, \quad \text{ins}\,\gamma_1 \cap C_k = \emptyset,\ k \ne j, \quad \text{ins}\,\gamma_1 \cap \overline{B}_{\varepsilon_1} = \emptyset,$$

and put $G = \text{ins}\,\gamma_1 \setminus \overline{\text{ins}\,\gamma_2} (\supset \gamma)$.

As $v^{(i)}(z,\varepsilon)$ is harmonic in $z \in G$, it admits a representation

$$v^{(i)}(z,\varepsilon) = \int_{\gamma_1\cup\gamma_2} p(z,\zeta) v^{(i)}(\zeta,\varepsilon) s(d\zeta), \quad z \in G,$$

for the Poisson kernel $p(z,\zeta)$ of G according to (1.1.26). In particular

$$\frac{\partial}{\partial \mathbf{n}_z} v^{(i)}(z,\varepsilon) = \int_{\gamma_1\cup\gamma_2} \frac{\partial}{\partial \mathbf{n}_z} p(z,\zeta) v^{(i)}(\zeta,\varepsilon) s(d\zeta), \quad z \in \gamma,$$

for the outer normal $\mathbf{n}_z$ at $z \in \gamma$. Since $\sup_{z\in\gamma,\, \zeta\in\gamma_1\cup\gamma_2} \left|\frac{\partial}{\partial \mathbf{n}_z} p(z,\zeta)\right| =: c_5 < \infty$, it follows from (A.7.22) that

$$\left|\int_\gamma \frac{\partial}{\partial \mathbf{n}_z} v^{(i)}(z,\varepsilon) s(dz)\right| \le c_5\, s(\gamma) \int_{\gamma_1\cup\gamma_2} |v^{(i)}(\zeta,\varepsilon)| s(d\zeta) \le c_6\varepsilon,$$

for some $c_6 < \infty$. Hence we get (A.7.23) from (A.7.22) on account of

$$a^{\varepsilon}_{ij} - a_{ij} = \int_{\gamma} \frac{\partial}{\partial \mathbf{n}_z} v^{(i)}(z,\varepsilon) s(dz) = \int_{\gamma_1 \cup \gamma_2} \left[\int_{\gamma} \frac{\partial}{\partial \mathbf{n}_z} p(z,\zeta) s(dz) \right] v^{(i)}(\zeta,\varepsilon) s(\zeta).$$

3. It holds that, for each $1 \le i,j \le N$ and any $\varepsilon \in (0,\varepsilon_0)$,

$$\begin{cases} b^{\varepsilon}_{ij} = b_{ij} + d_{ij}(\varepsilon)\cdot \varepsilon \quad \text{where} \\ d_{ij}(\varepsilon) \text{ is a continuous function of } \varepsilon \text{ with } \sup_{\varepsilon\in(0,\varepsilon_0)} |d_{ij}(\varepsilon)| < \infty. \end{cases} \tag{A.7.24}$$

Define matrices

$$\mathring{A}_{\varepsilon} = (a^{\varepsilon}_{ij}), \quad \mathring{A} = (a_{ij}), \quad C_{\varepsilon} = (-c_{ij}(\varepsilon)), \quad B_{\varepsilon} = (b^{\varepsilon}_{ij}) \quad \text{and} \quad B = (b_{ij}).$$

By (A.7.23), $\mathring{A}_{\varepsilon} = \mathring{A} - \varepsilon C_{\varepsilon} = \mathring{A}(I - \varepsilon B C_{\varepsilon})$ and so $B_{\varepsilon} = (I - \varepsilon \widetilde{C}_{\varepsilon})^{-1} B$ with $\widetilde{C}_{\varepsilon} := BC_{\varepsilon}$. Denote by $\widetilde{c}_{ij}(\varepsilon)$, $1 \le i,j \le N$, the entries of $\widetilde{C}_{\varepsilon}$. We then see by (A.7.23) that

$$\max_{1\le i,j\le N} \sup_{\varepsilon\in(0,\varepsilon_0)} |\widetilde{c}_{ij}(\varepsilon)| =: \widetilde{c} < \infty,$$

and that the absolute value of each entry of $\widetilde{C}^n_{\varepsilon}$ is dominated by $N^{n-1}\widetilde{c}^n$ for $n \in \mathbb{N}$. Therefore for $\varepsilon \in (0, 1/(N\widetilde{c}+1))$,

$$B_{\varepsilon} = B + \varepsilon \widehat{C}_{\varepsilon} \cdot B \quad \text{where } \widehat{C}_{\varepsilon} := \sum_{k=1}^{\infty} \varepsilon^{k-1} \widetilde{C}^k_{\varepsilon},$$

and each entry $\widehat{c}_{ij}(\varepsilon)$ of $\widehat{C}_{\varepsilon}$ is absolutely convergent with

$$\sup_{\varepsilon\in(0,1/(N\widetilde{c}+1))} |\widehat{c}_{ij}(\varepsilon)| \le \sum_{k=1}^{\infty} \varepsilon^{k-1} N^{k-1} \widetilde{c}^k \le \widetilde{c}(N\widetilde{c}+1).$$

Accordingly, (A.7.24) holds with $\sup_{\varepsilon\in(0,1/(N\widetilde{c}+1))} |d_i(\varepsilon)| < \infty$ in place of $\sup_{\varepsilon\in(0,\varepsilon_0)} |d_{ij}(\varepsilon)| < \infty$. On the other hand, each entry of A_{ε} is continuous in ε and $\det A_{\varepsilon}$ never vanishes. Hence b^{ε}_{ij} is continuous in ε and $d_{ij}(\varepsilon) = (b^{\varepsilon}_{ij} - b_{ij})/\varepsilon$ is bounded in $\varepsilon \in (1/(N\widetilde{c}+1), \varepsilon_0)$.

4. It holds for any $\theta \in (0,\pi)$ and $\varepsilon \in (0,\varepsilon_0)$ that

$$\begin{cases} -\frac{1}{2}\frac{\partial}{\partial \mathbf{n}_{\zeta}} \varphi^{(j)}_{\varepsilon}(\zeta)\big|_{\zeta=\varepsilon e^{i\theta}} = \left(-\frac{\partial}{\partial \mathbf{n}_{\zeta}} \varphi^{(j)}(\zeta)\big|_{\zeta=0} + c_5(\theta,\varepsilon)\cdot\varepsilon \right) \sin\theta \quad \text{with} \\ \sup_{\theta\in(0,\pi),\, \varepsilon\in(0,\varepsilon_0)} |c_5(\theta,\varepsilon)| < \infty. \end{cases} \tag{A.7.25}$$

In proving (A.7.10) of Lemma A.7.5, we have considered the ABM $\mathbf{Z}^{D^j}$ on $D^j = D \cup C_j$ and its Green function $G^{D^j}(z,z')$. Analogously, we put $D^j_{\varepsilon} = D_{\varepsilon} \cup C_j (= \mathbb{H}_{\varepsilon} \setminus \bigcup_{k\ne j} C_k)$ and consider the ABM $\mathbf{Z}^{D^j_{\varepsilon}}$ on D^j_{ε} and its Green function $G^{D^j_{\varepsilon}}(z,z')$. Then $\varphi^{(j)}_{\varepsilon}(z) = \mathbb{P}^{D^j_{\varepsilon}}_z(\sigma_{C_j} < \infty)$, $z \in D^j_{\varepsilon}$. By the strong Markov property of $\mathbf{Z}^{D^j}$, we have

$$\varphi^{(j)}_{\varepsilon}(z) = \varphi^{(j)}(z) - \mathbb{E}^{D^j}_z \left[\varphi^{(j)}(Z_{\sigma_{\partial B_{\varepsilon}\cap\mathbb{H}}}); \sigma_{\partial B_{\varepsilon}\cap\mathbb{H}} < \infty \right] \quad \text{for } z \in D^j_{\varepsilon},$$

and

$$G^{D^j_\varepsilon}(z,z') = G^{D^j}(z,z') - \mathbb{E}^{D^j}_z\left[G^{D^j}(Z_{\sigma_{\partial B_\varepsilon \cap \mathbb{H}}}, z'); \sigma_{\partial B_\varepsilon \cap \mathbb{H}} < \infty\right],$$
$$z, z' \in D^j_\varepsilon, \quad z \neq z'.$$

Therefore, we can deduce from (A.7.10) that

$$\varphi^{(j)}_\varepsilon(z) = \int_{C_j} G^{D^j_\varepsilon}(z,z')\nu_j(dz'), \quad z \in D^j_\varepsilon, \tag{A.7.26}$$

for the same measure ν_j as in (A.7.10).

Thus we have from (A.7.4), (A.7.10) with D^j in place of D, and from (A.7.6), (A.7.26) with D^j_ε in place of D_ε, respectively, that

$$-\frac{1}{2}\frac{\partial}{\partial \mathbf{n}_\zeta}\varphi^{(j)}(\zeta) = \int_{C_j} K_{D^j}(z',\zeta)\nu_j(dz'), \quad \zeta \in \partial\mathbb{H}, \tag{A.7.27}$$

$$-\frac{1}{2}\frac{\partial}{\partial \mathbf{n}_\zeta}\varphi^{(j)}_\varepsilon(\zeta) = \int_{C_j} K_{D^j_\varepsilon}(z',\zeta)\nu_j(dz'), \quad \zeta \in \partial(\mathbb{H} \setminus B_\varepsilon). \tag{A.7.28}$$

Consequently, we get (A.7.25) from (A.7.27), (A.7.28) and (A.7.8) with D^j_ε and D^j in place of D_ε and D, respectively.

5. It follows from **1**, **3**, **4** that

$$\varphi^{(i)}_\varepsilon(z)\, b^\varepsilon_{ij}\, \frac{\partial}{\partial \mathbf{n}_\zeta}\varphi^{(j)}_\varepsilon(\zeta)\big|_{\zeta=\varepsilon e^{i\theta}} - 2\varphi^{(i)}(z)\, b_{ij}\, \frac{\partial}{\partial \mathbf{n}_\zeta}\varphi^{(j)}(\zeta)\big|_{\zeta=0} \sin\theta$$
$$= c_6(\theta,z,\varepsilon)\frac{\varepsilon}{|z|} + \varphi^{(i)}(z)c_7(\theta,\varepsilon)\cdot\varepsilon, \tag{A.7.29}$$

with

$$\sup_{\theta\in(0,\pi),\ |z|>\varepsilon_1,\ \varepsilon\in(0,\varepsilon_0)} |c_6(\theta,z,\varepsilon)| < \infty \quad \text{and} \quad \sup_{\theta\in(0,\pi),\ \varepsilon\in(0,\varepsilon_0)} |c_7(\theta,\varepsilon)| < \infty.$$

To estimate the second term of (A.7.29), take $\rho_0 > 0$ with $B_{\rho_0} \supset K$. We then see from Proposition 1.1.2 that, for $z \in \mathbb{H} \setminus B_{2\rho_0}$,

$$\varphi^{(i)}(z) \le \mathbb{P}^{\mathbb{H}}_z\left(\sigma_{\partial B_{\rho_0}\cap\mathbb{H}} < \infty\right) \le \frac{2\rho_0(1+c_{1/2})}{|z|},$$

and so $\sup_{|z|>\varepsilon_1} |z|\varphi^{(i)}(z) \le 2\rho_0(1+c_{1/2})$. Thus we arrive at the desired (A.7.20) and (A.7.21) on account of (A.7.8), (A.7.9), (A.7.13) and (A.7.18). □

A.8 Comparison of half-plane capacities

We fix a standard slit domain $D = \mathbb{H} \setminus K$ with $K = \bigcup_{j=1}^N C_j$. Recall that for $r > 0$, $B_r := \{z \in \mathbb{C} : |z| < r\}$. For $T > 0$, we consider an increasing family $\{F_t; t \in (0,T]\}$ of $\mathbb{H}$-hulls with $F_T \subset D$ such that there is a positive increasing function ε_t, $t \in (0,T]$, satisfying

$$\lim_{t\to 0} \varepsilon_t = 0, \quad B_{\varepsilon_T} \subset D \quad \text{and} \quad F_t \subset B_{\varepsilon_t} \text{ for } t \in (0,T]. \tag{A.8.1}$$

Let a_t be the half-plane capacity of the $\mathbb{H}$-hull F_t relative to D introduced in Section 2.1.1. Let g_t^0 be the unique Riemann map from $\mathbb{H} \setminus F_t$ onto $\mathbb{H}$ satisfying the hydrodynamic normalization $g_t^0(z) = z + \dfrac{a_t^0}{z} + o(1/|z|)$ near infinity. Clearly, $a_t^0 = \lim_{z\to\infty} z(g_t^0(z) - z)$.

Theorem A.8.1. $\lim_{t\downarrow 0} a_t/t$ *exists if and only if* $\lim_{t\downarrow 0} a_t^0/t$ *exists. If both limits exist, they have the same value.*

Proof. We give a proof by making use of Theorems 2.2.4, 2.2.5 and A.7.7.

1. Along with the ABM $\mathbf{Z}^{\mathbb{H}} = (Z_t^{\mathbb{H}}, \zeta^{\mathbb{H}}, \mathbb{P}_z^{\mathbb{H}})$ on $\mathbb{H}$, we consider BMD $\mathbf{Z}^* = (Z_t^*, \zeta^*, \mathbb{P}_z^*)$ on $D^* = D \cup \{c_1^*, \ldots, c_N^*\}$ obtained from $\mathbf{Z}^{\mathbb{H}}$ by rendering each slit C_i into a single point c_i^*. The Poisson kernel of $\mathbf{Z}^*$ will be denoted by $K_D^*(z,\zeta)$. We write $S = \partial B_{\varepsilon_T} \cap \mathbb{H}$ and define for $t \in (0,T)$,

$$M_1(t) = \int_0^\pi \mathbb{E}^{\mathbb{H}}_{\varepsilon_t e^{i\theta}}\left[\Im Z^{\mathbb{H}}_{\sigma_{F_t}}; \sigma_{F_t} < \infty\right] \sin\theta \, d\theta,$$

$$M_1^*(t) = \int_0^\pi \mathbb{E}^{*}_{\varepsilon_t e^{i\theta}}\left[\Im Z^{*}_{\sigma_{F_t}}; \sigma_{F_t} < \infty\right] \sin\theta \, d\theta,$$

$$M_2(t) = \int_0^\pi \mathbb{E}^{\mathbb{H}}_{\varepsilon_t e^{i\theta}}\left[\Im Z^{\mathbb{H}}_{\sigma_{F_t}}; \sigma_{F_t} < \sigma_S\right] \sin\theta \, d\theta,$$

$$M_2^*(t) = \int_0^\pi \mathbb{E}^{*}_{\varepsilon_t e^{i\theta}}\left[\Im Z^{*}_{\sigma_{F_t}}; \sigma_{F_t} < \sigma_S\right] \sin\theta \, d\theta.$$

Let $g_t(z)$ be the canonical map from $D \setminus F_t$ and let $h_t(z) = \Im(z - g_t(z))$. By virtue of Theorem 2.2.4, we then have

$$h_t(z) = \mathbb{E}_z^*\left[\Im Z^*_{\sigma_{F_t}}; \sigma_{F_t} < \infty\right], \quad z \in D \setminus F_t, \tag{A.8.2}$$

and, for a fixed $\rho_0 > 0$ with $B_{\rho_0} \supset K$,

$$a_t = \frac{2\rho_0}{\pi}\int_0^\pi h_t(\rho_0 e^{i\theta}) \sin\theta \, d\theta. \tag{A.8.3}$$

Since $K_D^*(z,0)$ is $\mathbf{Z}^*$-harmonic on D^*, it is $\mathbf{Z}^{\mathbb{H}}$-harmonic on $\mathbb{H} \setminus \overline{B}_{\rho_0}$. By Proposition 1.7.4, $\lim_{z\in\mathbb{H},\, z\to\zeta} K_D^*(z,0) = 0$ for any $\zeta \in \partial\mathbb{H} \setminus \{0\}$ and $\lim_{z\to\infty} K_D^*(z,0) = 0$. Consequently, by making a similar consideration to the proof of Lemma 1.1.4 and using Lemma 1.1.3, we are led to

$$K_D^*(z,0) = \mathbb{E}_z^{\mathbb{H}}\left[K_D^*(Z^{\mathbb{H}}_{\sigma_{\partial B_{\rho_0}\cap\mathbb{H}}}, 0)\right], \quad z \in \mathbb{H} \setminus \overline{B}_{\rho_0}.$$

Hence we get from Proposition 1.1.2

$$K_D^*(z,0) = \frac{2\rho_0}{\pi}\frac{\Im z}{|z|^2}\left[\int_0^\pi K_D^*(\rho_0 e^{i\theta},0)\sin\theta d\theta\right](1 + O(1/|z|)). \tag{A.8.4}$$

Define $K_D^*(\infty,0) = \lim_{y\uparrow\infty} yK_D^*(iy,0)$. We then obtain from the above that

$$\begin{aligned} K_D^*(\infty,0) &= \frac{2\rho_0}{\pi}\int_0^\pi K_D^*(\rho_0 e^{i\theta},0)\sin\theta d\theta, \\ K_D^*(z,0) &= \frac{\Im z}{|z|^2}K_D^*(\infty,0) + O(1/|z|^2). \end{aligned} \tag{A.8.5}$$

Notice that (A.8.5) holds not only for a standard slit domain D but also for a more general domain $D = \mathbb{H} \setminus \bigcup_{j=1}^{N} A_j$ where $\{A_j\}$ are mutually disjoint compact continua contained in $\mathbb{H}$.

For $z \in D$ with $|z| \geq \varepsilon_T$ and $t \in (0, T)$, $\mathbb{P}_z^* \left[\sigma_{\partial B_{\varepsilon_t}} < \sigma_{F_t}\right] = 1$, and consequently we get from (A.8.2) and the strong Markov property of $\mathbf{Z}^*$ that

$$h_t(z) = \mathbb{E}_z^* \left[h_t(Z^*_{\sigma_{\partial B_{\varepsilon_t} \cap \mathbb{H}}}); \sigma_{\partial B_{\varepsilon_t} \cap \mathbb{H}} < \infty \right]. \tag{A.8.6}$$

Hence (A.8.3) and Theorem A.7.7 with ε_t in place of ε and (A.8.5) lead us to

$$\begin{aligned} a_t &= \frac{2\rho_0}{\pi} \int_0^{\pi} \mathbb{E}^*_{\rho_0 e^{i\theta_1}} \left[h_t(Z^*_{\sigma_{\partial B_{\varepsilon_t} \cap \mathbb{H}}}); \sigma_{\partial B_{\varepsilon_t} \cap \mathbb{H}} < \infty \right] \sin\theta_1 d\theta_1 \\ &= \frac{2\varepsilon_t \rho_0}{\pi} \int_0^{\pi} \int_0^{\pi} \left[2K^*_D(\rho_0 e^{i\theta_1}, 0) + O(\varepsilon_t) \right] \sin\theta_2 h_t(\varepsilon_t e^{i\theta_2}) d\theta_2 \sin\theta_1 d\theta_1 \\ &= 2\varepsilon_t M_1^*(t) \left[K^*_D(\infty, 0) + O(\varepsilon_t) \right]. \end{aligned} \tag{A.8.7}$$

2. We claim that

$$K^*_D(\infty, 0) = 1/\pi. \tag{A.8.8}$$

To this end, consider the conformal map $\phi(z) = -1/z$ from $\mathbb{H}$ onto $\mathbb{H}$. $\phi\big|_D$ is then a conformal map from D onto $\widehat{D} = \mathbb{H} \setminus \bigcup_{j=1}^{N} \phi(C_j)$. Let $K_{\widehat{D}^*}(z, \xi)$ be the BMD-Poisson kernel of $\widehat{D}^*$.

We first show for $K^*_{\widehat{D}}(\infty, 0) = \lim_{y \to \infty} y K^*_{\widehat{D}}(iy, 0)$ that

$$K^*_{\widehat{D}}(\infty, 0) = 1/\pi. \tag{A.8.9}$$

Let $\widehat{\Psi}(z, \xi)$ be the BMD-complex Poisson kernel on $\widehat{D}^*$, which has the properties $\Im \widehat{\Psi}(z, \xi) = K^*_{\widehat{D}}(z, \xi)$ and $\lim_{z \to \infty} \widehat{\Psi}(z, \xi) = 0$. Let b be a half of the BMD-domain constant; that is, $b = \lim_{z \to 0} (\pi \widehat{\Psi}(z, 0) + 1/z)$, which is a real number. The BMD-complex Poisson kernel and the BMD-domain constant have been formulated in Section 1.9 and Section 5.4 for any standard slit domain, but the same formulations are possible for the multiply connected domain of the type $\widehat{D}$.

Define $\varphi_D(z) = \pi \widehat{\Psi}(-1/z, 0) - b$. Then $\Im \varphi_D(z) = \pi K^*_{\widehat{D}}(\phi(z), 0)$ is constant on each slit C_j and $\lim_{z \to \infty} (\varphi_D(z) - z) = \lim_{w \to 0} (\pi \widehat{\Psi}(w, 0) + 1/w) - b = 0$. Therefore, in the same way as the proof of Theorem 2.2.3, we can see that φ_D is a conformal map from D onto a standard slit domain, and consequently a canonical map from D. Since the identity mapping on D is also a canonical map, $z = \varphi_D(z)$ holds by uniqueness so that $y = \pi K^*_{\widehat{D}}(i/y, 0)$. On the other hand, we see from (A.8.5) for $\widehat{D}^*$ and $z = i/y$ that $K^*_{\widehat{D}}(i/y, 0) = y K^*_{\widehat{D}}(\infty, 0) + O(y^2)$, and accordingly $y = \pi y K^*_{\widehat{D}}(\infty, 0) + O(y^2)$, $y \to 0$, yielding (A.8.9).

We next prove

$$K^*_D(\infty, 0) = K^*_{\widehat{D}}(\infty, 0), \tag{A.8.10}$$

which together with (A.8.9) gives (A.8.8). Let $G^*_D(z,z')$ (resp. $G^*_{\widehat{D}}(w,w')$) be the Green function (0-order resolvent density) of BMD on D^* (resp. $\widehat{D}^*$). Take $\rho_0>0$ with $B_{\rho_0}\supset K$. For sufficiently small $\varepsilon>0$, we then have similarly to (A.8.4)

$$G^*_D(z,i\varepsilon)=\frac{2\rho_0}{\pi}\frac{\Im z}{|z|^2}\left[\int_0^\pi G^*_D(\rho_0e^{i\theta},i\varepsilon)\sin\theta d\theta\right](1+O(1/|z|))\quad\text{as } |z|\to\infty,$$

and so

$$yG^*_D(iy,i\varepsilon)=\frac{2\rho_0}{\pi}\int_0^\pi G^*_D(\rho_0e^{i\theta},i\varepsilon)\sin\theta d\theta+O(1/y)\quad\text{for } y>\rho_0.$$

This combined with (A.8.5) implies

$$\lim_{\varepsilon\downarrow 0}\lim_{y\to\infty}\frac{yG^*_D(iy,i\varepsilon)}{2\varepsilon}=\frac{2\rho_0}{\pi}\int_0^\pi K^*_D(\rho_0e^{i\theta},0)\sin\theta d\theta=K^*_D(\infty,0).\tag{A.8.11}$$

On the other hand, by using the conformal invariance of BMD Green function $G^*_{\widehat{D}}(w,w')=G^*_D(\phi^{-1}w,\phi^{-1}w')$ for $w,w'\in\widehat{D}^*$ from Theorem 1.8.1 and the symmetry of $G^*_D(z,z')$, we are led to

$$\begin{aligned}K^*_{\widehat{D}}(\infty,0)&=\lim_{y\to\infty}\lim_{\varepsilon\downarrow 0}\frac{yG^*_{\widehat{D}}(iy,i\varepsilon)}{2\varepsilon}\\&=\lim_{y\to\infty}\lim_{\varepsilon\downarrow 0}\frac{yG^*_D(i/y,i/\varepsilon)}{2\varepsilon}=\lim_{y\to\infty}\lim_{\varepsilon\downarrow 0}\frac{yG^*_D(i/\varepsilon,i/y)}{2\varepsilon}.\end{aligned}$$

By substituting $1/\varepsilon=\widetilde{y}$ and $1/y=\widetilde{\varepsilon}$ in the last expression, we get

$$K^*_{\widehat{D}}(\infty,0)=\lim_{\widetilde{\varepsilon}\downarrow 0}\lim_{\widetilde{y}\to\infty}\frac{\widetilde{y}}{2\widetilde{\varepsilon}}G^*_D(i\widetilde{y},i\widetilde{\varepsilon}),$$

which is equal to $K^*_D(\infty,0)$ by (A.8.11).

3. From (A.8.7) and (A.8.8), we finally get

$$a_t=\frac{2}{\pi}\varepsilon_tM^*_1(t)[1+O(\varepsilon_t)]\quad\text{as } \varepsilon_t\to 0.\tag{A.8.12}$$

An analogous formula holds for a^0_t in view of (2.2.21) in Theorem 2.2.5:

$$a^0_t=\frac{2}{\pi}\varepsilon_tM_1(t).\tag{A.8.13}$$

We now use Theorem A.7.7 again to verify that

$$\lim_{t\downarrow 0}\frac{\varepsilon_tM^*_1(t)}{t}\quad\text{exists if and only if}\quad\lim_{t\downarrow 0}\frac{\varepsilon_tM^*_2(t)}{t}\quad\text{exists,}\tag{A.8.14}$$

and, in this case, they are equal.

Indeed, we have from (A.8.2)

$$h_t(\varepsilon_te^{i\theta})=\mathbb{E}^*_{\varepsilon_te^{i\theta}}\left[\Im Z^*_{\sigma_{F_t}};\sigma_{F_t}<\sigma_S\right]+\mathbb{E}^{\mathbb{H}}_{\varepsilon_te^{i\theta}}\left[h_t(Z_{\sigma_S});\sigma_S<\infty\right],$$

and so

$$M^*_1(t)=M^*_2(t)+\int_0^\pi\mathbb{E}^{\mathbb{H}}_{\varepsilon_te^{i\theta}}\left[h_t(Z_{\sigma_S});\sigma_S<\infty\right]\sin\theta d\theta.\tag{A.8.15}$$

Further, we obtain from (A.8.6) combined with Theorem A.7.7

$$h_t(z) = 2\varepsilon_t K_D^*(z,0) M_1^*(t) \left(1 + O(\varepsilon_t)\right), \quad z \in S. \tag{A.8.16}$$

If $\lim_{t\downarrow 0}(\varepsilon_t M_1^*(t)/t) = \gamma$ exists, then $h_t(z)/t$ is uniformly bounded in $t > 0$ and $z \in S$ by (A.8.16), and so $\lim_{t\downarrow 0}(\varepsilon_t M_2^*(t)/t) = \gamma$ by (A.8.15). Conversely, suppose $\lim_{t\downarrow 0}(\varepsilon_t M_2^*(t)/t) = \gamma'$ exists. Since $M_1^*(t) - M_2^*(t) \leq C\varepsilon_t M_1^*(t)$ for some constant $C > 0$ from (A.8.15) and (A.8.16), we get $M_1^*(t) \leq 2M_2^*(t)$ for sufficiently small $t > 0$. Hence $\limsup_{t\downarrow 0}(\varepsilon_t M_1^*(t)/t) < \infty$ and we conclude that $\lim_{t\downarrow 0}(\varepsilon_t M_1^*(t)/t) = \gamma'$ just as above.

We can use (A.7.8) in exactly the same way to verify that

$$\lim_{t\downarrow 0} \frac{\varepsilon_t M_1(t)}{t} \quad \text{exists if and only if} \quad \lim_{t\downarrow 0} \frac{\varepsilon_t M_2(t)}{t} \quad \text{exists,} \tag{A.8.17}$$

and, in this case, they are equal. As $M_2^*(t) = M_2(t)$, the desired statement of Theorem A.8.1 follows from (A.8.12), (A.8.13), (A.8.14) and (A.8.17). □

Notes

Chapter 1 is mainly based on [Chen (2012); Chen, Fukushima and Rohde (2016)]. The main source for Chapter 2 up to Section 2.5 is [Chen, Fukushima and Rohde (2016)], and the main source for Section 2.6, Chapter 3 and Chapter 4 is [Chen and Fukushima (2018)].

The contents of Sections 5.1 and 5.2 are based on [Murayama (2019)], while the contents of Sections 5.4 and 5.5 are based on [Chen and Fukushima (2018); Chen, Fukushima and Suzuki (2017)].

The self-contained construction of a Riemann map in Theorem 2.1.1 for the simply connected domain $\mathbb{H} \setminus F$ appears here for the first time. We learned from G. Lawler (private communication) that he also has a self-contained probabilistic proof of the Riemann mapping theorem but without using the degree theorem. The Komatu-Loewner left-differential equation (2.1.23) was first derived by [Bauer and Friedrich (2008)] under an analogous consideration to [Komatu (1950)] where a circularly slit annulus was treated. [Bauer and Friedrich (2008)] also made a first derivation of the equation (3.1.3) for the slit motion $\mathbf{s}(t)$. [Bauer and Friedrich (2008)] went on further to observe that the random process $(\xi(t), \mathbf{s}(t))$ induced by a random Jordan arc satisfying domain Markov property and conformal invariance in the sense of (4.1.6) and (4.1.7) is a Markov process with a certain spatial homogeneity.

Lawler (2006) indicated a probabilistic expression of $\Im g_F(z)$ in terms of the excursion reflected Brownian motion (ERBM). See Remark 1.8.3 for ERBM. Theorem 2.2.1 and its proof in Section A.4 of the Appendix replace ERBM by BMD in the derivation of a probabilistic representation of $\Im g_F(z)$. Such a probabilistic representation of $\Im g_F(z)$ plays a crucial role in addressing not only the continuity problem in Section 2.3 but also the measurability and adaptedness issues in Sections 4.1.2 and 5.4.2.

Sections A.4 to A.6 of the Appendix are based on [Chen, Fukushima and Rohde (2016)], while Sections A.7 and A.8 are based on [Chen and Fukushima (2018)]. Results in Sections A.7 and A.8 were first formulated in [Drenning (2011)] for ERBM in place of BMD. The comparison theorem in Section A.8 is used crucially in the determination of the half-plane capacity of the image hulls by a univalent map in Proposition 5.3.6.

Bibliography

L. V. Ahlfors (1972), *Complex Analysis.* McGraw-Hill.

R. O. Bauer and R. M. Friedrich (2006), On radial stochastic Loewner evolution in multiply connected domains. *J. Funct. Anal.* **237**, 565–588.

R. O. Bauer and R. M. Friedrich (2008), On chordal and bilateral SLE in multiply connected domains. *Math. Z.* **258**, 241–265.

C. Boehm and W. Lauf (2014), A Komatu-Loewner equation for multiple slits. *Computational Methods in Function Theory* **14**, 639–660, Springer.

R. M. Blumenthal and R. K. Getoor (1968), *Markov Processes and Potential Theory.* Dover 2007, republication of 1968 edition (Academic Press).

Z.-Q. Chen (2012), Browniam Motion with Darning. Lecture notes for talks given at RIMS, Kyoto University.

Z.-Q. Chen and M. Fukushima (2012), *Symmetric Markov Processes, Time Change and Boundary Theory.* Princeton University Press.

Z.-Q. Chen and M. Fukushima (2015), One-point reflection. *Stochastic Process Appl.* **125**, 1368–1393.

Z.-Q. Chen and M. Fukushima (2018), Stochastic Komatu-Loewner evolutions and BMD domain constant. *Stochastic Process. Appl.* **128**, 545–594.

Z.-Q. Chen, M. Fukushima and S. Rhode (2016), Chordal Komatu-Loewner equation and Brownian motion with darning in multiply connected domains. *Trans. Amer. Math. Soc.* **368**, 4065–4114.

Z.-Q. Chen, M. Fukushima and H. Suzuki (2017), Stochastic Komatu-Loewner evolutions and SLEs. *Stochastic Process. Appl.* **127**, 2068–2087.

E. A. Coddington and N. Levinson (1955), *Theory of Ordinary Differntial Equations,* McGraw-Hill.

J. B. Conway (1978), *Functions of One Complex Variable I.* Springer.

J. B. Conway (1995), *Functions of One Complex Variable II.* Springer.

P. Courrége and P. Priouret (1964), Temps d'arret d'une fonction aléatoire: propriétés de décomposition. *C. R. Acad. Sci. Paris* **259**, 3933–3935.

S. Drenning (2011), *Excursion reflected Brownian motions and Loewner equations in multiply connected domains.* ProQuest LLC, Ann Arbor, MI. Thesis (Ph.D.) The University of Chicago, arXiv:1112.4123.

P. L. Duren (1983), *Univalent Functions.* Springer.

E. B. Dynkin (1960), *Theory of Markov Processes.* Pergamon Press.

E. B. Dynkin (1965), *Markov Processes,* Vol. I, Vol. II. Springer.

M. Fukushima (1969), On boundary conditions for multi-dimensional Brownian motions with symmetric resolvent densities. *J. Math. Soc. Japan* **21**, 58–93.

M. Fukushima (2020), Komatu-Loewner differential equations. *SUGAKU Expositions* **33**, 239–260, American Mathematical Society.

M. Fukushima and H. Kaneko (2014), On Villat's kernels and BMD Schwarz kernels in Komatu-Loewner equations. In: *Stochastic Analysis and Applications 2014*, Springer Proc. in Math. and Stat. Vol.100 (Eds) D. Crisan, B. Hambly, T. Zariphopoulous, 327–348.

M. Fukushima, Y. Oshima and M. Takeda (2011), *Dirichlet Forms and Symmetric Markov Processes*, De Gruyter, 1994, Second Extended Edition.

M. Fukushima and H. Tanaka (2005), Poisson point processes attached to symmetric diffusions. *Ann. Inst. Henri Poincaré Probab. Statist.* **41**, 419–459.

A. R. Galmarino (1963), A test for Markov times. *Rev. Un. Mat. Argentina* **21**, 173-178.

P. R. Garabedian (1964), *Partial Differential Equations.* AMS Chelsia, 2007, republication of 1964 edition.

J. B. Garnett and D. E. Marshal (2005), *Harmonic Measure.* Cambridge University Press.

G.M. Goluzin (1951), On the parametric representation of functions univalent in a ring, (in Russian). *Math. Sbornik N.S.* **29(71)**, 469–476.

G.M. Goluzin (1969), *Geometric Theory of Functions of a Complex Variable*, American Mathematical Society Translations 26, Providence.

G. Grimmett (2006), *The Random-Cluster Model*, Springer.

P. Hartman (1964), *Ordinary Differential Equations.* John Wiley.

L. L. Helms (2009), *Potential Theory.* Springer.

N. Ikeda, M. Nagasawa and S. Watanabe (1966), A construction of Markov processes by piecing out. *Proc. Japan Acad.* **42**, 370-375.

N. Ikeda and S. Watanabe (1981), *Stochastic Differential Equations and Diffusion Processes.* North-Holland/Kodansha.

K. Itô (1960), *Lectures on Stochastic Processes.* Tata Institute of Fundamental Research, Bombay.

K. Itô (1970), Poisson point processes attached to Markov processes. *Proc. Sixth Berkeley Symp. Math. Stat. Probab.* **3**, 225–239.

K. Itô and H.P. McKean,Jr. (1965), *Diffusion Processes and their Sample Paths.* Springer.

I. Karatzas and S. E. Shreve (1998), *Brownian Motion and Stochastic Calculus*, 2nd ed., Springer.

Y. Komatu (1943), Untersuchungen über konforme Abbildung von zweifach zusammenhängenden Gebieten. *Proc. Phys.-Math. Soc. Japan* **25**, 1–42.

Y. Komatu (1950), On conformal slit mapping of multiply-connected domains. *Proc. Japan Acad.* **26**, 26–31.

H. Kunita (1990), *Stochastic Flows and Stochastic Differential Equations.* Cambridge University Press.

G. F. Lawler (2005), *Conformally Invariant Processes in the Plane.* Mathematical Surveys and Monographs, AMS.

G. F. Lawler (2006), The Laplacian-b random walk and the Schramm-Loewner evolution. *Illinois J. Math.* **50**, 701–746 (Special volume in memory of Joseph Doob).

G. F. Lawler (2009), Conformal invariance and $2D$ statistical physics, *Bulletin Amer. Math. Soc.* **46**, 35–54.

G. F. Lawler, O. Schramm and W. Werner (2001), Values of Brownian intersection exponents. I: Half-plane exponents. *Acta Math.* **187**, 237–273.

G. F. Lawler, O. Schramm and W. Werner (2003), Conformal restriction: the chordal case. *J. Amer. Math. Soc.* **16**, 917–955.

K. Löwner (1923), Untersuchungen über schlichte konforme Abbildungen des Einheitskreises I. *Math. Ann.* **89**, 103–121.

J. Milnor (1965), *Topology from the Differentiable Viewpoint.* Princeton University Press.

T. Murayama (2019), Chordal Komatu-Loewner equation for a family of continuously growing hulls. *Stochastic Process Appl.* **129**, 2968–2990.

T. Murayama (2020), On the slit motion obeying chordal Komatu-Loewner equation with finite explosion time, *J. Evol. Equ.* **20**, 233–255.

M. A. Newman (1961), *Elements of the Topology of Plane Sets of Points.* Cambridge University Press.

C. Pommerenke (1975), *Univalent Functions.* Vandenhoeck & Ruprecht, Göttingen.

S. C. Port and C. J. Stone (1978), *Brownian Motion and Classical Potential Theory.* Academic Press.

D. Revuz and M. Yor (1999), *Continuous Martingales and Brownian Motion.* Springer.

L. C. G. Rogers and D. Williams (1979), *Diffusions, Markov Processes and Martingales,* Vol. 1. Cambridge University Press.

S. Rohde and O. Schramm (2005), Basic properties of SLE. *Ann. Math.* **161**, 879–920.

O. Schramm (2000), Scaling limits of loop-erased random walks and uniform spanning trees. *Israel J. Math.* **118**, 221–288.

E. M. Stein (1970), *Singular Integrals and Differentiability Properties of Functions.* Princeton University Press.

M. Tsuji (1959), *Potential Theory in Modern Function Theory.* Marzen, Tokyo.

W. Werner (2004), *Random Planar Curves and Schramm-Loewner Evolutions.* Lecture Notes in Math. **1840**, Springer.

D. Zhan (2004), Stochastic Loewner evolution in doubly connected domains. *Probab. Theory Relat. Fields* **129**, 340–380.

Index

$\mathbb{H}$-hull, v, 59

BMD, v, 18
  domain constant, viii, 180
Brownian motion
  absorbed, 3
  planar, 1
  with darning, vi, 17, 18

canonical map
  from $\mathbb{H} \setminus F$, 62
  from $D \setminus F$, 62, 69
capacity
  $\mathcal{E}_1^*$-, 22
  $\mathcal{E}_1^{*,O}$-, 23
  $\mathcal{E}_1^E$-, 20
complex Poisson kernel
  of the BMD, vi, 56
  of $\mathbf{Z}^{\mathbb{H}}$, 57
condition **(L)**, 145
continuum, 7

degree, 208
diffusion, 18
Dirichlet form, 19
  quasi-regular, 26
  regular, 20
  strongly local, 22
domain Markov property, 130
driving function, 165

energy measure, 20, 29
excursion reflected Brownian
  motion, 53

exit distribution, 2
  α-order, 193
exit time, 1, 186, 193

Feller-Dynkin diffusion, 140
filtration, 147
first passage time relation, 3
flux, 37
fundamental identity for Logarithmic
  potential, 3

Green function, 5, 9
  of the BMD, 45

half-plane capacity
  for Komatu-Loewner evolution, 124
  of $\mathbb{H}$-hull, 62
  reparametrization, 103
harmonic, 2
  $\mathbf{Z}^*$-, 39
  BMD-, 39
harmonic basis, 8
hitting time, 1
homogeneous
  in horizontal direction, 113
  with degree -1, 113
  with degree 0, 113
Hunt process, 24
  symmetric, 24
hydrodynamic normalization, 62

infinitesimal generator, 37

kernel, 157
  convergence, 162

Komatu-Loewner
differential equation for
the slits, 109, 111
equation, vii, 103
intrinsic chordal, 103
evolution, vii, 124
stochastic, viii, 145, 151

Lipschitz
continuity of complex Poisson kernel,
77, 89
continuous, 85
locality, 39
of SLE_6, 191
locally
bounded, 3, 158
compact, 24
uniform convergence, 158
Loewner
equation, v
evolution, v
stochastic, vi, 145

mean value property, 2
multiply connected domain, 8
$(N+1)$-connected domain, 7

normal family, 158
null chain, 63

parallel slit plane, 62
period, 8, 41
Poisson integral formula, 9
Poisson kernel, 9
of the BMD, 45
polar, 2
$\mathcal{E}$-, 24, 26
$\mathcal{E}^E$-, 20
ν-, 26
m-, 29
prime end, 63
proper map, 207
properly associated with, 24
properly exceptional, 24

quasi-continuous, 20
quasi-everywhere, 20
quasi-homeomorphic, 27

regular point, 2
resolvent, 30
0-order, 31, 38
resolvent density
0-order, 3, 5
α-order, 3
Riemann mapping theorem, 59

SKLE, 145, 151
$SKLE_{\alpha,b}$, 151
SLE, 145, 153
standard process, 26, 27
standard slit domain, vi, 7, 61
stochastic differential equation, 143
symmetric Markov process, 18

CPSIA information can be obtained
at www.ICGtesting.com
Printed in the USA
JSHW050008130223
37638JS00003B/228